Essentials of Meteorology

An Invitation to the Atmosphere

Clouds forming downwind of Mt. Washington, New Hampshire. (Photo by the author)

Essentials of Meteorology

An Invitation to the Atmosphere

C. Donald Ahrens
Modesto Junior College

West Publishing Company
Minneapolis/St. Paul   New York   Los Angeles   San Francisco

West's Commitment to the Environment

Production, Prepress, Printing and Binding by West Publishing Company.

COPYRIGHT © 1993 By WEST PUBLISHING COMPANY
610 Opperman Drive
P.O. Box 64526
St. Paul, MN 55164-0526

Library of Congress Cataloging-in-Publication Data

Ahrens, C. Donald.
    Essentials of meteorology / C. Donald Ahrens.
        p.   cm.
    Includes index.
    ISBN 0-314-01245-1 (pbk.)
    1. Meteorology.   I. Title.
QC861.2.A29   1993
551.5—dc20   92-41121
                                              CIP

Design and production: Janet Bollow
Copy editing: Stuart Kenter
Illustrations: Joe Medeiros, George Kelvin, Barbara Barnett, Folium, House of Graphics, and Alexander Productions
Cover Photograph: © 1992 Richard Lee Kaylin
Composition: Janet Hansen, Alphatype

Photo Credits

Chapter 1

Fig. 1.1   NASA photo.
Fig. 1.2   Photo by author.
Fig. 1.4   Photo by Ron Tingley.
Fig. 1   Photo by author.
Fig. 2   Courtesy of NASA.
Fig. 1.10   NOAA photo.

Chapter 2

Figs. 2.2, 2.11, 3   Photos by author.
Fig. 2.14   NASA photo.
Fig. 2.18   Courtesy of the American Museum of Natural History.
Fig. 2.22   Photo by author.

Chapter 3

Figs. 3.5, 3.6   Photos by author.
Fig. 3.15   Photo by Ross DePaola.

Chapter 4

Figs. 1, 4.7   Photos by author.
Fig. 4.9   Photo by Elizabeth Beaver Burnett.
Figs. 4.10, 4.11   Photos by author.
Fig. 4.12   NOAA.
Figs. 4.13, 4.14, 4.15, 4.16, 4.17, 4.18, 4.19, 4.20, 4.21, 4.22, 4.23   Photos by author.
Fig. 4.24   Courtesy T. Ansel Toney.
Fig. 4.26   Photo by J. L. Medeiros.
Figs. 4.27, 4.28, 4.29   Photos by author.
Fig. 4.30   Ben Fogle, NCAR.

Chapter 5

Fig. 5.4   Photo by J. L. Medeiros.
Figs. 5.6, 5.11   Photos by author.
Fig. 5.19   Photo by Ross DePaola.
Fig. 5.20   Photo by author.
Figs. 5.21, 5.23   NOAA photos.
Fig. 5.24   National Center for Atmospheric Research/National Science Foundation.
Fig. 5.28   Photo courtesy of Alden Electronics, Inc., Westboro, MA.

Chapter 6

Figs. 3, 6.23, 6.26   Photos by author.

Chapter 7

Fig. 2   Courtesy of NCAR/NSF.
Fig. 7.5   Photo by author.
Fig. 7.6   Courtesy of T. Ansel Toney.
Figs. 7.9, 7.11   Photos by author.
Fig. 7.13   Courtesy of Sherwood B. Idso.

Chapter 8

Fig. 8.4   NOAA photo.
Figs. 8.8, 8.13   Photos by author.
Fig. 8.20   NOAA photo.
Figs. 4, 5   NASA photos.

Chapter 9

Figs. 9.2, 2, 3   Photos by author.
Figs. 9.5, 9.7a, 9.7b, 9.8, 9.9   NOAA photos.

(Continued following Index)

Contents

Chapter 5
Cloud Development
and Precipitation   105

Chapter 6
Air Pressure and Winds   131

Chapter 12
Air Pollution   295

Chapter 13
Climate Change   315

Chapter 14
Global Climate   339

▲▼▲

Chapter 15
Light, Color, and Atmospheric
Optics   373

Preface

In recent years, weather and climate have become front page news—from greenhouse warming to the cooling brought on by the eruption of Mt. Pinatubo. The dynamic nature of the atmosphere seems to demand our attention and understanding more these days than ever before. Almost daily, there are articles in the newspaper describing some weather event or impending climate change. For this reason, and the fact that weather influences our daily lives in so many ways, interest in meteorology (the study of the atmosphere) has been growing in popularity. With this interest in mind, this book was created to stimulate curiosity in the reader and answer questions about weather that arise in our day-to-day lives, without the intimidation that equations and mathematics can sometimes produce.

▲▼▲

About this Book

Essentials of Meteorology is written for those students taking an introductory course on the atmospheric environment. The main emphasis of the text is to convey meteorological concepts in a visual, practical, and non-mathematical manner. Although introductory in nature, the text maintains scientific integrity and includes up-to-date information on important topics, such as ozone depletion and global warming. Discussion on recent weather events, such as the destruction wrought by hurricanes Andrew and Iniki, are also included. Moreover, the book requires no special prerequisites in either science or mathematics.

Written with the student in mind, the book covers topics directly related to our everyday experiences with weather and stresses the understanding and application of meteorological principles. The book also emphasizes watching the weather so that it becomes "alive," and readers can immediately apply text material to the world around them. To assist with this endeavor, a color cloud chart appears toward the back of the text. The chart can be separated from the book and used as a learning tool when observing the sky.

To strengthen points and clarify concepts, illustrations are rendered in full color throughout the book. Color photographs were carefully selected to illustrate features, stimulate interest, and show how exciting weather can be.

Organization of the text into 15 chapters is designed to provide maximum flexibility to instructors of weather and climate courses. Thus, the chapters can be covered in any desired order. For example, Chapter 12, "Air Pollution," and Chapter 15, "Atmospheric Optics," are both self-contained and can be covered earlier if so desired. Instructors, then, are able to tailor this text to their particular needs.

This book basically follows a traditional approach. After an introductory chapter on the origin, composition, and structure of the atmosphere, it covers solar energy, air temperature, humidity, clouds, precipitation, and winds. Then comes a chapter on air masses, fronts, and middle-latitude storms. Weather prediction and severe storms are next. A chapter on air pollution is followed by climate change and global climates. The final chapter deals with atmospheric optics.

Each chapter contains at least two focus sections, which expand on material in the main text or explore a subject closely related to what is being covered. Focus sections fall into one of five distinct categories: observations, applications, issues, instruments, and special topics. Some include material that is not always found in introductory meteorology textbooks, subjects such as the ozone hole, storms on other planets, TV weathercasting, and cloud seeding. Others help bridge theory and practice.

Set apart as "Did You Know" headings in each chapter is weather information that may not be commonly known, yet pertains to the topic under discussion. Designed to bring the reader into the text, most of these weather highlights relate to some interesting weather fact or astonishing event.

Each chapter incorporates a number of other effective learning aids:

- ▲ A major topic outline begins each chapter
- ▲ Interesting introductory pieces draw the reader naturally into the main text
- ▲ Important terms are boldfaced, with their definitions appearing in the glossary or in the text
- ▲ Key phrases are italicized
- ▲ English equivalents of metric units are immediately provided in parenthesis
- ▲ Summaries at the end of each chapter review the chapter's main ideas
- ▲ A list of key terms following each chapter allow students to review and reinforce their knowledge of the chief concepts they encountered
- ▲ Review questions check how well students assimilate the material

Eight appendices conclude the book. Some are more technical than the main text, such as Appendix B, "Equations and Constants." Others can be used in observing the weather, such as Appendix G, "The Beaufort Wind Scale." In addition, at the end of the book, a compilation of supplementary reading material is presented, as is an extensive glossary.

▲▼▲
Supplemental Material

Available with this book is an instructor's manual with test bank by Charles Weidman of the University of Arizona, a slide package, overhead transparencies, meteorology software (for demonstration and laboratory purposes), and meteorology videos. A computerized test bank is available as Westest for IBM and IBM-compatible, Macintosh, and Apple microcomputers. Also available is a study guide prepared by the author. For additional information on this package, contact your West sales representative.

▲▼▲
Acknowledgements

My thanks go to the many individuals who have contributed to the production of this book. Special thanks to my friends and colleagues who were kind enough to allow me to use their beautiful photographs and to Kay Behrens, Nancy Newsom, and Laurie Rydelius for their careful reading of the entire book. A special debt of gratitude goes to Janet Bollow, who put a tremendous effort into ensuring an attractive product by transforming black and white illustrations into beautiful color art. A thank you to Stuart Kenter for his conscientious editing and suggestions. I am most grateful to my wife Suzy, for her patience, understanding, and invaluable assistance during this project.

My special thanks to the many professional people at West Publishing who gave support, especially Clyde Perlee, Denise Simon, Steve Schonebaum, and Denis Ralling. Finally, for their comments and suggestions, I am indebted to those colleagues, who were kind enough to review all or part of the manuscript, including:

Gerald C. Brothen, El Camino College

Dennis M. Driscoll, Texas A & M University

Rex J. Hess, University of Utah

Max Malmquist, Anoka-Ramsey Community College

Walter W. Moore, Sullivan County Community College

Charles D. Weidman, University of Arizona

Bob Weisman, St. Cloud State University

▲▼▲
To the Student

Learning about the atmosphere can be an enjoyable experience, especially if you become involved. This book is intended to give you some insight into the workings of the atmosphere. But, for a real appreciation of your atmospheric environment, you must go outside and observe. Mountains take millions of years to form, while a cumulus cloud can develop into a raging thunderstorm in less than an hour. To help with your observations, a color cloud chart is bound toward the back of the book for easy reference. Remove it, and keep it with you. And remember, all of the information in this book is out there—please, take the time to look.

Don Ahrens

Essentials of Meteorology

An Invitation to the Atmosphere

Earth viewed by *Apollo 8* astronauts, as their spacecraft came from behind the moon. (Photo: NASA)

Chapter 1

Introduction: The Earth and Its Atmosphere

Contents

▲▼▲

I well remember a brilliant red balloon which kept me completely happy for a whole afternoon, until, while I was playing, a clumsy movement allowed it to escape. Spellbound, I gazed after it as it drifted silently away, gently swaying, growing smaller and smaller until it was only a red point in a blue sky. At that moment I realized, for the first time, the vastness above us: a huge space without visible limits. It was an apparent void, full of secrets, exerting an inexplicable power over all the earth's inhabitants. I believe that many people, consciously or unconsciously, have been filled with awe by the immensity of the atmosphere. All our knowledge about the air, gathered over hundreds of years, has not diminished this feeling.

Theo Loebsack, *Our Atmosphere*

▲

Our *atmosphere* is a delicate life-giving blanket of air that surrounds the earth. In one way or another, it influences everything we see and hear—it is intimately connected to our lives. Air is with us from birth and we cannot detach ourselves from its presence. In the open air, we can travel for thousands of miles in any horizontal direction, but should we move a mere five miles above the surface, we would suffocate. We may be able to survive without food for a few weeks, or without water for a few days, but, without our atmosphere, we would not survive more than a few minutes. Just as fish are confined to an environment of water, so we are confined to an ocean of air. Anywhere we go, it must go with us.

The earth without an atmosphere would have no lakes or oceans. There would be no sounds, no clouds, no red sunsets. The beautiful pageantry of the sky would be absent. It would be unimaginably cold at night and unbearably hot during the day. All things on the earth would be at the mercy of an intense sun beating down upon a planet utterly parched.

Living on the surface of the earth, we have adapted so completely to our environment of air that we sometimes forget how truly remarkable this substance is. Even though air is tasteless, odorless, and (most of the time) invisible, it protects us from the scorching rays of the sun and provides us with a mixture of gases that allows life to flourish. Because we cannot see, smell, or taste air, it may seem surprising that between your eyes and the pages of this book are trillions of air molecules.

Some of these may have been in a cloud only yesterday, or over another continent last week, or perhaps part of the life-giving breath of a person who lived hundreds of years ago.

Overview of the Earth's Atmosphere

The earth's **atmosphere** is a thin, gaseous envelope comprised mostly of nitrogen (N_2) and oxygen (O_2), with small amounts of other gases, such as water vapor (H_2O) and carbon dioxide (CO_2). Nested in the atmosphere are clouds of liquid water and ice crystals.

The thin blue area near the horizon in Fig. 1.1 represents the most dense part of the atmosphere. Although our atmosphere extends upward for many hundreds of kilometers, almost 99 percent of the atmosphere lies within a mere thirty kilometers of the earth's surface. This thin blanket of air constantly shields the surface and its inhabitants from the sun's dangerous ultraviolet radiant energy, as well as from the onslaught of material from interplanetary space. There is no definite upper limit to the atmosphere; rather, it becomes thinner and thinner, eventually merging with empty space, which surrounds all the planets.

Composition of the Atmosphere   Table 1.1 shows the various gases present in a volume of air near the earth's surface. Notice that nitrogen occupies about 78

Figure 1.1
The earth's atmosphere as viewed from space.

percent and oxygen about 21 percent of the total volume. If all the other gases are removed, these percentages for nitrogen and oxygen hold fairly constant up to an elevation of about eighty kilometers (or fifty miles).

At the surface, there is a balance between destruction (output) and production (input) of these gases. For example, nitrogen is removed from the atmosphere primarily by biological processes that involve soil bacteria. It is returned to the atmosphere mainly through the decaying of plant and animal matter. Oxygen, on the other hand, is removed from the atmosphere when organic matter decays and when oxygen combines with other substances, producing oxides. It is also taken from the atmosphere during breathing, as the lungs take in oxygen and release carbon dioxide. The addition of oxygen to the atmosphere occurs during photosynthesis, as plants, in the presence of sunlight, combine carbon dioxide and water to produce sugar and oxygen.

The concentration of the invisible gas **water vapor**, however, varies greatly from place to place, and from time to time. Close to the surface in warm, tropical locations, water vapor may account for up to 4 percent of the atmospheric gases, whereas in colder polar areas, its concentration may dwindle to a mere fraction of a percent. (See Table 1.1.) Water vapor molecules are, of course, invisible. They become visible only when they transform into larger liquid or solid particles, such as cloud droplets and ice crystals. The chang-

ing of water vapor into liquid water is called *condensation*, whereas the process of liquid water becoming water vapor is called *evaporation*. In the lower atmosphere, water is everywhere. It is the only substance that exists as a gas, a liquid, and a solid at those temperatures and pressures normally found near the earth's surface. (See Fig. 1.2.)

Water vapor is an *extremely* important element of the atmosphere. Not only does it form into both liquid and solid cloud particles that grow in size and fall to earth as precipitation, but it also releases large amounts of heat—called *latent heat*—when it changes from vapor into liquid water or ice. Latent heat is an important source of atmospheric energy, especially for storms, such as thunderstorms and hurricanes. Moreover, water vapor is a potent greenhouse gas because it strongly absorbs a portion of the earth's outgoing radiant energy (somewhat like the glass of a greenhouse prevents the heat inside from escaping and mixing with the outside air). Thus, water vapor plays a significant role in the earth's heat-energy balance.

Table 1.1   Composition of the Atmosphere Near the Earth's Surface

Permanent Gases			Variable Gases			
Gas	*Symbol*	*Percent (by Volume) Dry Air*	*Gas (and Particles)*	*Symbol*	*Percent (by Volume)*	*Parts per Million (ppm)**
Nitrogen	N_2	78.08	Water vapor	H_2O	0 to 4	
Oxygen	O_2	20.95	Carbon dioxide	CO_2	0.035	355
Argon	Ar	0.93	Methane	CH_4	0.00017	1.7
Neon	Ne	0.0018	Nitrous oxide	N_2O	0.00003	0.3
Helium	He	0.0005	Ozone	O_3	0.000004	0.04
Hydrogen	H_2	0.00005	Particles (dust, soot, etc.)		0.000001	0.01
Xenon	Xe	0.000009	Chlorofluorocarbons (CFCs)		0.00000001	0.0001

*For CO_2, 355 parts per million means that out of every million air molecules 355 are CO_2 molecules.

Figure 1.2
The earth's atmosphere is a rich mixture of many gases, with clouds of condensed water vapor and ice crystals. Here, water evaporates from the ocean's surface. Rising air current then transforms the invisible water vapor into visible, puffy, cumulus clouds.

Carbon dioxide (CO_2), a natural component of the atmosphere, occupies a small (but important) percent of a volume of air, about 0.035 percent. Carbon dioxide enters the atmosphere mainly from the decay of vegetation, but it also comes from volcanic eruptions, the exhalations of animal life, from the burning of fossil fuels (such as coal, oil, and natural gas), and from deforestation. The removal of CO_2 from the atmosphere takes place during *photosynthesis*, as plants consume CO_2 to produce green matter. The CO_2 is then stored in roots, branches, and leaves. The oceans act as a huge reservoir for CO_2, as phytoplankton (tiny, drifting plants) in surface water fix CO_2 into organic tissues. Carbon dioxide that dissolves directly into surface water mixes downward and circulates through greater depths. Estimates are that the oceans hold more than fifty times the total atmospheric CO_2 content.

Figure 1.3 reveals that the atmospheric concentration of CO_2 has risen more than 10 percent since 1958, when it was first measured at Mauna Loa Obser-

vatory in Hawaii. This increase appears to be due mainly to the burning of fossil fuels. However, deforestation also plays a role as cut timber, either burned or left to rot, releases CO_2 directly into the air, perhaps accounting for about 20 percent of the observed increase. Scientists speculate that atmospheric levels of CO_2 have increased by as much as 25 percent since the early 1800s. With CO_2 levels presently increasing by about 0.4 percent annually, scientists now estimate that the concentration of CO_2 will likely double from its current value (355 ppm) sometime toward the end of the twenty-first century.

Carbon dioxide is another important greenhouse gas because it, like water vapor, traps a portion of the earth's outgoing infrared energy. Consequently, as the atmospheric concentration of CO_2 increases, so could the average global surface air temperature. Most of the mathematical model experiments that predict future atmospheric conditions estimate that a doubling of CO_2 will result in a *global warming* of surface air between

2°C and 5°C (about 3°F to 9°F).* Such warming could reduce the quantity and quality of water resources (especially in the western United States) as the global air currents that guide the major storm systems across the earth begin to shift from their "normal" paths. (Chapter 13 examines the global warming phenomenon in depth.)

Carbon dioxide and water vapor are not the only greenhouse gases. Recently, others have been gaining notoriety, primarily because they, too, are becoming more concentrated. Such gases include *methane* (CH_4), *nitrous oxide* (N_2O), and *chlorofluorocarbons* (CFCs).†

Levels of methane, for example, have been rising over the past decade, increasing by about one-half of one percent per year. Most methane appears to derive from the breakdown of plant material by certain bacteria in rice paddies, wet oxygen-poor soil, and biochemical reactions in the stomachs of cows. Just why methane should be increasing so rapidly is currently under study. Levels of nitrous oxide—commonly known as laughing gas—have been rising annually at the rate of about one quarter of a percent. Nitrous ox-

---

*The abbreviation °C is used when measuring temperature in degrees Celsius, and °F is the abbreviation for degrees Fahrenheit. More information about temperature scales is given in Appendix A and in Chapter 2.
†Because these gases (including CO_2) occupy only a small fraction of a percent in a volume of air near the surface, they are referred to collectively as *trace gases*.

---

ide forms in the soil through a chemical process involving bacteria and certain microbes. Ultraviolet light from the sun destroys it.

Chlorofluorocarbons represent a group of greenhouse gases that also have been increasing in concentration. These had been the most widely used propellants in spray cans. Today, however, they are mainly used as refrigerants, as propellants for the blowing of plastic-foam insulation, and as solvents for cleaning electronic microcircuits. Although their average concentration in a volume of air is quite small (see Table 1.1), they are the most rapidly increasing greenhouse gas at a rate of 4 to 11 percent per year. Moreover, CFCs not only have the potential for raising global temperature, they also play a part in destroying the gas ozone.

At the surface, *ozone* (O_3) is the primary ingredient of *photochemical smog*,* which irritates the eyes and

---

*Originally the word *smog* meant the combining of smoke and fog. Today, however, the word refers to the type of smog that forms in large cities, such as Los Angeles, California. Because this type of smog forms when chemical reactions take place in the presence of sunlight, it is termed *photochemical smog*.

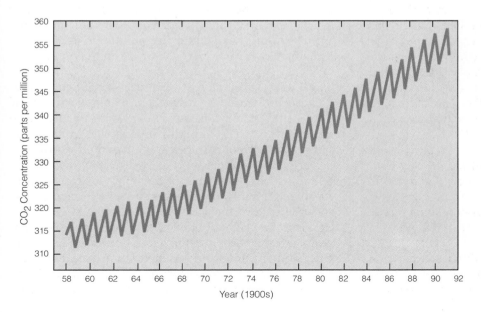

Figure 1.3
Measurements of CO_2 in parts per million (ppm) from 1958 through 1991 at Mauna Loa Observatory. Higher readings occur in winter when plants die and release CO_2 to the atmosphere. Lower readings occur in summer when more abundant vegetation absorbs CO_2 from the atmosphere.

throat and damages vegetation. But the majority of atmospheric ozone (about 97 percent) is found in the upper atmosphere—in the stratosphere—where it is formed naturally, as oxygen atoms combine with oxygen molecules. Here, the concentration of ozone averages less than 0.002 percent by volume. This small quantity is important, however, because it shields plants, animals, and humans from the sun's harmful ultraviolet rays. It is paradoxical that ozone, which damages plant life in a polluted environment, provides a natural protective shield in the upper atmosphere so that plants on the surface may survive. We will shortly look at how CFCs injected into the stratosphere may upset the ozone balance, and how the protective ozone shield itself may be gradually decreasing.

Impurities from both natural and human sources are also present in the atmosphere. Wind picks up dust and soil from the earth's surface and carries it aloft. Small saltwater drops from ocean waves are swept into the air. Upon evaporating, these drops leave microscopic salt particles suspended in the atmosphere.

Figure 1.4
Erupting volcanoes, such as Mount St. Helens, send tons of particles into the atmosphere, along with vast amounts of water vapor and carbon dioxide.

Smoke from forest fires is often carried high above the earth, and volcanoes spew many tons of fine ash particles and gases into our air (Fig. 1.4). Collectively, these tiny solid or liquid suspended particles of various composition are called **aerosols**.

Some natural impurities found in the atmosphere are quite beneficial. Small, floating particles, for instance, act as surfaces on which water vapor condenses to form clouds. However, most human-made impurities (and some natural ones) are a nuisance, as well as a health hazard. These we call **pollutants**. For example, automobile engines emit copious amounts of *nitrogen dioxide* (NO_2), *carbon monoxide* (CO), and *hydrocarbons*. Nitrogen dioxide gas often gives the atmosphere a dirty, light-brown color. In sunlight, it reacts with hydrocarbons and other gases to produce ozone. Carbon monoxide is a major pollutant of city air. Although colorless and odorless, this poisonous gas forms during the incomplete combustion of carbon-containing fuel. Hence, over 75 percent of carbon monoxide in urban areas comes from road vehicles.

The burning of sulfur-containing fuels (such as coal and oil) release the colorless gas *sulfur dioxide* (SO_2) into the air. When the atmosphere is sufficiently moist, the SO_2 may transform into tiny dilute drops of sulfuric acid. Rain containing sulfuric acid corrodes metals and painted surfaces and turns freshwater lakes acidic. *Acid rain* (thoroughly discussed in Chapter 12) is a major environmental problem, especially downwind from major industrial areas. In addition, high concentrations of SO_2 produce serious respiratory problems in humans, such as bronchitis and emphysema, and have an adverse effect on plant life. (More information on these and other pollutants is given in Chapter 12.)

The Early Atmosphere   The atmosphere that originally surrounded the earth was probably much different from the air we breathe today. The earth's first atmosphere (some 4.5 billion years ago) was most likely *hydrogen* and *helium*—the two most abundant gases found in the universe—as well as hydrogen compounds, such as methane and ammonia. Most scientists feel that this early atmosphere escaped into space from the earth's hot surface. However, a second, more dense atmosphere gradually enveloped the earth as gases from molten rock within its hot interior escaped through volcanoes and steam vents. We assume that volcanoes spewed out the same gases then as they do today: mostly water vapor (about 80 percent), carbon dioxide (about 10 percent), and up to a few percent nitrogen. These gases probably created the earth's second atmosphere.

As hundreds of millions of years passed, our atmosphere gradually cooled. The constant outpouring of gases from the hot interior—known as **outgassing**—provided a rich supply of water vapor, which formed into clouds. Rains fell upon the earth for many thousands of years, forming the rivers, lakes, and oceans of the world. During this time, large amounts of CO_2 were dissolved in the oceans. Through chemical and biological processes, much of the CO_2 became locked up in carbonate sedimentary rocks, such as limestone. With much of the water vapor already condensed and the concentration of CO_2 dwindling, the atmosphere gradually became rich in nitrogen (N_2), which is usually not chemically active.

It appears that oxygen (O_2), the second most abundant gas in today's atmosphere, probably began an extremely slow increase in concentration as energetic rays from the sun split water vapor (H_2O) into hydrogen and oxygen. The hydrogen, being lighter, probably rose and escaped into space, while the oxygen remained in the atmosphere.

This slow increase in oxygen may have provided enough of this gas for primitive plants to evolve, perhaps 2 to 3 billion years ago. Or the plants may have evolved in an almost oxygen-free (anaerobic) environment. At any rate, plant growth greatly enriched our atmosphere with oxygen. The reason for this is that, during photosynthesis, plants, in the presence of sunlight, combine carbon dioxide and water to produce oxygen. Hence, after plants evolved, the atmospheric oxygen content increased more rapidly, probably reaching its present composition about several hundred million years ago.

▲▼▲

Vertical Structure of the Atmosphere

A vertical profile of the atmosphere reveals that it can be divided into a series of layers. Each layer may be defined in a number of ways: by the manner in which the air temperature varies through it, by the gases that comprise it, or even by its electrical properties. At any rate, before we examine these various atmospheric layers, we need to look at the vertical profile of two important variables: air pressure and air density.

A Brief Look at Air Pressure and Air Density   Air molecules (as well as everything else) are held near the earth by gravity. This strong, invisible force pulling down on the air above squeezes (compresses) air molecules closer together, which causes their number in a given volume to increase. The more air above a level, the greater the squeezing effect or compression. Since *air density* is the number of air molecules in a given space (volume), it follows that air density is greatest at the surface and decreases as we move up into the atmosphere. Notice in Fig. 1.5 that, owing to the fact that the air near the surface is compressed, air density normally decreases rapidly at first, then more slowly as we move farther away from the surface.

Air molecules have weight. In fact, air is surprisingly heavy. The weight of all the air around the earth is a staggering 5600 trillion tons. The weight of the air molecules exerts a force upon the earth. The amount of force exerted over an area of surface is called *atmospheric pressure* or, simply, **air pressure**. The pressure at any level in the atmosphere may be measured in terms of the total weight of the air above any point. As we climb in elevation, fewer air molecules are above us; hence, *atmospheric pressure always decreases*

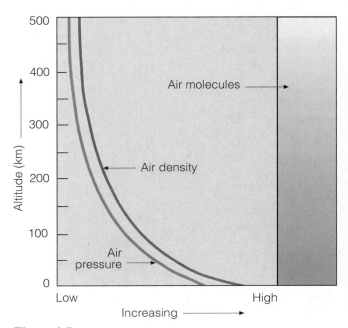

Figure 1.5
Both air pressure and air density decrease with increasing altitude.

Did you know?
On September 5, 1862, English meteorologist James Glaisher and a pilot named Coxwell ascended in a hot air balloon to collect atmospheric data. As the pair rose above 29,000 feet, the low air density and lack of oxygen caused Glaisher to become unconscious and Coxwell so paralyzed that he could only operate the control valve with his teeth.

with increasing height. Like air density, air pressure decreases rapidly at first, then more slowly at higher levels. (See Fig. 1.5.)

If we weigh a column of air 1 square inch in cross section, extending from the average height of the ocean surface (sea level) to the "top" of the atmosphere, it would weigh very nearly 14.7 pounds. Thus, normal atmospheric pressure near sea level is close to 14.7 pounds per square inch. If more molecules are packed into the column, it becomes more dense, the air weighs more, and the surface pressure goes up. On the other

hand, when fewer molecules are in the column, the air weighs less, and the surface pressure goes down. So, a change in air density can bring about a change in air pressure.

Pounds per square inch is, of course, one way of expressing air pressure. However, the most common unit presently found on surface weather maps is the *millibar* (mb). Another common pressure unit used in aviation and on television and radio weather broadcasts is inches of mercury (Hg). At sea level, the *average* or *standard value* for atmospheric pressure is

$$1013.25 \text{ mb} = 29.92 \text{ in. Hg.}$$

Figure 1.6 illustrates how rapidly air pressure decreases with height. Near sea level, atmospheric pressure decreases rapidly, whereas at high levels it decreases more slowly. With a sea level pressure near 1000 millibars, we can see in Fig. 1.6 that, at an altitude of only 5.5 kilometers (km) or 3.5 miles (mi), the air pressure is about 500 millibars, or half of the sea level pressure, meaning that, if you were at a mere 18,000 feet (ft) above the surface, you would be above one-half of all the molecules in the atmosphere.

At an elevation approaching the summit of Mount Everest (about 9 km or 29,000 ft—the highest mountain peak on earth), the air pressure would be about 300 millibars. The summit is above nearly 70 percent of all the molecules in the atmosphere. At an altitude of about 50 km, the air pressure is about 1 millibar, which means that 99.9 percent of all the molecules are below this level. Yet the atmosphere extends upwards for many hundreds of kilometers, gradually becoming thinner and thinner until it ultimately merges with outer space.

Layers of the Atmosphere   We have seen that both air pressure and density decrease with height above the earth—rapidly at first, then more slowly. Air temperature, however, has a more complicated vertical profile. Look closely at Fig. 1.7 and notice that air temperature normally decreases from the earth's surface up to an altitude of about 11 kilometers, which is nearly 36,000 ft, or 7 mi. This decrease in air temperature with increasing height is due primarily to the fact (investigated further in Chapter 2) that sunlight warms the earth's surface, and the surface, in turn, warms the air above it. The rate at which the air temperature decreases with height is called the temperature **lapse rate**. The *average* (or *standard*) *lapse rate* in this region of the lower

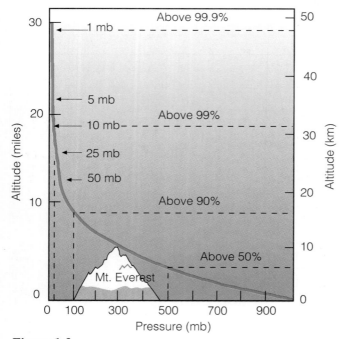

Figure 1.6
Atmospheric pressure decreases rapidly with height. Climbing to an altitude of only 5.5 km, where the pressure is 500 mb, would put you above one-half of the atmosphere's molecules.

atmosphere is about 6.5 degrees Celsius (°C) for every 1000 meters or about 3.6 degrees Fahrenheit for every 1000 feet rise in elevation. Keep in mind that these values are only averages. On some days, the air becomes colder more quickly as we move upward. This would increase or steepen the lapse rate. On other days, the air temperature would decrease more slowly with height, and the lapse rate would be less. So the lapse rate fluctuates, varying from day to day and season to season.

This part of the atmosphere (from the surface up to about 11 kilometers) contains all of the weather we

Did you know?
When flying in a jet aircraft at 30,000 feet, the air temperature just outside your window is normally more than 100°F colder than the air at the surface, directly below you.

are familiar with on earth. Also, this region is kept well-stirred by rising and descending air currents. Here, it is common for air molecules to circulate through a depth of more than 10 kilometers in just a few days. This region of circulating air extending upward from

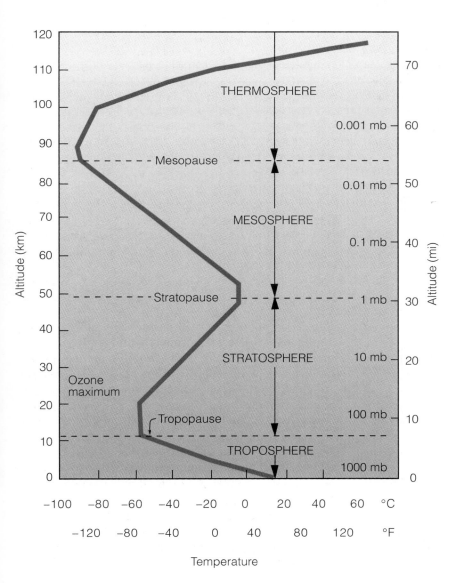

Figure 1.7
Layers of the atmosphere as related to the average profile of air temperature above the earth's surface. Heavy line illustrates how the average temperature varies in each layer.

Focus on Instruments
The Radiosonde

Figure 1
The radiosonde with parachute and balloon.

The vertical distribution of temperature, pressure, and humidity up to an altitude of about 30 km or 100,000 ft can be obtained with an instrument called a *radiosonde*. The radiosonde is a small, lightweight box equipped with weather instruments and a radio transmitter. It is attached to a cord that has a parachute and a gas-filled balloon tied tightly at the end. As the balloon rises, the attached radiosonde measures air temperature with a small electrical thermometer—a thermistor—located just outside the box. It measures humidity electrically by sending an electric current across a carbon-coated plate. Air pressure is obtained by a small barometer located inside the box. All of this information is transmitted to the surface by radio. Here, a computer rapidly reconverts the various frequencies into values of temperature, pressure, and moisture. Special tracking equipment at the surface may also be used to provide a vertical profile of winds.

When plotted on a graph, the vertical distribution of temperature, humidity, and wind is called a *sounding*. Eventually, the balloon bursts and the radiosonde returns to earth, its descent being slowed by its parachute.

At most sites, radiosondes are released twice a day, usually at the time that corresponds to midnight and noon in Greenwich, England. Releasing radiosondes is an expensive operation because many of the instruments are never retrieved, and many of those that are retrieved are often in poor working condition. For this reason, only the more affluent nations can afford the luxury of having a dense radiosonde network.

---

the earth's surface to where the air stops becoming colder with height is called the **troposphere**—from the Greek *tropein*, meaning to turn or change. (The instrument used to measure the vertical profile of air temperature in the atmosphere up to an elevation sometimes exceeding 30 km (100,000 ft) is the **radiosonde**. More information on this instrument is given in the Focus section above.)

Notice in Fig. 1.7 that just above 11 kilometers the air temperature normally stops decreasing with height. Here the lapse rate is zero. This region, where the air temperature remains constant with height, is referred to as an *isothermal* (equal temperature) zone. The bottom of this zone marks the top of the troposphere and the beginning of another layer, the **stratosphere**. The boundary separating the troposphere from the strato-

sphere is called the **tropopause**. The height of the tropopause varies. It is normally found at higher elevations over equatorial regions, and it decreases in elevation as we travel poleward. Generally, the tropopause is higher in summer and lower in winter at all latitudes. In some regions, the tropopause "breaks" and is difficult to locate and, here, scientists have observed tropospheric air mixing with stratospheric air and vice versa. These breaks also mark the position of *jet streams*—high winds that meander in a narrow channel, like an old river, often at speeds exceeding 100 knots.*

---

*A knot is a nautical mile per hour. One knot is equal to 1.15 miles per hour (mi/hr), or 1.9 kilometers per hour (km/hr).

From Fig. 1.7 we can see that, in the stratosphere at an altitude of about 20 km (12 mi), the air temperature begins to increase with height. Such an *increase* in measured air temperature with height is called a **temperature inversion**. The inversion region, along with the lower isothermal layer, tends to keep the vertical currents of the troposphere from spreading into the stratosphere. The inversion also tends to reduce the amount of vertical motion in the stratosphere itself; hence, it is a stratified layer.

Even though the air temperature is increasing with height, the air at an altitude of 30 kilometers is extremely cold, averaging less than –46°C. At this level above polar latitudes, air temperatures can change dramatically from one week to the next, as a *sudden warming* can raise the temperature in one week by more than 50°C. Such a rapid warming is probably due to sinking air associated with circulation changes that occur in late winter or early spring.

The reason for the inversion in the stratosphere is that the gas ozone plays a major part in heating the air at this altitude. Recall that ozone is important because it absorbs energetic ultraviolet (UV) solar energy. Some of this absorbed energy warms the stratosphere, which explains why there is an inversion. If ozone were not present, the air probably would become colder with height, as it does in the troposphere.

Above the stratosphere lies the **mesosphere** (middle sphere). The air here is extremely thin and the atmospheric pressure is quite low (again, refer back to Fig. 1.7). Even though the percentage of nitrogen and oxygen in the mesosphere is about the same as it was at the earth's surface, a breath of mesospheric air contains far fewer oxygen molecules than a breath of tropospheric air. At this level, without proper oxygen-breathing equipment, the brain would soon become oxygen-starved—a condition known as *hypoxia*—and suffocation would result. With an average temperature of –90°C, the top of the mesosphere represents the coldest part of our atmosphere.

The "hot layer" above the mesosphere is the **thermosphere**. Here oxygen molecules (O_2) absorb energetic solar rays, warming the air. In the thermosphere, there are relatively few atoms and molecules. Consequently, the absorption of a small amount of energetic solar energy can cause a large increase in air temperature. (See Fig. 1.8.)

Even though the temperature in the thermosphere is exceedingly high, a person shielded from the sun would not necessarily feel hot. The reason for this fact is that there are too few molecules in this region of the atmosphere to bump against something (exposed skin, for example) and transfer enough heat to it to make it feel warm. The low density of the thermosphere also means that an air molecule will move an average distance of over half a mile before colliding with another molecule. A similar air molecule at the earth's surface will move an average distance of less than one millionth of an inch before it collides with another molecule.

At the top of the thermosphere, about 500 km (300 mi) above the earth's surface, molecules can move greater distances before they collide with other molecules. Here, many of the lighter, faster-moving molecules traveling in the right direction actually escape the earth's gravitational pull. The region where atoms and molecules shoot off into space is sometimes referred to as the *exosphere*, which represents the upper limit of our atmosphere.

Up to this point, we have examined the atmospheric layers based on the vertical profile of temperature. The atmosphere, however, may also be divided into layers based on its composition. For example, the composition of the atmosphere begins to slowly

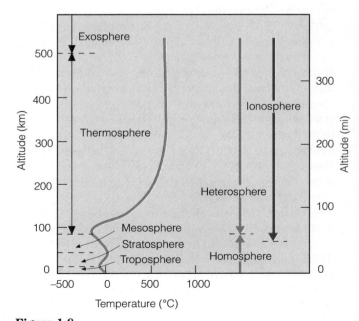

Figure 1.8
Layers of the atmosphere based on temperature (red line), composition (green line), and electrical properties (blue line).

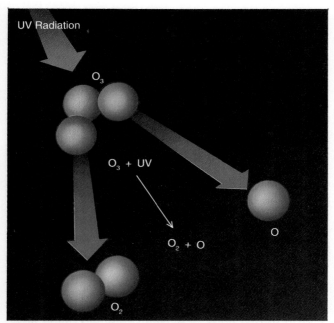

Figure 1.9
An ozone molecule absorbing ultraviolet radiation can become molecular and atomic oxygen.

change in the lower part of the thermosphere. Below the thermosphere, the composition of air remains fairly uniform (78 percent nitrogen, 21 percent oxygen) by turbulent mixing. This lower well-mixed region is known as the *homosphere* (Fig. 1.8). In the thermosphere, collisions between atoms and molecules are infrequent, and the air is unable to keep itself stirred. As a result, diffusion takes over as heavier atoms and molecules (such as oxygen and nitrogen) tend to settle to the bottom of the layer, while lighter gases (such as hydrogen and helium) float to the top. The region from about the base of the thermosphere to the top of the atmosphere is often called the *heterosphere*.

The Ozone Dilemma in the Stratosphere   Earlier in this chapter we learned that the majority of atmospheric ozone is found in the stratosphere. Even near 25 kilometers where ozone is most dense, its concentration is quite small—there are only about 12 ozone molecules for every million air molecules.* Although thin, this layer of ozone is significant, for it shields

earth's inhabitants from harmful amounts of ultraviolet solar radiation. This protection is fortunate because ultraviolet radiation has enough energy to cause skin cancer in humans.

If the concentration of stratospheric ozone were to decrease, the following might occur:

1. an increase in the number of cases of skin cancer
2. an adverse impact on crops and animals due to an increase in energetic ultraviolet radiation
3. a cooling of the stratosphere that could affect stratospheric wind patterns, possibly inducing some form of climate change at the surface

Ozone (O_3) forms naturally in the stratosphere by the combining of atomic oxygen (O) with molecular oxygen (O_2) in the presence of another molecule. Ozone is destroyed naturally by absorbing ultraviolet solar radiation (Fig. 1.9). Ozone is also destroyed by colliding with other gases, such as nitric oxide, nitrogen dioxide, and even other ozone molecules.

It appears that human activities are altering the amount of stratospheric ozone. This possibility was first brought to light in the early 1970s as Congress pondered over whether or not the United States should build a supersonic jet transport. One of the gases emitted from the engines of this aircraft is nitric oxide. Although the aircraft was designed to fly in the stratosphere below the level of maximum ozone, it was feared that the nitric oxide would eventually work its way upward, where it would have an adverse effect on the ozone. This factor was one of many considered when Congress decided to halt the development of the United States' version of the supersonic transport.

More recently, concerns involve emissions of chemicals at the earth's surface, such as *chlorofluorocarbons* (CFCs). Until the late 1970s, when the United States banned all nonessential uses of these chemicals, they were the most widely used propellants in spray cans. In the troposphere, these gases are quite safe, being nonflammable, nontoxic, and unable to chemically combine with other substances.* Hence, these gases slowly diffuse upward without being destroyed. They apparently enter the stratosphere (1) near breaks in the tropopause, especially in the vicinity of jet streams; and (2) in building thunderstorms that penetrate the lower stratosphere.

---

*Here, the composition of air is about the same as it is near the earth's surface: mainly 78 percent nitrogen and 21 percent oxygen.

*Recall, however, that CFCs do act as strong greenhouse gases in the troposphere.

Once CFC molecules enter the middle stratosphere, ultraviolet solar energy that is normally absorbed by ozone breaks them up, releasing *chlorine* in the process (and chlorine rapidly destroys ozone). In fact, estimates are that a single chlorine atom removes about 100,000 ozone molecules before it is taken out of action by combining with other substances.

Since the average lifetime of a CFC molecule in the stratosphere is about 100 years, any increase in the concentration of CFCs is long lasting and a genuine threat to the concentration of ozone. Given this fact and the additional knowledge that CFCs contribute to the earth's greenhouse effect, an international agreement signed in 1987—the *Montreal Protocol*—established a timetable for diminishing CFC emissions and the use of bromine compounds (halons), which destroy ozone at a rate ten times greater than do chlorine compounds. This agreement calls for a 50 percent cutback in CFC production (based on 1986 levels) by the year 1999. How successful this reduction will be depends on how well the signing nations comply and, of course, on the behavior of those who did not sign. The protocol calls for periodic scientific reviews for updating its requirements.

The question of just how much ozone is being depleted by chlorine is now under investigation. More than 5 billion kilograms of CFCs have already been released into the troposphere and will diffuse upward during the next few decades. In a 1991 study, an international panel of over eighty scientists concluded that the ozone layer thinned by about 3 percent during the summer from 1979 to 1991 over heavily populated areas of the Northern Hemisphere.

To further complicate the picture, theoretical calculations of ozone chemistry in the early 1980s suggested that an increase in CO_2 levels in the troposphere (resulting from fossil fuel and wood combustion) would raise the temperature in the troposphere, but lower it in the stratosphere. A cooler stratosphere would slow the process of ozone destruction.

It now appears, however, that ozone is being destroyed more quickly than expected. Scientists point to the fact that ozone concentrations over springtime Antarctica have plummeted at an alarming rate. This sharp drop in ozone is known as the **ozone hole**. (See the Focus section on p. 14.)

The Ionosphere The **ionosphere** is not really a layer, but rather an electrified region within the upper atmosphere where fairly large concentrations of ions and free electrons exist. Ions are atoms and molecules that have lost (or gained) one or more electrons. Atoms lose electrons and become positively charged when they cannot absorb all of the energy transferred to them by a colliding energetic particle or the sun's energy.

The lower region of the ionosphere is usually about 60 kilometers above the earth's surface. From here (60 km), the ionosphere extends upward to the top of the atmosphere. Hence, the bulk of the ionosphere is in the thermosphere. (See Fig. 1.8 on p. 11.)

The ionosphere plays a major role in radio communications. The lower part reflects standard AM radio waves back to earth, but at the same time it seriously weakens them as they are reflected. At night, though, the lower part of the ionosphere gradually disappears, and AM radio waves are able to penetrate higher into the ionosphere, where the waves are reflected back to earth. Because there is, at night, little absorbtion of radio waves in the higher reaches of the ionosphere, such waves bounce repeatedly from the ionosphere to the earth's surface and back to the ionosphere again. In this way, standard AM radio waves are able to travel for many hundreds of kilometers at night.

We have, in the last several sections, been examining our atmosphere from a vertical perspective. We will now turn our attention to weather events that take place in the lower atmosphere. As you read the remainder of this chapter, keep in mind that the content serves as a broad overview of material to come in later chapters, and that many of the concepts and ideas you encounter are designed to familiarize you with items you might read about in a newspaper or magazine, or see on television.

▲▼▲

Weather and Climate

When we talk about the **weather**, we are talking about the condition of the atmosphere at any particular time and place. Weather—which is always changing—is comprised of the elements of:

1. *air temperature*
2. *air pressure*
3. *humidity*
4. *clouds*
5. *precipitation*
6. *visibility*
7. *wind*

Focus on a Special Topic
The Ozone Hole

In 1974, two chemists from the University of California at Irvine—F. Sherwood Rowland and Mario J. Molina—warned that increasing levels of CFCs would eventually deplete stratospheric ozone on a global scale. Their studies suggested that ozone depletion would occur gradually and would perhaps not be detectable for many years to come. It was surprising, then, when British researchers identified a year-to-year decline in stratospheric ozone over Antarctica. Their findings, corroborated later by satellites and balloon-borne instruments, showed that since the late 1970s ozone concentrations have diminished each year during the months of September and October. Ozone diminished by as much as 40 percent in 1984 and shrank to the point of almost total depletion in certain regions of the lower stratosphere in 1987, in 1989, and again in 1991. This decrease in stratospheric ozone over Antarctica is known as the *ozone hole*. (See Fig. 2.)

In years of severe depletion, the ozone hole covers almost twice the area of the Antarctic continent. However, the results of research done with high-altitude aircraft flights during 1987 showed significant amounts of ozone being destroyed outside the main hole, suggesting that the area of ozone loss may be considerably greater than once thought. Furthermore, this same research revealed that ozone destruction begins earlier in the year than expected.

To understand the causes behind the ozone hole, scientists in 1986 organized the first *National Ozone Expedition*, NOZE-1, which set up a fully instrumented observing station near McMurdo Sound, Antarctica. During 1987, with the aid of instrumented aircraft, NOZE-2 got underway. The findings from these research programs

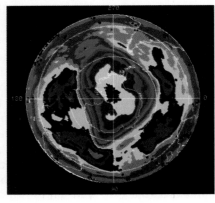

Figure 2
Ozone distribution over the Southern Hemisphere on October 5, 1989, as measured by the Total Ozone Mapping Spectrometer (TOMS) aboard the *NIMBUS-7* satellite. Notice that the area of lowest ozone concentration (purple shades) is larger than Antarctica, and that the ozone hole is nearly centered over the South Pole.

helped scientists put together the pieces of the ozone puzzle.

The stratosphere above Antarctica has one of the world's highest ozone concentrations. Most of this ozone forms over the tropics and is brought to the Antarctic by stratospheric winds. During September and October (spring in the Southern Hemisphere), a belt of stratospheric winds called the *polar vortex* encircles the Antarctic region near 66°S latitude, essentially isolating the cold Antarctic stratospheric air from the warmer air of the middle latitudes. During the long, dark Antarctic winter, temperatures inside the vortex can drop to −85°C (−121°F). This frigid air allows for the formation of *polar stratospheric clouds*. These ice clouds are critical in facilitating chemical interactions among nitrogen, hydrogen, and chlo-

rine atoms, the end product of which is the destruction of ozone.

In 1986, the NOZE-1 study detected unusually high levels of chlorine compounds in the stratosphere, and, in 1987, the instrumented aircraft of NOZE-2 measured enormous increases in chlorine compounds when it entered the polar vortex. These findings, in conjunction with other chemical discoveries, allowed scientists to pinpoint chlorine from CFCs as the main cause of the ozone hole.

Chemistry alone, however, does not explain the entire ozone depletion problem in Antarctica. It does not, for example, explain the early decline in ozone readings just as the ozone hole begins to form. Perhaps dynamic events play a role in this phenomenon, such as the mixing of stratospheric ozone-rich air with tropospheric ozone-poor air from below.

In the Polar Arctic, airborne instruments during 1989 measured high levels of ozone-destroying chlorine compounds in the stratosphere—levels 100 times higher than those over the United States. However, satellites monitoring ozone could detect no real hole. Apparently, several factors inhibit massive ozone loss in the Arctic. For one thing, in the stratosphere, the circulation of air over the Arctic differs from that over the Antarctic. Then, too, the Arctic stratosphere is too warm for the widespread development of clouds that help activate chlorine molecules.

We still have much to learn about stratospheric ozone and the processes that both form and destroy it. Presently, atmospheric studies are providing more information so that a more complete assessment of the ozone problem will become available in the future.

If we measure and observe these **weather elements** over a specified interval of time, say, for many years, we would obtain the "average weather" or the **climate** of a particular region. Climate, therefore, represents the accumulation of daily and seasonal weather events over a long period of time. The concept of climate is much more than this for it also includes the extremes of weather—the heat waves of summer and the cold spells of winter—that occur in a particular region. The *frequency* of these extremes is what helps us distinguish among climates that have similar averages.

If we were able to watch the earth for many thousands of years, even the climate would change. We would see rivers of ice moving down stream-cut valleys and huge glaciers—sheets of moving snow and ice—spreading their icy fingers over large portions of North America. Advancing slowly from Canada, a single glacier might extend as far south as Kansas and Illinois, with ice several thousands of meters thick covering the region now occupied by Chicago. Over an interval of two million years or so, we would see the ice advance and retreat several times. Of course, for this phenomenon to happen, the average temperature of North America would have to decrease and then rise in a cyclic manner.

Suppose we could photograph the earth once every thousand years for many hundreds of millions of years. In time-lapse film sequence, these photos would show that not only is the climate altering, but the whole earth itself is changing as well: Mountains would rise up only to be torn down by erosion; isolated puffs of smoke and steam would appear as volcanoes spew hot gases and fine dust into the atmosphere; and the entire surface of the earth would undergo a gradual transformation as some ocean basins widen and others shrink.*

In summary, the earth and its atmosphere are dynamic systems that are constantly changing. While major transformations of the earth's surface are completed only after long spans of time, the state of the atmosphere can change in a matter of minutes. Hence, a watchful eye turned skyward will be able to observe many of these changes.

Up to this point, we have looked at the concepts of weather and climate without discussing the word **meteorology**. What does this term actually mean and where did it originate?

*The movement of the ocean floor and continents is explained in the widely acclaimed theory of *plate tectonics*, formerly called the theory of continental drift.

Meteorology—the Science of the Atmosphere

Meteorology is the study of the atmosphere and its phenomena. The term itself goes back to the Greek philosopher Aristotle who, about 340 B.C., wrote a book on natural philosophy entitled *Meteorologica*. This work represented the sum of knowledge on weather and climate at that time, as well as material on astronomy, geography, and chemistry. Some of the topics covered included clouds, rain, snow, wind, hail, thunder, and hurricanes. In those days, all substances that fell from the sky, and anything seen in the air, were called meteors, hence the term *meteorology*. Today, we differentiate between those meteors that come from extraterrestrial sources outside our atmosphere (meteoroids) and particles of water and ice observed in the atmosphere (hydrometeors).

In *Meteorologica*, Aristotle attempted to explain atmospheric phenomena in a philosophical and speculative manner. Even though many of his speculations were found to be erroneous, Aristotle's ideas were accepted without reservation for almost two thousand years. In fact, the birth of meteorology as a genuine natural science did not take place until the invention of weather instruments, such as the thermometer at the end of the sixteenth century, the barometer (for measuring air pressure) in 1643, and the hygrometer (for measuring humidity) in the late 1700s. With observations from instruments available, attempts were then made to explain certain weather phenomena employing scientific experimentation and the physical laws that were being developed at the time.

As more and better instruments were developed in the 1800s, the science of meteorology progressed. The invention of the telegraph in 1843 allowed for the transmission of routine weather observations. Ideas about winds and storms became, at least partially, understood, and crude weather maps were drawn. Around 1920, the concepts of air masses and weather fronts were formulated in Norway. By the 1940s, daily upper-air balloon observations of temperature, humidity, and pressure gave a three-dimensional view of the atmosphere.

Meteorology took another step forward in the 1950s, when high-speed computers were developed to solve the equations that describe the behavior of the atmosphere. Along with drawing current weather maps, the computers were used to predict the state of the atmosphere at some desired time in the future. In 1960, *Tiros I*, the first weather satellite, was launched, ushering in space-age meteorology. Subsequent satellites provided a wide range of useful information, ranging from day and night time-lapse photographs of clouds and storms to pictures that depict swirling ribbons of water vapor flowing around the globe. Even today, more sophisticated satellites are being developed to supply computers with a far greater network of data, so that more accurate forecasts—perhaps up to a week or more—will be available in the future.

A Satellite's View of the Weather   A good view of the weather can be seen from a weather satellite. Figure 1.10 is a satellite photograph showing a portion of the Pacific Ocean and the North American continent. The photograph was obtained from a *geostationary satellite* situated about 36,000 km (22,300 mi) above the earth. At this elevation, the satellite travels at the same rate as the earth spins, which allows it to remain positioned above the same spot so it can continuously monitor what is taking place beneath it.

Storms of All Sizes   Probably the most dramatic spectacle in Fig. 1.10 is the whirling cloud masses of all shapes and sizes. The clouds appear white because sunlight is reflected back to space from their tops. The dark areas show where skies are clear. The largest of the organized cloud masses are the sprawling storms. One such storm shows as an extensive band of clouds, over twelve hundred miles long, west of the Great Lakes. This **middle latitude cyclonic storm** system (or *extratropical cyclone*) forms outside the tropics and, in the Northern Hemisphere, has winds spinning counterclockwise about its center, which is presently over Minnesota.

A slightly smaller but more vigorous storm is located over the Pacific Ocean. This tropical storm system, with its swirling band of rotating clouds and surface winds in excess of 64 knots or 74 miles per hour (mi/hr), is known as a **hurricane**. The diameter of the hurricane is about 800 km (500 mi). The tiny dot at its center is called the *eye*. In the eye, winds are light and skies are generally clear. Around the eye, however,

is an extensive region where heavy rain and high surface winds are reaching peak gusts of 100 knots.

Smaller storms are seen as bright spots over the Gulf of Mexico. These spots represent clusters of towering *cumulus* clouds that have grown into **thunderstorms**; that is, tall churning clouds accompanied by lightning, thunder, strong gusty winds, and heavy rain. If you look closely at Fig. 1.10, you will see similar cloud forms in many regions. There were probably thousands of thunderstorms occurring throughout the world at that very moment. Although they cannot be seen individually, there are even some thunderstorms embedded in the cloud mass west of the Great Lakes. Later in the day on which this photograph was taken, a few of these storms spawned the most violent disturbance in the atmosphere—the **tornado**.

A tornado is an intense rotating column of air that extends downward from the base of a thunderstorm. Sometimes called *twisters* or *cyclones*, they may appear as ropes or as a large circular cylinder. The majority are less than a kilometer wide and many are smaller than a football field. Tornado winds may exceed 200 knots but most probably peak at less than 125 knots. Some tornadoes never reach the ground, and often appear to hang from the base of a parent cloud as a rapidly rotating funnel. Often, they dip down then rise up before disappearing.

A Look at a Weather Map   We can obtain a better picture of the middle latitude storm system by examining a simplified surface weather map for the same day that the satellite picture was taken. The weight of the air above different regions varies and, hence, so does the atmospheric pressure. In Fig. 1.11 the letter L on the map indicates a region of low atmospheric pressure, often called a *low*, which marks the center of the middle latitude storm. The two letters H on the map represent regions of high atmospheric pressure, called *highs*, or *anticyclones*. The circles on the map represent individual weather stations. The **wind** is the horizontal movement of air. The **wind direction**—the direction from which the wind is blowing—is given by lines that parallel the wind and extend outward from the center of the station. The wind speed is indicated by barbs.

Notice how the wind blows around the highs and low. The horizontal pressure differences cause the air to move from higher pressure toward lower pressure. Because of the earth's rotation, the winds are bent in such a way that, in the Northern Hemisphere, they blow *clockwise* and *outward* from the center of the highs, and *counterclockwise* and *inward* toward the center of the low. (We will examine this concept more completely in Chapter 6.)

As the surface air spins into the low, it flows together and rises, much like toothpaste does when its open tube is squeezed. The rising air cools, and the moisture in the air condenses into clouds. Notice in Fig. 1.11 that the area of precipitation (the shaded green area) in the vicinity of the low corresponds to an extensive cloudy region in the satellite photo (Fig. 1.10).

Also notice by comparing Figs. 1.10 and 1.11 that, in the regions of high pressure, skies are generally clear. As the surface air flows outward away from the center of a high, air sinking from above must replace the laterally spreading air. Since sinking air does not usually produce clouds, we find generally clear skies and fair weather associated with the regions of high pressure.

The swirling air around the areas of high and low pressure are the major weather producers for the middle latitudes. Look at the middle latitude storm and the surface temperatures in Fig. 1.11 and notice that, to the southeast of the storm, southerly winds from the Gulf of Mexico are bringing warm, humid air northward over much of the southeastern portion of the nation. On the storm's western side, cool, dry northerly breezes combine with sinking air to create generally clear weather over the Rocky Mountains. The boundary that separates the warm and cool air appears as a heavy, dark line on the map—a **front**, across which there is a sharp change in temperature, humidity, and wind direction.

Where the cool air from Canada replaces the warmer air from the Gulf of Mexico, a *cold front* is drawn in blue, with arrowheads showing its direction of movement. Where the warm Gulf air is replacing cooler air to the north, a *warm front* is drawn in red, with half circles showing its direction of movement. Where cold air is replacing cool air, an *occluded front* is drawn in purple, with alternating arrowheads and half circles to show how it is moving. Along each of the fronts, warm air is rising, producing clouds and precipitation. In the satellite photo (Fig. 1.10), the occluded front and the cold front appear as an elongated, curling cloud band that stretches from the low pressure area over Minnesota into the northern part of Texas.

Notice in Fig. 1.11 that the weather front is to the west of Chicago. As the westerly winds aloft push the

front eastward, a person on the outskirts of Chicago might observe the approaching front as a line of towering thunderstorms similar to those in Fig. 1.12. In a few hours, Chicago should experience heavy showers with thunder, lightning, and gusty winds as the front passes. All of this, however, should give way to clearing skies and surface winds from the west or northwest after the front has moved on by.

Observing storm systems, we see that not only do they move but they constantly change. Steered by the upper-level westerly winds, the middle latitude storm in Fig. 1.11 intensifies into a larger storm, which moves eastward, carrying its clouds and weather with it. In advance of this system, a sunny day in Ohio will gradually cloud over and yield heavy showers and thunderstorms by nightfall. Behind the storm, cool, dry northerly winds rushing into eastern Colorado cause an overcast sky to give way to clearing conditions. Further south, the thunderstorms presently over the Gulf of Mexico (Fig. 1.10) expand a little, then dissipate as new storms appear over water and land areas. To the west, the hurricane over the Pacific Ocean drifts northwestward and encounters cooler water. Here, away from its warm energy source, it loses its punch; winds taper off, and the storm soon turns into an unorganized mass of clouds and tropical moisture.

Weather and Climate in Our Lives   Weather and climate play a major role in our lives. Weather, for example, often dictates the type of clothing we wear, while climate influences the type of clothing we buy. Climate determines when to plant crops as well as what type of crops can be planted. Weather determines if these same crops will grow to maturity. Although weather and climate affect our lives in many ways, perhaps their most immediate effect is on our comfort. In order to survive the cold of winter and heat of summer, we build homes, heat them, air condition them, insulate them—only to find that when we leave our shelter, we are at the mercy of the weather elements.

Figure 1.10
This satellite picture (taken in visible, reflected light) shows a variety of cloud patterns and storms in the earth's atmosphere.

Even when we are dressed for the weather properly, wind, humidity, and precipitation can change our perception of how cold or warm it feels. On a cold, windy day the effects of *wind chill* tell us that it feels much colder than it really is, and, if not properly dressed, we run the risk of *frostbite* or even *hypother-*

mia (the rapid, progressive mental and physical collapse that accompanies the lowering of human body temperature). On a hot, humid day we normally feel uncomfortably warm and blame it on the humidity. If we become too warm, our bodies overheat and *heat exhaustion* or *heat stroke* may result. Those most likely

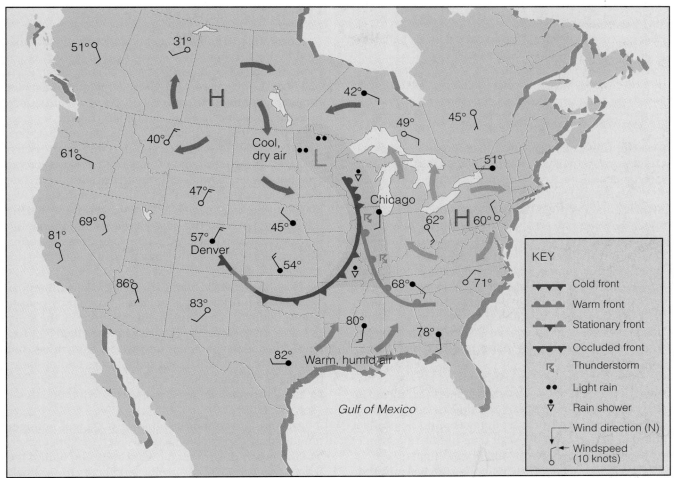

Figure 1.11
Simplified surface weather map that correlates with the satellite picture shown in Fig. 1.10.
The shaded green area represents precipitation. Air temperatures are in °F.

Figure 1.12
Thunderstorms developing along an approaching cold front.

Did you know?
During a severe drought in the southwestern United States in 1971, most of the 200,000 bats that lived in the Carlsbad Caverns of New Mexico fled their caves, where there was no food or water. Massive flocks could be seen searching for food in the town of Carlsbad and along the Pecos River.

to suffer these maladies are the elderly with impaired circulatory systems and infants, whose heat regulatory mechanisms are not yet fully developed.

Weather affects how we feel in other ways, too. Arthritic pain is most likely to occur when rising humidity is accompanied by falling pressures. In ways not well understood, weather does seem to affect our health. The incidence of heart attacks shows a statistical peak after the passage of warm fronts, when rain and wind are common and after the passage of cold fronts, when an abrupt change takes place as showery precipitation is accompanied by cold, gusty winds. Headaches are common on days when we are forced to squint, often due to hazy skies or a thin, bright overcast layer of high clouds.

For some people, a warm, dry wind blowing downslope (a *chinook wind*) adversely affects their behavior (they often become irritable and depressed). Just how and why these winds impact humans physiologically is not well understood. We will take up the question of why these winds are warm and dry in Chapter 7.

When the weather turns colder or warmer than normal it influences the lives and pocketbooks of many people. For example, the cool summer of 1992 over the eastern two-thirds of North America saved people billions of dollars in air conditioning costs. On the other side of the coin, the bitter cold winter of 1986–1987 over Europe killed many hundreds of people and caused fuel rationing as demands for fuel exceeded supplies.

Major cold spells accompanied by heavy snow and ice can play havoc by snarling commuter traffic, curtailing airport services, closing schools, and downing power lines, thereby cutting off electricity to thousands of customers. One such storm during February, 1983, buried many eastern cities, and left Philadelphia paralyzed with a snow depth of twenty-one inches. When the frigid air settles into the deep south, many millions of dollars worth of temperature-sensitive fruits and vegetables may be ruined, the consequence of which

often shows up as higher produce prices in the supermarket.

Prolonged dry spells, especially when accompanied by high temperatures, can lead to a shortage of food and, in some places, widespread starvation. Parts of Africa, for example, have been in the grip of a major drought and famine for several decades. Organizations and major benefits have saved millions of starving people by providing outside aid. In 1986, the southeastern section of the United States suffered through its worst drought as searing summer temperatures wilted crops, causing losses in excess of one billion dollars. When the climate turns hot and dry, animals suffer too. Over 500,000 chickens perished in Georgia alone during a two-day period at the peak of the summer heat. Severe drought also has an effect on water reserves, often forcing communities to ration water and restrict its use.

Each year the violent side of weather influences the lives of millions. It is amazing how many people whose family roots are in the Midwest know the story of someone who was severely injured or killed by a tornado. Tornadoes have not only taken many lives, but annually they cause damage to buildings and property totaling in the hundreds of millions of dollars, as a single large tornado can level an entire section of a town. (See Fig. 1.13.)

Although the gentle rains of a typical summer thunderstorm are welcome over much of North America, the heavy downpours, high winds, and hail of the *severe thunderstorms* are not. Cloudbursts from slowly moving violent thunderstorms can provide too much rain too quickly, creating *flash floods* as small streams become raging rivers composed of mud and sand entangled with uprooted plants and trees. Strong downdrafts originating inside a severe thunderstorm (a *downburst*) create turbulent winds that are capable of destroying crops and inflicting damage upon surface structures. Several airline crashes have been attributed to the turbulent *wind shear* zone within the downburst. Annually, hail damages crops worth millions of dollars, and lightning starts fires that destroy many thousands of acres of valuable timber.

Even the quiet side of weather has its influence. When winds die down, and humid air becomes more tranquil, fog may form. Heavy fog can restrict visibility at airports, causing flight delays and cancellations. Every winter, deadly fog-related auto accidents occur along our busy highways and turnpikes. But, fog has a positive side too, especially during a dry spell, as fog moisture collects on tree branches and drips to the ground, where it provides water for the tree's root system.

Weather and climate have become so much a part of our lives that the first thing many of us do in the morning is to listen to the local weather forecast. For this reason, many radio and television newscasts have their own "weather person" to present weather information and give daily forecasts. More and more of these people are professionally trained in meteorology, and many stations require that the weather caster obtain a seal of approval from the American Meteorological Society (AMS), or a certificate from the National Weather Association (NWA). To make their weather presentation as up-to-the-minute as possible, an increasing number of stations are taking advantage of the information provided by the National Weather Service (NWS), such as computerized weather forecasts, time-lapse satellite pictures, and color radar displays. A few stations are even purchasing their own Doppler radar units to aid in the forecasting of severe local storms.

For many years now, educational television stations have aired a fifteen-minute weather program, Monday through Friday, entitled "AM Weather"—a program in which national weather service meteorologists present forecasts, summaries, and aviation weather. On cable television, the "Weather Channel" has a staff of trained professionals who give weather information twenty-four hours a day. Finally, the National Oceanic and Atmospheric Administration (NOAA), in cooperation with the National Weather Service, sponsors

Figure 1.13
Tornadoes annually inflict widespread damage and cause the loss of many lives.

weather radio broadcasts at selected locations across the United States. Known as *NOAA weather radio* (and transmitted at VHF–FM frequencies), this service provides continuous weather information and regional forecasts (as well as special weather advisories, including watches and warnings) for over 90 percent of the nation.

Summary

This chapter provides a brief overview of the earth's atmosphere. Our atmosphere is one rich in nitrogen and oxygen as well as smaller amounts of other gases and particles, some of them being extremely important, such as water vapor, carbon dioxide, and ozone. We examined the earth's early atmosphere and found it to be much different from the air we breathe today.

We then looked at the various layers of the atmosphere and found that stratospheric ozone, which protects us from the sun's harmful ultraviolet rays, may be decreasing in concentration as gases such as chlorofluorocarbons in the stratosphere break apart and release ozone-destroying chlorine. We found that the coldest part of our atmosphere is in the mesosphere, that the warmest is in the thermosphere, and that all the weather we have come to know exists in the troposphere.

We looked briefly at the weather map and a satellite photo and observed that dispersed throughout the atmosphere are storms and clouds of all sizes and shapes. The movement, intensification, and weakening of these systems, as well as the dynamic nature of air itself, produce a variety of weather events that we describe in terms of weather elements. The sum total of weather over a long period of time is what we call climate. Although sudden changes in weather may occur in a moment, climatic change takes place gradually over many years. The study of the atmosphere and all of its related phenomena is called meteorology, a term whose origin dates back to the days of Aristotle. Finally, we discussed some of many ways weather and climate influence our lives.

Key Terms

The following terms are listed in the order they appear in the text. Define each. Doing so will aid you in reviewing the material covered in this chapter.

atmosphere	troposphere	ozone hole	hurricane
water vapor	radiosonde	ionosphere	thunderstorm
aerosols	stratosphere	weather	tornado
pollutants	tropopause	weather elements	wind
outgassing	temperature inversion	climate	wind direction
air pressure	mesosphere	meteorology	front
lapse rate	thermosphere	middle latitude cyclonic storm	

Review Questions

1. What are the four most abundant gases in today's atmosphere?
2. Explain how the atmosphere "protects" inhabitants at the earth's surface.
3. Basically, how do the three states of water differ?
4. What are some of the important roles that water plays in our atmosphere?
5. What are some of the aerosols in our atmosphere?
6. How has the earth's atmosphere changed over time?

7. (a) Explain the concept of air pressure in terms of weight of air above some level.
   (b) Why does air pressure always decrease with height?
8. What atmospheric layer contains all of our weather?
9. On the basis of temperature, list the layers of the atmosphere from the lowest layer to the highest.
10. Briefly describe how the air temperature changes from the earth's surface to the lower thermosphere.

11. Explain how scientists believe the Antarctic "ozone hole" forms.
12. (a) How are CFCs related to the destruction of stratospheric ozone?
    (b) If all the ozone in the stratosphere were destroyed, what possible effects might this have on the earth's atmosphere and its inhabitants?
13. Define temperature inversion.
14. What is the ionosphere and where is it located?
15. List the common weather elements.
16. How does weather differ from climate?
17. Define *meteorology* and discuss the origin of this word.
18. Rank the following storms in size from largest to smallest: hurricane, tornado, middle latitude cyclonic storm, thunderstorm.
19. When someone says that "the wind direction today is north," what does that mean?
20. Weather in the middle latitudes tends to move in what general direction?
21. Describe some of the ways weather and climate influence the lives of people.

Shorter days and longer nights (created by a tilted earth revolving about the sun) promote cooler weather, hazy days, and a hint of autumn. (Photo by author)

Chapter 2

Warming the Earth and the Atmosphere

Contents

▲▼▲

. . . The sun doesn't rise or fall: it doesn't move, it just sits there, and we rotate in front of it. Dawn means that we are rotating around into sight of it, while dusk means we have turned another 180 degrees and are being carried into the shadow zone. The sun never "goes away from the sky." It's still there sharing the same sky with us; it's simply that there is a chunk of opaque earth between us and the sun which prevents our seeing it. Everyone knows that, but I really see it now. No longer do I drive down a highway and wish the blinding sun would set; instead I wish we could speed up our rotation a bit and swing around into the shadows more quickly.

Michael Collins, *Carrying the Fire*   ▲

2 As you sit quietly reading this book, you are part of a moving experience. The earth is speeding around the sun at thousands of miles per hour while, at the same time, it is spinning on its axis. When we look down upon the North Pole, we see that the direction of spin is counterclockwise, meaning that we are moving toward the east at hundreds of miles per hour. We normally don't think of it in that way, but, of course, this is what causes the sun, moon, and stars to rise in the east and set in the west. In fact, it is these motions coupled with energy from the sun, striking a tilted planet, that causes our seasons. But, as we will see later, the sun's energy is not distributed evenly over the earth, as tropical regions receive more energy than polar regions. It is this energy imbalance that drives our atmosphere into the dynamic patterns we experience as wind and weather.

Therefore, we will begin this chapter by examining the concept of energy and heat transfer. Then we will see how our atmosphere warms and cools. Finally, we will examine how the earth's motions and the sun's energy work together to produce the seasons.

▲▼▲

Temperature and Heat Transfer

Temperature is the quantity that tells us how hot or cold something is relative to some set standard value. But we can look at temperature in another way.

We know that air is a mixture of countless billions of atoms and molecules. If they could be seen, they would appear to be moving about in all directions, freely darting, twisting, spinning, and colliding with one another like an angry swarm of bees. Close to the earth's surface, each individual molecule would travel about a thousand times its diameter before colliding with another molecule. Moreover, we would see that all the atoms and molecules are not moving at the same speed, as some are moving faster than others. The energy associated with this motion is called **kinetic energy**, the energy of motion. The temperature of the air (or any substance) is a measure of its average kinetic energy. Simply stated, **temperature** *is a measure of the average speed of the atoms and molecules*, where higher temperatures correspond to faster average speeds.

If we slowly cool air, its atoms and molecules would move slower and slower until the air reaches a temperature of −273°C (−459°F), which is the lowest temperature possible. At this temperature, called *absolute zero*, the atoms and molecules would possess a minimum amount of energy and theoretically no thermal motion. At absolute zero, we can begin a temperature scale called the *absolute* or **Kelvin scale**, after Lord Kelvin (1824–1907), a famous British scientist who first introduced it. (If you are unfamiliar with this temperature scale, or with the Celsius or Fahrenheit scales, read the Focus section on p. 27.)

The atmosphere contains internal (thermal) energy, which is the energy it possesses due to its temperature. **Heat**, on the other hand, *is energy in the process of being transferred from one object to another because of the temperature difference between them*. After heat is transferred, it is stored as internal energy. In the atmosphere, heat is transferred by *conduction, convection*, and *radiation*. We will examine these mechanisms of energy transfer after we look at the important concept of latent heat.

Latent Heat—The Hidden Warmth   The heat energy required to change a substance from one state to another is called **latent heat**. But why is this heat referred to as "latent"? To answer this question, we will begin with something familiar to most of us—the cooling produced by evaporating water.

Suppose we microscopically examine a small drop of pure water. At the drop's surface, molecules are constantly escaping (evaporating). Because the more energetic, faster-moving molecules escape most easily, the average motion of all the molecules left behind decreases as each additional molecule evaporates. Since temperature is a measure of average molecular motion, the slower motion suggests a lower water temperature. *Evaporation is, therefore, a cooling process.* Stated another way, evaporation is a cooling process because the energy needed to evaporate the water—that is, to change its phase from a liquid to a gas—may come from the water or other sources, including the air.

The energy lost by liquid water during evaporation can be thought of as carried away by, and "locked up" within, the water vapor molecule. The energy is thus in a "stored" or "hidden" condition and is, therefore, called *latent heat*. It is latent (hidden) in that the temperature of the substance changing from liquid to vapor is still the same. However, the heat energy will reappear as **sensible heat** (the heat we can feel and measure with a thermometer) when the vapor condenses back

Focus on Instruments
Temperature Scales

The Kelvin scale starts at absolute zero. Since it contains no negative numbers, it is quite convenient for scientific calculations. Two other temperature scales commonly used today are the Fahrenheit and Celsius (formerly centigrade). The **Fahrenheit scale** was developed in the early 1700s by the physicist G. Daniel Fahrenheit, who assigned the number 32 to the temperature at which water freezes, and the number 212 to the temperature at which water boils. The zero point was simply the lowest temperature that he obtained with a mixture of ice, water, and salt. Between the freezing and boiling points are 180 equal divisions, each of which is called a *degree*. A thermometer calibrated with this scale is referred to as a Fahrenheit thermometer, for it measures an object's temperature in degrees Fahrenheit (°F).

The **Celsius scale** was introduced later in the eighteenth century. The number 0 (zero) on this scale is assigned to the temperature at which water freezes, and the number 100 to the temperature at which water boils. The space between freezing and boiling is divided into 100 equal degrees. Therefore, each degree Celsius (°C) is 180/100 or 1.8 times bigger than a degree Fahrenheit. Put another way, an increase in temperature of 1°C equals an increase of 1.8°F.

On the Kelvin scale, degrees Kelvin are called *Kelvins* (abbreviated K). Each degree on the Kelvin scale is exactly the same size as a degree Celsius, and a temperature of 0 K is equal to –273°C. Converting from °C to K can be made by simply adding 273 to the Celsius temperature, as

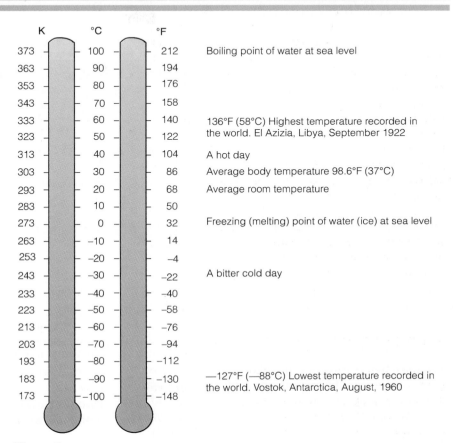

Figure 1
Comparison of Fahrenheit, Celsius, and Kelvin scales.

$$K = °C + 273.$$

Figure 1 compares the Fahrenheit, Celsius, and Kelvin scales. Converting a temperature from one scale to another can be done by simply reading the corresponding temperature from the adjacent scale. Thus, 303 on the Kelvin scale is the equivalent of 30°C and 86°F.*

*A more complete table of conversions is given in Appendix A.

In most of the world, temperature readings are taken in °C. In the United States, however, temperatures above the surface are taken in °C, while temperatures at the surface are typically read in °F. Currently, then, temperatures on upper-level maps are plotted in °C, while, on surface weather maps, they are in °F. Since both scales are in use, temperature readings in this book will, in most cases, be given in °C followed by their equivalent in °F.

Figure 2.1
Heat energy absorbed
and released.

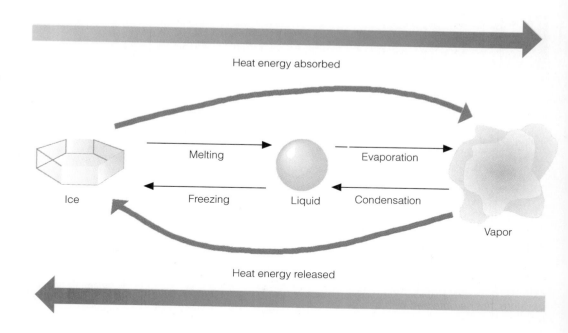

into liquid water. Therefore, *condensation* (the opposite of evaporation) *is a warming process.*

The heat energy released when water vapor condenses to form liquid droplets is called *latent heat of condensation.* Conversely, the heat energy used to change liquid into vapor at the same temperature is called *latent heat of evaporation* (vaporization). Nearly 600 calories* are required to evaporate a single gram of water at room temperature. With many hundreds of grams of water evaporating from the body, it is no wonder that after a shower we feel cold before drying off.

Figure 2.1 summarizes the concepts examined so far. When the change of state is from left to right, heat is absorbed by the substance and taken away from the environment. When the change of state is from right to left, heat energy is given up by the substance and added to the environment.

Latent heat is an important source of atmospheric energy. Once vapor molecules become separated from the earth's surface, they are swept away by the wind, like dust before a broom. Rising to high altitudes where the air is cold, the vapor changes into liquid and ice cloud particles. During these processes, a tremendous amount of heat energy is released into the environment. (See Fig. 2.2.)

---

*By definition, a calorie is the amount of heat required to raise the temperature of 1 gram of water from 14.5°C to 15.5°C.

Water vapor evaporated from warm, tropical water can be carried into polar regions, where it condenses and gives up its heat energy. Thus, as we will see, evaporation-transportation-condensation is an extremely important mechanism for the relocation of heat energy (as well as water) in the atmosphere.

Conduction   The transfer of heat from molecule to molecule within a substance is called **conduction**. Hold one end of a metal straight pin between your fingers and place a flaming candle under the other end. (See Fig. 2.3.) Because of the energy they absorb from the flame, the molecules in the pin vibrate faster. These molecules collide with neighboring molecules, causing them to move faster. These, in turn, collide with their neighbors, and so on until the molecules at the finger-held end of the pin begin to vibrate rapidly. These fast-moving molecules eventually cause the molecules of your finger to vibrate more quickly. Heat is now being transferred from the pin to your finger, and both the pin and your finger feel hot. If enough heat is transferred, you will drop the pin. The transmission of heat from one end of the pin to the other, and from the pin to your finger, occurs by conduction. Heat transferred in this fashion always flows from *warmer* to *colder* regions. Generally, the greater the temperature difference, the more rapid the heat transfer.

Figure 2.2
Every time a cloud forms, it warms the atmosphere. Inside this developing thunderstorm a vast amount of stored heat energy (latent heat) is given up to the air, as invisible water vapor becomes countless billions of water droplets and ice crystals. In fact, for the duration of this storm alone, more heat energy is released inside this cloud than is unleashed by a small nuclear bomb.

When materials can easily pass energy from one molecule to another, they are considered to be good conductors of heat. How well they conduct heat depends upon how their molecules are structurally bonded together. Table 2.1 shows that solids, such as metals, are good heat conductors. It is often difficult, therefore, to judge the temperature of metal objects. For example, if you grab a metal pipe at room temperature, it will seem to be much colder than it actually is because the metal conducts heat away from the hand quite rapidly. Conversely, air is an extremely poor conductor of heat, which is why most insulating materials have a large number of air spaces trapped within them. Air is such a poor heat conductor that, in calm weather, the hot ground only warms a shallow layer of air a few inches thick by conduction. Yet, air can carry this energy rapidly from one region to another. How then does this phenomenon happen?

Convection　The transfer of heat by the mass movement of a fluid (such as water and air) is called **convection**. This type of heat transfer takes place in

Table 2.1	
Heat Conductivity* of Various Substances	
Material	**Heat Conductivity (Cal per sec per cm per °C)**
Still air	0.0000614 (at 20°C)
Dry soil	0.0006
Water	0.00143
Snow	0.0015 (density 0.5 g/cm³)
Wet soil	0.0050
Ice	0.0053 (at 0°C)
Granite	0.0065
Iron	0.161
Silver	1.006

*Heat (thermal) conductivity describes a substance's ability to conduct heat as a consequence of molecular motion.

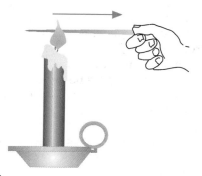

Figure 2.3
The transfer of heat from the hot end of the metal pin to the cool end by molecular contact is called *conduction*.

liquids and gases because they can move freely and it is
possible to set up currents within them.

Convection happens naturally in the atmosphere.
On a warm, sunny day certain areas of the earth's surface absorb more heat from the sun than others; as a result, the air near the earth's surface is heated somewhat
unevenly. Air molecules adjacent to these hot surfaces
bounce against them, thereby gaining some extra energy by conduction. The heated air expands and becomes less dense than the surrounding cooler air. The
expanded warm air is buoyed upward and rises. In this
manner, large bubbles of warm air rise and transfer
heat energy upward. Cooler, heavier air flows toward
the surface to replace the rising air. This cooler air becomes heated in turn, rises, and the cycle is repeated.
In meteorology, this vertical exchange of heat is called
convection and the rising air bubbles are known as
thermals (Fig. 2.4).

The rising air expands, cools, and gradually
spreads outward. It then slowly begins to sink. Near the
surface, it moves back into the heated region, replacing
the rising air. In this way, a *convective circulation*, or
thermal "cell," is produced in the atmosphere.

Although the entire process of heated air rising,
spreading out, sinking, and finally flowing back toward
its original location is known as a convective circulation, meteorologists usually restrict the term *convec-*
tion to the process of the rising and sinking part of the
circulation.

The horizontally moving part of the circulation
(called *wind*) carries properties of the air in that particular area with it. The transfer of these properties by
horizontally moving air is called **advection**. For example, wind blowing across a body of water will "pick up"
water vapor from the evaporating surface and transport it elsewhere in the atmosphere. If the air cools, the
water vapor may condense into cloud droplets and release latent heat. In a sense, then, heat is advected (carried) by the water vapor as it is swept along with the
wind. Earlier, we saw that this is an important way to
redistribute heat energy in the atmosphere.

In the convective circulation you may have noticed
that as the warm surface air was rising, it was also cooling. In our atmosphere, *any air that rises will expand*
and cool, and *any air that sinks is compressed and*
warms. This important concept is detailed in the Focus
section on p. 31.

Radiation   On a summer day, you may have noticed
how warm and flushed your face feels as you stand facing the sun. Sunlight travels through the surrounding
air with little effect upon the air itself. Your face, however, absorbs this energy and converts it to thermal energy. Thus, sunlight warms your face without actually
warming the air. The energy transferred from the sun to
your face is called **radiant energy** or **radiation**. It
travels in the form of waves that release energy when
they are absorbed by an object. Because these waves
have magnetic and electrical properties, we call them
electromagnetic waves. Electromagnetic waves do
not need molecules to propagate them. In a vacuum,

Figure 2.4
The development of a thermal. A thermal is a rising bubble of air that carries
heat energy upward by *convection*.

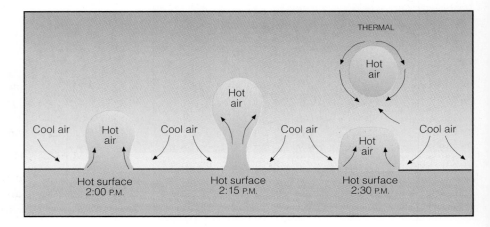

Focus on a Special Topic
Rising Air Cools and Sinking Air Warms

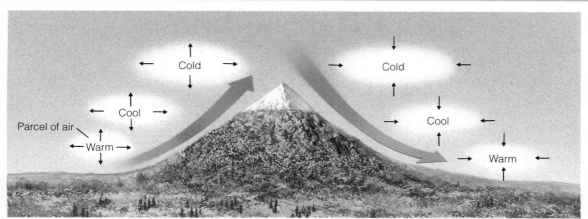

Figure 2
Rising air expands and cools; sinking air is compressed and warms.

To understand why rising air cools and sinking air warms we need to examine some air. Suppose we place air in an imaginary thin, elastic wrap about the size of a large balloon. (See Fig. 2.) This invisible balloonlike "blob" is called a *parcel*. The air parcel can expand and contract freely, but neither external air nor heat is able to mix with the air inside. By the same token, as the parcel moves, it does not break apart, but remains as a single unit.

At the earth's surface, the parcel has the same temperature and pressure as the air surrounding it. Sup-pose we lift the parcel. Recall from Chapter 1 that air pressure always decreases as we move up into the atmosphere. Consequently, as the parcel rises, it enters a region where the surrounding air pressure is lower. To equalize the pressure, the parcel molecules inside push the parcel walls outward, expanding it. Because there is no other energy source, the air molecules inside use some of their own energy to expand the parcel. This energy loss shows up as slower molecular speeds, which represent a lower parcel temperature. Hence, *any air that rises always expands and cools*.

If the parcel is lowered to the earth, it returns to a region where the air pressure is higher. The higher outside pressure squeezes (compresses) the parcel back to its original (smaller) shape. Because air molecules have a faster rebound velocity after striking the sides of a collapsing parcel, the average speed of the molecules inside goes up. (A ping-pong ball moves faster after striking a paddle that is moving toward it.) This increase in molecular speed represents a warmer parcel temperature. Therefore *any air that sinks (subsides), warms by compression*.

they travel at a constant speed of nearly 300,000 km (186,000 mi) per second—the speed of light.

Figure 2.5 shows some of the different wavelengths of radiation. Notice that the *wavelength* is the distance measured along a wave from one crest to another. Also notice that some of the waves have exceedingly short lengths. For example, radiation that we can see (visible light) has an average wavelength of less than one-millionth of a meter—a distance nearly one-

hundredth the diameter of a human hair. To measure these short lengths, we introduce a new unit of measurement called a **micrometer** (abbreviated μm), which is equal to one-millionth of a meter (m):

$$1 \text{ micrometer} = 0.000001 \text{ m} = 10^{-6} \text{ m}.$$

In Fig. 2.5, we can see that the average wavelength of visible light is about 0.0000005 meters, which is the same as 0.5 micrometers. To give you a common

Did you know?
The large ears of a jackrabbit are efficient emitters of infrared energy. Its ears help the rabbit survive the heat of a summer's day by radiating a great deal of infrared energy to the cooler sky above.

object for comparison, the average height of a letter on this page is about 2000 micrometers, or 2 millimeters.

We can also see in Fig. 2.5 that the longer waves carry less energy than do the shorter waves. When comparing the energy carried by various waves, it is useful to give electromagnetic radiation characteristics of particles in order to explain some of the wave's behavior. We can actually think of radiation as streams of particles or **photons** that are discrete packets of energy.

An ultraviolet photon carries more energy than an infrared photon. In fact, certain ultraviolet photons have enough energy to produce sunburns and penetrate skin tissue, sometimes causing skin cancer. As we discussed in Chapter 1, it is ozone in the stratosphere that protects us from the vast majority of these harmful rays.

To better understand the concept of radiation, here are a few important concepts and facts to remember:

1. All things (whose temperature is above absolute zero), no matter how big or small, emit radiation. The air, your body, flowers, trees, the earth, the stars are all radiating a wide range of electromagnetic waves. The energy originates from rapidly vibrating electrons, billions of which exist in every object.
2. The wavelengths of radiation that an object emits depend primarily on the object's temperature. The higher the object's temperature, the shorter are the wavelengths of emitted radiation. By the same token, as an object's temperature increases, its peak emission of radiation shifts toward shorter wavelengths.
3. Objects that have a high temperature emit radiation at a greater rate or intensity than objects with a lower temperature. Thus, as the temperature of an object increases, more total radiation (over a given surface area) is emitted each second.

Objects at a high temperature (above about 500°C) radiate waves with many lengths, but some of them are short enough to stimulate the sensation of color. We actually see these objects glow red. Objects cooler than this radiate at wavelengths that are too long for us to

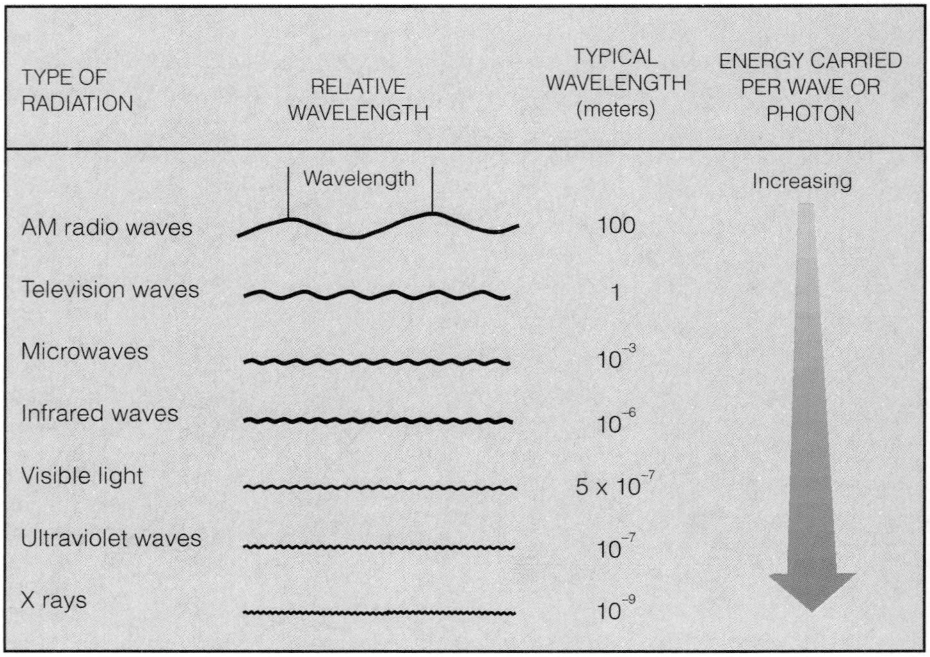

Figure 2.5
Radiation can be characterized according to its wavelength.

TYPE OF RADIATION	RELATIVE WAVELENGTH	TYPICAL WAVELENGTH (meters)	ENERGY CARRIED PER WAVE OR PHOTON
	Wavelength		Increasing
AM radio waves		100	
Television waves		1	
Microwaves		10^{-3}	
Infrared waves		10^{-6}	
Visible light		5×10^{-7}	
Ultraviolet waves		10^{-7}	
X rays		10^{-9}	

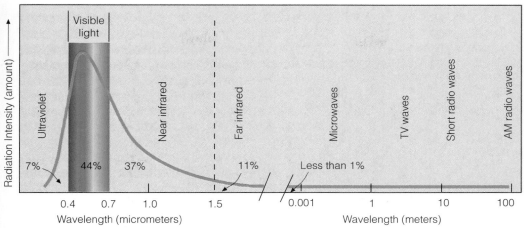

Figure 2.6
The sun's electromagnetic spectrum and some of the descriptive names of each region. The numbers underneath the curve approximate the percent of energy the sun radiates in various regions.

see. The page of this book, for example, is radiating electromagnetic waves. But because its temperature is only around 20°C (68°F), the waves emitted are much too long to stimulate vision. We are able to see the page, however, because light waves from other sources (such as light bulbs or the sun) are being *reflected* (bounced) off the paper. If this book were carried into a completely dark room, it would continue to radiate, but the pages would appear black because there are no visible light waves in the room to reflect off the page.

The sun emits radiation at almost all wavelengths, but because its surface is hot—6000 K (10,500°F)—it radiates the majority of its energy at relatively short wavelengths. If we look at the amount of radiation given off by the sun at each wavelength, we obtain the sun's *electromagnetic spectrum*. A portion of this spectrum is shown in Fig. 2.6.

Notice that the sun emits a maximum amount of radiation at wavelengths near 0.5 micrometers. Since our eyes are sensitive to radiation between 0.4 and 0.7 micrometers, these waves reach the eye and stimulate the sensation of color. This portion of the spectrum is therefore referred to as the **visible region**, and the light that reaches our eye is called *visible light*. The color violet is the shortest wavelength of visible light. Wavelengths shorter than violet (0.4 micrometers) are **ultraviolet (UV)**. The longest wavelengths of visible light correspond to the color red. Wavelengths longer than red (0.7 micrometers) are called **infrared (IR)**.

Whereas the hot sun emits only a part of its energy in the infrared portion of the spectrum, the relatively cool earth emits almost all of its energy at infrared wavelengths. In fact, the earth, with an average surface temperature near 288K (15°C, or 59°F), radiates nearly all its energy between 5 and 25 micrometers, with a peak intensity in the infrared region near 10 micrometers. (See Fig. 2.7.) Since the sun radiates the majority

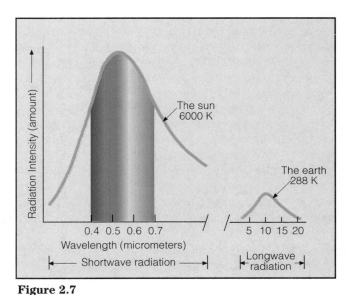

Figure 2.7
The hotter sun not only radiates more energy than the cooler earth, but it also radiates the majority of its energy at much shorter wavelengths.

of its energy at much shorter wavelengths than does the earth, solar radiation is often called *shortwave radiation*, whereas the earth's radiation is referred to as *longwave radiation*.

▲▽▲

Balancing Act—Absorption, Emission, and Equilibrium

If the earth and all things on it are continually radiating energy, why doesn't everything get progressively colder? The answer is that all objects not only radiate energy, they absorb it as well. If an object radiates more energy than it absorbs, it becomes colder; if it absorbs more energy than it emits, it becomes warmer. On a sunny day, the earth's surface warms by absorbing more energy from the sun and the atmosphere than it radiates, whereas at night the earth cools by radiating more energy than it absorbs from its surroundings. When an object emits and absorbs energy at equal rates, its temperature remains constant.

The rate at which something radiates and absorbs energy depends strongly on its surface characteristics, such as color, texture, and moisture, as well as temperature. For example, a black object in direct sunlight is a good absorber of radiation. It converts energy from the sun into internal energy, and its temperature ordinarily increases. You need only walk barefoot on a black asphalt road on a summer afternoon to experience this. At night, the blacktop road will cool quickly by emitting infrared energy and, by early morning, it may be cooler than surrounding surfaces.

Any object that is a perfect absorber (that is, absorbs all the radiation that strikes it) and a perfect emitter (emits the maximum radiation possible at its given temperature) is called a **black body**. Black bodies do not have to be colored black, they simply must absorb and emit all possible radiation. Since the earth's surface and the sun absorb and radiate with nearly 100 percent efficiency for their respective temperatures, they both behave as black bodies.

When we look at the earth from space, we see that half of it is in sunlight, the other half is in darkness. The outpouring of solar energy constantly bathes the earth with radiation, while the earth, in turn, constantly emits infrared radiation. If we assume that there is no other method of transferring heat, then, when the rate of ab-

sorption of solar radiation equals the rate of emission of infrared earth radiation, a state of *radiative equilibrium* is achieved. The average temperature at which this occurs is called the **radiative equilibrium temperature**. At this temperature, the earth (behaving as a black body) is absorbing solar radiation and emitting infrared radiation at equal rates, and its average temperature does not change. As the earth is about 150 million km (93 million mi) from the sun, the earth's radiative equilibrium temperature is about 255 K (−18°C, 0°F). But this temperature is *much* lower than the earth's observed average surface temperature of 288 K (15°C, 59°F). Why is there such a large difference?

The answer lies in the fact that the earth's atmosphere absorbs and emits infrared radiation. Unlike the earth, the atmosphere does *not* behave like a black body, as it absorbs some wavelengths of radiation and is transparent to others. Objects that selectively absorb and emit radiation, such as gases in our atmosphere, are known as **selective absorbers**.

Selective Absorbers and the Atmospheric Greenhouse Effect   There are many selective absorbers in our environment. Snow, for example, is a good absorber of infrared radiation but a poor absorber of sunlight. Objects that selectively absorb radiation usually selectively emit radiation at the same wavelength. Snow is therefore a good emitter of infrared energy. At night, a snow surface usually emits much more infrared energy than it absorbs from its surroundings. This large loss of infrared radiation (coupled with the insulating qualities of snow) causes the air above a snow surface on a clear, winter night to become extremely cold.

Figure 2.8 shows some of the most important selectively absorbing gases in our atmosphere. Notice that both water vapor (H_2O) and carbon dioxide (CO_2) are strong absorbers of infrared radiation and poor absorbers of visible solar radiation. Other, less important selective absorbers include nitrous oxide (N_2O), methane (CH_4), and ozone (O_3), which is most abundant in the stratosphere. (See Chapter 1.) As these gases absorb infrared radiation emitted from the earth's surface, they gain kinetic energy (energy of motion). The gas molecules share this energy by colliding with neighboring air molecules, such as oxygen and nitrogen (both of which are poor absorbers of infrared energy). These collisions increase the average kinetic energy of the air, which results in an increase in air temperature.

Figure 2.8
Absorption of radiation by gases in the atmosphere.

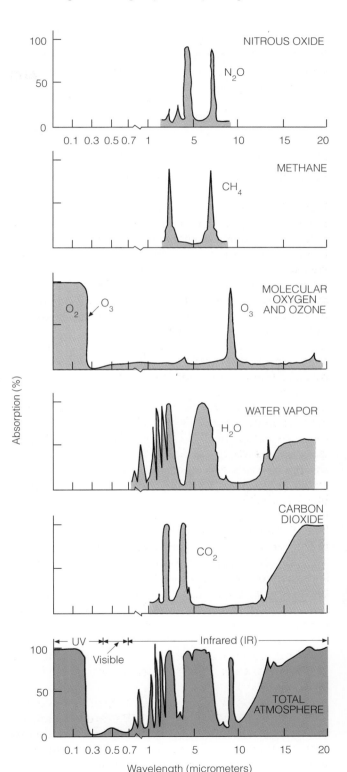

Thus, some of the infrared energy emitted from the earth's surface warms the lower atmosphere.

Besides being selective absorbers, water vapor and CO_2 selectively emit radiation at infrared wavelengths.* This radiation travels away from these gases in all directions. A portion of this energy is radiated toward the earth's surface and absorbed, thus heating the ground. The earth, in turn, reradiates infrared energy upward, where it is absorbed and warms the lower atmosphere. In this way, water vapor and CO_2 absorb and reradiate infrared energy and act as an insulating layer around the earth, keeping part of the earth's infrared radiation from escaping rapidly into space. Consequently, the earth's surface and the lower atmosphere are much warmer than they would be if these selectively absorbing gases were not present. In fact, as we saw earlier, the earth's mean radiative equilibrium temperature without CO_2 and water vapor would be around –18°C (0°F), or about 33°C (59°F) lower than at present.

The absorption characteristics of water vapor, CO_2, and other gases such as methane and nitrous oxide (see Fig. 2.8), were, at one time, thought to be similar to the glass of a florist's greenhouse. In a greenhouse, the glass allows visible radiation to come in, but inhibits to some degree the passage of outgoing infrared radiation. For this reason, the behavior of the water vapor and CO_2 in the atmosphere is popularly called the **greenhouse effect**. However, studies have shown that the warm air inside a greenhouse is probably caused more by the air's inability to circulate and mix with the cooler outside air, rather than by the entrapment of infrared energy. Because of these findings, some scientists insist that the greenhouse effect should be called the *atmosphere effect*. To accommodate everyone, we will use the term *atmospheric greenhouse effect* when describing the role that water vapor and CO_2 play in keeping the earth's mean surface temperature higher than it otherwise would be.

Look again at Fig. 2.8 and observe that, in the bottom diagram, there is a region between about 8 and 11 micrometers where neither water vapor nor CO_2 read-

---

*Nitrous oxide, methane, and ozone also emit infrared radiation, but their concentration in the atmosphere is much smaller than water vapor and carbon dioxide (see Table 1.1, p. 3.)

ily absorb infrared radiation. Because these wavelengths of emitted energy pass upward through the atmosphere and out into space, the wavelength range (between 8 and 11 micrometers) is known as the **atmospheric window**. At night, clouds can enhance the atmospheric greenhouse effect. Tiny liquid cloud droplets are selective absorbers in that they are good absorbers of infrared radiation but poor absorbers of visible solar radiation. Clouds even absorb the wavelengths between 8 and 11 micrometers, which are otherwise "passed up" by water vapor and CO_2. Thus, they have the effect of enhancing the atmospheric greenhouse effect by closing the atmospheric window.

Clouds are also excellent emitters of infrared radiation. Their tops radiate infrared energy upward and their bases radiate energy back to the earth's surface where it is reabsorbed and, in a sense, reradiated back to the clouds. This process keeps calm, cloudy nights warmer than calm, clear ones. If the clouds remain into the next day, they prevent much of the sunlight from reaching the ground by reflecting it back to space. Since the ground does not heat up as much as it would in full sunshine, cloudy, calm days are normally cooler than clear, calm days. Hence, the presence of clouds tends to keep nighttime temperatures higher and daytime temperatures lower.

In summary, the atmospheric greenhouse effect occurs because water vapor, CO_2, and other trace gases are selective absorbers. They allow most of the sun's radiation to reach the surface, but they absorb a good portion of the earth's infrared radiation, preventing it from escaping into space. (See Fig. 2.9.)

Enhancement of the Greenhouse Effect In spite of the inaccuracies that plague temperature measurements, several studies suggest that for the past 100 years or so, the earth's surface has been undergoing a slight warming trend of about 0.6°C (about 1°F). Sci-

Figure 2.9
Sunlight warms the earth's surface only during the day, while the surface *constantly* emits infrared radiation upward during the day and at night. (a) Near the surface, water vapor, CO_2, and other trace gases absorb some of this infrared energy and reradiate it back to the ground—the atmospheric greenhouse effect. This provides warmth, like an insulating blanket, which helps to keep the earth's average surface temperature near 15°C (59°F), which is comfortable. (b) Without water vapor, CO_2, and other greenhouse gases, the earth's surface would constantly emit infrared energy, but it would not receive infrared energy from its lower atmosphere. Hence, there would be no atmospheric greenhouse effect and the earth's average surface temperature would be much lower than it is presently.

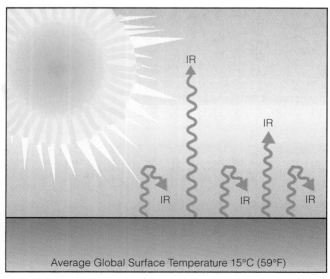

(a) Earth's atmosphere *with* H_2O and CO_2

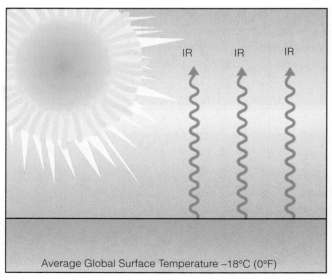

(b) Earth's atmosphere *without* H_2O and CO_2

entific computer models, called *General Circulation Models* (GCMs), predict that if such a warming should continue unabated, we are irrevocably committed to some measure of climate change, notably a shifting of the world's wind patterns that steer the rain-producing storms across the globe.

Some scientists believe that the main cause for the warming is the greenhouse gas CO_2, whose concentration has been increasing primarily because of the burning of fossil fuels and deforestation. However, in recent years, increasing concentration of other greenhouse gases, such as *methane* (CH_4), *nitrous oxide* (N_2O), and *chlorofluorocarbons* (CFCs), has collectively been shown to have an effect equal to CO_2. Look at Fig. 2.8 and notice that both CH_4 and N_2O absorb strongly at infrared wavelengths. Moreover, a particular CFC (CFC-12) absorbs in the region of the atmospheric window between 8 and 11 micrometers. Thus, in terms of its absorption impact on infrared radiation, the addition of a single CFC-12 molecule to the atmosphere is equivalent to adding 10,000 molecules of CO_2.

Presently, the concentration of CO_2 in a volume of air near the surface is about 0.035 percent. Some computer models predict that doubling this amount will cause the earth's average surface temperature to rise between about 2° and 5°C. How can doubling such a small quantity of CO_2 and adding miniscule amounts of other greenhouse gases bring about such a large temperature increase?

Mathematical climate models predict that rising ocean temperatures will cause an increase in evaporation rates. The added water vapor—the primary greenhouse gas—will enhance the atmospheric greenhouse effect and double the temperature rise, in what is known as a *positive feedback*. But there are other feedbacks to consider.*

The two potentially largest and least understood feedbacks in the climate system are the clouds and the oceans. Clouds can change area, depth, and radiation properties simultaneously with climatic changes. The net effect of all these changes is not well understood. Oceans, on the other hand, cover 70 percent of the planet. The response of ocean circulations, ocean temperatures, and sea ice to global warming will determine

the global pattern and speed of climate change. Unfortunately, it is not now known how quickly or in what direction each of these will respond.

Satellite data from the *Earth Radiation Budget Experiment* (ERBE) suggest that clouds overall appear to cool the earth's climate, as they reflect and radiate away more energy than they retain. (The earth would be warmer if clouds were not present.) So an increase in global cloudiness (if it were to occur) might offset some of the global warming brought on by an enhanced atmospheric greenhouse effect. Therefore, if clouds were to act on the climate system in this manner, they would provide a *negative feedback* on climate change.

Uncertainties unquestionably exist about the impact that increasing levels of CO_2 and other trace gases will have on enhancing the atmospheric greenhouse effect. Nonetheless, many (but not all) scientific studies suggest that increasing the concentration of these gases in our atmosphere will lead to global-scale climatic change within the next century. Such change could adversely affect water resources and agricultural productivity. (We will examine this topic further in Chapter 13, when we cover climatic change in more detail.)

Warming the Air from Below   On a clear day, solar energy passes through the lower atmosphere with little effect upon the air. Ultimately it reaches the surface, warming it (Fig. 2.10). Air molecules in contact with the heated surface bounce against it, gain energy by *conduction*, then shoot upward like freshly popped kernels of corn, carrying their energy with them. Because the air near the ground is very dense, these molecules only travel a short distance before they collide with other molecules. During the collision, these more rapidly moving molecules share their energy with less energetic molecules, raising the average temperature of the air. But air is such a poor heat conductor that this process is only important within a few centimeters of the ground.

As the surface air warms, it actually becomes less dense than the air directly above it. The warmer air rises and the cooler air sinks, setting up thermals, or *free convection cells* that transfer heat upward and distribute it through a deeper layer of air. The rising air expands and cools, and, if sufficiently moist, the water vapor condenses into cloud droplets, releasing latent heat that warms the air. Meanwhile, the earth constantly emits infrared energy, which is being absorbed

---

*A feedback is a process whereby an initial change in a process will tend to either reinforce the process (positive feedback) or weaken the process (negative feedback).

and re-emitted by greenhouse gases like water vapor and CO_2. Since the concentration of water vapor decreases rapidly above the earth, most of the absorption occurs in a layer near the surface. Hence, the lower atmosphere is mainly heated from below.

Incoming Solar Energy

As the sun's radiant energy travels through space essentially nothing interferes with it until it reaches the atmosphere. At the top of the atmosphere, solar energy received on a surface perpendicular to the sun's rays appears to remain fairly constant at nearly two calories on each square centimeter each minute—a value called the *solar constant*.*

*The solar constant is not really "constant." Recent satellite measurements suggest the solar constant varies slightly as the sun's radiant output varies.

Upon entering the atmosphere, a number of interactions take place between the atmosphere and solar radiation. For example, some of the energy is absorbed by gases, such as ozone, in the upper atmosphere. Moreover, when sunlight strikes very small objects, such as air molecules and dust particles, the light itself is deflected in all directions—forward, sideways, and backwards. The distribution of light in this manner is called **scattering**. (Scattered light is also called *diffuse light*.) Because air molecules are much smaller than the wavelengths of visible light, they are more effective scatterers of the shorter (blue) wavelengths than the longer (red) wavelengths. Hence, when we look away from the direct beam of sunlight, blue light strikes our eyes from all directions, turning the daytime sky blue. At midday, all the wavelengths of visible light from the sun strike our eyes, and the sun is perceived as white. At sunrise and sunset, when the white beam of sunlight must pass through a thick portion of the atmosphere, scattering by air molecules removes the blue light, leaving the longer wavelengths of red, orange, and yellow to pass on through, creating the image of a ruddy or yellowish sun (Fig. 2.11).

Sunlight can be *reflected* from clouds and from the earth's surface. **Albedo** is the percent of radiation returning from a surface compared to that which strikes it. Albedo, then, represents the *reflectivity* of the sur-

Figure 2.10
Air in the lower atmosphere is heated from below. Sunlight warms the ground, and the air above is warmed by conduction, convection, and radiation. Further warming occurs during condensation as latent heat is given up to the air inside the cloud.

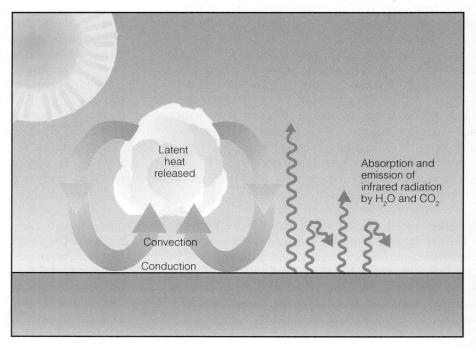

Latent heat released

Absorption and emission of infrared radiation by H_2O and CO_2

Convection

Conduction

face. In Table 2.2 notice that thick clouds have a higher albedo than thin clouds. On the average, the albedo of clouds is near 60 percent. When solar energy strikes a surface covered with snow, up to 95 percent of the sunlight may be reflected. However, notice in Table 2.2 that a water surface reflects only a small amount of sunlight, about 10 percent for an entire day. Since the earth's surface is almost three-fourths water and the continents are dotted with trees and vegetation, the albedo of the earth's relatively dark surface is only about 4 percent. Consequently, in Fig. 2.12, we can see that averaged for an entire year, the earth and its atmosphere redirect about 30 percent of the sun's incoming radiation back to space, which gives the earth and its atmosphere a combined albedo of 30 percent.

The Earth's Annual Energy Balance   Although the average temperature at any one place may vary considerably from year to year, the earth's overall average equilibrium temperature changes only slightly from one year to the next. This fact indicates that, each year, the earth and its atmosphere combined must send off into space just as much energy as they receive from the sun. The same type of energy balance must exist between the earth's surface and the atmosphere. That is, each year, the earth's surface must return to the atmosphere the same amount of energy that it absorbs. If this did not occur, the earth's average surface temperature would change. How do the earth and its atmosphere maintain this yearly energy balance?

Suppose 100 units of solar energy reach the top of the earth's atmosphere. We already know from Fig. 2.12 that, on the average, clouds, the earth, and the atmosphere reflect and scatter 30 units back to space and that the atmosphere and clouds together absorb 19 units, which leaves 51 units of direct and indirect (diffuse) solar radiation to be absorbed at the earth's surface.

Figure 2.13 shows approximately what happens to the solar radiation that is absorbed by the surface and the atmosphere. Out of 51 units reaching the surface, a large amount (23 units) is used to evaporate water and about 7 units are lost through conduction and convection, which leaves 21 units to be radiated away as infrared energy. Look closely at Fig. 2.13 and notice that the earth's surface actually radiates upward a whopping 117 units. It does so because, although it receives solar radiation only during the day, it constantly emits infrared energy both during the day and at night.

Figure 2.11
A brilliant red sunset produced by the process of scattering.

Table 2.2	Typical Albedo of Various Surfaces
Surface or Object	**Albedo (percent)**
Fresh snow	75 to 95
Clouds (thick)	60 to 90
Clouds (thin)	30 to 50
Venus	78
Ice	30 to 40
Sand	15 to 45
Earth and atmosphere	30
Mars	17
Grassy field	10 to 30
Dry, plowed field	5 to 20
Water	10*
Forest	3 to 10
Moon	7
*Daily average.	

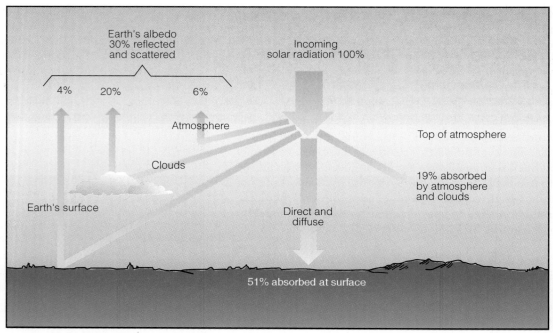

Figure 2.12
On the average, of all the solar energy that reaches the earth's atmosphere annually, about 30 percent is reflected and scattered back to space giving the earth and its atmosphere an albedo of 30 percent. Of the remaining solar energy about 19 percent is absorbed by the atmosphere and clouds and 51 percent is absorbed at the surface.

Additionally, the atmosphere above only allows a small fraction of this energy (6 units) to pass through into space. The majority of it (111 units) is absorbed mainly by the greenhouse gases water vapor and CO_2, and by clouds. Much of this energy (96 units) is then reradiated back to earth, producing the atmospheric greenhouse effect. In all these exchanges, notice that the energy lost at the earth's surface (147 units) is exactly balanced by the energy gained there (147 units).

A similar balance exists between the earth's surface and its atmosphere. Again in Fig. 2.13 observe that the energy gained by the atmosphere (160 units) balances the energy lost. Moreover, averaged for an entire year, the solar energy received at the earth's surface (51 units) and that absorbed by the earth's atmosphere (19 units) balances the infrared energy lost to space by the earth's surface (6 units) and its atmosphere (64 units).

And so, the earth and the atmosphere absorb energy from the sun, as well as from each other. In all of the energy exchanges, a delicate balance is maintained.

Essentially, there is no yearly gain or loss of total energy, and the average temperature of the earth and the atmosphere remains fairly constant from one year to the next. This equilibrium does not imply that the earth's average temperature does not change, but that the changes are small from year to year (usually less than one-tenth of a degree Celsius) and become significant only when measured over many years.

Before we turn our attention to how incoming solar energy produces the earth's seasons, we need to examine how solar energy, in the form of particles, produces a dazzling light show known as the *aurora*.

Solar Energy and the Aurora   From the sun and its tenuous atmosphere comes a continuous discharge of particles. This discharge happens because, at extremely high temperatures, gases become stripped of electrons by violent collisions and acquire enough speed to escape the gravitational pull of the sun. As these charged particles (ions and electrons) travel through space, they are known as the *solar wind*.

When the solar wind moves close enough to the earth, it interacts with the earth's magnetic field, disturbing it. This disturbance causes energetic solar wind particles to enter the upper atmosphere, where they collide with atmospheric gases. These gases then become excited and emit visible radiation (light), which causes the sky to glow like a neon light, thus producing the **aurora** (Fig. 2.14).

The actual appearance of the aurora can vary. Sometimes it appears as a faint white or red glow, lasting from a few minutes to a few hours. The light may move across the sky as a yellow-green arc much wider than a rainbow; or it may faintly decorate the sky with flickering draperies of blue, green, and purple light that constantly change in form and location, as if blown by a gentle breeze.

In the Northern Hemisphere, the aurora is called the *aurora borealis*, or northern lights; its counterpart

in the Southern Hemisphere is the *aurora australis*, or southern lights.

The aurora is most frequently seen in the polar regions, where the earth's magnetic field lines emerge from the earth. But during active sun periods when there are numerous sunspots and giant flares (solar eruptions), large quantities of particles travel outward away from the sun at high speeds (hundreds of miles a second). These energetic particles are able to penetrate unusually deep into the earth's magnetic field, where they provide sufficient energy to produce auroral displays. During these conditions in North America, we see the aurora much farther south than usual.

Why the Earth Has Seasons   The earth revolves completely around the sun in an elliptical path (not quite a circle) in slightly longer than 365 days (one year). As the earth revolves around the sun, it spins on

Figure 2.13
The earth-atmosphere energy balance. Numbers represent approximations based on surface observations and satellite data. While the actual value of each process may vary by several percent, it is the relative size of the numbers that is important.

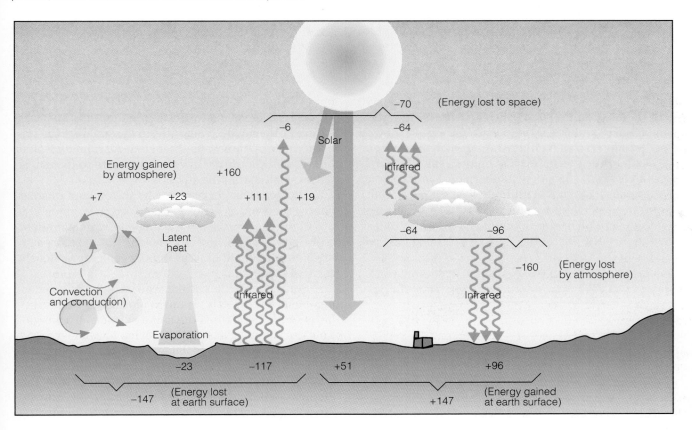

Figure 2.14
Aurora australis, the southern lights, photographed between Antarctica and Australia by space shuttle astronauts. The aurora forms as energetic particles from the sun interact with the earth's atmosphere.

its own axis, completing one spin in 24 hours (one day). The average distance from the earth to the sun is 150 million km (93 million mi). Because the earth's orbit is an ellipse instead of a circle, the actual distance from the earth to the sun varies during the year. The earth comes closer to the sun in January (147 million km) than it does in July (152 million km). (See Fig. 2.15.) From this we might conclude that our warmest weather should occur in January and our coldest weather in July. But, in the Northern Hemisphere, we

normally experience cold weather in January when we are closer to the sun and warm weather in July when we are farther away. If nearness to the sun were the primary cause of the seasons then, indeed, January would be warmer than July. However, nearness to the sun is only a small part of the story.

Our seasons are regulated by the amount of solar energy received at the earth's surface. This amount is determined primarily by the angle at which sunlight strikes the surface, and by how long the sun shines on any latitude (daylight hours). Let's look more closely at these factors.

Solar energy that strikes the earth's surface perpendicularly (directly) is much more intense than solar energy that strikes the same surface at an angle. Think of shining a flashlight straight at a wall—you get a small, circular spot of light (Fig. 2.16). Now, tip the flashlight and notice how the spot of light spreads over a larger area. The same principle holds for sunlight. Sunlight striking the earth at an angle spreads out and must heat a larger region than sunlight impinging

January   July

|←——— 147 million km ———→|←——— 152 million km ———→|

Figure 2.15
The elliptical path (highly exaggerated) of the earth about the sun brings the earth slightly closer to the sun in January than in July.

directly on the earth. Everything else being equal, an area experiencing more direct solar rays will receive more heat than the same size area being struck by sunlight at an angle. In addition, the more the sun's rays are slanted from the perpendicular, the more atmosphere they must penetrate. And the more atmosphere they penetrate, the more they can be scattered and absorbed (attenuated). As a consequence, when the sun is high in the sky, it can heat the ground to a much higher temperature than when it is low on the horizon.

The second important factor determining how warm the earth's surface becomes is the length of time the sun shines each day. Longer daylight hours, of course, mean that more energy is available from sunlight. In a given location, more solar energy reaches the earth's surface on a clear, long day than on a day that is clear but much shorter. Hence, more surface heating takes place.

From a casual observation, we know that summer days have more daylight hours than winter days. Also, the noontime summer sun is higher in the sky than is the noontime winter sun. Both of these events occur because our spinning planet is inclined on its axis (tilted) as it revolves around the sun. As Fig. 2.17 illustrates, the angle of tilt is 23½° from the perpendicular drawn to the plane of the earth's orbit. The earth's axis points to the same direction in space all year long; thus, the Northern Hemisphere is tilted toward the sun in summer (June), and away from the sun in winter (December).

Seasons in the Northern Hemisphere Notice in Fig. 2.17 that on June 22, the northern half of the world is directed toward the sun. At noon on this day, solar

Did you know?
If the tilt of the earth should increase dramatically, summers would be warmer and winters colder in the middle latitudes of both hemispheres.

rays beat down upon the Northern Hemisphere more directly than during any other time of year. The sun is at its highest position in the noonday sky, directly above 23½° north (N) latitude (Tropic of Cancer). If you were standing at this latitude on June 22, the sun at noon would be directly overhead. This day, called the **summer solstice**, is the astronomical first day of summer in the Northern Hemisphere.*

Study Fig. 2.17 closely and notice that, as the earth spins on its axis, the side facing the sun is in sunshine and the other side is in darkness. Thus, half of the globe is always illuminated. If the earth's axis were not tilted, the noonday sun would always be directly overhead at the equator, and there would be 12 hours of daylight and 12 hours of darkness at each latitude every day of the year. However, the earth is tilted. Since the Northern Hemisphere faces towards the sun on June 22, each latitude in the Northern Hemisphere will have more than 12 hours of daylight. The farther north we go, the longer are the daylight hours. When we reach the Arctic Circle (66½°N), daylight lasts for 24 hours, as the sun does not set. Notice in Fig. 2.17 how the region above 66½°N never gets into the "shadow" zone as the

*As we will see later in this chapter, the seasons are reversed in the Southern Hemisphere. Hence, in the Southern Hemisphere, this same day is the winter solstice, or the astronomical first day of winter.

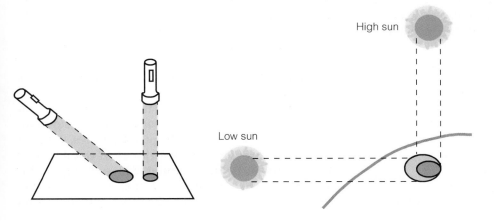

Figure 2.16
Sunlight that strikes a surface at an angle is spread over a larger area than sunlight that strikes the surface directly. Oblique sun rays deliver less energy (are less intense) to a surface than direct sun rays.

High sun

Low sun

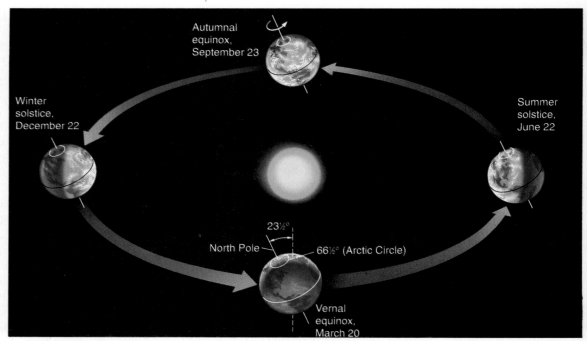

Figure 2.17
As the earth revolves about the sun, it is tilted on its axis by an angle of 23½°. The earth's
axis always points to the same area in space (as viewed from a distant star). Thus, in June,
when the Northern Hemisphere is tipped toward the sun, more direct sunlight and long hours
of daylight cause warmer weather than in December, when the Northern Hemisphere is
tipped away from the sun. (Diagram, of course, is not to scale.)

earth spins. At the North Pole, the sun actually rises
above the horizon on March 20 and has six months until it sets on September 23. No wonder this region is
called the "Land of the Midnight Sun"! (See Fig. 2.18.)

Even though in the far north the sun is above the
horizon for many hours during the summer (see Table
2.3), the surface air there is not warmer than the air farther south, where days are appreciably shorter. The
reason for this is shown in Fig. 2.19. When incoming solar radiation (called *insolation*) enters the atmosphere,
fine dust, air molecules and clouds reflect and scatter it,
and some of it is absorbed by atmospheric gases. Generally, the greater the thickness of atmosphere that sunlight must penetrate, the greater are the chances that

it will be either reflected or absorbed by the atmosphere. During the summer in far northern latitudes,
the sun is never very high above the horizon, so its radiant energy must pass through a thick portion of atmosphere before it reaches the earth's surface. Some of
the solar energy that does reach the surface melts
frozen soil or is reflected by snow or ice. And, that
which is absorbed is spread over a large area. So, even
though northern cities may experience long hours of
sunlight they are not warmer than cities farther south.
Overall, they receive less radiation at the surface, and
what radiation they do receive does not effectively heat
the surface.

Look at Fig. 2.17 again and notice that, by September 23, the earth will have moved so that the sun is directly above the equator. Except at the poles, the days
and nights throughout the world are of equal length.
This day is called the **autumnal** (fall) **equinox**, and it
marks the astronomical beginning of fall in the Northern Hemisphere. At the North Pole, the sun appears on

Did you know?
The word *autumn* is derived from the Latin *autumnum*,
meaning "the fairest season of the year."

Figure 2.18
Land of the Midnight Sun. These eight exposures of the sun were taken in northern Greenland (latitude 78°N) during late July between about 11:00 P.M. and 1:00 A.M..

the horizon for 24 hours, due to the bending of light by
the atmosphere. The following day (or at least within
several days), the sun disappears from view, not to rise
again for a long, cold six months. Throughout the
northern half of the world on each successive day,
there are fewer hours of daylight, and the noon sun is
slightly lower in the sky. Less direct sunlight and
shorter hours of daylight spell cooler weather for the
Northern Hemisphere. Reduced sunlight, lower air temperatures, and cooling breezes stimulate the beautiful
pageantry of fall colors. (Does the first frost cause the
beautiful colors? If you would like to know, read the
Focus section on p. 46.)

In some years around the middle of autumn, there
is an unseasonably warm spell, especially in the eastern
two-thirds of the United States. This warm period, referred to as **Indian summer**,* may last from several
days up to a week or more. It usually occurs when a

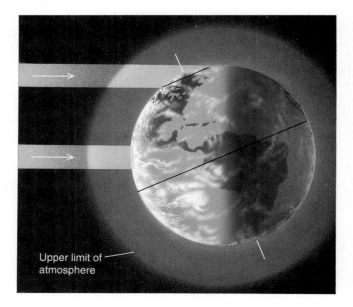

Upper limit of
atmosphere

Figure 2.19
During the Northern Hemisphere summer, sunlight that reaches
the earth's surface in far northern latitudes has passed through a
thicker layer of absorbing, scattering, and reflecting atmosphere
than sunlight that reaches the earth's surface farther south.

*The origin of the term is uncertain, as it dates back to the eighteenth
century. It may have originally referred to the good weather that
allowed the Indians time to harvest their crops. Normally, a period
of cool autumn weather must precede the warm weather period to be
called Indian summer.

Focus on an Observation
Does the First Frost Cause the Leaves to Change Color in Autumn?

Contrary to what many people believe, it is not the first frost that causes the leaves of deciduous trees to change color. The colors are actually in the leaves during the summer, but we do not see them because the green pigments—known as *chlorophylls*—are far more dominant.

Chlorophylls absorb and use the energy of red and violet wavelengths of sunlight to manufacture the simple sugars and starches that trees use as food. Most of the green wavelengths are not absorbed; instead, they are reflected. This, of course, causes the leaf to appear green.

The leaves remain green as long as the tree is able to replenish the chlorophylls it uses up. About several weeks before the first frost, however, a chemical change begins in the leaf, as shorter days and cooler nights cause leaf activity to diminish. As the chlorophylls decrease, the green slowly disappears. Other pigments, such as the yellow and orange carotenoid, which were hidden by the

Figure 3
The pageantry of fall colors along a country road in Vermont.

green of the leaf, now begin to show through. Additionally, the cooler weather stimulates the manufacture of other pigments, such as the anthocyanins, which turn the leaves red or purple. The combination of these pigments can produce a fiery display of autumn color. (See Fig. 3.)

The weather most suitable for an impressive display of fall colors is warm, sunny days followed by clear, cool nights, with temperatures dropping below 7°C (45°F), but remaining above freezing.

---

large high pressure area stalls near the southeast coast. The clockwise flow of air around this system moves warm air from the Gulf of Mexico into the central or eastern half of the nation. The warm, gentle breezes and smoke from a variety of sources respectively make for mild, hazy days. The warm weather ends abruptly when an outbreak of polar air reminds us that winter is not far away.

On December 22 (three months after the autumnal equinox), the Northern Hemisphere is tilted as far away from the sun as it will be all year. (See Fig. 2.17.) Nights are long and days are short. Notice in Table 2.3 that daylight decreases from 12 hours at the equator to 0 (zero) at latitudes above 66½°N (the arctic circle).

This is the shortest day of the year, called the **winter solstice**—the astronomical beginning of winter in the northern world. On this day, the sun shines directly above latitude 23½°S (Tropic of Capricorn). In the northern half of the world, the sun is at its lowest position in the noon sky. Its rays pass through a thick section of atmosphere and spread over a large area on the surface.

With so little incident sunlight, the earth's surface cools quickly. A blanket of clean snow covering the ground aids in the cooling. In northern Canada and Alaska, arctic air rapidly becomes extremely cold as it lies poised, ready to do battle with the milder air to the south. Periodically, this cold arctic air pushes down

Table 2.3 Length of Time from Sunrise to Sunset for Various Latitudes on Different Dates				
	Northern Hemisphere (Read Down)			
Latitude	*March 20*	*June 22*	*Sept. 23*	*Dec. 22*
0°	12 hr	12.0 hr	12 hr	12.0 hr
10°	12 hr	12.6 hr	12 hr	11.4 hr
20°	12 hr	13.2 hr	12 hr	10.8 hr
30°	12 hr	13.9 hr	12 hr	10.1 hr
40°	12 hr	14.9 hr	12 hr	9.1 hr
50°	12 hr	16.3 hr	12 hr	7.7 hr
60°	12 hr	18.4 hr	12 hr	5.6 hr
70°	12 hr	2 months	12 hr	0 hr
80°	12 hr	4 months	12 hr	0 hr
90°	12 hr	6 months	12 hr	0 hr
Latitude	*Sept 23*	*Dec. 22*	*March 20*	*June 22*
	Southern Hemisphere (Read Up)			

ward the poles, and cold air and water toward the equator. Thus, the transfer of heat energy by atmospheric and oceanic circulations prevents low latitudes from steadily becoming warmer and high latitudes from steadily growing colder. These circulations are extremely important to weather and climate, and will be treated more completely in Chapter 7.

Seasons in the Southern Hemisphere On June 22, the Southern Hemisphere is adjusting to an entirely different season. Because this part of the world is now tilted away from the sun, nights are long, days are short, and solar rays come in at an angle. All of these factors keep air temperatures fairly low. The June solstice marks the astronomical beginning of winter in the Southern Hemisphere. In this part of the world, summer will not "officially" begin until the sun is over the Tropic of Capricorn (23½°S)—remember that this occurs on December 22. So, when it is winter and June in the Southern Hemisphere, it is summer and June in the Northern Hemisphere. If you are tired of the hot June

into the northern United States, producing a rapid drop in temperature called a *cold wave*, which occasionally reaches far into the south. Sometimes, these cold spells arrive well before the winter solstice—the "official" first day of winter—bringing with them heavy snow and blustery winds. (More information on this "official" first day of winter is given in the Focus section on p. 48.)

Three months past the winter solstice marks the astronomical arrival of spring, which is called the **vernal** (spring) **equinox**. The date is March 20 and, once again, the noonday sun is shining directly on the equator, days and nights throughout the world are of equal length, and, at the North Pole, the sun rises above the horizon after a long six month absence.

Up to this point, we have seen that the seasons are controlled by solar energy striking our tilted planet, as it makes its annual voyage around the sun. This tilt of the earth causes a seasonal variation in both the length of daylight and the intensity of sunlight that reaches the surface. Because of these facts, high latitudes tend to lose more energy to space each year than they receive from the sun, while low latitudes tend to gain more energy during the course of a year than they lose. From Fig. 2.20 we can see that only at middle latitudes near 37° does the amount of energy received each year balance the amount lost. From this situation, we might conclude that polar regions are growing colder each year, while tropical regions are becoming warmer. But this does not happen. To compensate for these gains and losses of energy, winds in the atmosphere and currents in the oceans circulate warm air and water to-

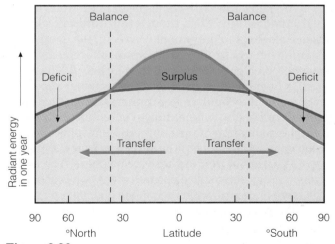

Figure 2.20
The average annual incoming solar energy (red line) absorbed by the earth and the atmosphere, along with the average annual outgoing energy (blue line) emitted by the earth and the atmosphere.

Focus on an Issue
Is December 22 Really the First Day of Winter?

On December 22 (or 21, depending on the year) after nearly a month of cold weather, and perhaps a snow-storm or two, someone on the radio or television has the audacity to proclaim that "today is the first official day of winter." If during the last several weeks it was not winter, then what season was it?

Actually, December 22 marks the *astronomical* first day of winter in the Northern Hemisphere (NH), just as June 22 marks the astronomical first day of summer (NH). The earth is tilted on its axis by 23½° as it revolves around the sun. This fact causes the sun (as we view it from earth) to move in the sky from a point where it is di-

rectly above 23½° South latitude on December 22, to a point where it is directly above 23½° North latitude on June 22. The astronomical first day of spring (NH) occurs around March 20 as the sun crosses the equator moving northward and, likewise, the astronomical first day of autumn (NH) occurs around September 23 as the sun crosses the equator moving southward.

In the middle latitudes, summer is defined as the warmest season and winter the coldest season. If the year is divided into four seasons with each season consisting of three months, then the *meteorological definition* of summer over much of the Northern

Hemisphere would be the three warmest months of June, July, and August. Winter would be the three coldest months of December, January, and February. Autumn would be September, October, and November—the transition between summer and winter. And spring would be March, April, and May—the transition between winter and summer.

So, the next time you hear some-one remark on December 22 that "winter officially begins today," re-member that this is the astronomical definition of the first day of winter. According to the meteorological defini-tion, winter has been around for several weeks.

---

weather in your Northern Hemisphere city, travel to the winter half of the world and enjoy the cooler weather. The tilt of the earth as it revolves around the sun makes all this possible.

We know the earth comes nearer to the sun in January than in July. Even though this difference in distance amounts to only about 3 percent, the energy that strikes the top of the earth's atmosphere is almost 7 percent greater on January 3 than on July 4. These statistics might lead us to believe that summer should be warmer in the Southern Hemisphere than in the Northern Hemisphere, which, however, is not the case. A close examination of the Southern Hemisphere reveals that nearly 81 percent of the surface is water compared to 61 percent in the Northern Hemisphere. The added solar energy due to the closeness of the sun is absorbed by large bodies of water, becoming well mixed and circulated within them. This keeps the average summer (January) temperatures in the Southern Hemisphere cooler than summer (July) temperatures in the Northern Hemisphere. Because of water's large heat capacity,

it also tends to keep winters in the Southern Hemisphere warmer than we might expect.*

Local Seasonal Variations   Figure 2.21 shows how the sun's position changes in the middle latitudes of the Northern Hemisphere during the course of one year. Note that, during the winter, the sun rises in the southeast and sets in the southwest. During the summer, it rises in the northeast, reaches a much higher position in the sky at noon, and sets in the northwest. Clearly, objects facing south will receive more sunlight during a year than those facing north. This fact becomes strikingly apparent in hilly or mountainous country.

Hills that face south receive more sunshine and, hence, become warmer than the partially shielded north-facing hills. Higher temperatures usually mean greater rates of evaporation and slightly drier soil conditions. Thus, south-facing hillsides are usually warmer

---

*For a comparison of January and July temperatures see Figs. 3.7 and 3.8, pp. 62 and 63.

and drier as compared to north-facing slopes at the same elevation. In many areas of the far west, only sparse vegetation grows on south-facing slopes, while, on the same hill, dense vegetation grows on the cool, moist hills that face north. (See Fig. 2.22.)

In the mountains, snow usually lingers on the ground for a longer time on north slopes than on the warmer south slopes. For this reason, ski runs are built facing north wherever possible. Also, homes and cabins built on the north side of a hill usually have a steep pitched roof, as well as a reinforced deck to withstand the added weight of snow from successive winter storms.

The seasonal change in the sun's position during the year can have an effect on the vegetation around the home. In winter, a large two-story home can shade its own north side, keeping it much cooler than its south side. Trees that require warm, sunny weather should be planted on the south side, where sunlight reflected from the house can even add to the warmth.

The design of a home can be important in reducing heating and cooling costs. Large windows should face south, allowing sunshine to penetrate the home in winter. To block out excess sunlight during the summer, a small eave or overhang should be built. A kitchen with windows facing east will let in enough warm morning sunlight to help heat this area. Because the west side warms rapidly in the afternoon, rooms having small windows (such as garages) should be placed here to act as a thermal buffer. Deciduous trees planted on the

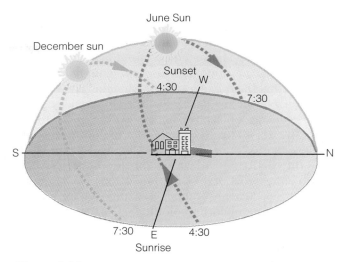

Figure 2.21
The changing position of the sun, as observed in middle latitudes in the Northern Hemisphere.

west side of a home provide shade in the summer. In winter, they drop their leaves, allowing the winter sunshine to warm the house. If you like the bedroom slightly cooler than the rest of the home, face it toward the north. Let nature help with the heating and air conditioning. Proper house design, orientation, and landscaping can help cut the demand for electricity, as well as for natural gas and fossil fuels, which are rapidly being depleted.

Figure 2.22
In areas where small temperature changes can cause major changes in soil moisture, sparse vegetation on the south-facing slopes will often contrast with lush vegetation on the north-facing slopes.

Summary

In this chapter, we looked at the concepts of heat and temperature and learned that latent heat is an important source of atmospheric heat energy. We also learned that the transfer of heat can take place by conduction, convection, and radiation—the transfer of energy by means of electromagnetic waves.

The hot sun emits most of its radiation as short-wave radiation. A portion of this energy heats the earth, and the earth, in turn, warms the air above. The cool earth emits most of its radiation as longwave infrared energy. Selective absorbers in the atmosphere, such as water vapor and carbon dioxide, absorb some of the earth's infrared radiation and reradiate a portion of it back to the surface, where it warms the surface, producing the atmospheric greenhouse effect. The average equilibrium temperature of the earth and the atmosphere remains fairly constant from one year to the next because the amount of energy they absorb each year is equal to the amount of energy they lose.

Finally, we examined the seasons and found that the earth has seasons because it is tilted on its axis as it revolves around the sun. The tilt of the earth causes a seasonal variation in both the length of daylight and the intensity of sunlight that reaches the surface.

Key Terms

The following terms are listed in the order they appear in the text. Define each. Doing so will aid you in reviewing the material covered in this chapter.

kinetic energy	convection	infrared radiation (IR)	aurora
temperature	thermals	black body	summer solstice
Kelvin scale	advection	radiative equilibrium	autumnal equinox
Fahrenheit scale	radiant energy (radiation)	temperature	Indian summer
Celsius scale	electromagnetic waves	selective absorbers	winter solstice
heat	micrometer	greenhouse effect	vernal equinox
latent heat	photons	atmospheric window	
sensible heat	visible region	scattering	
conduction	ultraviolet radiation (UV)	albedo	

Review Questions

1. Distinguish between temperature and heat.
2. Explain how heat is transferred in our atmosphere by:
   (a) conduction
   (b) convection
   (c) radiation
3. What is latent heat? How is latent heat an important source of atmospheric energy?
4. Why is the Kelvin temperature scale often used in scientific calculations?
5. How does the amount of radiation emitted by the earth differ from that emitted by the sun?
6. How do the wavelengths of most of the radiation emitted by the sun differ from those emitted by the surface of the earth?
7. When a body reaches a radiative equilibrium temperature, what is taking place?
8. Why are carbon dioxide and water vapor called selective absorbers?
9. Explain how the earth's atmospheric greenhouse effect works.
10. What gases appear to be responsible for the enhancement of the earth's greenhouse effect?
11. Why does the albedo of the earth and its atmosphere average about 30 percent?

12. Explain how the atmosphere near the earth's surface is warmed from below.
13. What causes the aurora?
14. In the Northern Hemisphere, why are summers warmer than winters even though the earth is actually closer to the sun in winter?
15. What are the main factors that determine seasonal temperature variations?
16. During the Northern Hemisphere's summer, the daylight hours in northern latitudes are longer than in middle latitudes. Explain why northern latitudes are not warmer.
17. Explain why the vegetation on the north-facing side of a hill is frequently different from the vegetation on the south-facing side of the same hill.

A coating of ice protects these almond trees from damaging low temperatures as an early spring freeze drops air temperatures well below freezing. (Courtesy of the *Modesto Bee*, photograph by Ted Benson.)

Chapter 3

Air Temperature

Contents

The sun shining full upon the field, the soil of which was sandy, the mouth of a heated oven seemed to me to be a trifle hotter than this ploughed field; it was almost impossible to breathe . . . The weather was almost too hot to live in, and the British troops in the orchard were forced by the heat to shelter themselves from it under trees . . . I presume everyone has heard of the heat that day, but none can realize it that did not feel it. Fighting is hot work in cool weather, how much more so in such weather as it was on the twenty-eighth of June 1778.

David M. Ludlum, *The Weather Factor*

Air temperature is an important weather element. Not only does it dictate how we should dress for the day, but the careful recording and application of temperature data are tremendously important to all. For without accurate information of this type, the work of farmers, weather analysts, construction workers, and many others would be a great deal more difficult. Therefore, we begin this chapter by examining the daily variation in air temperature. Here, we will answer such questions as why the warmest time of the day is normally in the afternoon, and why the coldest is usually in the early morning. And why calm, clear nights are usually colder than windy, clear nights. After we examine the factors that cause temperatures to vary from one place to another, we will look at daily, monthly, and yearly temperature averages and ranges with an eye toward practical applications for everyday living. Near the end of the chapter, we will see how air temperature is measured and how the wind can change our perception of air temperature.

▲▽▲

Daily Temperature Variations

In Chapter 2, we learned how the sun's energy coupled with the motions of the earth produce the seasons. In a way, each sunny day is like a tiny season as the air goes through a daily cycle of warming and cooling. The air warms during the morning hours, as the sun gradually rises higher in the sky, spreading a blanket of heat energy over the ground. The sun reaches its highest point around noon, after which it begins its slow journey toward the western horizon. It is around noon when the earth's surface receives the most intense solar rays. However, somewhat surprisingly, noontime is usually not the warmest part of the day. Rather, the air continues to be heated, often reaching a maximum temperature later in the afternoon. To find out why this *lag in temperature* occurs, we need to examine a shallow layer of air in contact with the ground.

Daytime Warming   As the sun rises in the morning, sunlight warms the ground, and the ground warms the air in contact with it by conduction. However, air is such a poor heat conductor that this process only takes place within a few inches of the ground. As the sun rises higher in the sky, the air in contact with the ground becomes even warmer, and, on a windless day, a substantial temperature difference usually exists just above the ground. This explains why joggers on a clear, windless, hot summer afternoon may experience air temperatures of over 50°C (122°F) at their feet and only 35°C (95°F) at their waist. (See Fig. 3.1.)

Near the surface, convection begins, and rising air bubbles (thermals) help to redistribute heat. In calm weather, these thermals are small and do not effectively mix the air near the surface. Thus, large vertical temperature differences are able to exist. On windy days, however, turbulent eddies are able to mix hot, surface air with the cooler air above. This form of mechanical stirring, sometimes called *forced convection*, helps the thermals to transfer heat away from the surface more efficiently. Therefore, on windy days the temperature difference between the surface air and the air directly above is not as great as it is on calm days.

We can now see why the warmest part of the day is usually in the afternoon. Around noon, the sun's rays are most intense. However, even though incoming solar radiation decreases in intensity after noon, it still exceeds outgoing heat energy from the surface for a time. This yields an energy surplus for two to four hours after noon and substantially contributes to a lag between the time of maximum solar heating and the time of maximum air temperature several feet above the surface.

The exact time of the highest temperature reading varies somewhat. Where the summer sky remains

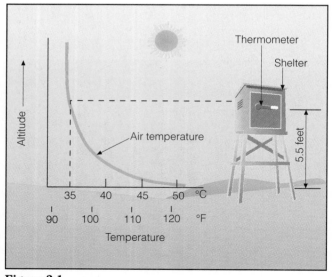

Figure 3.1
On a sunny, calm day, the air near the surface can be substantially warmer than the air several feet above the surface.

cloud-free all afternoon, the maximum temperature may occur sometime between 3:00 and 5:00 P.M. Where there is afternoon cloudiness or haze, the temperature maximum occurs an hour or two earlier. If clouds persist throughout the day, the overall daytime temperatures are usually lower, as clouds reflect a great deal of incoming sunlight.

Adjacent to large bodies of water, cool air moving inland may modify the rhythm of temperature change such that the warmest part of the day occurs at noon or before. In winter, atmospheric storms circulating warm air northward can even cause the highest temperature to occur at night.

Just how warm the air becomes depends on such factors as the type of soil, its moisture content, and vegetation cover. When the soil is a poor heat conductor (as loosely packed sand is), heat energy does not readily transfer into the ground. This allows the surface layer to reach a higher temperature, availing more energy to warm the air above. On the other hand, if the soil is moist or covered with vegetation, much of the available energy evaporates water, leaving less to heat the air. As you might expect, the highest summer temperatures usually occur over desert regions, where clear skies coupled with low humidities and meager vegetation permit the surface and the air above to warm up rapidly.

Where the air is humid, haze and cloudiness lower the maximum temperature by preventing some of the sun's rays from reaching the ground. In humid Atlanta, Georgia, the average maximum temperature for July is 30.5°C (87°F). In contrast, Phoenix, Arizona—in the desert southwest at the same latitude as Atlanta—experiences an average July maximum of 40.5°C (105°F). (Additional information on high daytime temperatures is given in the Focus section on p. 56.)

Nighttime Cooling   As the sun lowers, its energy is spread over a larger area, which reduces the heat available to warm the ground. Sometime in late afternoon or early evening, the earth's surface and air above begin to lose more energy than they receive (Fig. 3.2); hence, they start to cool.

Both the ground and air above cool by radiating infrared energy, a process called **radiational cooling**. The ground, being a much better radiator than air, is able to cool more quickly. Consequently, shortly after sunset, the earth's surface is slightly cooler than the air directly above it. The surface air transfers some en-

ergy to the ground by conduction, which the ground, in turn, quickly radiates away.

As the night progresses, the ground and the air in contact with it continue to cool more rapidly than the air a few feet higher. The warmer upper air does transfer *some* heat downward, a process that is slow due to the air's poor thermal conductivity. Therefore, by late night or early morning, the coldest air is next to the ground, with slightly warmer air above. (See Fig. 3.3.)

This measured increase in air temperature just above the ground is known as a **radiation inversion** because it forms mainly through radiational cooling of the surface. Because radiation inversions occur on most clear, calm nights, they are also called *nocturnal inversions.*

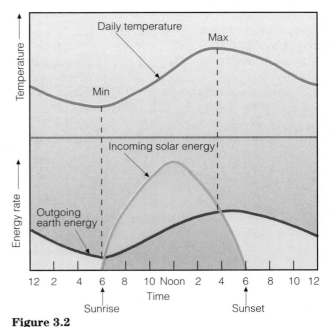

Figure 3.2
The daily variation in air temperature is controlled by incoming energy (primarily from the sun) and outgoing energy from the earth's surface. Where incoming energy exceeds outgoing energy (orange shade), the air temperature rises. Where outgoing energy exceeds incoming energy (blue shade) the air temperature falls.

Focus on a Special Topic
Record High Temperatures

Most people are aware of the extreme heat that exists during the summer in the desert southwest of the United States. But how hot does it get there? On July 10, 1913, Greenland Ranch in Death Valley, California, reported the highest temperature ever observed in North America: 57°C (134°F). Here, air temperatures are persistently hot throughout the summer, with the average maximum for July being 47°C (116°F). During the summer of 1917, there was an incredible period of 43 consecutive days when the maximum temperature reached 120°F or higher.

Probably the hottest urban area in the United States is Yuma, Arizona. Located along the California–Arizona border, Yuma's high temperature during July averages 42°C (108°F). In 1937, the high reached 100°F or more for 101 consecutive days.

In a more humid climate, the maximum temperature rarely climbs above 41°C (106°F). However, during the record heat wave of 1936, the air temperature reached 121°F near Alton, Kansas. And during the heat wave of 1983, which destroyed about $7 billion in crops and increased the nation's air-conditioning bill by an estimated $1 billion, Fayetteville reported North Carolina's all-time record high temperature when the mercury hit 110°F.

These readings, however, do not hold a candle to the hottest place in the world. That distinction probably belongs to Dallol, Ethiopia. Dallol is located south of the Red Sea, near lati-tude 12°N, in the hot, dry Danakil Depression. A prospecting company kept weather records at Dallol from 1960 to 1966. During this time, the average daily maximum temperature exceeded 38°C (100°F) every month of the year, except during December and January, when the average maximum lowered to 98°F and 97°F, respectively. On many days, the air temperature exceeded 120°F. The average annual temperature for the six years at Dallol was 34°C (94°F). In comparison, the average annual tem-perature in Yuma is 74°F and at Death Valley, 76°F. The highest temperature reading on earth (under standard conditions) occurred northeast of Dallol at El Azizia, Libya (32°N), when, on September 13, 1922, the temperature reached a scorching 58°C (136°F). Table 1 gives record high temperatures throughout the world.

Table 1   Some Record High Temperatures Throughout the World

Location (Latitude)	Record High Temperature (°C)	(°F)	Record for:	Date
El Azizia, Libya (32°N)	58	136	The world	September 13, 1922
Death Valley, Calif. (36°N)	57	134	Western Hemisphere	July 10, 1913
Tirat Tsvi, Israel (32°N)	54	129	Middle East	June 21, 1942
Cloncurry, Queensland (21°S)	53	128	Australia	January 16, 1889
Seville, Spain (37°N)	50	122	Europe	August 4, 1881
Rivadavia, Argentina (35°S)	49	120	South America	December 11, 1905
Midale, Saskatchewan (49°N)	45	113	Canada	July 5, 1937
Fort Yukon, Alaska (66°N)	38	100	Alaska	June 27, 1915
Pahala, Hawaii (19°N)	38	100	Hawaii	April 27, 1931
Esparanza, Antarctica (63°S)	14	58	Antarctica	October 20, 1956

Cold Air Near the Surface   A strong radiation inversion occurs when the air near the ground is much colder than the air higher up. Ideal conditions for a strong inversion and, hence, very low nighttime temperatures exist when the air is calm, the night is long, and the air is fairly dry and cloud-free. Let's examine these ingredients one by one.

A windless night is essential for a strong radiation inversion because a stiff breeze tends to mix the colder air at the surface with the warmer air above. This mixing, along with the cooling of the warmer air as it comes in contact with the cold ground, causes a vertical temperature profile that is almost isothermal (constant temperature) in a layer several feet thick. In the absence of wind, the cooler, more dense surface air does not readily mix with the warmer, less dense air above, and the inversion is more strongly developed as illustrated in Fig. 3.3.

A long night also contributes to a strong inversion. Generally, the longer the night, the longer the time of radiational cooling and the better are the chances that the air near the ground will be much colder than the air above. Consequently, winter nights provide the best conditions for a strong radiation inversion, other factors being equal.

Finally, radiation inversions are more likely with a clear sky and dry air. Under these conditions, the ground is able to radiate its energy to outer space and thereby cool rapidly. However, with cloudy weather and moist air, much of the outgoing infrared energy is absorbed and reradiated to the surface, retarding the rate of cooling. Also, on moist nights, condensation in the form of fog or dew will release latent heat, which warms the air. So, radiation inversions may occur on any night. But, during long winter nights, when the air is still, cloud-free, and relatively dry, these inversions can become strong and deep.

It should now be apparent that how cold the night air becomes depends primarily on the length of the night, the moisture content of the air, cloudiness, and the wind. Even though wind may initially bring cold air into a region, the coldest nights usually occur when the air is clear and relatively calm. (Additional information on very low nighttime temperatures is given in the Focus section on p. 58.)

Look back at Fig. 3.2 (p. 55) and observe that the lowest temperature on any given day is usually observed around sunrise. However, the cooling of the

ground and surface air may even continue beyond sunrise for a half hour or so, as outgoing energy can exceed incoming energy. This happens because light from the early morning sun passes through a thick section of atmosphere and strikes the ground at a low angle. Consequently, the sun's energy does not effectively warm the surface. Surface heating may be reduced further when the ground is moist and available energy is used for evaporation. Hence, the lowest temperature may occur shortly after the sun has risen.

Cold, heavy surface air slowly drains downhill during the night and eventually settles in low-lying basins and valleys. Valley bottoms are thus colder than the surrounding hillsides. (See Fig. 3.4.) In middle latitudes, these warmer hillsides, called **thermal belts**, are less likely to experience freezing temperatures than

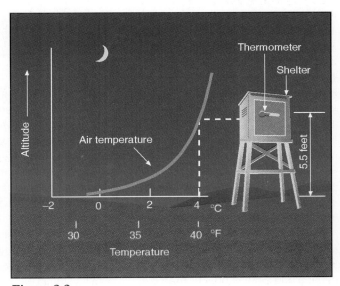

Figure 3.3
On a clear, calm night, the air near the surface can be much colder than the air several feet above. The increase in air temperature with increasing height above the surface is called a *radiation temperature inversion*.

Focus on a Special Topic
Record Low Temperatures

One city in the United States that experiences very low temperatures is International Falls, Minnesota, where the average temperature for January is −16°C (3°F). Located several hundred miles to the south, Minneapolis–St. Paul, with an average temperature of −9°C (16°F) for the three winter months, is the coldest major urban area in the nation. For duration of extreme cold, Minneapolis reported 186 consecutive hours of temperatures below 0°F during the winter of 1911–1912. Within the forty-eight adjacent states, however, the record for the longest duration of severe cold belongs to Langdon, North Dakota, where the thermometer remained below 0°F for 41 consecutive days during the winter of 1936. The official record for the lowest temperature in the forty-eight adjacent states belongs to Rogers Pass, Montana, where on the morning of January 20, 1954, the mercury dropped to −57°C (−70°F). The lowest official temperature for Alaska, −62°C (−80°F), occurred at Prospect Creek on January 23, 1971.

The coldest areas in North America are found in the Yukon and Northwest Territory of Canada. Resolute, Canada (latitude 75°N), has an average temperature of −32°C (−26°F) for the month of January.

The lowest temperatures and coldest winters in the Northern Hemisphere are found in the interior of Siberia and Greenland. For example, the average January temperature in Yakutsk, Siberia (latitude 62°N), is −43°C (−46°F). There, the mean temperature for the entire year is a bitter

cold 12°F. At Eismitte, Greenland, the average temperature for February (the coldest month) is −47°C (−53°F), with the mean annual temperature being a frigid −30°C (−22°F). Even though these temperatures are extremely low, they do not come close to the coldest area of the world: the Antarctic.

At the geographical South Pole, over nine thousand feet above sea level, where the Amundsen-Scott scientific station has been keeping records for more than twenty-five years, the average temperature for the month of July (winter) is −59°C (−74°F) and the mean annual temper-

ature is −49°C (−57°F). The lowest temperature ever recorded there was −83°C (−117°F). This temperature occurred under clear skies with a light wind on the morning of June 23, 1983. Cold as it was, it was not the record low for the world. That belongs to the Russian station at Vostok, Antarctica (latitude 72°S), where the temperature plummeted to −88°C (−127°F) on August 24, 1960. (See Table 2 for record low temperatures throughout the world.)

Table 2   Some Record Low Temperatures Throughout the World

Location (Latitude)	Record Low Temperature (°C)	(°F)	Record for:	Date
Vostok, Antarctica (72°S)	−88	−127	The world	August 24, 1960
Verkhoyansk, Russia (67°N)	−68	−90	Northern Hemisphere	February 7, 1892
Northice, Greenland (72°N)	−66	−87	Greenland	January 9, 1954
Snag, Yukon (62°N)	−63	−81	North America	February 3, 1947
Prospect Creek, Alaska (66°N)	−62	−80	Alaska	January 23, 1971
Rogers Pass, Montana (47°N)	−57	−70	U.S. (excluding Alaska)	January 20, 1954
Sarmiento, Argentina (34°S)	−33	−27	South America	June 1, 1907
Ifrane, Morocco (33°N)	−24	−11	Africa	February 11, 1935
Charlotte Pass, Australia (36°S)	−22	−8	Australia	July 22, 1949
Mt. Haleakala, Hawaii (20°N)	−10	14	Hawaii	January 2, 1961

the valley below. This encourages farmers to plant on hillsides those trees unable to survive the valley's low temperature.

On the valley floor, the cold, dense air is unable to rise. Smoke and other pollutants trapped in this heavy air restrict visibility. Therefore, valley bottoms are not only colder, but are also more frequently polluted than nearby hillsides. Even when the land is only gently sloped, cold air settles into lower-lying areas, such as river basins and floodplains. Because the flat flood-plains are agriculturally rich areas, cold air drainage often forces farmers to seek protection for their crops.

Protecting Crops from the Cold Night Air   On cold nights, many plants may be damaged by low tempera-tures. To protect small plants or shrubs, cover them with straw, cloth, or plastic sheeting. This prevents ground heat from being radiated away to the colder surroundings. If you are a household gardener con-cerned about outside flowers and plants during cold weather, simply wrap them in plastic or cover each with a paper cup.

Did you know?
When the air temperature reached its all time record low of −127°F on the Antarctic Plateau at Vostok Station, a drop of saliva from the lips of a person taking an observation would have frozen before reaching the ground.

Fruit trees are particularly vulnerable to cold weather in the spring when they are blossoming. The protection of such trees presents a serious problem to the farmer. Since the lowest temperatures on a clear, still night occur near the surface, the lower branches of a tree are the most susceptible to damage. Therefore, increasing the air temperature close to the ground may prevent damage. One way this can be done is to use **orchard heaters**, or "smudge pots," which warm the air around them by setting up convection currents close to the ground. (See Fig. 3.5.)

Another way to protect trees is to mix the cold air at the ground with the warmer air above, thus raising the temperature of the air next to the ground. Such

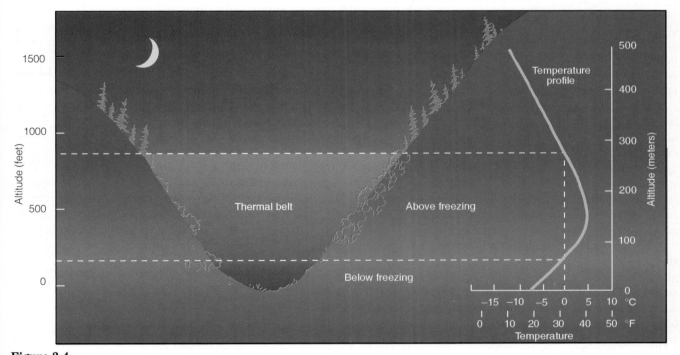

Figure 3.4
On cold, clear nights, the settling of cold air into valleys makes them colder than surrounding hillsides. The region where the air temperature is above freezing is known as a *thermal belt.*

Figure 3.5
Orchard heaters circulate the air by setting up convection currents.

Figure 3.6
Wind machines mix cooler surface air with warmer air above.

mixing can be accomplished by using **wind machines** (Fig. 3.6), which are power-driven fans that resemble airplane propellers. Farmers without their own wind machines can rent air mixers in the form of helicopters. Although helicopters are effective in mixing the air, they are expensive to operate.

If sufficient water is available, trees can be protected by irrigation. On potentially cold nights, the orchard may be flooded. Because water has a high heat capacity, it cools more slowly than dry soil. Consequently, the surface does not become as cold as it would if it were dry. Furthermore, wet soil has a higher thermal conductivity than dry soil. Hence, in wet soil heat is conducted upward from subsurface soil more rapidly, which helps to keep the surface warmer.

So far, we have discussed protecting trees against the cold air near the ground during a radiation inversion. Farmers often face another nighttime cooling problem. For instance, when subfreezing air blows into a region, the coldest air is not found at the surface; the air actually becomes colder with height. This condition is known as a **freeze** or *advection frost*. A single freeze in California or Florida can cause several million dollars damage to citrus crops.

Protecting an orchard from the damaging cold air blown by the wind can be a problem. Wind machines

will not help because they would only mix cold air at the surface with the colder air above. Orchard heaters and irrigation are of little value as they would only protect the branches just above the ground. However, there is one form of protection that does work: An orchard's sprinkling system may be turned on so that it emits a fine spray of water. In the cold air, the water freezes around the branches and buds, coating them with a thin veneer of ice. As long as the spraying continues, the latent heat—given off as the water changes into ice—keeps the ice temperature at 0°C (32°F). The ice acts as a protective coating against the subfreezing air by keeping the buds (or fruit) at a temperature higher than their damaging point. Care must be taken since too much ice can cause the branches to break. The fruit may be saved from the cold air, while the tree itself may be damaged by too much protection.

The Controls of Temperature

The main factors that cause variations in temperature from one place to another are called the **controls of temperature**. In the previous chapter, we saw that the greatest factor in determining temperature is the amount of solar radiation that reaches the surface. This, of course, is determined by the length of daylight hours and the intensity of incoming solar radiation. Both of these factors are a function of latitude, hence latitude is considered an important control of temperature. The main controls are listed below.

1. latitude
2. land and water
3. ocean currents
4. elevation

We can obtain a better picture of these controls by examining Figs. 3.7 and 3.8, which show the average monthly temperatures throughout the world for January and July. The lines on the map are **isotherms**—lines connecting places that have the same temperature. Because air temperature normally decreases with height, cities at very high elevations are much colder than their sea level counterparts. Consequently, the isotherms in Figs. 3.7 and 3.8 are corrected to read at the same horizontal level (sea level) by adding to each station above sea level an amount of temperature that would correspond to an average temperature change with height.*

Figures 3.7 and 3.8 show the importance of latitude on temperature. Note that, on the average, temperatures decrease poleward from the tropics and subtropics in both January and July. However, because there is a greater variation in solar radiation between low and high latitudes in winter than in summer, the isotherms in January are closer together (a tighter gradient) than they are in July. This means that if you travel from New Orleans to Detroit in January, you are more likely to experience greater temperature variations than if you make the same trip in July. Notice also in Figs. 3.7 and 3.8 that the isotherms do not run horizontally; rather, in many places they bend, especially where they approach an ocean-continent boundary.

On the January map, the temperatures are much lower in the middle of continents than they are at the same latitude near the oceans; on the July map, the reverse is true. The reason for these temperature variations can be attributed to the unequal heating and cooling properties of land and water. For one thing, so-

lar energy reaching land is absorbed in a thin layer of soil; reaching water, it penetrates deeply. Because water is able to circulate, it distributes its heat through a much deeper layer. Also, some of the solar energy striking the water is used to evaporate it rather than heat it.

Another important reason for the temperature contrasts is that water has a higher *specific heat* than land. The **specific heat** of a substance *is the amount of heat needed to raise the temperature of one gram of a substance by one degree Celsius*. It takes a great deal more heat (about five times more) to raise the temperature of a given amount of water by one degree than it does to raise the temperature of the same amount of soil or rock by one degree. Consequently, water has a much higher specific heat than either of these substances. Water not only heats more slowly than land, it cools more slowly as well, and so the oceans act like huge heat reservoirs. Thus, mid-ocean surface temperatures change relatively little from summer to winter compared to the much larger annual temperature changes over the middle of continents.

Along the margin of continents, ocean currents often influence air temperatures. For example, along the eastern margins, warm ocean currents transport warm water poleward, while, along the western margins, they transport cold water equatorward. Some coastal areas also experience upwelling, which brings even colder water to the surface. (See Chapter 7.)

Even large lakes can modify the temperature around them. In summer, the Great Lakes remain cooler than the land. As a result, refreshing breezes blow inland, bringing relief from the sometimes sweltering heat. As winter approaches, the water cools more slowly than the land. The first blast of cold air from Canada is modified as it crosses the lakes, and so the first freeze is delayed on the eastern shores of Lake Michigan.

▲▼▲

Air Temperature Data

In the previous sections we considered how air temperature varies on a daily basis and from one place to another. We will now focus on the ways temperature data are organized and used.

Daily, Monthly, and Yearly Temperatures The greatest variation in daily temperature occurs at the earth's surface. In fact, the difference between the daily

*The amount of change is usually less than the normal temperature lapse rate of 3.6°F per 1000 feet (6.5°C per 1000 meters). The reason is that the normal lapse rate is computed for altitudes above the earth's surface in the "free" atmosphere. In the less dense air at high elevations, the absorption of solar radiation by the ground causes an overall slightly higher temperature than that of the free atmosphere at the same level.

Figure 3.7
Average air temperature near sea level in January (°F).

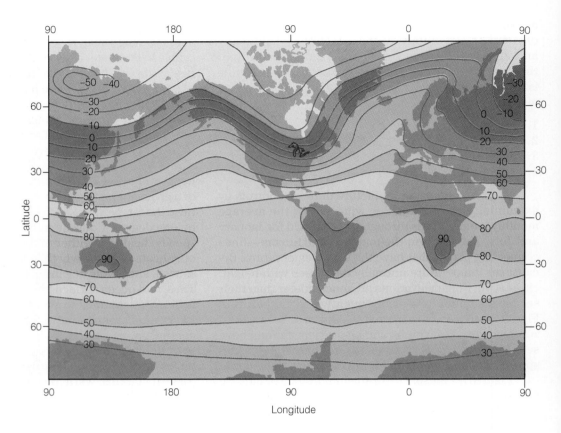

maximum and minimum temperature—called the **daily (diurnal) range of temperature**—is greatest next to the ground and becomes progressively smaller as we move upward. This daily variation in temperature is also much larger on clear days than on cloudy ones.

The largest diurnal range of temperature occurs on high deserts, where the air is fairly dry, often cloud-free, and there is little water vapor to reradiate much infrared energy back to the surface. By day, clear summer skies allow the sun's energy to quickly warm the ground which, in turn, warms the air above to a temperature sometimes exceeding 35°C (95°F). At night, the ground cools rapidly by radiating infrared energy to space, and the minimum temperature in these regions

occasionally dips below 5°C (41°F), thus giving a daily temperature range of more than 30°C (54°F).

In humid regions, the diurnal temperature range is usually small. Here, haze and clouds lower the maximum temperature by preventing some of the sun's energy from reaching the surface. At night, the moist air keeps the minimum temperature high by absorbing the earth's infrared radiation and reradiating a portion of it to the ground. An example of a humid city with a small summer diurnal temperature range is Charleston, South Carolina, where the average July maximum temperature is 32°C (90°F), the average minimum is 22°C (72°F), and the diurnal range is only 10°C (8°F).

Cities near large bodies of water typically have smaller diurnal temperature ranges than cities further inland. This is caused in part by the additional water vapor in the air and by the fact that water warms and cools much more slowly than land.

The average of the highest and lowest temperature for a 24-hour period is known as the **mean daily temperature**. Most newspapers list the mean daily temperature along with the highest and lowest temper-

Did you know?
One of the greatest temperature ranges ever recorded (100°F) occurred at Browning, Montana, when on January 23, 1916, the air temperature plummeted from 44°F to –56°F in less than 24 hours.

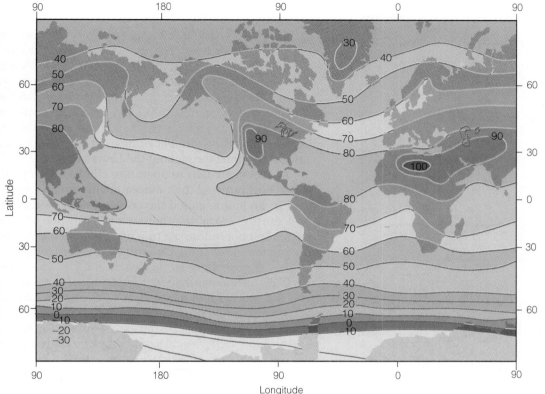

Figure 3.8
Average air temperature
near sea level in July (°F).

atures for the preceding day. The average of the mean daily temperatures for a particular date averaged for a 30-year period gives the average (or "*normal*") temperatures for that date. The average temperature for each month is the average of the daily mean temperatures for that month.

At any location, the difference between the average temperature of the warmest and coldest months is called the **annual range of temperature**. Usually the largest annual ranges occur over land, the smallest over water. Hence, inland cities have larger annual ranges than coastal cities. Near the equator (because daylight length varies little and the sun is always high in the noon sky), annual temperature ranges are small, usually less than 3°C (5°F). Quito, Ecuador—on the equator at an elevation of nine thousand feet—experiences an annual range of less than 1°F. In middle and high latitudes, large seasonal variations in the amount of sunlight reaching the surface produce large temperature contrasts between winter and summer. Here, annual ranges are large, especially in the middle of a continent. Yakutsk, in northeastern Siberia near the

Arctic Circle, has an extremely large annual temperature range of 62°C (112°F).

The average temperature of any station for the entire year is the **mean annual temperature**, which represents the average of the twelve monthly average temperatures. When two cities have the same mean annual temperature, it might first seem that their temperatures throughout the year are quite similar. However, often this is not the case. For example, San Francisco, California, and Richmond, Virginia, are at the same latitude (37°N). Both have similar hours of daylight during the year; both have the same mean annual temperature—14°C (57°F). Here, the similarities end. The temperature differences between the two cities are apparent to anyone who has traveled to San Francisco during the summer with a suitcase full of clothes suitable for summer weather in Richmond.

Figure 3.9 summarizes the average temperatures for San Francisco and Richmond. Notice that the coldest month for both cities is January. Even though January in Richmond averages only 8°C (14°F) colder than January in San Francisco, people in Richmond awaken

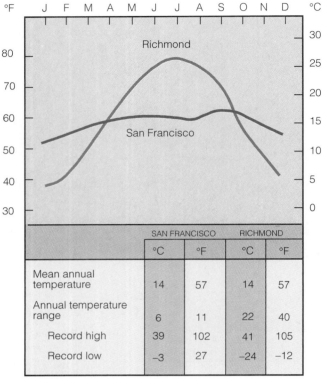

Figure 3.9
Temperature data for San Francisco, California (37°N) and
Richmond, Virginia (37°N)—two cities with the same mean annual
temperature.

	SAN FRANCISCO		RICHMOND	
	°C	°F	°C	°F
Mean annual temperature	14	57	14	57
Annual temperature range	6	11	22	40
Record high	39	102	41	105
Record low	−3	27	−24	−12

to an average January minimum temperature of –6°C
(21°F), which is much colder than the lowest temperature ever recorded in San Francisco. Trees that thrive
in San Francisco's weather would find it difficult surviving a winter in Richmond. So, even though San Francisco and Richmond have the same mean annual
temperature, the behavior and range of their temperatures differ greatly.

The Use of Temperature Data   An application of
daily temperature developed by heating engineers in
estimating energy needs is the **heating degree-day**.
The heating-degree day is based on the assumption that
people will begin to use their furnaces when the mean
daily temperature drops below 65°F. Therefore, heating
degree-days are determined by subtracting the mean
temperature for the day from 65°F. Thus, if the mean

temperature for a day is 64°F, there would be 1 heating
degree-day on this day.*

On days when the mean temperature is above
65°F, there are no heating degree-days. Hence, the
lower the average daily temperature, the more heating
degree-days and the greater the predicted consumption
of fuel. When the number of heating degree-days for a
whole year is calculated, the heating fuel requirements
for any location can be estimated. Figure 3.10 shows
the yearly average number of heating degree-days in
various locations throughout the United States.

As the mean daily temperature climbs above 65°F,
people begin to cool their indoor environment. Consequently, an index, called the **cooling degree-day**, is
used during warm weather to estimate the energy
needed to cool indoor air to a comfortable level. The
forecast of mean daily temperature is converted to
cooling degree-days by subtracting 65°F from the
mean. The remaining value is the number of cooling degree-days for that day. For example, a day with a mean
temperature of 70°F would correspond to (70–65), or
5 cooling degree-days. High values indicate warm
weather and high power production for cooling. (See
Fig. 3.11.)

Knowledge of the number of cooling degree-days
in an area allows a builder to plan the size and type of
equipment that should be installed to provide adequate
air conditioning. Also, the forecasting of cooling degree-days during the summer gives power companies
a way of predicting the energy demand during peak energy periods. A composite of heating plus cooling degree-days would give a practical indication of the
energy requirements over the year.

Farmers use an index, called **growing degree-
days**, as a guide to planting and for determining the
approximate dates when a crop will be ready for harvesting. A growing degree-day for a particular crop is
defined as a day on which the mean daily temperature
is one degree above the *base temperature* (also known
as the *zero temperature*)—the minimum temperature
required for growth of that crop. For sweet corn, the
base temperature is 50°F and, for peas, it is 40°F.

On a summer day in Iowa, the mean temperature
might be 80°F. From Table 3.1, we can see that, on this
day, sweet corn would accumulate (80–50), or 30 grow-

---

*In the United States, the National Weather Service and the Depart-
ment of Agriculture use degrees Fahrenheit in their computations.

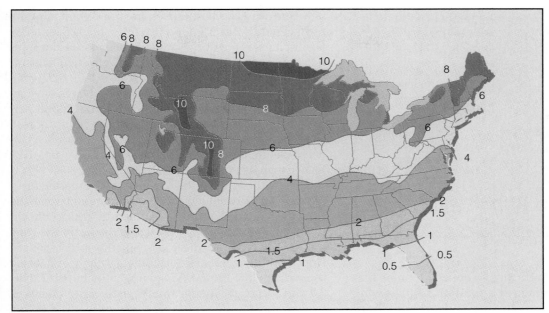

Figure 3.10
Mean annual total heating degree-days in thousands of °F, where
the number 4 on the map represents 4000 (base 65°F).

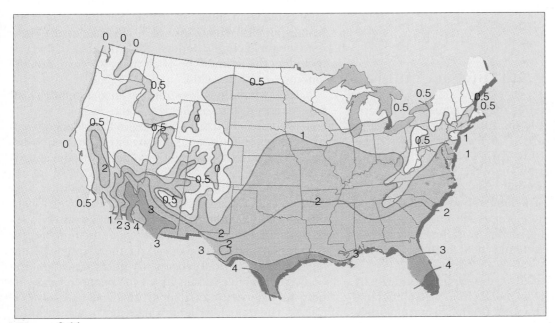

Figure 3.11
Mean annual total cooling degree-days in thousands of °F, where
the number 1 on the map represents 1000 (base 65°F).

Table 3.1 Estimated Growing Degree-Days for Certain Agricultural Crops to Reach Maturity

Crop (Variety, Location)	Base Temperature (°F)	Growing Degree-Days to Maturity
Beans (Snap, South Carolina	50	1200–1300
Corn (Sweet, Indiana)	50	2200–2800
Cotton (Delta Smooth Leaf, Arkansas)	60	1900–2500
Peas (Early, Indiana)	40	1100–1200
Rice (Vegold, Arkansas)	60	1700–2100
Wheat (Indiana)	40	2100–2400

ing degree-days. Theoretically, sweet corn can be harvested when it accumulates a total of 2200 growing degree-days. So, if sweet corn is planted in early April and each day thereafter averages about 20 growing degree-days, the corn would be ready for harvest about 110 days later, or around the middle of July. Although moisture and other conditions are not taken into account, growing degree-days nevertheless serve as a useful guide in forecasting approximate dates of crop maturity.

Air Temperature and Human Comfort

Probably everyone realizes that the same air temperature can feel differently on different occasions. For example, a temperature of 70°F on a clear, windless March afternoon in New York City can almost feel balmy after a long, hard winter. Yet, this same temperature may feel uncomfortably cool on a summer afternoon in a stiff breeze. The human body's perception of temperature obviously changes with varying atmospheric conditions. The reason for these changes is related to how we exchange heat energy with our environment.

The body stabilizes its temperature primarily by converting food into heat (*metabolism*). To maintain a constant temperature, the heat produced and absorbed by the body must be equal to the heat it loses to its surroundings. There is, therefore, a constant exchange of heat—especially at the surface of the skin—between the body and the environment.

One way the body loses heat is by emitting infrared energy. But we not only emit radiant energy, we absorb it as well. Another way the body loses and gains heat is by conduction and convection, which transfer heat to and from the body by air motions. On a cold day, a thin layer of warm air molecules forms close to the skin, protecting it from the surrounding cooler air and from the rapid transfer of heat. Thus, in cold weather, when the air is calm, the temperature we perceive—called the *sensible temperature*—is often higher than a thermometer might indicate.

Once the wind starts to blow, the insulating layer of warm air is swept away, and heat is rapidly removed from the skin by the constant bombardment of cold air. When all other factors are the same, the faster the wind blows, the greater the heat loss, and the colder we feel. How cold the wind makes us feel is usually expressed as a **wind-chill factor**. The wind-chill charts (Tables 3.2 and 3.3) translate the ability of the air to take heat away from the human body with wind (its cooling power) into a wind-chill equivalent temperature with no wind. For example, notice that, in Table 3.2, an air temperature of 20°F with a wind speed of 30 miles per hour produces a wind-chill equivalent temperature of –18°F. This means that exposed skin would lose as much heat in one minute in air with a temperature of 20°F and a wind speed of 30 miles per hour as it would in calm air with a temperature of –18°F. Of course, how cold we feel actually depends on a number of factors, including the fit and type of clothing we wear, and the amount of exposed skin.

High winds, in below-freezing air, can remove heat from exposed skin so quickly that the skin may actually freeze and discolor. The freezing of skin, called *frostbite*, usually occurs on the body extremities first because they are the greatest distance from the source of body heat.

In cold weather, wet skin can be a factor in how cold we feel. A cold, rainy day (drizzly, or even foggy) often feels colder than a "dry" one because water on exposed skin conducts heat away from the body better than air does. In fact, in cold, wet, and windy weather a person may actually lose body heat faster than the body can produce it. This may even occur in relatively mild weather with air temperatures as high as 10°C (50°F). The rapid loss of body heat may lower the body temperature below its normal level and bring on a condition known as **hypothermia**—the rapid, progressive mental and physical collapse that accompanies the lowering of human body temperature.

The first symptom of hypothermia is exhaustion. If exposure continues, judgment and reasoning power begin to disappear. Prolonged exposure, especially at temperatures near or below freezing, produces stupor, collapse, and death when the internal body temperature drops to about 26°C (79°F).

In cold weather, heat is more easily dissipated through the skin. To counteract this rapid heat loss, the peripheral blood vessels of the body constrict, cutting off the flow of blood to the outer layers of the skin. In hot weather, the blood vessels enlarge, allowing a greater loss of heat energy to the surroundings. In addition to this we perspire. As evaporation occurs, the skin cools. When the air contains a great deal of water vapor and it is close to being saturated, perspiration does not readily evaporate from the skin. Less evaporational cooling causes most people to feel hotter than it really is, and a number of people start to complain about the "heat and humidity." (A closer look at how

Did you know?
The coldest Boston Marathon was in 1925, when the air temperature throughout most of the race hovered in the mid thirties (°F) and a penetrating north wind of 20 miles per hour kept the wind chill near 10°F.

we feel in hot weather will be given in Chapter 4, after we have examined the concepts of relative humidity and wet-bulb temperature.)

Measuring Air Temperature

Thermometers were developed in the late sixteenth century to measure air temperature. The most commonly used thermometers for measuring surface air

---

Table 3.2   Wind Chill Equivalent Temperature (°F)
A 20-Mile-per-Hour Wind Combined with an Air Temperature of 10°F
Produces a Wind Chill Equivalent Temperature of –24°F

							Air Temperature (°F)											
		35	30	25	20	15	10	5	0	–5	–10	–15	–20	–25	–30	–35	–40	–45
	4	35	30	25	20	15	10	5	0	–5	–10	–15	–20	–25	–30	–35	–40	–45
	5	32	27	22	16	11	6	0	–5	–10	–15	–21	–26	–31	–36	–42	–47	–52
	10	22	16	10	3	–3	–9	–15	–22	–27	–34	–40	–46	–52	–58	–64	–71	–77
Wind speed (mi/hr)	**15**	16	9	2	–5	–11	–18	–25	–31	–38	–45	–51	–58	–65	–72	–78	–85	–92
	20	12	4	–3	–10	–17	–24	–31	–39	–46	–53	–60	–67	–74	–81	–88	–95	–103
	25	8	1	–7	–15	–22	–29	–36	–44	–51	–59	–66	–74	–81	–88	–96	–103	–110
	30	6	–2	–10	–18	–25	–33	–41	–49	–56	–64	–71	–79	–86	–93	–101	–109	–116
	35	4	–4	–12	–20	–27	–35	–43	–52	–58	–67	–74	–82	–89	–97	–105	–113	–120
	40	3	–5	–13	–21	–29	–37	–45	–53	–60	–69	–76	–84	–92	–100	–107	–115	–123
	45	2	–6	–14	–22	–30	–38	–46	–54	–62	–70	–78	–85	–93	–102	–109	–117	–125

Table 3.3   Wind Chill Equivalent Temperature (°C)

							Air Temperature (°C)								
		8	4	0	–4	–8	–12	–16	–20	–24	–28	–32	–36	–40	–44
	Calm	8	4	0	–4	–8	–12	–16	–20	–24	–28	–32	–36	–40	–44
	10	5	0	–4	–8	–13	–17	–22	–26	–31	–35	–40	–44	–49	–53
Wind Speed (km/hr)	**20**	0	–5	–10	–15	–21	–26	–31	–36	–42	–47	–52	–57	–63	–68
	30	–3	–8	–14	–20	–25	–31	–37	–43	–48	–54	–60	–65	–71	–77
	40	–5	–11	–17	–23	–29	–35	–41	–47	–53	–59	–65	–71	–77	–83
	50	–6	–12	–18	–25	–31	–37	–43	–49	–56	–62	–68	–74	–80	–87
	60	–7	–13	–19	–26	–32	–39	–45	–51	–58	–64	–70	–77	–83	–89

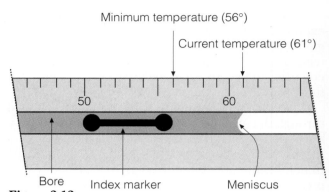

Minimum temperature (56°)

Current temperature (61°)

50   60

Bore   Index marker   Meniscus

Figure 3.13
A section of a minimum thermometer showing both the current air temperature and the minimum temperature.

temperature are **liquid-in-glass thermometers**, for they are easy to read and inexpensive to construct.

These thermometers have a glass bulb attached to a sealed, graduated tube about ten inches long. A very small opening, or bore, extends from the bulb to the end of the tube. A liquid in the bulb (usually mercury or red-colored alcohol) is free to move from the bulb up through the bore and into the tube. When the air temperature increases, the liquid in the bulb expands and rises up the tube. When the air temperature decreases, the liquid contracts and moves down the tube. Hence, the length of the liquid in the tube represents the air temperature. Because the bore is very narrow, a small temperature change will show up as a relatively large change in the length of the liquid column.

Maximum and minimum thermometers are liquid-in-glass thermometers used exclusively for determining daily maximum and minimum temperatures. The **maximum thermometer** looks like any other liquid-in-glass thermometer with one exception: It has a small constriction within the bore just above the bulb (Fig. 3.12). As the air temperature increases, the mercury expands and freely moves past the constriction up the tube, until the maximum temperature occurs. However, as the air temperature begins to drop, the small constriction prevents the mercury from flowing back into the bulb. Thus, the end of the stationary mercury column indicates the maximum temperature for the day. The mercury will stay at this position until either the air

warms to a higher reading or the thermometer is reset by whirling it on a special holder and pivot. Usually, the whirling is sufficient to push the mercury back into the bulb past the constriction until the end of the column indicates the present air temperature.*

A **minimum thermometer** measures the lowest temperature reached during a given period. Most minimum thermometers use alcohol as a liquid, since it freezes at a much lower temperature than mercury. The minimum thermometer is similar to other liquid-in-glass thermometers except that it contains a small barbell-shaped index marker in the bore (Fig. 3.13). The index marker is about 1 inch long and is free to slide back and forth within the liquid. It cannot move out of the liquid because the surface tension at the end of the liquid column (the *meniscus*) holds it in.

A minimum thermometer is mounted horizontally. As the air temperature drops, the contracting liquid moves back into the bulb and brings the index marker down the bore with it. When the air temperature stops decreasing, the liquid and the index marker stop moving down the bore. As the air warms, the alcohol expands and moves freely up the tube past the stationary index marker. Because the index marker does not move as the air warms, the minimum temperature is read by observing the upper end of the marker.

To reset a minimum thermometer, simply tip it upside down. This allows the index marker to slide to the

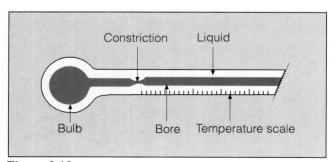

Constriction   Liquid

Bulb   Bore   Temperature scale

Figure 3.12
A section of a maximum thermometer.

---

*Thermometers that measure body temperature are maximum thermometers. It should be apparent why they are shaken both before and after you take your temperature.

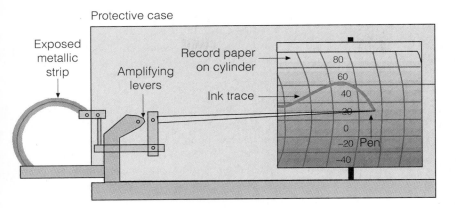

Figure 3.14
The thermograph with a bimetallic thermometer.

upper end of the alcohol column, which is indicating the current air temperature. The thermometer is then remounted horizontally, so that the marker will move toward the bulb as the air temperature decreases.

Highly accurate temperature measurements may be made with **electrical thermometers**, such as the *thermistor* and the *electrical resistance thermometer*. Both of these instruments do not actually measure air temperature but rather the electrical resistance of some material. Since the resistance of the particular material changes as the temperature changes, a meter can measure the resistance and be calibrated to represent air temperature.

Air temperature may also be obtained with instruments called *infrared sensors*, or **radiometers**. Radiometers do not measure temperature directly; rather, they measure emitted radiation (usually infrared). By measuring both the intensity of radiant energy and the wavelength of maximum emission of a particular gas (either water vapor or carbon dioxide), radiometers in orbiting satellites are now able to estimate the air temperature at selected levels in the atmosphere.

A **bimetallic thermometer** consists of two different pieces of metal (usually iron and brass) welded together to form a single strip. As the temperature changes, one metal expands more than the other, causing the strip to bend. The small amount of bending is amplified through a system of levers to a pointer on a calibrated scale. The bimetallic thermometer is usually the temperature-sensing part of the **thermograph**, an instrument that measures and records temperature. (See Fig. 3.14.)

Thermometers and other instruments are usually housed in an **instrument shelter** (Fig. 3.15). The shelter completely encloses the instruments, protecting them from rain, snow, and the sun's direct rays. It is painted white to reflect sunlight and has louvered sides, so that air is free to flow through it. This helps to keep

Figure 3.15
An instrument shelter protects the instruments inside from the weather elements.

Focus on Instruments
Thermometers Should Be Read in the Shade

When we measure air temperature with a common liquid thermometer, an incredible number of air molecules bombard the bulb, transferring energy either to or away from it. When the air is warmer than the thermometer, the liquid gains energy, expands, and rises up the tube; the opposite will happen when the air is colder than the thermometer. The liquid stops rising (or falling) when equilibrium be-tween incoming and outgoing energy is established. At this point, we can read the temperature by observing the height of the liquid in the tube.

It is *impossible* to measure air temperature accurately in direct sunlight because the thermometer absorbs radiant energy from the sun in addition to energy from the air molecules. The thermometer gains energy at a much faster rate than it can radiate it away, and the liquid keeps expanding and rising until there is equilibrium be-tween incoming and outgoing energy. Because of the direct absorption of solar energy, the level of the liquid in the thermometer indicates a tempera-ture much higher than the actual air temperature. Hence, a thermometer must be kept in a shady place to measure the temperature of the air accurately.

the air inside the shelter at the same temperature as the air outside.

The thermometers inside a standard shelter are mounted about five feet above the ground. Because air temperatures vary considerably above different types of surfaces, shelters are usually placed over grass to en-sure that the air temperature is measured at the same elevation over the same type of surface. Unfortunately, some shelters are placed on asphalt, others sit on con-crete, while others are located on the tops of tall build-ings, making it difficult to compare air temperature measurements from different locations. In fact, if either the maximum or minimum air temperature in your area seems suspiciously different from those of nearby towns, find out where the instrument shelter is situ-ated. (If you have wondered whether it is possible to measure air temperature accurately in the sun with a common liquid thermometer, read the Focus section above.)

Summary

The daily variation in air temperature near the earth's surface is controlled mainly by the input of energy from the sun and the output of energy from the surface. On a clear, calm day, the surface air warms, as long as heat input (mainly sunlight) exceeds heat output (mainly convection and radiated infrared energy). The surface air cools at night, as long as heat output exceeds input. Because the ground at night cools more quickly than the air above, the coldest air is normally found at the surface where a radiation inversion usually forms. When the air temperature in agricultural areas drops to dangerously low readings, fruit trees and grape vine-yards can be protected from the cold by a variety of means, from mixing the air to spraying the trees and vines with water.

The greatest daily variation in air temperature oc-curs at the earth's surface. Both the diurnal and annual range of temperature are greater in dry climates than in humid ones. Even though two cities may have similar average annual temperatures, the range and extreme of their temperatures can differ greatly. Temperature in-formation influences our lives in many ways, from de-ciding what clothes to take on a trip to providing critical information for energy-use predictions and agri-cultural planning. We reviewed some of the many types of thermometers in use: maximum, minimum, bimetal-lic, electrical, radiometer. Those designed to measure air temperatures near the surface are housed in instru-ment shelters to protect them from direct sunlight and precipitation.

Key Terms

The following terms are listed in the order they appear in the text. Define each. Doing so will aid you in reviewing the material covered in this chapter.

radiational cooling
radiation inversion
thermal belt
orchard heater
wind machine
freeze
controls of temperature

isotherm
specific heat
daily (diurnal) range of
    temperature
mean daily temperature
annual range of temperature
mean annual temperature

heating degree-day
cooling degree-day
growing degree-day
wind-chill factor
hypothermia
liquid-in-glass
    thermometer

maximum thermometer
minimum thermometer
electrical thermometer
radiometer
bimetallic thermometer
thermograph
instrument shelter

Review Questions

1. Explain why the warmest time of the day is usually in the afternoon, even though the sun's rays are most direct at noon.
2. On a calm, sunny day, why is the air next to the ground normally much warmer than the air several feet above?
3. Explain how incoming energy and outgoing energy regulate the daily variation in air temperature.
4. Draw a vertical profile of air temperature from the ground to an elevation of 3 meters (10 feet) on a clear windless (a) afternoon and (b) early morning just before sunrise. Explain why the temperature curves are different.
5. Explain how radiational cooling produces a radiation temperature inversion.
6. What weather conditions are best suited for the formation of a cold night and a strong radiation inversion?
7. What are thermal belts?
8. List some of the measures farmers use to protect their crops against the cold. Explain the physical principle behind each method.
9. Why are the lower branches of trees most susceptible to damage from low temperatures?
10. Describe each of the controls of temperature.
11. Look at Fig. 3.7 (temperature map for January) and explain why the isotherms dip southward (equatorward) over the Northern Hemisphere continents.
12. During the winter, frost can form on the ground when the minimum thermometer indicates a low temperature above freezing. Explain.

13. Why do the first freeze in autumn and the last freeze in spring occur in bottomlands?
14. Explain why the daily range of temperature is normally greater (a) in dry regions than in humid regions and (b) on clear days than on cloudy days.
15. Why are the largest annual range of temperatures normally observed over continents away from large bodies of water?
16. Two cities have the same mean annual temperature. Explain why this fact does not mean that their temperatures throughout the year are similar.
17. What is a heating degree-day? A cooling degree-day? How are these units calculated?
18. During a cold, calm, sunny day, why do we usually feel warmer than a thermometer indicates?
19. (a) Assume the wind is blowing at 30 mi/hr and the air temperature is 5°F. Determine the wind chill equivalent temperature in Table 3.2, p. 67; (b) under the conditions listed in (a) above, explain why an ordinary thermometer would measure a temperature of 5°F.
20. What atmospheric conditions can bring on hypothermia?
21. Explain why the minimum thermometer is the one with a small barbell-shaped index marker in the bore.
22. Briefly describe how the following thermometers measure air temperature:
    (a) liquid-in-glass
    (b) bimetallic
    (c) electrical
    (d) radiometer

Drought-resistant vegetation struggles to survive another day on the arid plateau of Nevada. Yet, on any summer day, there is actually more water vapor in the air of this desert than there is in the air of a wet New England snowstorm. (Photo by author)

Chapter 4

Humidity, Condensation, and Clouds

Contents

Sometimes it rains and still fails to moisten the desert—the falling water evaporates halfway down between cloud and earth. Then you see curtains of blue rain dangling out of reach in the sky while the living things wither below for want of water. Torture by tantalizing, hope without fulfillment. And the clouds disperse and dissipate into nothingness. . . . The sun climbed noon-high, the heat grew thick and heavy on our brains, the dust clouded our eyes and mixed with our sweat. My canteen is nearly empty and I'm afraid to drink what little water is left—there may never be anymore. I'd like to cave in for a while, crawl under yonder cottonwood and die peacefully in the shade, drinking dust.

Edward Abbey, *Desert Solitaire—A Season in the Wilderness*

4 We know from Chapter 1 that, in our atmosphere, the concentration of the invisible gas water vapor is normally less than a few percent of all the atmospheric molecules. Yet water vapor is exceedingly important, for it transforms into cloud particles—particles that grow in size and fall to the earth as precipitation. The term *humidity* is used to describe the amount of water vapor in the air. To most of us, a moist day suggests high humidity. However, there is usually more water vapor in the hot, "dry" air of the Sahara Desert than in the cold, "damp" polar air in New England, which raises an interesting question: Does the desert air have a higher humidity? As we will see later in this chapter, the answer to this question is both yes and no, depending on the type of humidity we mean.

So that we may better understand the concept of humidity, we will begin this chapter by examining the circulation of water in the atmosphere. Then, we will look at the different ways to express humidity. Near the end of the chapter, we will investigate the various forms of condensation, including dew, fog, and clouds.

▲▽▲

Circulation of Water in the Atmosphere

Within the atmosphere, there is an unending circulation of water. Since the oceans occupy over 70 percent of the earth's surface, we can think of this circulation as beginning over the ocean. Here, the sun's energy transforms enormous quantities of liquid water into water vapor in a process called **evaporation**. Winds then transport the moist air to other regions, where the water vapor changes back into liquid, forming clouds, in a process called **condensation**. Under certain conditions, the liquid (or solid) cloud particles may grow in size and fall to the surface as **precipitation**—rain, snow, or hail. If the precipitation falls into an ocean, the water is ready to begin its cycle again. If, on the other hand, the precipitation falls on a continent, a great deal of the water returns to the ocean in a complex journey. This cycle of moving and transforming water molecules from liquid to vapor and back to liquid again is called the **hydrologic** (water) **cycle**. In the most simplistic form of this cycle, water molecules travel from ocean to atmosphere to land and then back to the ocean.

Figure 4.1 illustrates the complexities of the hydrologic cycle. For example, before falling rain ever reaches the ground, a portion of it evaporates back into the air. Some of the precipitation may be intercepted by vegetation, where it evaporates or drips to the ground long after a storm has ended. Once on the surface, a portion of the water soaks into the ground by percolating downward through small openings in the soil and rock, forming groundwater that can be tapped by wells. What does not soak in collects in puddles of standing water or runs off into streams and rivers, which find their way back to the ocean. Even the underground water moves slowly and eventually surfaces, only to evaporate or be carried seaward by rivers.

Over land, a considerable amount of vapor is added to the atmosphere through evaporation from the soil, lakes, and streams. Even plants give up moisture by a process called *transpiration*. The water absorbed by a plant's root system moves upward through the stem and emerges from the plant through numerous small openings on the underside of the leaf. In all, evaporation and transpiration from continental areas amount to only about 15 percent of the nearly 1.5 billion billion gallons of water vapor that annually evaporate into the atmosphere; the remaining 85 percent evaporates from the oceans. The total mass of water vapor stored in the atmosphere at any moment adds up to only a little over a week's supply of the world's precipitation. Since this amount varies only slightly from day to day, the hydrologic cycle is exceedingly efficient in circulating water in the atmosphere.

▲▽▲

Evaporation, Condensation, and Saturation

To obtain a slightly different picture of water in the atmosphere, suppose we examine water in a beaker similar to the one shown in Fig. 4.2(a). If we were able to magnify the surface water about a billion times, we would see water molecules fairly close together, jiggling, bouncing, and moving about. We would also see that the molecules are not all moving at the same speed—some are moving much faster than others. Recall from Chapter 2 that the *temperature* of the water is a measure of the average speed of its molecules. At the surface, molecules with enough speed (and traveling in the right direction) would occasionally break away from the liquid surface and enter into the air above. These molecules, changing from the *liquid state into*

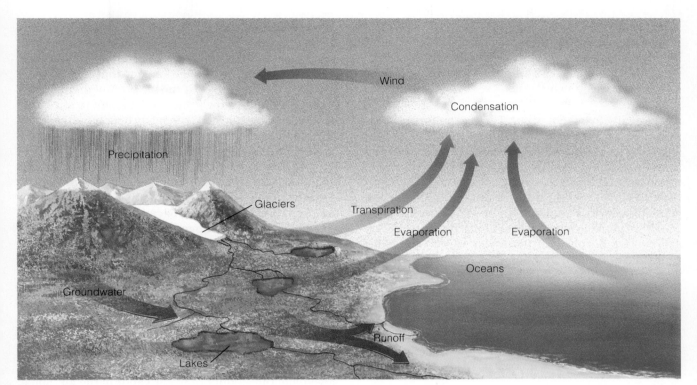

Figure 4.1
The hydrologic cycle.

the vapor state, are evaporating. While some water molecules are leaving the liquid, others are returning. Those returning are condensing as they are changing from a *vapor state to a liquid state*.

When a cover is placed over the beaker (Fig. 4.2b), after a while the total number of molecules escaping from the liquid (evaporating) would be balanced by the number returning (condensing). When this condition exists, the air is said to be *saturated* with water vapor. For every molecule that evaporates, one must condense, and no net loss of liquid or vapor molecules results.

If we remove the cover and blow across the top of the water, some of the vapor molecules already in the air above would be blown away, creating a difference between the actual number of vapor molecules and the total number required for **saturation**. This would help prevent saturation from occurring and would allow for a greater amount of evaporation. Wind, therefore, enhances evaporation.

The temperature of the water also influences evaporation. All else being equal, warm water will evaporate

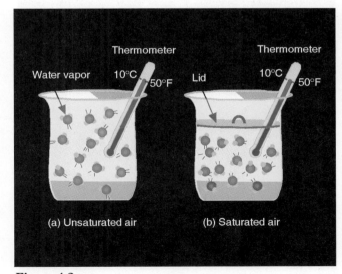

Figure 4.2
In beaker A, water molecules at the surface of the water are evaporating (changing from liquid into vapor) and condensing (changing from vapor into liquid). When evaporation and condensation are in balance (beaker B), the air above the liquid is saturated. (For clarity, only water vapor molecules are illustrated.)

more readily than cool water. The reason for this phenomenon is that, when heated, the water molecules will speed up. At higher temperatures, a greater fraction of the molecules have sufficient speed to break through the surface tension of the water and zip off into the air above. Consequently, the warmer the water, the greater the rate of evaporation.

If we could examine the air above the water in Fig. 4.2a we would observe the water vapor molecules freely darting about and bumping into each other as well as neighboring molecules of oxygen and nitrogen. We would also observe that mixed in with all of the air molecules are microscopic bits of dust, smoke, and salt from ocean spray. Near the surface, on an ordinary day, a volume of air about the size of your index finger may contain thousands of these particles. Since many of

these serve as surfaces on which water vapor may condense, they are called **condensation nuclei**. In the warm air above the water, fast-moving vapor molecules strike the nuclei with such impact that they simply bounce away. However, if the air is chilled, the molecules move more slowly and are more apt to stick and condense to the nuclei. When many billions of these vapor molecules condense onto the nuclei, tiny liquid cloud droplets form.

We can see then that condensation is more likely to happen as the air cools and the speed of the vapor molecules decreases. As the air temperature increases, condensation is less likely because most of the molecules have sufficient speed (sufficient energy) to remain as a vapor. As we will see in this and other chapters, *condensation occurs primarily when the air is cooled.*

▲▼▲

Humidity

Humidity refers to any one of a number of ways of specifying the amount of water vapor in the air. Since there are several ways to express atmospheric water vapor content, there are several meanings for the concept of humidity.

Suppose, for example, we enclose a volume of air in an imaginary, thin, elastic container—a *parcel*—as illustrated in Fig. 4.3. If we extract the water vapor from the parcel, we would specify the humidity in the following ways:

1. We could compare the weight (mass) of the water vapor with the volume of air in the parcel and obtain the *water vapor density*, or *absolute humidity*.
2. We could compare the weight (mass) of the water vapor in the parcel with the total weight (mass) of all the air in the parcel (including vapor) and obtain the *specific humidity*.
3. Or, we could compare the weight (mass) of the water vapor in the parcel with the weight (mass) of the remaining dry air and obtain the *mixing ratio*.

Absolute humidity is normally expressed as grams of water vapor per cubic meter of air (g/m^3), whereas both specific humidity and mixing ratio are expressed as grams of water vapor per kilogram of air (g/kg).

Look at Fig. 4.3 and notice that we could also express the humidity of the air in terms of *water vapor*

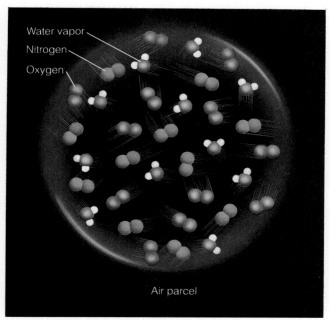

Figure 4.3
The humidity of the air inside the air parcel can be obtained by determining the density of water vapor, the weight (mass) of water vapor, or the pressure that the water vapor molecules exert inside the parcel.

pressure—the push (force) that the water vapor molecules are exerting against the inside walls of the parcel.

Vapor Pressure   Suppose the air parcel is near sea level and the air pressure inside the parcel is 1000 millibars (mb). The total air pressure inside the parcel is due to the collision of all the molecules against the walls of the parcel. In other words, the total pressure inside the parcel is equal to the sum of the pressures of the individual gases. Since the total pressure inside the parcel is 1000 millibars, and the gases inside include nitrogen (78 percent), oxygen (21 percent), and water vapor (1 percent), then the partial pressure exerted by nitrogen would be 780 millibars and that exerted by oxygen, 210 millibars. The partial pressure of water vapor, called the **actual vapor pressure**, would be only 10 millibars. It is evident, then, that because the number of water vapor molecules in any volume of air is small compared with the total number of air molecules in the volume, the actual vapor pressure is normally a small fraction of the total air pressure.

Everything else being equal, the more air molecules in a parcel, the greater the total air pressure. When you blow up a balloon, you increase its pressure by putting in more air. Similarly, an increase in the number of water vapor molecules will increase the total vapor pressure. Hence, the actual vapor pressure is a fairly good measure of the total amount of water vapor in the air: *High actual vapor pressure indicates large numbers of water vapor molecules, whereas low actual vapor pressure indicates comparatively small numbers of vapor molecules.*

Actual vapor pressure indicates the air's total water vapor content, whereas **saturation vapor pressure** describes how much water vapor is necessary to make the air saturated at any given temperature. Put another way, saturation vapor pressure is the pressure that the water vapor molecules would exert if the air were saturated with vapor at a given temperature. At higher air temperatures, it takes more water vapor to saturate the air. Saturation vapor pressure, then, depends primarily on air temperature. (See Fig. 4.4.)

Look back at the air in the saturated beaker in Fig. 4.2, p. 75, and notice that the temperature of the saturated air is 10°C (50°F). By looking at the saturation vapor pressure curve in Fig. 4.4, we can see that the saturation vapor pressure of this air is about 12 millibars.

Relative Humidity   While relative humidity is the most commonly used way of describing atmospheric

moisture, it is also, unfortunately, the most misunderstood. The concept of relative humidity may at first seem confusing because it does not indicate the actual amount of water vapor in the air. Instead, it tells us how close the air is to being saturated. The **relative humidity** is the *ratio of the amount of water vapor actually in the air compared to the maximum amount of water vapor required for saturation at that particular temperature (and pressure).* It is the *ratio* of the air's water vapor *content* to its *capacity*:

$$\text{Relative humidity} = \frac{\text{water vapor content}}{\text{water vapor capacity}}.$$

We can think of the actual vapor pressure as a measure of the air's actual water vapor content, and the saturation vapor pressure as a measure of air's total capacity

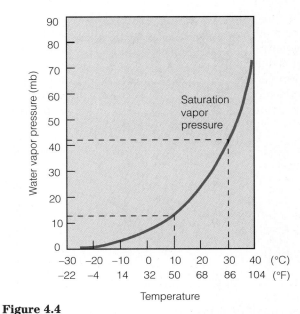

Figure 4.4
Saturation vapor pressure increases with increasing temperature. At a temperature of 10°C, the saturation vapor pressure is about 12 millibars, whereas at 30°C it is about 42 millibars.

Did you know?

It is possible for you to feel quite cool (and even begin to shiver) when the air temperature is 100°F or above as long as your skin is wet, the wind is blowing, and the wet-bulb temperature is considerably below your exposed skin's temperature.

for water vapor. Hence, the relative humidity can be expressed as:

$$RH = \frac{\text{actual vapor pressure}}{\text{saturation vapor pressure}} \times 100 \text{ percent.}$$

Relative humidity is given as a percent. Air with a 50 percent relative humidity actually contains one-half the amount required for saturation. Air with a 100 percent relative humidity is said to be saturated because it is filled to capacity with water vapor.

Let's again look at the two beakers in Fig. 4.2, p. 75. Since the air is saturated in beaker B, the relative humidity is obviously 100 percent. But what about the relative humidity of the unsaturated air in beaker A? Since the air temperature in beaker A is 10°C we know (from Fig. 4.4) that the saturation vapor pressure is 12 millibars. If the actual vapor pressure of the water vapor molecules in the beaker is 6 millibars, then the relative humidity of the air would be $^{6}/_{12} \times 100$ percent, or 50 percent. We can see that if more water vapor molecules are added to the air, the actual vapor pressure would increase and gradually approach the saturation vapor pressure. Hence, the relative humidity would increase. Likewise, if water vapor molecules were removed from the air, the actual vapor pressure

would decrease and so would the relative humidity. Consequently, as water vapor is added to the air (with no change in air temperature), the relative humidity increases, and, as water vapor is removed from the air, the relative humidity decreases.

It is also possible to change the relative humidity without changing the air's water vapor content, as a change in air temperature can bring about a change in relative humidity. This happens because a change in air temperature alters the air's saturation vapor pressure. We can see how this happens by examining Fig. 4.4. We have already seen that when the air temperature is 10°C, the saturation vapor pressure is 12 millibars. If the actual vapor pressure of this air is 6 millibars, then the relative humidity is 50 percent. However, if the temperature of the air increases to 30°C, the saturation vapor pressure increases to 42 millibars, and with no change in the air's water vapor content, the actual vapor pressure remains at 6 millibars. Hence, the relative humidity lowers to $^{6}/_{42}$ or 14 percent. Therefore, as the air temperature rises (with no change in water vapor content), the relative humidity decreases. As the air temperature drops, the relative humidity increases because the air is approaching saturation.

In many places, the air's total vapor content varies only slightly during an entire day, and so it is the changing air temperature that primarily regulates the daily variation in relative humidity (Fig. 4.5). As the air cools during the night, the relative humidity increases. Normally, the highest relative humidity occurs in the early morning, during the coolest part of the day. As the air warms during the day, the relative humidity decreases, with the lowest values usually occurring during the warmest part of the afternoon.

These changes in relative humidity are important in determining the amount of evaporation from vegetation and wet surfaces. If you water your lawn on a hot afternoon, when the relative humidity is low, much of the water will evaporate quickly from the lawn, instead of soaking into the ground. Watering the same lawn in the evening, when the relative humidity is higher, will cut down the evaporation and increase the effectiveness of the watering.

Very low relative humidities in a house can have an adverse effect on things living inside. For example, house plants have a difficult time surviving because the moisture from their leaves and the soil evaporates rapidly. People suffer, too, when the relative humidity is quite low. The rapid evaporation of moisture from ex-

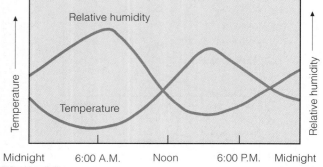

Figure 4.5

When the air is cool (morning), the relative humidity is high. When the air is warm (afternoon), the relative humidity is low.

posed flesh causes skin to crack, dry, flake, or itch. These low humidities also irritate the mucous membranes in the nose and throat, producing an "itchy" throat. Similarly, dry nasal passages permit inhaled bacteria to incubate, causing persistent infections.

The relative humidity in a home can be increased just by heating water and allowing it to evaporate into the air. This will raise the relative humidity to a more comfortable level. In modern homes, a humidifier, installed near the furnace, adds moisture to the air at a rate of about one gallon per room per day. The air, with its added moisture, is circulated throughout the home by a forced air heating system. In this way, all rooms get their fair share of moisture—not just the room where the vapor is added.

Relative Humidity and Human Discomfort On a hot, muggy day when the relative humidity is high, it is common to hear someone exclaim (often in exasperation), "It's not so much the heat, it's the humidity." Actually, this statement is valid. In warm weather the main source of body cooling is through evaporation of perspiration. When the air temperature is high and the relative humidity low, perspiration on the skin evaporates quickly, often making us feel that the air temperature is lower than it really is. However, when both the air temperature and relative humidity are high and the air is nearly saturated with water vapor, body moisture does not readily evaporate; instead, it collects on the skin as beads of perspiration. Less evaporation means less cooling, and so we usually feel warmer than we did with a similar air temperature, but a lower relative humidity.

A good measure of how cool the skin can become is the **wet-bulb temperature**—the lowest temperature that can be reached by evaporating water into the air. On a hot day when the wet-bulb temperature is low, rapid evaporation (and, hence, cooling) takes place at the skin's surface. As the wet-bulb temperature approaches the air temperature, less cooling occurs, and the skin temperature may begin to rise. When the wet-bulb temperature exceeds the skin's temperature, no net evaporation occurs, and the body temperature can rise quite rapidly. Fortunately, most of the time, the wet-bulb temperature is considerably below the temperature of the skin.

When the weather is hot and muggy, a number of heat-related problems may occur. For example, in hot weather when the human body temperature rises, the *hypothalamus* gland (a gland in the brain that regulates body temperature) activates the body's heat-regulating mechanism, and over ten million sweat glands wet the body with as much as two liters of liquid per hour. As this perspiration evaporates, rapid loss of water and salt can result in a chemical imbalance that may lead to painful *heat cramps*. Excessive water loss through perspiring coupled with an increasing body temperature may result in *heat exhaustion*—fatigue, headache, nausea, and even fainting. If one's body temperature rises above about 41°C (106°F), *heat stroke* can occur, resulting in complete failure of the circulatory functions. If the body temperature continues to rise, death may result.

In an effort to draw attention to this serious weather-related health hazard, an index called the **heat index (HI)**, is used by the National Weather Service. The index combines air temperature with relative humidity to determine an **apparent temperature**—what the air temperature "feels like" to the average person for various combinations of air temperature and relative humidity. For example, in Fig. 4.6, an air temperature of 100°F and a relative humidity of 60 percent produce an apparent temperature of 130°F. Heatstroke or sunstroke is imminent when the index reaches this level. (See Table 4.1.)

During hot, humid weather some people remark about how "heavy" or how dense, the air feels. Is hot,

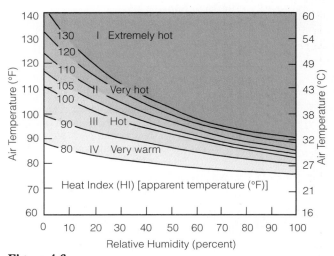

Figure 4.6
The Heat Index (HI). To calculate the apparent temperature, find the intersection of the air temperature and the relative humidity.

Focus on a Special Topic
Humid Air and Dry Air Do Not Weigh the Same

Does a volume of hot, humid air really weigh more than a similar size volume of hot, dry air? The answer is no! At the same temperature and at the same level in the atmosphere, hot, humid air is lighter (less dense) than hot, dry air. The reason for this fact is that a molecule of water vapor (H_2O) weighs appreciably less than a molecule of either nitrogen (N_2) or oxygen (O_2).

Consequently, in a given volume of air, as lighter water vapor molecules replace either nitrogen or oxygen mol-ecules one for one, the number of molecules in the volume does not change, but the total weight of the air becomes slightly less. Since air density is the mass (weight) of air in a vol-ume, the more humid air must be lighter than the drier air. Hence, hot, humid air at the surface is lighter (less dense) than hot, dry air.

This fact can have an important in-fluence in the weather. The lighter the air becomes, the more likely it is to rise. All other factors being equal, hot, humid (less dense) air will rise more readily than hot, dry (more dense) air. It is of course the water vapor in the rising air that changes into liquid cloud droplets and ice crystals, which, in turn, grow large enough to fall to the earth as precipitation.

Of lesser importance to weather but of greater importance to sports is the fact that a baseball will "carry" far-ther in less dense air. Consequently, without the influence of wind, a ball will travel slightly farther on a hot, hu-mid day in Atlanta's Fulton County sta-dium than it will on a hot, dry day.

---

humid air really more dense than hot, dry air? If you are interested in the answer, read the Focus section above.

Dew Point   Consider a volume of air whose temper-ature is 20°C and relative humidity is 100 percent. Sup-pose the air warms to 30°C, with no change in water va-por content or air pressure. The relative humidity drops, and the air is no longer saturated. To what tem-perature must the 30°C air be cooled so that it is once again saturated? Of course, the answer is 20°C. For this amount of moisture, 20°C is called the **dew-point tem-perature**, or, simply, the **dew point**. It represents *the temperature to which air would have to be cooled (with no change in air pressure or moisture content) for saturation to occur.* Since atmospheric pressure varies only slightly at the earth's surface, *the dew point is a good indicator of the air's actual water vapor content.* High dew points indicate high water vapor content; low dew points, low water vapor content. Ad-dition of water vapor to the air increases the dew point; removing water vapor lowers it.

The difference between air temperature and dew point can indicate whether the relative humidity is low or high. When the air temperature and dew point are far apart, the relative humidity is low; when they are close to the same value, the relative humidity is high. When the air temperature and dew point are equal, the air is saturated and the relative humidity is 100 percent. Even though the relative humidity may be 100 percent, the air, under certain conditions, may be considered "dry." More information on this is given in the Focus section on p. 81 entitled "Dry Air with a High Humidity."

Table 4.1	The Heat Index (HI)	
Category	Apparent Temperature (°F)	Heat Syndrome
I	130° or higher	Heatstroke or sunstroke *imminent*
II	105°–130°	Sunstroke, heat cramps, or heat exhaustion *likely*, heatstroke *possible* with prolonged exposure and physical activity
III	90°–105°	Sunstroke, heat cramps, and heat exhaustion *possible* with prolonged exposure and physical activity
IV	80°–90°	Fatigue *possible* with prolonged exposure and physical activity

Focus on a Special Topic
"Dry" Air with a High Humidity

Polar Air: Air temperature –2°C
Dew point –2°C

Desert Air: Air temperature 35°C
Dew point 4°C

Figure 1
The polar air has the highest relative humidity, whereas the desert air, with the highest dew point, contains the most water vapor.

At the beginning of the chapter, we raised the question as to which air has the higher humidity—hot, "dry" desert air or cold, "damp" polar air? To answer this question, look at the two photographs in Fig. 1. In the polar air, the air temperature and the dew point are the same, the air is saturated and the relative humidity is 100 percent. Meanwhile, in the desert air, there is a large spread between air temperature and dew point; the air is *not* close to being saturated and the

relative humidity is therefore low. So the polar air has a *higher relative humidity* than the desert air. However, notice that the dew point of the desert air is higher than the dew point of the polar air. Since dew point is a measure of the amount of water vapor in the air, the desert air must contain *more* water vapor. Hence, the water vapor density, or absolute humidity, is higher in the air of the desert.*

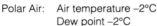

*The specific humidity and mixing ratio are also higher in the desert air.

Now we can see why the polar air is often described as being "dry" when the relative humidity is high (often close to 100 percent). In cold, polar air the dew point and air temperatures are usually close together. But the low dew-point temperature means that there is little water vapor in the air. Consequently, the air is described as "dry" even though the relative humidity is high.

Measuring Humidity   The common instrument used to obtain dew point and relative humidity is a **psychrometer**, which consists of two liquid-in-glass thermometers mounted side by side and attached to a piece of metal that has either a handle or chain at one end (Fig. 4.7). The thermometers are exactly alike except that one has a piece of cloth (wick) covering the bulb. The wick-covered thermometer—called the *wet bulb*—is dipped in clean water, whereas the other thermometer is kept dry. Both thermometers are ventilated for a few minutes, either by whirling the instrument (*sling psychrometer*), or by drawing air past it with an electric fan (*aspiration psychrometer*). Water evaporates from the wick and that thermometer cools. The drier the air, the greater the amount of evaporation and cooling. After a few minutes, the wick-covered thermometer will cool to the lowest value possible. Recall from an earlier section that this is the *wet-bulb temperature*—the lowest temperature that can be obtained by evaporating water into the air.

The dry thermometer (commonly called the *dry bulb*) gives the current air temperature, or *dry-bulb temperature*. The temperature difference between the dry bulb and the wet bulb is known as the *wet-bulb depression*. A large depression indicates that a great deal of water can evaporate into the air and that the relative humidity is low. A small depression indicates that little evaporation of water vapor is possible, so the air is close to saturation and the relative humidity is high. If there is no depression, the dry bulb and wet bulb are the same; the air is saturated and the relative humidity is 100 percent. (Tables used to compute relative humidity and dew point are given in Appendix D.)

Instruments that measure humidity are commonly called **hygrometers**. One type—called the *hair hygrometer*—uses human (or horse) hair to measure relative humidity. It is constructed on the principle that, as the relative humidity increases, the length of hair increases and, as the relative humidity decreases, so does the hair length. A number of strands of hair (with oils removed) are attached to a system of levers. A small change in hair length is magnified by a linkage system and transmitted to a dial (Fig. 4.8) calibrated to show relative humidity, which can then be read directly or recorded on a chart. (Often, the chart is attached to a clock-driven rotating drum that gives a continuous record of relative humidity.) Because the hair hygrometer is not as accurate as the psychrometer (especially at very high and very low relative humidities), it requires frequent calibration, principally in areas that experience large daily variations in relative humidity.

The *electrical hygrometer* is another instrument used to measure humidity. It consists of a flat plate coated with a film of carbon. An electric current is sent across the plate. As the moisture content of the air changes, the electrical resistance of the carbon coating changes. These changes are translated into relative humidity. This instrument is commonly used in the radiosonde, which gathers atmospheric data at various levels above the earth. Still another instrument—the *infrared hygrometer*—measures atmospheric humidity by measuring the amount of infrared energy absorbed by water vapor in a sample of air. Finally, the *dew cell* determines the amount of water vapor in the air by measuring the air's actual vapor pressure.

Over the last several sections we saw that, as the air cools, the air temperature approaches the dew-point temperature and the relative humidity increases. When the air temperature reaches the dew point, the air is saturated with water vapor and the relative humidity is 100 percent. Continued cooling, however, causes

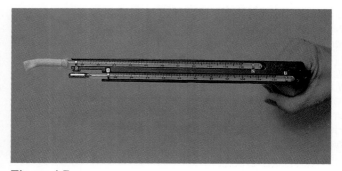

Figure 4.7
The sling psychrometer.

some of the water vapor to condense into liquid water. The cooling may take place in a thick portion of the atmosphere, or it may occur near the earth's surface. In the next section, we will examine condensation that forms near the ground.

▲▽▲

Dew and Frost

On clear, calm nights, objects near the earth's surface cool rapidly by emitting infrared radiation. The ground and objects on it often become much colder than the surrounding air. Air that comes in contact with these cold surfaces cools by conduction. Eventually, the air cools to the dew point. As surfaces (such as twigs, leaves, and blades of grass) cool below this temperature, water vapor begins to condense upon them, forming tiny visible specks of water called **dew**. If the air temperature should drop to freezing or below, the dew will freeze, becoming tiny beads of ice called *frozen dew*. Because the coolest air is usually at ground level, dew is more likely to form on blades of grass than on objects several feet above the surface. This thin coating of dew not only dampens bare feet, but it also is a valuable source of moisture for many plants during periods of low rainfall.

Dew is more likely to form on nights that are clear and calm than on nights that are cloudy and windy. Clear nights allow objects near the ground to cool rapidly, and calm winds mean that the coldest air will

be located at ground level. These atmospheric conditions are usually associated with large fair-weather, high-pressure systems. On the other hand, the cloudy, windy weather that inhibits rapid cooling near the ground and the forming of dew often signifies the approach of a rain-producing storm system. These observations inspired the following folk-rhyme:

> When the dew is on the grass,
> rain will never come to pass.
> When grass is dry at morning light,
> look for rain before the night!

Visible white frost forms on cold, clear, calm mornings when the dew-point temperature is at or below freezing. When the air temperature cools to the dew point (now called the *frost point*) and further cooling occurs, water vapor can change directly to ice without becoming a liquid first—a process called *deposition*.* The delicate, white crystals of ice that form in this manner are called *hoarfrost, white frost,* or simply **frost**. Frost has a treelike branching pattern that easily dis-

---

*When the ice changes back into vapor without melting, the process is called *sublimation*.

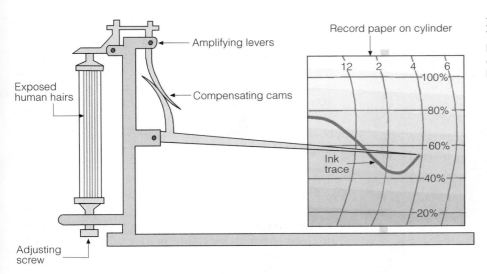

Exposed human hairs

Amplifying levers

Compensating cams

Record paper on cylinder

Ink trace

12 2 4 6
100%
80%
60%
40%
20%

Adjusting screw

Figure 4.8
The hair hygrometer measures relative humidity by amplifying and measuring changes in the length of human (or horse) hair.

tinguishes it from the nearly spherical beads of frozen dew. (See Fig. 4.9.)

In very dry weather, the air may become quite cold and drop below freezing without ever reaching the frost point, and no visible frost forms. *Freeze* and *black frost* are words denoting this situation—a situation that can severely damage certain crops. (See Chapter 3, pp. 59–60.)

As a deep layer of air cools during the night, its relative humidity increases. When the air's relative humidity reaches about 75 percent, some of its water vapor may begin to condense onto tiny floating particles of sea salt and other substances—condensation nuclei—that are *hygroscopic* ("water seeking") in that they allow water vapor to condense onto them when

the relative humidity is considerably below 100 percent. As water collects onto these nuclei, their size increases and the particles, although still small, are now large enough to scatter visible light in all directions, becoming **haze**—a layer of particles dispersed through a portion of the atmosphere. (See Fig. 4.10.)

As the relative humidity gradually approaches 100 percent, the haze particles grow larger, and condensation begins on the less-active nuclei. Now a large fraction of the available nuclei have water condensing onto them, causing the droplets to grow even bigger, until eventually they become visible to the naked eye. The increasing size and concentration of droplets further restrict visibility. When the visibility lowers to less than 1 kilometer (or 0.62 mile), and the air is wet with millions of tiny floating water droplets, the haze becomes a cloud resting near the ground, which we call **fog**.

▲▼▲

Fog

Fog, like any cloud, usually forms in one of two ways: (1) by cooling—air is cooled below its saturation point (dew point); and (2) by evaporation and mixing—water vapor is added to the air by evaporation, and the moist air mixes with relatively dry air. Once fog forms it is maintained by new fog droplets, which constantly form on available nuclei. In other words, the air must maintain its degree of saturation either by continual cooling or by evaporation and mixing of vapor into the air.

Fog produced by the earth's radiational cooling is called **radiation fog**, or *ground fog*. It forms best on clear nights when a shallow layer of moist air near the ground is overlain by drier air. Under these conditions, the ground cools rapidly since the shallow, moist layer does not absorb much of the earth's outgoing infrared radiation. As the ground cools, so does the air directly above it, and a surface inversion forms. The moist, lower layer (chilled rapidly by the cold ground) quickly becomes saturated, and fog forms. The longer the night, the longer the time of cooling and the greater the likelihood of fog. Hence, radiation fogs are most common over land in late fall and winter.

Another factor promoting the formation of radiation fog is a light breeze of less than five knots. Although radiation fog may form in calm air, slight air movement brings more of the moist air in direct contact with the cold ground and the transfer of heat

Figure 4.9
These are the delicate ice-crystal patterns that frost exhibits on a window during a cold winter morning.

Figure 4.10
The high relative humidity of the cold air above the lake is causing a layer of haze to form on a still winter morning.

occurs more rapidly. A strong breeze would prevent a radiation fog from forming by mixing the air near the surface with the drier air above. The ingredients of clear skies and light winds are associated with large high-pressure areas (anticyclones). Consequently, during the winter, when a high becomes stagnant over an area, radiation fog may form on consecutive days.

Because cold, heavy air drains downhill and collects in valley bottoms, we normally see radiation fog forming in low-lying areas. Hence, radiation fog is frequently called *valley fog*. The cold air and high moisture content in river valleys make them susceptible to radiation fog. Since radiation fog normally forms in lowlands, hills may be clear all day long, while adjacent valleys are fogged in (Fig. 4.11).

Radiation fogs are usually deepest around sunrise. Usually, however, a shallow fog layer will dissipate or *burn off* by afternoon. Of course, the fog does not "burn"; rather, sunlight penetrates the fog and warms the ground, causing the temperature of the air in con-

tact with the ground to increase. The warm air rises and mixes with the foggy air above, which increases the temperature of the foggy air. In the slightly warmer air, some of the fog droplets evaporate, allowing more sunlight to reach the ground, which produces more heating, and soon the fog completely disappears. If the fog layer is quite thick, it may not completely dissipate and a layer of low clouds (called *stratus*) covers the region. This type of fog is sometimes called *high fog*.

When warm, moist air moves over a sufficiently colder surface, the moist air may cool to its saturation point, forming **advection fog**. A good example of advection fog may be observed along the Pacific Coast during summer. The main reason fog forms in this region is that the surface water near the coast is much colder than the surface water farther offshore. Warm, moist air from the Pacific Ocean is carried (advected) by westerly winds over the cold coastal waters. Chilled from below, the air temperature drops to the dew point, and fog is produced. Advection fog, unlike radiation

Figure 4.11
Radiation fog nestled in a valley.

fog, always involves the movement of air, so when there is a stiff summer breeze in San Francisco, it's common to watch advection fog roll in past the Golden Gate Bridge (Fig. 4.12).

As summer winds carry the fog inland over the warmer land, the fog near the ground dissipates, leaving a sheet of low-lying gray clouds that block out the sun. Further inland, the air is sufficiently warm, so that even these low clouds evaporate and disappear.

Because they provide moisture to the coastal redwood trees, advection fogs are important to the scenic beauty of the Pacific Coast. Much of the fog moisture collected by the needles and branches of the redwoods drips to the ground, where it is utilized by the tree's shallow root system. Without the summer fog, coastal redwood trees would have trouble surviving the dry California summers. Hence, we find them nestled in the fog belt along the coast.

Advection fogs also prevail where two ocean currents with different temperatures flow next to one another. Such is the case in the Atlantic Ocean off the coast of Newfoundland, where the cold southward-flowing Labrador Current lies almost parallel to the warm northward-flowing Gulf Stream. Warm southerly air moving over the cold water produces fog in that region—so frequently that fog occurs on about two out of three days during summer.

Advection fog also forms over land. In winter, warm, moist air from the Gulf of Mexico moves northward over progressively colder and slightly elevated land. As the air cools to its saturation point, a fog forms in the southern or central United States. Because the cold ground is often the result of radiation cooling, fog that forms in this manner is sometimes called *advection-radiation fog*. During this same time of year, air moving across the warm Gulf Stream encounters the colder land of the British Isles and produces the thick fogs of England. Similarly, fog forms as marine air moves over an ice or snow surface. In extremely cold arctic air, ice crystals form instead of water droplets, producing an *ice fog*.

Fog that forms as moist air flows up along an elevated plain, hill, or mountain is called **upslope fog**. Typically, upslope fog forms during the winter and

Figure 4.12
Advection fog rolling in through the Golden Gate Bridge in San Francisco. As the fog moves inland, the air warms and the fog lifts above the surface. Eventually, the air becomes warm enough to totally evaporate the fog.

spring on the eastern side of the Rockies, where the eastward-sloping plains are nearly a kilometer higher than the land further east. Occasionally, cold air moves from the lower eastern plains westward. The air gradually rises, expands, becomes cooler, and—if sufficiently moist—a fog forms. Upslope fogs that form over an extensive area may last for many days.

So far, we have seen how the cooling of air produces fog. But remember that fog may also form by the mixing of two unsaturated masses of air. Fog that forms in this manner is usually called *evaporation fog* because evaporation initially enriches the air with water vapor. Probably, a more appropriate name for the fog is **evaporation (mixing) fog**. On a cold day, you may have unknowingly produced evaporation (mixing) fog. When moist air from your mouth or nose meets the cold air and mixes with it, the air becomes saturated, and a tiny cloud forms with each exhaled breath.

A common form of evaporation-mixing fog is the *steam fog*, which forms when cold air moves over warm water. This type of fog forms above a heated outside swimming pool in winter. As long as the water is warmer than the unsaturated air above, water will evaporate from the pool into the air. The increase in water vapor raises the dew point, and, if mixing is sufficient, the air above becomes saturated. The colder air directly above the water is heated from below and becomes warmer than the air directly above it. This warmer air rises and, from a distance, the rising condensing vapor appears as "steam."

It is common to see steam fog forming over lakes on autumn mornings, as cold air settles over water still warm from the long summer. On occasion, over the Great Lakes, columns of condensed vapor rise from the fog layer, forming whirling *steam devils*, which appear similar to the dust devils on land. If you travel to Yellowstone National Park, you will see steam fog forming above thermal ponds all year long (Fig. 4.13). Over the ocean in polar regions, steam fog is referred to as *arctic sea smoke*.

Steam fog may form above a wet surface on a sunny day. This is commonly observed after a rain shower as sunlight shines on a wet road, heats the asphalt, and quickly evaporates the water. This added vapor mixes with the air above, producing steam fog. Fog that forms in this manner is short-lived and disappears as the road surface dries.

A warm rain falling through a layer of cold, moist air can produce fog. As a warm raindrop falls into a cold layer of air, some of the water evaporates from the raindrop into the air. If the moist air mixes with the cooler air, fog forms. Fog of this type is often associated with warm air riding up and over a mass of colder surface air. The fog usually develops in the shallow layer of cold air just ahead of an approaching warm front or behind a cold front, which is why this type of evaporation fog is also known as *frontal fog*.

Up to this point, we have looked at the different forms of condensation that occur on or near the earth's surface. In particular, we learned that fog is simply many millions of tiny liquid droplets (or ice crystals) that form near the ground. In the following sections we will see how these same particles, forming well above the ground, are classified and identified as clouds. (Before we move on to the next section, you may wish to read the Focus section on foggy weather.)

Clouds

Clouds are aesthetically appealing and add excitement to the atmosphere. Without them, there would be no rain or snow, thunder or lightning, rainbows or halos. How monotonous if one had only a clear blue sky to look at. A cloud is a visible aggregate of tiny water droplets or ice crystals suspended in the air. Some are found only at high elevations, whereas others nearly touch the ground. Clouds can be thick or thin, big or little—they exist in a seemingly endless variety of forms. To impose order on this variety, we divide clouds into ten basic types. With a careful and practiced eye, you can become reasonably proficient in correctly identifying them.

Classification of Clouds   Although ancient astronomers named the major stellar constellations about

Figure 4.13
Even in summer, warm air rising above thermal pools in Yellowstone National Park condenses into a type of steam fog.

The foggiest regions in the United States are shown in Fig. 2. Notice that heavy fog is more prevalent in coastal margins (especially those regions lapped by cold ocean currents) than in the center of the continent. In fact, the foggiest spot near sea level in the United States is Cape Disappointment, Washington. Located at the mouth of the Columbia River, it averages 2556 hours of heavy fog each year. Anyone who travels to this spot hoping to enjoy the sun during August and September would find its name appropriate indeed.

Extremely limited visibility exists while driving at night in heavy fog with the high-beam lights on. The light scattered back to the driver's eyes from the fog droplets makes it difficult to see very far down the road. Along a gently sloping highway, the elevated sections may have excellent visibility, while in lower regions—only a few miles away—fog may cause poor visibility. Driving from the clear area into the fog on a major freeway can be extremely dangerous. In fact, every winter many people are involved in fog-related auto accidents. These usually occur when a car enters the fog and, because of the reduced visibility, the driver puts on the brakes to

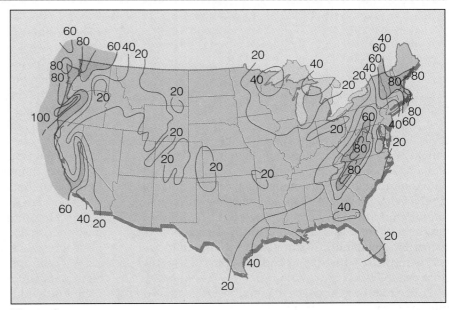

Figure 2
Average annual number of days with heavy fog throughout the United States.

slow down. The car behind then slams into the slowed vehicle, causing a chain-reaction accident with many cars involved.

Airports suspend flight operations when fog causes visibility to drop below a prescribed minimum. The resulting delays and cancellations become costly to the airline industry and irritate passengers. With fog-caused

problems such as these, it is no wonder that scientists have been seeking ways to disperse, or at least "thin," fog. Unfortunately, an inexpensive and practical method of dispersing warm fog—fog that forms at or above freezing temperatures—has yet to be discovered.

2000 years ago, clouds were not formally identified and classified until the early nineteenth century. The French naturalist Lamarck (1744–1829) proposed the first system for classifying clouds in 1802; however, his work did not receive wide acclaim. One year later, Luke Howard, an English naturalist, developed a cloud classification system that found general acceptance. In essence, Howard's innovative system employed Latin words to describe clouds as they appear to a ground observer. He named a sheetlike cloud *stratus* (Latin for "layer"); a puffy cloud *cumulus* ("heap"); a wispy cloud *cirrus* ("curl of hair"); and a rain cloud *nimbus* ("violent rain"). In Howard's system, these were the four basic cloud forms. Other clouds could be described by combining the basic types. For example, nimbostratus is a rain cloud that shows layering, whereas cumulonimbus is a rain cloud having pronounced vertical development.

Table 4.2
The Four Major Cloud Groups and Their Types

1. *High clouds*
   Cirrus (Ci)
   Cirrostratus (Cs)
   Cirrocumulus (Cc)
2. *Middle clouds*
   Altostratus (As)
   Altocumulus (Ac)
3. *Low clouds*
   Stratus (St)
   Stratocumulus (Sc)
   Nimbostratus (Ns)
4. *Clouds with vertical development*
   Cumulus (Cu)
   Cumulonimbus (Cb)

In 1887, Abercromby and Hildebrandsson expanded Howard's original system and published a classification system that, with only slight modification, is still used today. Ten principal cloud forms are divided into four primary cloud groups. Each group is identified by the height of the cloud's base above the surface: high clouds, middle clouds, and low clouds. The fourth group contains clouds showing more vertical than horizontal development. Within each group, cloud types are identified by their appearance. Table 4.2 lists these four groups and their cloud types.

The approximate base height of each cloud group is given in Table 4.3. Note that the altitude separating the high and middle cloud groups overlaps and varies with latitude. Large temperature changes cause most of this latitudinal variation. For example, high cirriform clouds are composed almost entirely of ice crystals. In subtropical regions, air temperatures low enough to freeze all liquid water usually occur only above about 20,000 feet. In polar regions, however, these same temperatures may be found at altitudes as low as 10,000 feet. Hence, while you may observe cirrus clouds at 12,000 feet over northern Alaska, you will not see them at that elevation above southern Florida.

Clouds cannot be accurately identified strictly on the basis of elevation. Other visual clues are necessary. Some of these are explained in the following section.

Cloud Identification

High Clouds   High clouds in middle and low latitudes generally form above 20,000 feet (or 6000 meters). Because the air at these elevations is quite cold and "dry," high clouds are composed almost exclusively of ice crystals and are also rather thin. High clouds usually appear white, except near sunrise and sunset, when the unscattered (red, orange, and yellow) components of sunlight are reflected from the underside of the clouds.

The most common high clouds are the **cirrus**, which are thin, wispy clouds blown by high winds into long streamers called *mares' tails*. Notice in Fig. 4.14 that they can look like a white, feathery patch with a faint wisp of a tail at one end. Cirrus clouds usually move across the sky from west to east, indicating the prevailing winds at their elevation.

Cirrocumulus clouds, seen less frequently than cirrus, appear as small, rounded, white puffs that may occur individually, or in long rows. (See Fig. 4.15.) When in rows, the cirrocumulus cloud has a rippling appearance that distinguishes it from the silky look of the cirrus and the sheetlike cirrostratus. Cirrocumulus seldom cover more than a small portion of the sky. The dappled cloud elements that reflect the red or yellow light of a setting sun make this one of the most beautiful of all clouds. The small ripples in the cirrocumulus strongly resemble the scales of a fish; hence, the expression "mackerel sky" commonly describes a sky full of cirrocumulus clouds.

The thin, sheetlike, high clouds that often cover the entire sky are **cirrostratus** (Fig. 4.16), which are so

---

Table 4.3   Approximate Height of Cloud Bases above the Surface for Various Locations

Cloud Group	Tropical Region	Middle Latitude Region	Polar Region
High	20,000 to 60,000 ft (6000 to 18,000 m)	16,000 to 43,000 ft (5000 to 13,000 m)	10,000 to 26,000 ft (3000 to 8000 m)
Middle	6500 to 26,000 ft (2000 to 8000 m)	6500 to 23,000 ft (2000 to 7000 m)	6500 to 13,000 ft (2000 to 4000 m)
Low	surface to 6500 ft (0 to 2000 m)	surface to 6500 ft (0 to 2000 m)	surface to 6500 ft (0 to 2000 m)

Figure 4.14
Cirrus clouds.

Figure 4.15
Cirrocumulus clouds.

thin that the sun and moon can be clearly seen through them. The ice crystals in these clouds bend the light passing through them and will often produce a halo. In fact, the veil of cirrostratus may be so thin that a halo is the only clue to its presence. Thick cirrostratus clouds give the sky a glary white appearance and frequently form ahead of an advancing storm; hence, they can be used to predict rain or snow within twelve to twenty-four hours, especially if they are followed by middle type clouds.

Middle Clouds   The middle clouds have bases between about 6500 and 23,000 feet (2000 and 7000 meters) in the middle latitudes. These clouds are com-

Figure 4.16
Cirrostratus clouds with a halo.

posed of water droplets and—when the temperature becomes low enough—some ice crystals.

Altocumulus clouds are middle clouds that appear as gray, puffy masses, sometimes rolled out in parallel waves or bands (Fig. 4.17). Usually, one part of the cloud is darker than another, which helps to separate it from the higher cirrocumulus. Also, the individual puffs of the altocumulus appear larger than those of the cirrocumulus. A layer of altocumulus may some-times be confused with altostratus; in case of doubt, clouds are called altocumulus if there are rounded masses or rolls present. Altocumulus clouds that look like "little castles" (*castellanus*) in the sky indicate the presence of rising air at cloud level. The appearance of these clouds on a warm, humid summer morning often portends thunderstorms by late afternoon.

The **altostratus** is a gray or blue-gray (never white) cloud that often covers the entire sky over an

Figure 4.17
Altocumulus clouds.

area that extends over many hundreds of square miles. In the thinner section of the cloud, the sun (or moon) may be dimly visible as a round disk, which is sometimes referred to as a "watery sun" (Fig. 4.18). Thick cirrostratus clouds are occasionally confused with thin altostratus clouds. The gray color, height, and dimness of the sun are good clues to identifying an altostratus. The fact that halos only occur with cirriform clouds also helps one distinguish them. Another way to separate the two is to look at the ground for shadows. If there are none, it is a good bet that the cloud is altostratus because cirrostratus are usually transparent enough to produce them. Altostratus clouds often form ahead of storms having widespread and relatively continuous precipitation.

Low Clouds   Low clouds, with their bases lying below 6500 feet (or 2000 meters) are almost always composed of water droplets; however, in cold weather, they may contain ice particles and snow.

The **nimbostratus** is a dark gray, "wet"-looking cloud layer associated with more or less continuously falling rain or snow (Fig. 4.19). The intensity of this precipitation is usually light or moderate—it is never of the heavy, showery variety. The base of the nimbostratus cloud is normally impossible to identify clearly and is easily confused with the altostratus. Thin nimbostra-

tus is usually darker gray than thick altostratus, and you cannot see the sun or moon through a layer of nimbostratus. Visibility below a nimbostratus cloud deck is usually quite poor because rain will evaporate and mix with the air in this region. If this air becomes saturated, a lower layer of clouds or fog may form beneath the original cloud base. Since these lower clouds drift rapidly with the wind, they form irregular shreds with a ragged appearance called *stratus fractus*, or *scud*.

A low, lumpy cloud layer is the **stratocumulus**. It appears in rows, in patches, or as rounded masses with blue sky visible between the individual cloud elements (Fig. 4.20). Often they appear near sunset as the spreading remains of a much larger cumulus cloud. The color of stratocumulus ranges from light to dark gray. It differs from altocumulus in that it has a lower base and larger individual cloud elements. (Compare Fig. 4.17 with Fig. 4.20.) To distinguish between the two, hold your hand at arm's length and point toward the cloud. Altocumulus cloud elements will generally be about the size of your thumbnail; stratocumulus cloud elements will usually be about the size of your fist. Rain or snow rarely fall from stratocumulus.

Stratus is a uniform grayish cloud that often covers the entire sky. It resembles a fog that does not reach the ground (Fig. 4.21). Actually, when a thick fog "lifts," the resulting cloud is a deck of low stratus. Normally,

Figure 4.19
The nimbostratus is the sheetlike cloud from which light rain is falling. The ragged-appearing cloud beneath the nimbostratus is *stratus fractus*, or *scud*.

Figure 4.20
Stratocumulus clouds. Notice that the rounded masses are larger than those of the altocumulus.

Figure 4.21
A layer of low-lying stratus clouds.

no precipitation falls from the stratus, but sometimes it is accompanied by a light mist or drizzle. This cloud commonly occurs over Pacific and Atlantic coastal waters in summer. A thick layer of stratus might be confused with nimbostratus, but the distinction between them can be made by observing the base of the cloud. Often, stratus has a more uniform base than does nimbostratus. Also, a deck of stratus may be confused with a layer of altostratus. However, if you remember that stratus clouds are lower and darker gray, the distinction can be made.

Clouds with Vertical Development   Familiar to almost everyone, the puffy **cumulus** cloud takes on a variety of shapes, but most often it looks like a piece of floating cotton with sharp outlines and a flat base (Fig. 4.22). The base appears white to light gray, and, on a humid day, may be only a few thousand feet above the ground and a half a mile or so wide. The top of the cloud—often in the form of rounded towers—denotes the limit of rising air and is usually not very high. These clouds can be distinguished from stratocumulus by the fact that cumulus clouds are detached (usually a great deal of blue sky between each cloud) whereas stratocumulus usually occur in groups or patches. Also, the cumulus has a dome- or tower-shaped top as opposed to the generally flat tops of the stratocumulus. Cumulus clouds that show only slight vertical growth (*cumulus*

humilis) are associated with fair weather; therefore, we call these clouds "fair weather cumulus." If the cumulus clouds are small and appear as broken fragments of a cloud with ragged edges, they are called *cumulus fractus*.

Harmless-looking cumulus often develop on warm summer mornings and, by afternoon, become much larger and more vertically developed. When the growing cumulus resembles a head of cauliflower, it becomes a *cumulus congestus*, or *towering cumulus*. Most often, it is a single large cloud, but, occasionally, several grow into each other, forming a line of towering clouds, as shown in Fig. 4.23. Precipitation that falls from a cumulus congestus is always showery.

If a cumulus congestus continues to grow vertically, it develops into a giant **cumulonimbus**—a thunderstorm cloud (Fig. 4.24). While its dark base may be no more than 1000 feet above the earth's surface, its top may extend upward to the tropopause, over 35,000 feet higher. A cumulonimbus can occur as an isolated cloud or as part of a line or "wall" of clouds.

Tremendous amounts of energy are released by the condensation of water vapor within a cumulonimbus and result in the development of violent up- and downdrafts, which may exceed fifty knots. The lower (warmer) part of the cloud is usually composed of only water droplets. Higher up in the cloud, water droplets and ice crystals both abound, while, toward the cold

Did you know?
On July 26, 1959, Colonel William A. Rankin took a wild ride inside a huge cumulonimbus cloud. Bailing out of his disabled military aircraft inside a thunderstorm at 47,000 feet, Rankin free-fell for about 10,000 feet. When his parachute opened, surging updrafts carried him higher into the cloud, where he was pelted by heavy rain and hail, and nearly struck by lightning.

top, there are only ice crystals. Swift winds at these higher altitudes can reshape the top of the cloud into a huge flattened anvil. These great thunderheads may contain all forms of precipitation—large raindrops, snowflakes, snow pellets, and sometimes hailstones—all of which can fall to earth in the form of heavy show-

ers. Lightning, thunder, and even violent tornadoes are associated with the cumulonimbus.

Cumulus congestus and cumulonimbus frequently look alike, making it difficult to distinguish between them. However, you can usually distinguish them by looking at the top of the cloud. If the sprouting upper part of the cloud is sharply defined and not fibrous, it is usually a cumulus congestus; conversely, if the top of the cloud loses its sharpness and becomes fibrous in texture, it is usually a cumulonimbus. Compare Fig. 4.23 with Fig. 4.24. The weather associated with these clouds also differs: lightning, thunder, and large hail only occur with cumulonimbus.

So far, we have discussed the ten primary cloud forms, summarized pictorially in Fig. 4.25. This figure, along with the cloud photographs and descriptions, should help you identify the more common cloud

Figure 4.22
Cumulus clouds. Small cumulus clouds such as these are sometimes called *fair weather cumulus*.

Figure 4.23
Cumulus congestus. This line of cumulus congestus clouds is building along Maryland's
eastern shore.

Figure 4.24
A cumulonimbus cloud. Strong upper-level winds blowing from right to left produce a well-defined anvil. Sunlight scattered by falling ice crystals produces the white (bright) area beneath the anvil. Notice the precipitation falling from the base of the cloud.

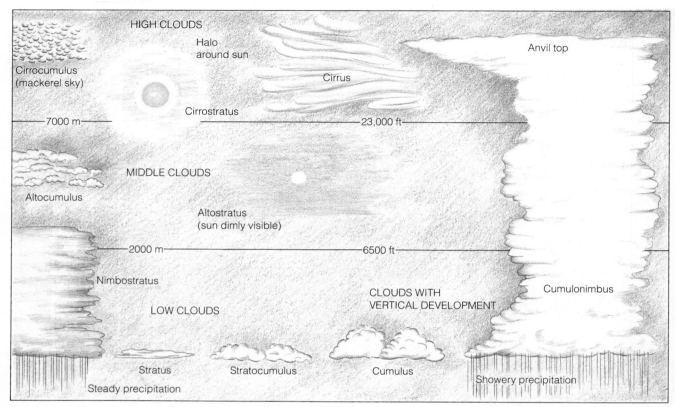

Figure 4.25
A generalized illustration of basic cloud types based on height above the surface and vertical development.

forms. Don't worry if you find it hard to estimate cloud heights. This is a difficult procedure, requiring much practice. You can use local objects (hills, mountains, tall buildings) of known height as references on which to base your height estimates.

To better describe a cloud's shape and form, a number of descriptive words may be used in conjunction with its name. We mentioned a few in the previous section; for example, a stratus cloud with a ragged appearance is a stratus fractus, and a cumulus cloud with marked vertical growth is a cumulus congestus. Table 4.4 lists some of the more common terms that are used in cloud identification.

Some Unusual Clouds   Although the ten basic cloud forms are the most frequently seen, there are some unusual clouds that deserve mentioning. For example,

moist air crossing a mountain barrier often forms into waves. The clouds that form in the wave crest usually have a lens shape and are, therefore, called **lenticular clouds** (Fig. 4.26). Frequently, they form one above the other like a stack of pancakes, and at a distance they may resemble a fleet of hovering spacecraft. Hence, it is no wonder a large number of UFO sightings take place when lenticular clouds are present.

Similar to the lenticular is the *cap cloud* or *pileus* that usually resembles a silken scarf capping the top of a sprouting cumulus cloud (Fig. 4.27). Pileus clouds form when moist winds are deflected up and over the top of a building cumulus congestus or cumulonimbus. If the air flowing over the top of the cloud condenses, a pileus often forms.

Most clouds form in rising air, but the mammatus forms in sinking air. **Mammatus clouds** derive their

Table 4.4   Common Terms Used in Identifying Clouds

Term	Latin Root and Meaning	Description
Lenticularis	(*lens, lenticula*, lentil)	Clouds having the shape of a lens; often elongated and usually with well-defined outlines. This term applies mainly to cirrocumulus, alto-cumulus, and stratocumulus
Fractus	(*frangere*, to break or fracture)	Clouds that have a ragged or torn appearance; applies only to stratus and cumulus
Humilis	(*humilis*, of small size)	Cumulus clouds with generally flattened bases and slight vertical growth
Congestus	(*congerere*, to bring together; to pile up)	Cumulus clouds of great vertical extent that from a distance may resemble a head of cauliflower
Undulatus	(*unda*, wave; having waves)	Clouds in patches, sheets, or layers showing undulations
Translucidus	(*translucere*, to shine through; transparent)	Clouds that cover a large part of the sky and are sufficiently translucent to reveal the position of the sun or moon
Mammatus	(*mamma*, mammary)	Baglike clouds that hang like a cow's udder on the underside of a cloud; may occur with cirrus, altocumulus, altostratus, stratocumulus, and cumulonimbus
Pileus	(*pileus*, cap)	A cloud in the form of a cap or hood above or attached to the upper part of a cumuliform cloud, particularly during its developing stage
Castellanus	(*castellum*, a castle)	Clouds that show vertical development and produce towerlike extensions, often in the shape of small castles

Figure 4.26
Lenticular clouds forming one on top of the other on the eastern side of the Sierra Nevada.

Figure 4.27
A pileus cloud forming above a developing cumulus cloud.

Figure 4.28
Mammatus clouds photographed after
sunset.

name from their appearance—baglike sacks that hang
beneath the cloud and resemble a cow's udder (Fig.
4.28). Although mammatus most frequently form on the
underside of cumulonimbus, they may develop beneath
cirrus, cirrocumulus, altostratus, altocumulus, and
stratocumulus.

Jet aircraft flying at high altitudes often produce a
cirruslike trail of condensed vapor called a *condensation trail* or **contrail** (Fig. 4.29). The condensation
may come directly from the water vapor added to the
air from engine exhaust. In this case, there must be sufficient mixing of the hot exhaust gases with the cold air

Figure 4.29
A contrail forming behind a jet aircraft.

Figure 4.30
The clouds in the upper two-thirds of this photograph are noctilucent clouds. They are usually observed at high latitudes, at altitudes between 75 km and 90 km above the earth's surface.

to produce saturation. Contrails evaporate rapidly when the relative humidity of the surrounding air is low. If the relative humidity is high, however, contrails may persist for many hours. Contrails may also form by a cooling process as the reduced pressure produced by air flowing over the wing causes the air to cool.

Aside from the cumulonimbus cloud that sometimes penetrates into the stratosphere, all of the clouds described so far are observed in the lower atmosphere—in the troposphere. Occasionally, however, clouds may be seen above the troposphere. For example, soft pearly looking clouds called *nacreous clouds*, or *mother-of-pearl clouds*, form in the stratosphere at altitudes above 30 km or 100,000 ft. They are best viewed in polar latitudes during the winter months when the sun, being just below the horizon, is able to il-

luminate them because of their high altitude. Their exact composition is not known, although they appear to be composed of water in either solid or liquid (super-cooled) form.

Wavy bluish-white clouds, so thin that stars shine brightly through them, may sometimes be seen in the upper mesosphere, at altitudes above 80 km (50 mi). The best place to view these clouds is in polar regions at twilight. At this time, because of their altitude, the clouds are still in sunshine. To a ground observer, they appear bright against a dark background and, for this reason, they are called *noctilucent clouds*, meaning "luminous night clouds." (See Fig. 4.30.) Scientists theorize that these clouds are composed of water that freezes on tiny meteoric dust particles. The water may actually originate in meteoroids that disintegrate when entering the upper atmosphere.

Summary

In this chapter, we examined some of the ways of describing humidity and found that relative humidity does not tell us how much water vapor is in the air but, rather, how close the air is to being saturated. A good indicator of the air's actual water vapor content is the dew-point temperature. When the air temperature and dew point are close together, the relative humidity is high, and, when they are far apart, the relative humidity is low.

When the air temperature drops below the dew point in a shallow layer of air near the surface, dew forms. If the dew freezes it becomes frozen dew. Visible white frost forms when the air cools to a below freezing dew-point temperature. As the air cools in a deeper layer near the surface, the relative humidity increases and water vapor begins to condense upon "water seeking" hygroscopic condensation nuclei, forming haze. As the relative humidity approaches 100 percent, the air can become filled with tiny liquid droplets (or ice crystals) called fog. Upon examining fog, we found that it forms in two primary ways: cooling the air and evaporating and mixing water vapor into the air.

Condensation above the earth's surface produces clouds. When clouds are classified according to their height and physical appearance, they are divided into four main groups: high, middle, low, and clouds with vertical development. Since each cloud has physical characteristics that distinguish it from all the others, careful observation normally leads to correct identification.

Key Terms

The following terms are listed in the order they appear in the text. Define each. Doing so will aid you in reviewing the material covered in this chapter.

evaporation	dew-point temperature	cirrocumulus clouds
condensation	(dew point)	cirrostratus clouds
precipitation	psychrometer	altocumulus clouds
hydrologic cycle	hygrometer	altostratus clouds
saturation	dew	nimbostratus clouds
condensation nuclei	frost	stratocumulus clouds
humidity	haze	stratus clouds
actual vapor pressure	fog	cumulus clouds
saturation vapor pressure	radiation fog	cumulonimbus clouds
relative humidity	advection fog	lenticular clouds
wet-bulb temperature	upslope fog	mammatus clouds
heat index (HI)	evaporation (mixing) fog	contrail
apparent temperature	cirrus clouds	

Review Questions

1. Briefly explain the movement of water in the hydrologic cycle.
2. Define:
   (a) evaporation
   (b) condensation
   (c) saturation.
3. What are condensation nuclei and why are they important in our atmosphere?
4. In a volume of air, how does the actual vapor pressure differ from the saturation vapor pressure? When are they the same?
5. What does saturation vapor pressure primarily depend upon?

6. (a) What does the relative humidity represent?
   (b) When the relative humidity is given, why is it also important to know the air temperature?
   (c) Explain two ways the relative humidity may be changed.
7. Why do hot and humid summer days usually feel hotter than hot and dry summer days?
8. Why is the wet-bulb temperature a good measure of how cool human skin can become?
9. (a) What is the dew-point temperature?
   (b) How is the difference between dew point and air temperature related to the relative humidity?

10. How can you obtain both the dew point and the relative humidity using a sling psychrometer?
11. Explain how dew, frozen dew, and visible frost form.
12. List the two primary ways in which fog forms.
13. Describe the conditions that are necessary for the formation of:
    (a) radiation fog
    (b) advection fog.
14. How does evaporation (mixing) fog form?
15. Clouds are most generally classified by height. List the major height categories and the cloud types associated with each.

16. How can you distinguish altostratus clouds from cirrostratus clouds?
17. Which clouds are associated with each of the following characteristics:
    (a) mackerel sky
    (b) lightning
    (c) halos
    (d) hailstones
    (e) mares' tails
    (f) anvil top
    (g) light continuous rain or snow

Heavy, wet snowflakes that fall through air with a temperature just below freezing cover everything with a blanket of white. (Photo by author)

Chapter 5

Cloud Development and Precipitation

Contents

The weather is an ever-playing drama before which we are a captive audience. With the lower atmosphere as the stage, air and water as the principal characters, and clouds for costumes, the weather's acts are presented continuously somewhere about the globe. The script is written by the sun; the production is directed by the earth's rotation; and, just as no theater scene is staged exactly the same way twice, each weather episode is played a little differently, each is marked with a bit of individuality.

Clyde Orr, Jr., *Between Earth and Space*

5 Clouds, spectacular features in the sky, add beauty and color to the natural landscape. Yet, clouds are important for nonaesthetic reasons, too. As they form, vast quantities of heat are released into the atmosphere. Clouds help regulate the earth's energy balance by reflecting and scattering solar radiation and by absorbing the earth's infrared energy. And, of course, without clouds there would be no precipitation. But clouds are also significant because they visually indicate the physical processes taking place in the atmosphere; to a trained observer, they are signposts in the sky. In the beginning of this chapter, we will look at the atmospheric processes these signposts point to, the first of which is atmospheric stability. Later, we will examine the different mechanisms responsible for the formation of most clouds. Toward the end of the chapter, we will peer into the tiny world of cloud droplets to see how rain, snow, and other types of precipitation form.

▲▼▲

Atmospheric Stability

We know that most clouds form as air rises, expands, and cools. But why does the air rise on some occasions and not on others? And why does the size and shape of clouds vary so much when the air does rise? To answer these questions, let's focus on the concept of atmospheric stability.

When we speak of atmospheric stability, we are referring to a condition of equilibrium. For example, rock A resting in the depression in Fig. 5.1 is in *stable* equilibrium. If the rock is pushed up along either side of the hill and then let go, it will quickly return to its original position. On the other hand, rock B, resting on the top

of the hill, is in a state of *unstable* equilibrium, as a slight push will set it moving away from its original position. Applying these concepts to the atmosphere, we can see that air is in stable equilibrium when, after being lifted or lowered, it tends to return to its original position—it resists upward and downward air motions. Air that is in unstable equilibrium will, when given a little push, move farther away from its original position—it favors vertical air currents.

In order to explore the behavior of rising and sinking air, we must first review some concepts we learned in Chapter 2. Recall that a balloonlike blob of air is called an *air parcel*. When an air parcel rises, it moves into a region where the air pressure surrounding it is lower. This situation allows the air molecules inside to push outward on the parcel walls, expanding it. As the air parcel expands, the air inside cools. If the same parcel is brought back to the surface, the increasing pressure around the parcel squeezes (compresses) it back to its original volume, and the air inside warms. If a parcel of air expands and cools, or compresses and warms, with no interchange of heat with its outside surroundings, this situation is called an **adiabatic process**. As long as the air in the parcel is unsaturated (the relative humidity is less than 100 percent), the rate of adiabatic cooling or warming remains constant and is about 10°C for every 1000 meters of change in elevation, or about 5.5°F for every 1000 feet. Since this rate of cooling or warming only applies to unsaturated air, it is called the **dry adiabatic rate**. (See Fig. 5.2.)

As the rising air cools, its relative humidity increases as the air temperature approaches the dew-point temperature. If the air cools to its dew-point temperature, the relative humidity becomes 100 percent. Further lifting results in condensation, a cloud forms, and latent heat is released into the rising air. Be-

Figure 5.1
When rock A is disturbed, it will return to its original position; rock B, however, will accelerate away from its original position.

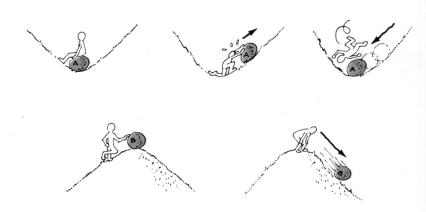

cause the heat added during condensation offsets some of the cooling due to expansion, the air no longer cools at the dry adiabatic rate but at a lesser rate called the **moist adiabatic rate**.* (Because latent heat is added to the rising saturated air, the process is not really adiabatic.) If a saturated parcel containing water droplets were to sink, it would compress and warm at the moist adiabatic rate because evaporation of the liquid droplets would offset the rate of compressional warming. Hence, the rate at which rising or sinking saturated air changes temperature—the moist adiabatic rate—is less than the dry adiabatic rate.

Unlike the dry adiabatic rate, the moist adiabatic rate is not constant, but varies greatly with temperature and, hence, with moisture content—as warm saturated air produces more liquid water than cold saturated air. The added condensation in warm, saturated air liberates more latent heat. Consequently, the moist adiabatic rate is much less than the dry adiabatic rate when the rising air is quite warm; however, the two rates are nearly the same when the rising air is very cold. Although the moist adiabatic rate does vary, we will use an average of 6°C per 1000 meters (3.3°F per 1000 feet) in most of our examples and calculations.

Determining Stability

We determine the stability of the air by comparing the temperature of a rising parcel to that of its surroundings. If the rising air is colder than its environment, it will be more dense† (heavier) and tend to sink back to its original level. In this case, the air is *stable* because it resists upward displacement. If the rising air is warmer and, therefore, less dense (lighter) than the surrounding air, it will continue to rise until it reaches the same temperature as its environment. This is an example of *unstable* air. To figure out the air's stability, we need to measure the temperature both of the rising air and of its environment at various levels above the earth.

---

*If condensed water or ice is removed from the rising saturated air, the cooling process is called *pseudoadiabatic*.

†When, at the same level in the atmosphere, we compare parcels of air that are equal in size but vary in temperature, we find that cold air parcels are more dense than warm air parcels; that is, in the cold parcel, there are more molecules that are crowded closer together.

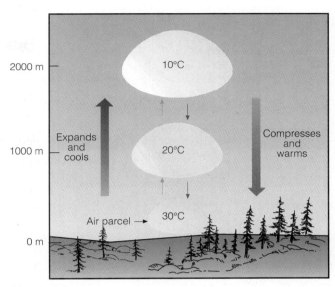

Figure 5.2
The dry adiabatic rate. As long as the air parcel remains unsaturated, it expands and cools by 10°C per 1000 meters; the sinking parcel compresses and warms by 10°C per 1000 meters.

Stable Air Suppose we release a balloon-borne instrument—a radiosonde—and it sends back temperature data as shown in Fig. 5.3. We measure the air temperature in the vertical and find that it decreases by 4°C for every 1000 meters. Remember from Chapter 1 that the rate at which the air temperature changes with elevation is called the *lapse rate*. Because this is the rate at which the air temperature surrounding us would be changing if we were to climb upward into the atmosphere, we refer to it as the **environmental lapse rate**.

Notice in Fig. 5.3a that (with an environmental lapse rate of 4°C per 1000 meters) a rising parcel of unsaturated, "dry" air is colder and heavier than the air surrounding it at all levels. Even if the parcel is initially saturated (Fig. 5.3b), as it rises it, too, would be colder than its environment at all levels. In both cases, the atmosphere is **absolutely stable** because the lifted parcel of air is colder and heavier than the air surrounding it. If released, the parcel would have a tendency to return to its original position.

Since stable air strongly resists upward vertical motion, it will, *if forced to rise*, tend to spread out horizontally. If clouds form in this rising air, they, too, will spread horizontally in relatively thin layers and

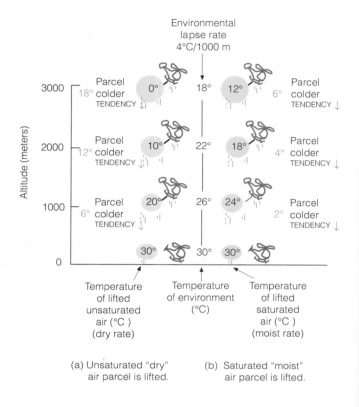

Figure 5.3
A stable atmosphere. An *absolutely stable atmosphere* exists when a rising air parcel is colder and heavier (i.e., more dense) than the air surrounding it. If given the chance (i.e., released), the air parcel in both situations would return to its original position, the surface.

Figure 5.4
Cold surface air, on this morning, produces a stable atmosphere that inhibits vertical air motions and allows the fog and haze to linger close to the ground.

usually have flat tops and bases. We might expect to see clouds—such as cirrostratus, altostratus, nimbostratus, or stratus—forming in stable air.

The atmosphere is stable when the environmental lapse rate is small; that is, when there is a relatively small difference in temperature between the surface air and the air aloft. Consequently, the atmosphere tends to become more stable as the air aloft warms or the surface air cools. The cooling of the surface air may be due to:

1. nighttime radiational cooling of the surface
2. an influx of cold air brought in by the wind
3. air moving over a cold surface

It should be apparent that, on any given day, the air is generally most stable in the early morning around sunrise, when the lowest air temperature is recorded.

The air aloft may warm as winds bring in warmer air or as the air slowly sinks over a large area. Recall that sinking (subsiding) air warms as it is compressed. The warming may produce an inversion, where the air aloft is actually warmer than the air at the surface. An inversion that forms by slow, sinking air is termed a *subsidence inversion*. Because inversions represent a very stable atmosphere, they act as a lid on vertical air motion. When an inversion exists near the ground, stratus, fog, haze, and pollutants are all kept close to the surface (Fig. 5.4).

Unstable Air   The atmosphere is unstable when the air temperature decreases rapidly as we move up into the atmosphere. For example, in Fig. 5.5, notice that the measured air temperature decreases by 11°C for every 1000 meters rise in elevation, which means that the environmental lapse rate is 11°C per 1000 meters. Also notice that a lifted parcel of unsaturated "dry" air in Fig. 5.5a, as well as a lifted parcel of saturated "moist" air in Fig. 5.5b, will, at each level above the surface, be warmer than the air surrounding them. Since, in both cases, the rising air is warmer and less dense than the air around them, once the parcels start upward, they will continue to rise on their own, away from the surface. Thus, we have an **absolutely unstable atmosphere**.

The atmosphere becomes more unstable as the environmental lapse rate steepens; that is, as the temperature of the air drops rapidly with increasing height. This circumstance may be brought on by either the air aloft becoming colder or the surface air becoming warmer (Fig. 5.6). The warming of the surface air may be due to:

1. daytime solar heating of the surface
2. an influx of warm air brought in by the wind
3. air moving over a warm surface

Generally, then, as the surface air warms during the day, the atmosphere becomes more unstable.

Suppose an unsaturated (but humid) air parcel is somehow forced to rise from the surface, as shown in Fig. 5.7. As the parcel rises, it expands, and cools at

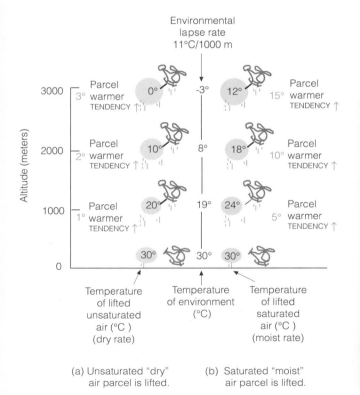

Figure 5.5
An unstable atmosphere. An *absolutely unstable atmosphere* exists when a rising air parcel is warmer and lighter (i.e., less dense) than the air surrounding it. If given the chance (i.e., released), the lifted parcel in both (a) and (b) would continue to move away from its original position.

Figure 5.6
Unstable air. The warmth from the forest fire heats the air, causing instability near the surface. Warm, less-dense air (and smoke) bubbles upward, expanding and cooling as it rises. Eventually the rising air cools to its dew point, condensation begins, and a cumulus cloud forms.

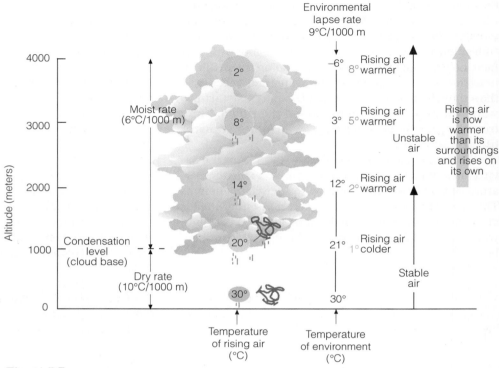

Figure 5.7
Conditionally unstable air. The atmosphere is conditionally unstable when unsaturated, stable air is lifted to a level where it becomes saturated and warmer than the air surrounding it. If the atmosphere remains unstable, vertical developing cumulus clouds can build to great heights.

the *dry adiabatic rate* until its air temperature cools to its dew point. At this level, the air is saturated, the relative humidity is 100 percent, and further lifting results in condensation and the formation of a cloud. The elevation above the surface where the cloud first forms is called the **condensation level**.

In Fig. 5.7, notice that above the condensation level, the rising saturated air cools at the *moist adiabatic rate*. Notice also that up to a level of 2000 meters, the rising, lifted air is colder than the air surrounding it. The atmosphere up to this level is *stable*. However, due to the release of latent heat, at 2000 meters and above, the rising air has actually become warmer than the air around it. Since the lifted air can rise on its own accord, the atmosphere is now *unstable*.

The atmospheric layer from the surface up to 4000 meters in Fig. 5.7 has gone from stable to unstable because the rising air was humid enough to become saturated, form a cloud, and release latent heat, which warms the air. Had the cloud not formed, the rising air

would have remained colder at each level than the air surrounding it. From the surface to 4000 meters, we have what is said to be a **conditionally unstable atmosphere**—the condition for instability being whether or not the rising air becomes saturated. Therefore, *conditional instability* means that, if unsaturated stable air is somehow lifted to a level where it becomes saturated, instability may result.

In Fig. 5.7, we can see that the environmental lapse rate is 9°C per 1000 meters. This value is between the dry adiabatic rate and the moist adiabatic rate. Consequently, conditional instability exists whenever the environmental lapse rate is between the dry and moist adiabatic rates.

At this point, it should be apparent that the stability of the air changes during the course of a day. In clear, calm weather around sunrise, surface air is normally colder than the air above it, a radiation inversion exists, and the air is quite stable, as indicated by smoke or haze lingering close to the ground. As the day progresses, sunlight warms the surface and the surface warms the air above. As the air temperature near the ground increases, the lower atmosphere gradually becomes more unstable, with maximum instability usually occurring during the hottest part of the day. On a humid summer afternoon this phenomenon can be witnessed by the development of cumulus clouds.

Up to now, we have looked briefly at stability as it relates to cloud development; that is, layered clouds tend to form in stable air, whereas cumuliform clouds tend to form in unstable air. The following section describes how atmospheric stability influences the physical mechanisms responsible for the development of individual cloud types.

Cloud Development and Stability

Most clouds form as air rises, expands, and cools. Basically, the following mechanisms are responsible for the development of the majority of clouds we observe:

(a) surface heating and free convection; (b) topography; (c) widespread ascent due to the flowing together (convergence) of surface air; and (d) uplift along weather fronts. (See Fig. 5.8.)

Convection and Clouds   Some areas of the earth's surface are better absorbers of sunlight than others and, therefore, heat up more quickly. The air in contact with these "hot spots" becomes warmer than its surroundings. A hot "bubble" of air—a *thermal*—breaks away from the warm surface and rises, expanding and cooling as it ascends. As the thermal rises, it mixes with the cooler, drier air around it and gradually loses its identity. Its upward movement now slows. Frequently, before it is completely diluted, subsequent rising thermals penetrate it and help the air rise a little higher. If the rising air cools to its saturation point, the moisture will condense, and the thermal becomes visible to us as a cumulus cloud.

Observe in Fig. 5.9 that the air motions are downward on the outside of the cumulus cloud. The downward motions are caused in part by evaporation around

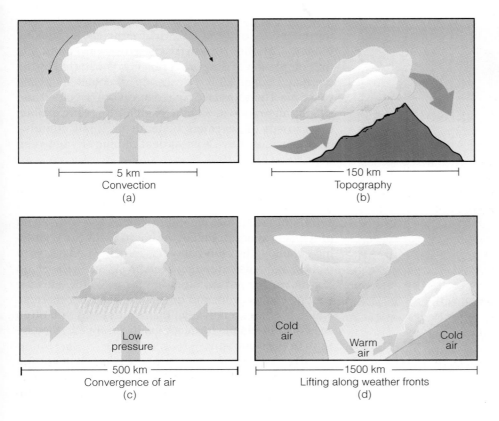

Figure 5.8
The primary ways clouds form: (a) surface heating and convection; (b) forced lifting along topographic barriers; (c) convergence of surface air; (d) forced lifting along weather fronts.

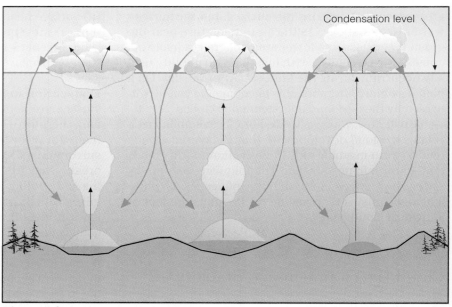

Figure 5.9
Cumulus clouds form as hot, invisible air bubbles detach themselves from the surface, then rise and cool to the condensation level. Below and within the cumulus clouds, the air is rising. Around the cloud, the air is sinking.

Figure 5.10
Cumulus clouds building over the warm Florida landscape. Each tiny cloud represents a region where thermals are rising from the surface. The clear areas between the clouds are regions where the air is sinking. Clouds are notably absent over the slightly cooler water.

the outer edge of the cloud, which cools the air, making it heavy. Another reason for the downward motion is the completion of the convection current started by the thermal. Cool air slowly descends to replace the rising warm air. Therefore, we have rising air in the cloud and sinking air around it. Since subsiding air greatly inhibits the growth of thermals beneath it, small cumulus clouds usually have a great deal of blue sky between them. (See Fig. 5.10.)

As the cumulus clouds grow, they shade the ground from the sun. This, of course, cuts off surface heating and upward convection. Without the continual supply of rising air, the cloud begins to erode as its droplets evaporate. Unlike the sharp outline of a growing cumulus, the cloud now has indistinct edges, with cloud fragments extending from its sides. As the cloud dissipates (or moves along with the wind), surface heating begins again and regenerates another thermal, which becomes a new cumulus. This is why you often see cumulus clouds form, gradually disappear, then re-form in the same spot.

The stability of the atmosphere plays an important part in determining the vertical growth of cumulus clouds. For example, if a stable layer (such as an inversion) exists near the top of the cumulus cloud, the cloud would have a difficult time rising much higher, and it would remain as a "fair-weather" cumulus cloud. However, if a deep unstable or conditionally unstable layer exists above the cloud, then the cloud may develop vertically into a towering cumulus congestus with a cauliflowerlike top. When the unstable air is several miles deep, the cumulus congestus may even develop into a cumulonimbus (Fig. 5.11).

Notice in Fig. 5.11 that the distant thunderstorm has a flat anvil-shaped top. The reason for this shape is due to the fact that the cloud has reached the stable stratosphere, and the rising air is unable to puncture very far into it. Consequently, the top of the cloud spreads horizontally as high winds at this altitude (usually above 10,000 meters or 34,000 feet) blow the cloud's upper icy part horizontally.

Topography and Clouds   Horizontally moving air obviously can not go through a large obstacle, such as a mountain, so the air must go over it. Forced lifting along a topographic barrier is called **orographic uplift**. Often, large masses of air rise when they approach a long chain of mountains such as the Sierra Nevada and Rockies. This lifting produces cooling, and if the air is humid, clouds form. Clouds produced in this manner are called *orographic clouds*.

An example of orographic uplift and cloud development is given in Fig. 5.12. Notice that, after having risen over the mountain, the air at the surface on the leeward (downwind) side is considerably warmer than it was at the surface on the windward (upwind) side. The higher air temperature on the leeward side is the result of latent heat being converted into sensible heat during condensation on the windward side. In fact, the rising air at the top of the mountain is considerably warmer than it would have been had condensation not occurred.

Notice also in Fig. 5.12 that the dew-point temperature of the air on the leeward side is lower than it was before the air was lifted over the mountain. The lower dew point and, hence, drier air on the leeward side is the result of water vapor condensing and then remaining as liquid cloud droplets and precipitation on the windward side. This region on the leeward side of a mountain, where precipitation is noticably low, and the air is often drier, is called a **rain shadow**.

Figure 5.11
Cumulus clouds developing into thunderstorms in unstable air over the Great Plains. Notice that the cumulonimbus in the distance, with the anvil top, has reached the stable stratosphere.

Although clouds are more prevalent on the windward side of mountains, they may, under certain atmospheric conditions, form on the leeward side as well. For example, stable air flowing over a mountain often moves in a series of waves that may extend for several hundred miles on the leeward side. Such waves often resemble the waves that form in a river downstream from a large boulder. Recall from Chapter 4 that wave clouds often have a characteristic lens shape and are called *lenticular clouds*.

The formation of lenticular clouds is shown in Fig. 5.13. As moist air rises on the upwind side of the wave, it cools and condenses, producing a cloud. On the downwind side, the air sinks and warms—the cloud evaporates. Viewed from the ground, the clouds appear motionless as the air rushes through them. When the air between the cloud-forming layers is too dry to produce clouds, lenticular clouds will form one above the other, sometimes extending into the stratosphere and appearing as a fleet of hovering spacecraft. (See Fig. 4.26, p. 99.)

Notice in Fig. 5.13 that beneath the lenticular cloud, a large swirling eddy forms. The rising part of the eddy may cool enough to produce *rotor clouds*. The air in the rotor is extremely turbulent and presents a major hazard to aircraft in the vicinity. Dangerous flying conditions also exist near the leeside of the mountain, where strong downward air motions are present.

Now, having examined the concept of stability and the formation of clouds, we are ready to see how minute cloud particles are transformed into rain and snow. The next section, therefore, takes a look at the processes that produce precipitation.

▲▽▲ Precipitation Processes

As we all know, cloudy weather does not necessarily mean that it will rain or snow. In fact, clouds may form, linger for many days, and never produce precipitation. In Eureka, California, the August daytime sky is overcast more than 50 percent of the time, yet the average precipitation there for August is merely one-tenth of an inch. How, then, do cloud droplets grow large enough to produce rain? And why do some clouds produce rain, but not others?

In Fig. 5.14, we can see that an ordinary cloud droplet is extremely small, having an average diameter

Figure 5.12
Orographic uplift, cloud development, and the formation of a rain shadow.

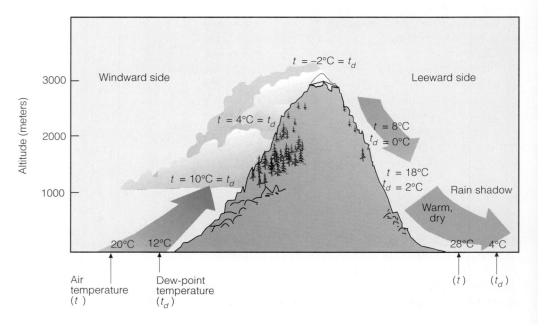

of 0.02 millimeters, which is less than one-thousandth of an inch. Also, notice in Fig. 5.14 that a typical cloud droplet is 100 times smaller than a typical raindrop. Clouds then are composed of many small droplets—too small to fall as rain. These minute droplets require only slight upward air currents to keep them suspended. Those droplets that do fall, descend slowly and evaporate in the drier air beneath the cloud.

In Chapter 4, we learned that condensation begins on tiny particles called *condensation nuclei*. The growth of cloud droplets by condensation is slow and, even under ideal conditions, it would take several days for this process alone to create a raindrop. It is evident, then, that the condensation process by itself is entirely too slow to produce rain. Yet, observations show that clouds can develop and begin to produce rain in less than an hour. Consequently, there must be some other process by which cloud droplets grow large and heavy enough to fall as precipitation.

Even though all the intricacies of how rain is produced are not yet fully understood, two important processes stand out: (1) the collision-coalescence process and (2) the ice crystal process.

Collision and Coalescence Process   In clouds that form at temperatures above freezing (*warm clouds*), collisions between droplets play a significant part in producing precipitation. To produce the many collisions necessary to form a raindrop, some cloud droplets must be larger than others. Larger drops could have formed on large condensation nuclei, such as salt particles. Random bumping and touching of droplets could also account for size variation. Recent studies, however, suggest that turbulent mixing between the cloud and its drier environment may actually play a major role in producing larger droplets. For example, dry air injected into a cloud near its top will evaporate cloud droplets until the mixed air reaches saturation. If this mixed blob of air rises and cools, the water vapor available is consequently shared by fewer droplets, which allows them to grow to a larger size than they would otherwise.

Large drops fall faster than small drops. Consequently, large drops are able to overtake and collide with smaller drops in their path. This merging of cloud droplets by collision is called **coalescence**. Laboratory studies show that collision does not always guarantee coalescence; sometimes, the droplets actually bounce apart during collision. Coalescence appears to be en-

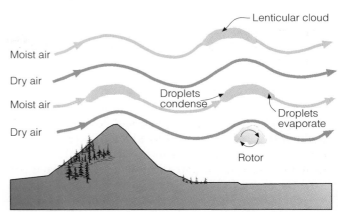

Figure 5.13
The formation of lenticular clouds.

hanced if colliding droplets have opposite (and, hence, attractive) electrical charges. Therefore, atmospheric electricity seems to play a role in the growth of cloud droplets and in the production of rain. Another important factor that determines cloud droplet growth by the collision process is the amount of time the droplet spends in the cloud. Since rising air currents slow the

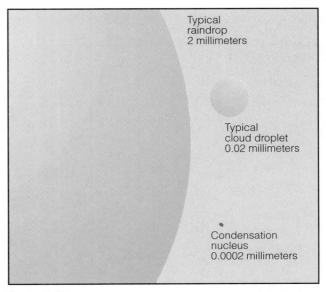

Figure 5.14
Relative sizes of raindrops, cloud droplets, and condensation nuclei.

rate at which droplets fall, a thick cloud with strong updrafts will maximize the time cloud droplets spend in a cloud and, hence, the size to which they can grow.

In tropical regions, where warm cumulus clouds build to great heights, strong convective updrafts frequently occur. In Fig. 5.15, suppose a cloud droplet is caught in a strong updraft. As the droplet rises, it collides with and captures smaller drops in its path, and grows until it reaches a size of about one millimeter. At this point, the updraft in the cloud is just able to balance the pull of gravity on the drop. Here, the drop remains suspended until it grows just a little bigger. Once the fall velocity of the drop is greater than the updraft velocity in the cloud, the drop slowly descends. As the drop falls, some of the smaller droplets get caught in the airstream around it, and are swept aside. Larger cloud droplets are captured by the falling drop, which then grows larger. By the time this drop reaches the bottom of the cloud, it will be a large raindrop with a diameter of over 5 millimeters. Because raindrops of this size fall faster and reach the ground first, they typically

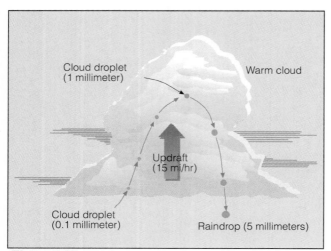

Figure 5.15
A cloud droplet rising then falling through a warm cumulus cloud can grow by collision and coalescence and emerge from the cloud as a large raindrop.

occur at the beginning of a rain shower originating in these warm, convective cumulus clouds.

So far, we have examined the way cloud droplets in warm clouds grow large enough by the collision-coalescence process to fall as raindrops. The most important factor in the production of raindrops is the cloud's liquid water content. In a cloud with sufficient water, other significant factors are:

1. the relative droplet size
2. the electric charge of the droplets and the electric field in the cloud
3. the cloud thickness
4. the updrafts of the cloud

Relatively thin stratus clouds with slow upward air currents are, at best, only able to produce drizzle (the lightest form of rain), whereas the towering cumulus clouds associated with rapidly rising air can cause heavy showers. Now, let's turn our attention to the ice-crystal process of rain formation.

Ice-Crystal Process   The ice-crystal process* of rain formation proposes that both ice crystals and liquid cloud droplets must co-exist in clouds at temperatures below freezing. Consequently, this process of rain formation is extremely important in middle and high latitudes, where clouds are able to extend upwards into regions where air temperatures are below freezing. Figure 5.16 illustrates a typical cumulonimbus cloud that has formed over the Great Plains of North America.

In the warm region of the cloud (below the freezing level) where only water droplets exist, we might expect to observe cloud droplets growing larger by the collision and coalescence process described in the previous section. Surprisingly, in the cold air just above the freezing level, almost all of the cloud droplets are still composed of liquid water. Water droplets existing at temperatures below freezing are referred to as **supercooled**. At higher levels, ice crystals become more numerous, but are still outnumbered by water droplets. Ice crystals exist overwhelmingly in the upper part of the cloud, where air temperatures drop to well below freezing. Why are there so few ice crystals in the middle of the cloud, even though temperatures there, too, are below freezing? Laboratory studies reveal that the

---

*The ice crystal process is also known as the *Bergeron process* after the Swedish meteorologist Tor Bergeron, who proposed that essentially all raindrops begin as ice crystals.

smaller the amount of pure water, the lower the temperature at which water freezes. Since cloud droplets are extremely small, it takes very low temperatures to turn them into ice.

Just as liquid cloud droplets form on condensation nuclei, ice crystals may form in subfreezing air if there are ice-forming particles present called **ice nuclei**. The number of ice-forming nuclei available in the atmosphere is small, especially at temperatures above 10°C (14°F). Although some uncertainty exists regarding the principal source of ice nuclei, it is known that certain clay minerals, bacteria in decaying plant leaf material, and ice crystals themselves are excellent ice nuclei. Moreover, particles serve as excellent ice-forming nuclei if their geometry resembles that of an ice crystal.

We can now understand why there are so few ice crystals in the subfreezing region of some clouds. Liquid cloud droplets may freeze, but only at very low temperatures. Ice nuclei may initiate the growth of ice crystals, but they do not abound in nature. Therefore, we are left with a cold cloud that contains many more liquid droplets than ice particles, even at low temperatures. Neither the tiny liquid nor solid particles are large enough to fall as precipitation. How, then, does the ice-crystal process produce rain and snow?

In subfreezing air of a cold cloud, many supercooled liquid droplets will surround each ice crystal. Suppose that the ice particle and liquid droplet in Fig. 5.17 are part of a cold, supercooled cloud. Observe that more water vapor molecules surround each supercooled water droplet than each ice particle. This surplus in the number of vapor molecules around the droplet causes water vapor to move (diffuse) from the droplet toward the ice crystal. The removal of vapor molecules above the liquid droplet causes it to evaporate and replenish the diminished supply of water vapor above it. This process provides a continuous source of moisture for the ice crystal, which absorbs the water vapor and grows rapidly. Hence, during the **ice crystal process**, *ice crystals grow larger at the expense of the surrounding water droplets.*

The ice crystals may now grow even larger. For example, in some clouds, ice crystals might collide with supercooled liquid droplets. Upon contact, the liquid droplets freeze into ice and stick to the ice crystal—a process called **accretion**, or *riming*. The icy matter (rime) that forms is called *graupel*. As the graupel falls, it may fracture or *splinter* into tiny ice particles when it collides with cloud droplets. These splinters may then go on themselves to become new graupel, which, in

Figure 5.16
The distribution of ice and water in a cumulonimbus cloud.

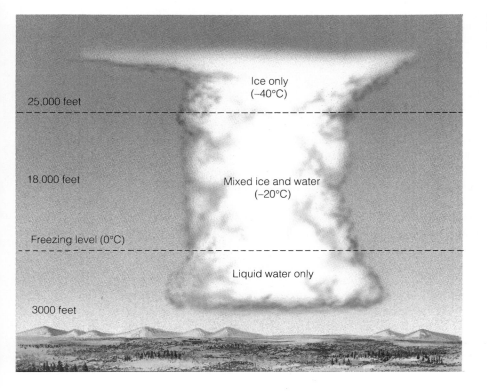

Ice only
(−40°C)

25,000 feet

18,000 feet

Mixed ice and water
(−20°C)

Freezing level (0°C)

Liquid water only

3000 feet

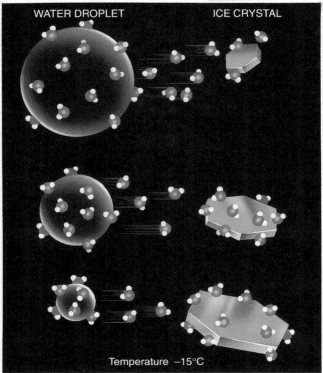

Figure 5.17
The ice-crystal process. The greater number of water vapor molecules around the liquid droplets causes the ice crystals to grow by diffusion as water molecules move from the liquid droplets toward the ice crystals. The ice crystals absorb the water vapor and grow larger, while the water droplets grow smaller.

turn, may produce more splinters. In colder clouds, the delicate ice crystals may collide with other crystals and fracture into smaller ice particles, or tiny seeds, which freeze hundreds of supercooled droplets on contact. In both cases a chain reaction may develop, producing many ice crystals. As they fall, they may collide and stick to one another, forming an aggregate of ice crystals called a *snowflake* (Fig. 5.18). If the snow-flake melts before reaching the ground, it continues its fall as a raindrop. Therefore, much of the rain falling in middle and northern latitudes—even in summer—begins as snow.

Cloud Seeding and Precipitation   The primary goal in many experiments concerning **cloud seeding** is to inject (or seed) a cloud with small particles that will act as nuclei, so that the cloud particles will grow large enough to fall to the surface as precipitation. The first

ingredient in any seeding project is, of course, the presence of clouds. (Seeding does not generate clouds.) However, at least a portion of the cloud (preferably the upper part) must be supercooled because cloud seeding uses the ice-crystal process to cause the cloud particles to grow.

Some of the first experiments in cloud seeding were conducted by Vincent Schaefer and Irving Langmuir during the late 1940s. To seed a cloud, they dropped crushed pellets of dry ice (solid carbon dioxide) from a plane. Because dry ice has a temperature of –78°C (–108°F), it acts as a cooling agent. Small pellets dropped into the cloud cool the air to the point where new liquid droplets are able to form. These new droplets (and the original ones) are now able to change into ice in an instant. The newly formed ice crystals then grow larger at the expense of the nearby liquid droplets and, upon reaching a sufficiently large size, fall as precipitation.

In 1947, Bernard Vonnegut demonstrated that silver iodide (AgI) could be used as a cloud-seeding agent. Because silver iodide has a crystalline structure similar to an ice crystal, it acts as an effective ice nucleus at below-freezing temperatures. Silver iodide causes ice crystals to form in two primary ways:

1. Ice crystals form when silver iodide crystals come in contact with supercooled droplets.
2. Ice crystals grow as water vapor deposits onto the silver iodide crystal.

Silver iodide is much easier to handle than dry ice, since it can be supplied to the cloud from burners located either on the ground or on the wing of a small aircraft. Although other substances, such as lead iodide and cupric sulfide, are also effective ice nuclei, silver iodide still remains the most commonly used substance in cloud-seeding projects. (Additional information on the controversial topic, the effectiveness of cloud seeding, is given in the Focus section on p. 120.)

Precipitation in Clouds   In cold, strongly convective clouds, precipitation may begin only minutes after the cloud forms and may be initiated by either the collision-coalescence or the ice-crystal process. Once either process begins, most precipitation growth is by accretion. Although precipitation is commonly absent in warm-layered clouds, it is often associated with such cold-layered clouds as nimbostratus and altostratus. This precipitation is thought to form principally by the

(a) Falling ice crystals may freeze supercooled droplets on contact (accretion), producing larger ice particles.

(b) Falling ice particles may collide and fracture into many tiny (secondary) ice particles.

(c) Falling ice crystals may collide and stick to other ice crystals (aggregation), producing snowflakes.

Figure 5.18
Ice particles in clouds.

ice-crystal process because the liquid water content of these clouds is generally lower than that in convective clouds, thus making the collision-coalescence process much less effective. Nimbostratus clouds are normally thick enough to extend to levels where air temperatures are quite low, and they usually last long enough for the ice-crystal process to initiate precipitation.

▲▼▲

Precipitation Types

Up to now, we have seen how cloud droplets are able to grow large enough to fall to the ground as rain or snow. While falling, raindrops and snowflakes may be altered by atmospheric conditions encountered beneath the cloud and transformed into other forms of precipitation that can profoundly influence our environment.

Rain   Most people consider **rain** to be any falling drop of liquid water. To the meteorologist, however, that falling drop must have a diameter equal to, or

greater than, 0.5 millimeters to be considered rain. Fine uniform drops of water whose diameters are smaller than this figure are called **drizzle**. Most drizzle falls from stratus clouds; however, small raindrops may fall through air that is unsaturated, partially evaporate, and reach the ground as drizzle. Occasionally, the rain falling from a cloud never reaches the surface because the low humidity causes rapid evaporation. As the drops become smaller, their rate of fall decreases, and they appear to hang in the air as a rain streamer. These evaporating streaks of precipitation are called **virga** (see Fig. 5.19).

Raindrops may also fall from a cloud and not reach the ground if they encounter the rapidly rising air of an updraft. If the updraft weakens or changes direction and becomes a downdraft, the suspended drops will fall to the ground as a sudden rain **shower**. The showers falling from cumuliform clouds are usually brief and sporadic, as the cloud moves overhead and then drifts on by. If the shower is excessively heavy, it is termed a *cloudburst*. Beneath a cumulonimbus cloud, which normally contains large convection currents, it is

Focus on an Issue
Does Cloud Seeding Enhance Precipitation?

Just how effective is artificial seeding with silver iodide in increasing precipitation? This is a much-debated question among meteorologists. First of all, it is difficult to evaluate the results of a cloud-seeding experiment. When a seeded cloud produces precipitation, the question always remains as to how much precipitation would have fallen had the cloud not been seeded.

Other factors must be considered when evaluating cloud-seeding experiments: the type of cloud, its temperature, moisture content, and droplet size distribution.

Although some experiments suggest that cloud seeding does not increase precipitation, others seem to indicate that seeding *under the right conditions* may enhance precipitation between 5 and 20 percent. And so the controversy continues.

Some cumulus clouds show an "explosive" growth after being seeded. The latent heat given off when the droplets freeze functions to warm the cloud, causing it to become more buoyant. It grows rapidly and becomes a longer-lasting cloud, which may produce more precipitation.

The business of cloud seeding can be a bit tricky, since over-seeding can produce too many ice crystals. When this occurs, the ice particles, being very small, do not fall as precipitation. Since few liquid droplets exist, the ice crystals cannot grow by the ice-crystal process; rather, they evaporate, leaving a clear area in a thin stratified cloud.

Warm clouds have also been seeded in an attempt to produce rain.

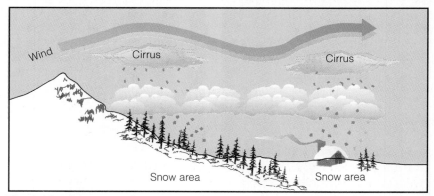

Figure 1
Natural seeding by cirrus clouds may form bands of precipitation downwind of a mountain chain.

Tiny water drops and particles of hygroscopic salt are injected into the base of the cloud. These particles when carried into the cloud by updrafts create large cloud droplets, which grow even larger by the collision-coalescence process. To date, the results obtained using this method are inconclusive.

Under certain conditions, clouds may be seeded naturally. For example, when cirriform clouds lie directly above a lower cloud deck, ice crystals may descend from the higher cloud and seed the cloud below. As the ice crystals mix into the lower cloud, supercooled droplets are converted to ice crystals, and the precipitation process is enhanced. When the cirrus clouds form waves downwind from a mountain chain (Fig. 1), bands of precipitation often form.

Cloud seeding may be inadvertent. Some industries emit large concentrations of condensation nuclei and ice nuclei into the air. Studies have shown that these particles are at least partly responsible for increasing precipitation in, and downwind of, cities. On the other hand, studies have also indicated that the burning of certain types of agricultural waste may produce smoke containing many condensation nuclei. These produce clouds that yield less precipitation because they contain numerous, but very small, droplets.

In summary, cloud seeding in certain instances may lead to more precipitation; in others, to less precipitation, and, in still others, to no change in precipitation amounts. Many of the questions about cloud seeding have yet to be resolved.

entirely possible that one side of a street may be dry (updraft side), while a heavy shower is occurring across the street (downdraft side). Continuous rain, on the other hand, usually falls from a layered cloud that covers a large area and has smaller vertical air currents. These are the conditions normally associated with nimbostratus clouds.

Raindrops that reach the earth's surface are seldom larger than about 6 millimeters, the reason being that the collisions (whether glancing or head-on) between raindrops tend to break them up into many smaller drops. Additionally, when raindrops grow too large they become unstable and break apart. (If you are curious as to the actual shape of a falling raindrop, read the Focus section on p. 122.)

After a rainstorm, visibility usually improves primarily because precipitation removes (scavenges) many of the suspended particles. When rain combines with gaseous pollutants, such as oxides of sulfur and nitrogen, it becomes acidic. *Acid rain*, which has an adverse effect on plants and water resources, is becoming a major problem in many industrialized regions of the world. (See Chapter 12.)

Snow We have learned that much of the precipitation reaching the ground actually begins as **snow**. In summer, the freezing level is usually high and the snowflakes falling from a cloud melt before reaching the surface. In winter, however, the freezing level is much lower, and falling snowflakes have a better chance of

survival. In fact, snowflakes can generally fall about 300 meters (or 1000 feet) below the freezing level before completely melting. When the warmer air beneath the cloud is relatively dry, the snowflakes partially melt. As the liquid water evaporates, it chills the snowflake, which retards its rate of melting. Consequently, in air that is relatively dry, snowflakes may reach the ground even when the air temperature is considerably above freezing.

Is it ever "too cold to snow"? Although many believe this expression, the fact remains that it is *never* too cold to snow. True, more water vapor will condense from warm saturated air than from cold saturated air. But, no matter how cold the air becomes, it always contains some water vapor that could produce snow. In fact, tiny ice crystals have been observed falling at temperatures as low as –47°C (–53°F). We usually associate extremely cold air with "no snow" because the coldest winter weather occurs on clear, calm nights—

Figure 5.19
In the drier air beneath these clouds, falling rain evaporates, producing streaks of precipitation called virga.

Focus on a Special Topic
Are Raindrops Tear-Shaped?

As rain falls, the drops take on a characteristic shape. Choose the shape in Fig. 2 that you feel most accurately describes that of a falling raindrop. Did you pick number 1? The tear-shaped drop has been depicted by artists for many years. Unfortunately, *raindrops are not tear-shaped*. Actually, the shape depends on the drop size. Raindrops less than 2 millimeters in diameter are nearly spherical and look like raindrop number 2. The attraction among the molecules of the liquid (surface tension) tends to

Figure 2
Which of the three drops drawn here represents the real shape of a falling raindrop?

squeeze the drop into a shape that has the smallest surface area for its total volume—a sphere.

Large raindrops, with diameters exceeding 2 millimeters, take on a different shape as they fall. Believe it or

not, they look like number 3, slightly elongated, flattened on the bottom, and rounded on top. As the larger drop falls, the air pressure against the drop is greatest on the bottom and least on the sides. The pressure of the air on the bottom flattens the drop, while the lower pressure on its sides allows it to expand a little. This shape has been described as everything from a falling parachute to a loaf of bread, or even a hamburger bun. You may call it what you wish, but remember: it is not tear-shaped.

---

conditions that normally prevail with strong high pressure areas that have few if any clouds.

When ice crystals and snowflakes fall from high cirrus clouds they are called **fall streaks**. Fall streaks behave in much the same way as virga—as the ice particles fall into drier air, they usually disappear as they change from ice into vapor (called *sublimation*). Because the wind at higher levels moves the cloud and ice particles horizontally more quickly than do the slower winds at lower levels, fall streaks often appear as dangling white streamers (Fig. 5.20).

Snowflakes falling through moist air that is slightly above freezing slowly melt as they descend. A thin film of water forms on the edge of the flakes, which acts like glue when other snowflakes come in contact with it. In this way, several flakes join to produce giant snowflakes that often measure an inch or more in diameter. These large, soggy snowflakes are associated with moist air and temperatures near freezing. However, when snowflakes fall through extremely cold air with

Did you know?
Some of the largest snowflakes ever seen—measuring about 5 inches in diameter—fell over parts of England on April 13, 1951, when the air temperature was 33°F.

a low moisture content, they do not readily stick together and small, powdery flakes of "dry" snow accumulate on the ground.

If you catch falling snowflakes on a dark object and examine them closely, you will see that the most common snowflake form is a fernlike branching shape called *dendrite*. As ice crystals fall through a cloud, they are constantly exposed to changing temperatures and moisture conditions. Since many ice crystals can join together to form a much larger snowflake, ice crystals may assume many complex patterns (Fig. 5.21).

Snow falling from developing cumulus clouds is often in the form of **flurries**. These are usually light showers that fall intermittently for short durations and produce only light accumulations. A more intense snow shower is called a **snow squall**. These brief but heavy falls of snow are comparable to summer rain showers and, like snow flurries, usually fall from cumuliform clouds. A more continuous snowfall (sometimes steadily, for several hours) accompanies nimbostratus and altostratus clouds.

When a strong wind is blowing at the surface, snow can be picked up and deposited into huge drifts. Drifting snow is usually accompanied by *blowing snow*; that is, snow lifted from the surface by the wind and blown about in such quantities that horizontal visibility is greatly restricted. The combination of drifting

Figure 5.20
The dangling white streamers of ice crystals beneath these cirrus clouds are known as *fall streaks*. The bending of the streaks is due to the changing wind speed with height.

Figure 5.21
The many patterns of dendrite snow crystals.

and blowing snow, after falling snow has ended, is called a *ground blizzard*. A true **blizzard** is a weather condition characterized by low temperatures and strong winds (greater than 30 knots) bearing large amounts of fine, dry, powdery particles of snow, which can reduce visibility to only a few feet.

Sleet and Freezing Rain   Consider the falling snowflake in Fig. 5.22. As it falls into warmer air, it begins to melt. When it falls through the deep subfreezing surface layer of air, the partially melted snowflake or cold raindrop turns back into ice, not as a snowflake, but as a tiny transparent (or translucent) *ice pellet* called **sleet**.* Generally, these ice pellets bounce when striking the ground and produce a tapping sound when they hit a window or piece of metal.

*Occasionally, the news media incorrectly use the term sleet to represent a mixture of rain and snow. This, however, is the British meaning.

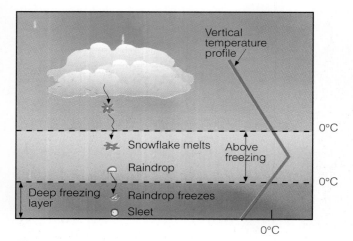

Figure 5.22
Sleet forms when a partially melted snowflake or a cold raindrop freezes into a pellet of ice before reaching the ground.

The cold surface layer beneath a cloud may be too shallow to freeze raindrops as they fall. In this case, they reach the surface as supercooled liquid drops. Upon striking a cold object, the drops spread out and almost immediately freeze, forming a thin veneer of ice. This form of precipitation is called **freezing rain**, *glaze*, or *silver thaw*. If the drops are quite small, the precipitation is called *freezing drizzle*.

Freezing rain can create a beautiful winter wonderland by coating everything with silvery, glistening ice. At the same time, highways turn into skating rinks for automobiles, and the destructive weight of the ice—which can be many tons on a single tree—breaks tree branches, power lines, and telephone cables. (See Fig. 5.23.) The area most frequently hit by these storms extends over a broad region from Texas into Minnesota and eastward into the middle Atlantic states and New England. Such storms are extremely rare in southern California and Florida. (For additional information on freezing rain and its effect on aircraft, read the Focus section on p. 125.)

Snow Grains and Snow Pellets   **Snow grains** are small, opaque grains of ice, the solid equivalent of drizzle. They fall in small quantities from stratus clouds, and never in the form of a shower. Upon striking a hard surface, they neither bounce nor shatter. **Snow pellets**, on the other hand, are white, opaque grains of ice about the size of an average raindrop. They are sometimes confused with snow grains. The distinction is easily made, however, by remembering that, unlike snow grains, snow pellets are brittle, crunchy, and bounce (or break apart) upon hitting a hard surface. They usually fall as showers, especially from cumulus congestus clouds. Snow pellets form as ice crystals collide with supercooled water droplets that freeze into a spherical aggregate of icy matter (rime) containing many air spaces. When the ice particles accumulate so much rime that it can no longer be recognized as an ice crystal (or snowflake), it is called *graupel*. If the graupel reaches the surface as a light round clump of snowlike ice, we call it a snow pellet.

Focus on an Application
Aircraft Icing

Consider an aircraft flying through an area of freezing rain. As the large, supercooled drops strike the leading edge of the wing, they break apart and form a film of water, which quickly freezes into a solid sheet of ice. This smooth, transparent ice—called *clear ice*—is similar to the glaze that coats trees during ice storms. Clear ice can build up quickly; it is heavy and difficult to remove, even with modern de-icers.

When an aircraft flies through a cloud composed of tiny, supercooled liquid droplets, *rime ice* may form. Rime ice forms when some of the cloud droplets strike the wing and freeze before they have time to spread, thus leaving a rough and brittle coating of ice on the wing. Because the small, frozen droplets trap air between them, rime ice usually appears white. Even though rime ice redistributes the flow of air over the wing more than clear ice does, it is lighter in weight and is more easily removed with de-icers.

Because the raindrops and cloud droplets in most clouds vary in size, a mixture of clear and rime ice usually forms on aircraft. Also, because concentrations of liquid water tend to be greatest in warm air, icing is usually heaviest and most severe when the air temperature is just below freezing.

A major hazard to aviation, icing reduces aircraft efficiency by increasing weight. Icing has other adverse effects, depending on where it forms. On a wing or fuselage, ice can disrupt the air flow and decrease the plane's flying capability. When ice forms in the air intake of the engine, it robs the engine of air, causing a reduction in power. Icing may also affect the operation of brakes, landing gear, and instruments. Because of the hazards of ice on an aircraft, its wings are usually sprayed with a type of antifreeze before taking off during cold, inclement weather.

Figure 5.23
A heavy coating of freezing rain during this ice storm caused tree limbs to break and power lines to sag.

Did you know?
What is believed to be the world's heaviest hailstone (weighing in at nearly 11 pounds) fell in the Guangxi region of China on May 1, 1986.

Hail **Hailstones** are pieces of ice either transparent or partially opaque, ranging in size from that of small peas to that of golf balls or larger. Some are round, others take on irregular shapes. The largest authenticated hailstone in the United States fell on Coffeyville, Kansas, in September, 1970. (See Fig. 5.24.) This giant weighed over one and one-half pounds and had a measured diameter of five and one-half inches. Canada's record hailstone fell on Cedoux, Saskatchewan, during August, 1973. It weighed over one-half pound and measured about four inches in diameter. Needless to say, large hailstones are quite destructive as they can break windows, dent cars, batter roofs of homes, and cause extensive damage to livestock and crops. In fact, a single hailstorm can destroy a farmer's crop in a matter of minutes.

Estimates are that, in the United States alone, hail accounts for some $700 million in damage annually. Although hailstones are potentially lethal, only two fatalities due to falling hail have been documented in the United States during this century.

Hail is produced in a cumulonimbus cloud when graupel, large frozen raindrops, or just about any particles (even insects) act as embryos that grow by accumulating supercooled liquid droplets—accretion. For a hailstone to grow to a golfball size, it must remain in the cloud for between five and ten minutes. Violent, upsurging air currents within the cloud carry small embryos high above the freezing level. When the updrafts are tilted, the embryos are swept laterally through the cloud. As the embryos pass through regions of varying liquid water content, a coating of ice forms around them and they grow larger and larger. When the ice particles are appreciable size, they become too large and heavy to be supported by the rising air, and they then begin to fall as hail. As they slowly descend, the hailstones may get caught in a violent updraft (Fig. 5.25) only to be carried upward once again to repeat the cycle. Or, they may fall through the cloud and begin to melt in the warmer air below. Small hailstones often melt before reaching the ground, but, in the violent thunderstorms of summer, hailstones may grow large enough to reach the ground before completely melting. Strangely, then, the largest form of frozen precipitation occurs during the warmest time of the year.

As the cumulonimbus cloud moves along, it may deposit its hail in a long, narrow band known as a *hailstreak*. If the cloud should remain almost stationary for a period of time, substantial accumulation of hail is

Figure 5.24
The Coffeyville hailstone. This giant hailstone—the largest ever reported in the United States—fell on the community of Coffeyville, Kansas, on September 3, 1970. The layered structure of the hailstone reveals that it traveled through a cloud of varying water content and temperature.

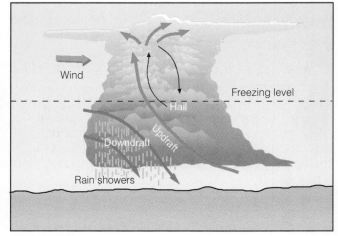

Figure 5.25
The formation of hailstones. The violent updrafts in this cumulonimbus cloud keep ice particles suspended in the cloud. The ice particles collide with supercooled liquid droplets, which freeze on contact. The ice particle eventually grows large enough and heavy enough to fall toward the ground as a hailstone.

possible. For example, in June, 1984, a devastating hailstorm lasting over an hour dumped knee-deep hail on the suburbs of Denver, Colorado. In addition to its destructive effect, accumulation of hail on a roadway is a hazard to traffic. For example, four people lost their lives near Soda Springs, California, in a 15-vehicle pileup on a hail-covered freeway in September, 1989.

Because hailstones are so damaging, various methods have been tried to prevent them from forming in thunderstorms. One method employs the seeding of clouds with large quantities of silver iodide. These nuclei freeze supercooled water droplets and convert them into ice crystals. The ice crystals grow larger as they come in contact with additional supercooled cloud droplets. In time, the ice crystals grow large enough to be called graupel, which then becomes a hailstone embryo. Large numbers of embryos are produced by seeding in hopes that competition for the remaining supercooled droplets may be so great that none of the embryos would be able to grow into large and destructive hailstones. Russian scientists claim great success in suppressing hail using ice nuclei, such as silver iodide and lead iodide. However, their experiments are performed in such a way that statistical evaluation is not possible. In the United States, the results of most hail-suppression experiments are still inconclusive.

Measuring Precipitation

Instruments   A **standard rain gauge** is commonly used to measure rainfall. This instrument consists of a funnel-shaped collector attached to a long measuring tube (Fig. 5.26). The cross-sectional area of the collector is ten times that of the tube. Hence, rain falling into the collector is amplified tenfold in the tube, permitting measurements of great precision—to as low as one-hundredth of an inch. An amount less than this is called a **trace**.

Another instrument that measures rainfall is the *tipping bucket rain gauge*. In Fig. 5.27, notice that this gauge has a receiving funnel leading to two small metal collectors (buckets). The bucket beneath the funnel collects the rain water. When it accumulates the equivalent of one-hundredth of an inch of rain, the weight of the water causes it to tip and empty itself. The second bucket immediately moves under the funnel to catch

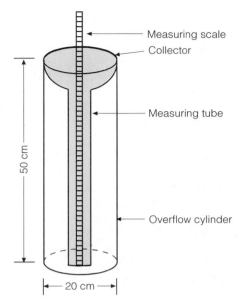

Figure 5.26
Components of the standard rain gauge.

the water. When it fills, it also tips and empties itself, while the original bucket moves back beneath the funnel. Each time a bucket tips, an electric contact is made, causing a pen to register a mark on a remote recording chart. Adding up the total number of marks gives the rainfall for a certain time period.

Remote recording of precipitation can also be made with a *weighing-type rain gauge*. With this gauge, precipitation is caught in a cylinder and accumulates in a bucket. The bucket sits on a sensitive weighing platform. Special gears translate the accumulated weight of rain or snow into millimeters or inches of precipitation. The precipitation totals are recorded

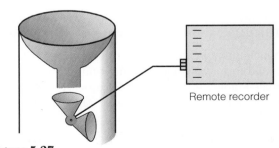

Figure 5.27
The tipping bucket rain gauge. Each time the bucket fills with 0.01 inch of rain, it tips, sending an electric signal to the remote recorder.

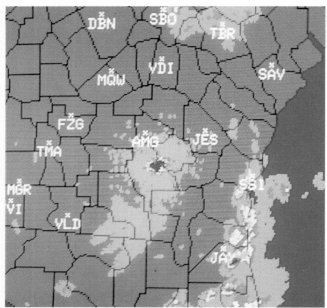

Figure 5.28
A color radar image of precipitation over southern Georgia. (Green represents lighter precipitation while yellow and red represent heavier precipitation.)

by a pen on chart paper, which covers a clock-driven drum. By using special electronic equipment, this information can be transmitted from rain gauges in remote areas to satellites or land-based stations, thus providing precipitation totals from previously inaccessible regions.

The depth of snow in a region is determined by measuring its depth at three or more representative areas. The amount of snowfall is defined as the average of these measurements. Snow depth may also be measured by removing the collector and inner cylinder of a standard rain gauge and allowing snow to accumulate in the outer tube. Generally, about ten inches of snow will melt down to about one inch of water, given a typical fresh snowpack a **water equivalent** of 10: 1. This ratio, however, will vary greatly, depending on the texture and packing of the snow.

Weather Radar and Precipitation Radar (*ra*dio *d*etection *a*nd *r*anging) has become an essential tool of the atmospheric scientist, for it gathers information about storms and precipitation in previously inaccessible regions. Atmospheric scientists use weather radar to examine the inside of a cloud much like physicians use X-rays to examine the inside of a human body. Essentially, the radar unit consists of a transmitter that sends out short, powerful microwave pulses. When this energy encounters a foreign object—called a *target*—a fraction of the energy is scattered back toward the transmitter and is detected by a receiver. The returning signal is amplified and displayed on a screen, producing an image or "echo" from the target. The elapsed time between transmission and reception indicates the target's distance.

The brightness of the echo is directly related to the amount (intensity) of rain falling in the cloud. So, the radar screen shows not only where precipitation is occurring, but also how intense it is. In recent years, the radar image has been displayed using various colors to denote the intensity of precipitation within the range of the radar unit (Fig. 5.28). Some television weather shows use this observation technique and call it "color radar."

Not only does radar provide information on precipitation, it also allows scientists to peer into a tornado-generating thunderstorm. We will investigate the special type of radar—called *Doppler radar*—used for this purpose in Chapter 10 when we consider the formation of severe thunderstorms and tornadoes.

Summary

In this chapter, we tied together the concepts of stability, cloud formation, and precipitation. We learned that because stable air tends to resist upward vertical motions, clouds forming in a stable atmosphere often spread horizontally and have a stratified appearance. Stable air may be caused by either the surface air being cooled or the air aloft being warmed.

An unstable atmosphere tends to favor vertical air currents and produce cumuliform clouds. Instability may be brought on by either the surface air being warmed or the air aloft being cooled. In a conditionally unstable atmosphere, rising unsaturated air may be lifted to a level where condensation begins, latent heat is released, and instability results.

We looked at cloud droplets and found that, individually, they are too small and light to reach the ground as rain. They can grow in size as large cloud droplets, falling through a cloud, collide and merge with smaller droplets in their path. In clouds where the temperature is below freezing, ice crystals can grow larger at the expense of the surrounding liquid droplets. As an ice crystal begins to fall, it may grow larger by colliding with liquid droplets, which freeze on contact. In an attempt to coax more precipitation from them, some clouds are seeded.

We examined the various forms of precipitation, from raindrops that freeze on impact (producing freezing rain) to raindrops that freeze into tiny ice pellets called sleet. We learned that strong updrafts in a cumulonimbus cloud may carry ice particles high above the freezing level, where they acquire a further coating of ice and form destructive hailstones. We looked at instruments and found that although the rain gauge is still the most commonly used method of measuring precipitation, radar has recently become an important tool for determining precipitation intensity.

Key Terms

The following terms are listed in the order they appear in the text. Define each. Doing so will aid you in reviewing the material covered in this chapter.

adiabatic process	orographic uplift	rain	sleet
dry adiabatic rate	rain shadow	drizzle	freezing rain
moist adiabatic rate	coalescence	virga	snow grains
environmental lapse rate	supercooled (water	shower (rain)	snow pellets
absolutely stable atmosphere	droplet)	snow	hailstones
absolutely unstable atmosphere	ice nuclei	fall streaks	standard rain gauge
condensation level	ice-crystal process	flurries (of snow)	trace (of precipitation)
conditionally unstable	accretion	snow squall	water equivalent
atmosphere	cloud seeding	blizzard	radar

Review Questions

1. Define each of the following:
   (a) dry adiabatic rate
   (b) moist adiabatic rate
   (c) environmental lapse rate
2. Why are the moist and dry adiabatic rates of cooling different?
3. How can the atmosphere be made more stable? More unstable?
4. If the atmosphere is conditionally unstable, what does this mean?
5. Explain why an inversion represents an extremely stable atmosphere.
6. What type of clouds would you most likely expect to see in stable air? What about unstable air?
7. Why are cumulus clouds more frequently observed during the afternoon?
8. There are usually large spaces of blue sky between cumulus clouds. Explain why this is so.
9. Why do most thunderstorms have flat tops?
10. List the four primary ways in which clouds form.
11. Explain why rain shadows form on the leeward side of mountains.
12. On which side of a mountain (windward or leeward) would lenticular clouds most likely form?

13. What is the primary difference between a cloud droplet and a raindrop?
14. Describe how the process of collision-and-coalescence produces rain.
15. How does the ice-crystal process produce precipitation? What is the *main* premise behind this process?
16. Explain the main principal behind cloud seeding.
17. How does rain differ from drizzle?
18. Why do heavy showers usually fall from cumuliform clouds? Why does steady precipitation normally fall from stratiform clouds?
19. Why is it *never* too cold to snow?
20. How would you be able to distinguish between virga and fall streaks?
21. What is the difference between freezing rain and sleet?
22. How do the atmospheric conditions that produce sleet differ from those that produce hail?
23. Describe how a standard rain gauge measures precipitation.
24. How does radar measure the intensity of precipitation?

Sharp variations in air pressure produce strong winds that sweep wispy cirrus clouds across the sky. (Photo: Ross DePaola)

Chapter 6

Air Pressure and Winds

Contents

▲▼▲

December 19, 1980, was a cool day in Lynn, Massachusetts, but not cool enough to dampen the spirits of more than 2000 people who gathered in Central Square—all hoping to catch at least one of the 1500 dollar bills that would be dropped from a small airplane at noon. Right on schedule, the aircraft circled the city and dumped the money onto the people below. However, to the dismay of the onlookers, a westerly wind caught the currency before it reached the ground and carried it out over the cold Atlantic Ocean. Had the pilot or the sponsoring leather manufacturer examined the weather charts beforehand, they might have been able to predict that the wind would ruin their advertising scheme.

▲

6 This opening scenario raises two questions: (1) Why does the wind blow? and (2) How can one tell its direction by looking at weather charts? Chapter 1 has already answered the first question: Air moves in response to horizontal differences in pressure. This happens when we open a vacuum-packed can—air rushes from the higher pressure region outside the can toward the region of lower pressure inside. In the atmosphere, the wind blows in an attempt to equalize imbalances in air pressure. Does this mean that the wind always blows directly from high to low pressure? Not really, because the movement of air is controlled not only by pressure differences but by other forces as well. In this chapter, we will consider the forces that influence atmospheric motions at the surface and aloft. Through studying these forces, we will be able to tell how the wind should blow in a particular region by examining surface and upper-air charts.

Because the wind is controlled by variations in air pressure, we will look at this weather element first.

▲▼▲
Atmospheric Pressure

In Chapter 1, we learned several important concepts about atmospheric pressure. One stated that **air pressure** is simply the weight of air above a given level. As we climb in elevation above the earth's surface, there are fewer air molecules above us; hence, atmospheric pressure always decreases with increasing height. Another concept we learned was that most of our atmosphere is crowded close to the earth's surface, which causes air pressure to decrease with height, rapidly at first, then more slowly at higher altitudes. Keep these concepts in mind as you read the remainder of this section.

To help eliminate some of the complexities of the atmosphere, scientists construct *models*. Figure 6.1 shows a simple atmospheric model—a column of air, extending well up into the atmosphere. In the column, the dots represent air molecules. Our model assumes several things:

1. that the air molecules are not crowded close to the surface and, hence, the air density remains constant from the surface up to the top of the column
2. the width of the column does not change

Suppose we somehow force more air into the column in Fig. 6.1. What would happen? If the air temperature in the column does not change, the added air would make the column more dense, and the added weight of the air in the column would increase the surface air pressure. Likewise, if a great deal of air were removed from the column, the surface air pressure would decrease. So the surface air pressure can be changed by changing the amount of air above the surface.

Suppose the two air columns in Figure 6.2a are located at the same elevation and have identical surface air pressures. This condition, of course, means that there must be the same number of molecules (same mass of air) in each column above both cities. Further suppose that the surface air pressure for both cities remains the same, while the air above city 1 cools and the air above city 2 warms (Figure 6.2b).

As the air in column 1 cools, the molecules move more slowly and crowd closer together—the air becomes more dense. In the warm air above city 2, the molecules move faster and spread farther apart—the air becomes less dense. If the width of the columns does not change (and if we assume an invisible barrier exists between the columns) then to keep the surface pressure from varying, the total number of molecules above each city must remain the same. In the more

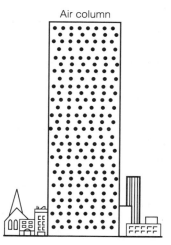

Figure 6.1
A model of the atmosphere where air density remains constant with height. The air pressure at the surface is related to the number of molecules above. When air of the same temperature is stuffed into the column, the surface air pressure rises. When air is removed from the column, the surface pressure falls.

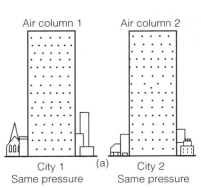
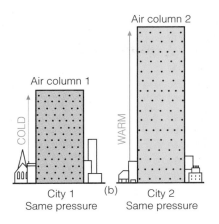
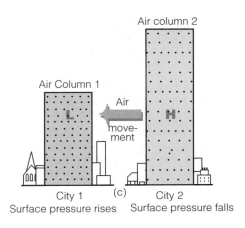

Figure 6.2
It takes a shorter column of cold air to exert the same pressure as a taller column of warm air. Because of this fact, cold air aloft is associated with low pressure and warm air aloft with high pressure. The pressure differences aloft create a force that causes the air to move from a region of higher pressure toward a region of lower pressure. The removal of air from column 2 causes its surface pressure to drop, whereas the addition of air into column 1 causes its surface pressure to rise. (The difference in height between the two columns is greatly exaggerated.)

dense, cold air above city 1, the column shrinks, while the column rises in the less dense, warm air above city 2.

We now have a cold, shorter column of air above city 1 and a warm, taller air column above city 2. From this situation, we can conclude that *it takes a shorter column of cold, dense air to exert the same surface pressure as a taller column of warm, less dense air.* This concept has a great deal of meteorological significance.

Atmospheric pressure decreases more rapidly with elevation in the cold column of air. In the cold air above city 1 (Fig. 6.2b), move up the column and observe how quickly you pass through the densely packed molecules. This activity indicates a rapid change in pressure. In the warmer, less dense air, the pressure does not decrease as rapidly with height, simply because you climb above fewer molecules in the same vertical distance.

In Fig. 6.2c move up the warm column until you come to the letter H. Now move up the cold column the same distance until you reach the letter L. Notice that there are more molecules above the letter H in the warm column than above the letter L in the cold column. The fact that the number of molecules above any level is a measure of the atmospheric pressure leads to an important concept: *Warm air aloft is normally associated with high atmospheric pressure, and cold air aloft is associated with low atmospheric pressure.*

In Fig. 6.2c, the horizontal difference in temperature creates a horizontal difference in pressure. The pressure difference establishes a force (called the *pressure gradient force*) that causes the air to move from higher pressure toward lower pressure. Consequently, if we remove the invisible barrier between the two columns and allow the air aloft to move horizontally, the air will move from column 2 toward column 1. As the air aloft leaves column 2, the weight of the air in the column decreases, and so does the surface air pressure. Meanwhile, the accumulation of air in column 1 causes the surface air pressure to increase.

In summary, heating or cooling a column of air can establish horizontal variations in pressure that cause the air to move. The net accumulation of air above the surface causes the surface air pressure to rise, whereas a decrease in the amount of air above the surface causes the surface air pressure to fall.

▲▼▲
Measuring Air Pressure

Up to this point, we have described air pressure as the weight of the atmosphere above any level. We can also define air pressure as the *force exerted by the air molecules over a given area.* Billions of air molecules constantly push on the human body. This force is exerted equally in all directions. We are not crushed by the

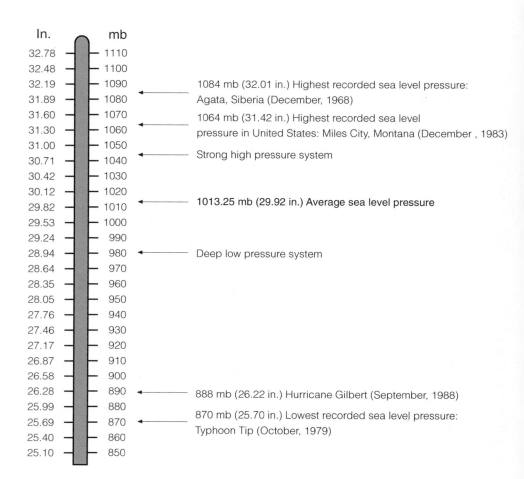

Figure 6.3
Atmospheric pressure in inches of mercury and in millibars.

In. / mb

1084 mb (32.01 in.) Highest recorded sea level pressure: Agata, Siberia (December, 1968)

1064 mb (31.42 in.) Highest recorded sea level pressure in United States: Miles City, Montana (December , 1983)

Strong high pressure system

1013.25 mb (29.92 in.) Average sea level pressure

Deep low pressure system

888 mb (26.22 in.) Hurricane Gilbert (September, 1988)

870 mb (25.70 in.) Lowest recorded sea level pressure: Typhoon Tip (October, 1979)

force because billions of molecules inside the body push outward just as hard.

Even though we do not actually feel the constant bombardment of air, we can detect quick changes in it. For example, if we climb rapidly in elevation our ears may "pop." This experience happens because air collisions outside the eardrum lessen as the air pressure decreases. The popping comes about as air collisions between the inside and outside of the ear equalize.

Instruments that detect and measure pressure changes are called *barometers*, which literally means an instrument that measures bars. In meteorology, the *bar* is a unit of pressure that describes a force over a given area.* Because the bar is a relatively large unit,

and because surface pressure changes are normally small, the unit of pressure most commonly found on surface weather maps is the **millibar** (mb), where one millibar is equal to one-thousandth of a bar. A common pressure unit used in aviation and on television and radio weather broadcasts is *inches of mercury* (Hg). At sea level, the *average* or **standard** value for **atmospheric pressure** is

$$1013.25 \text{ mb} = 29.92 \text{ in. Hg} = 76 \text{ cm.}$$

Figure 6.3 compares pressure readings in millibars and in inches of mercury. An understanding of how the unit "inches of mercury" is obtained is found in the following section on barometers.

Barometers   Because we measure atmospheric pressure with an instrument called a **barometer**, atmospheric pressure is also referred to as *barometric pressure*. Evangelista Torricelli, a student of Galileo, in-

*By definition, a bar is a force of 100,000 newtons acting on a surface area of 1 square meter. A *newton* is the amount of force required to move an object with a mass of 1 kilogram so that it increases its speed at a rate of 1 meter per second each second.

vented the **mercury barometer** in 1643. His barometer, similar to those used today, consisted of a long glass tube open at one end and closed at the other (Fig. 6.4). Removing air from the tube and covering the open end, Torricelli immersed the lower portion into a dish of mercury. He removed the cover, and the mercury rose up the tube to nearly thirty inches above the level in the dish. Torricelli correctly concluded that the column of mercury in the tube was balancing the weight of the air above the dish, and, hence, its height was a measure of atmospheric pressure.

The most common type of home barometer—the **aneroid barometer**—contains no fluid. Inside this instrument is a small, flexible metal box called an aneroid cell. Before the cell is tightly sealed, air is partially removed, so that small changes in external air pressure cause the cell to expand or contract. The size of the cell is calibrated to represent different pressures, and any change in its size is amplified by levers and transmitted to an indicating arm, which points to the current atmospheric pressure (Fig. 6.5).

Notice that the aneroid barometer often has descriptive weather-related words printed above specific pressure values. These adjectives indicate the most likely weather conditions when the needle is pointing to that particular pressure reading. Generally, the higher the reading, the more likely clear weather will occur, and the lower the reading, the better are the chances for inclement weather. This is because surface high pressure areas are associated with sinking air and normally fair weather, whereas surface low pressure areas are associated with rising air and usually cloudy, wet weather.

The *altimeter* and *barograph* are two types of aneroid barometers. Altimeters are aneroid barometers that measure pressure, but are calibrated to indicate altitude. Barographs are recording aneroid barometers. Basically, the barograph consists of a pen attached to an indicating arm that marks a continuous record of pressure on chart paper. The chart paper is attached to a drum rotated slowly by an internal mechanical clock (Fig. 6.6).

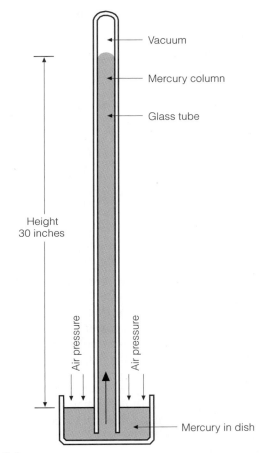

Figure 6.4
The mercury barometer. The height of the mercury column is a measure of atmospheric pressure.

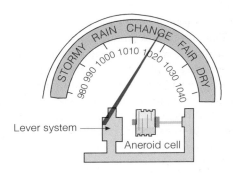

Figure 6.5
The aneroid barometer.

Figure 6.6
A recording barograph.

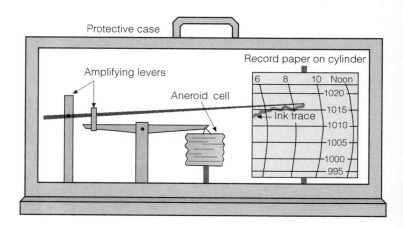

Pressure Readings The seemingly simple task of reading the height of the mercury column to obtain the air pressure is actually not all that simple. Being a fluid, mercury is sensitive to changes in temperature; it will expand when heated and contract when cooled. Consequently, to obtain accurate pressure readings without the influence of temperature, all mercury barometers are corrected as if they were read at the same temperature. Because the earth is not a perfect sphere, the force of gravity is not a constant. Since small, gravity differences influence the height of the mercury column, they must be considered when reading the barometer. Finally, each barometer has its own "built-in" error, called *instrument error*, which is caused, in part, by the surface tension of the mercury against the glass tube. After being corrected for temperature, gravity, and instrument error, the barometer reading at a particular location and elevation is termed **station pressure**.

Figure 6.7a gives the station pressure measured at four locations only a few hundred miles apart. The different station pressures of the four cities is due primarily to the cities being at different altitudes. This fact becomes even clearer when we realize that atmospheric pressure changes much more quickly when we move upward than it does when we move sideways. A small vertical difference between two observation sites can yield a large difference in station pressure. Thus, to properly monitor horizontal changes in pressure, barometer readings must be corrected for altitude.

Altitude corrections are made so that a barometer reading taken at one elevation can be compared with a barometer reading taken at another. Station pressure observations are normally adjusted to a level of mean sea level—the level representing the average surface of the ocean, which yields a reading called **sea level pressure**.

Near the earth's surface, atmospheric pressure decreases on the average by about 10 millibars for every 100 meters increase in elevation (about 1 inch of mercury for each 1000-foot rise). Notice in Fig. 6.7a that city A has a station pressure of 952 mb. Notice also that city A is 600 meters above sea level. Adding 10 millibars per 100 meters to its station pressure yields a sea level pressure of 1012 mb (Fig. 6.7b). After all the station pressures are adjusted to sea level (Fig. 6.7b), we are able to see the horizontal variations in sea level pressure—something we were not able to see from the station pressures alone in Fig. 6.7a.

When more pressure data are added (Fig. 6.7c), the chart can be analyzed and the pressure pattern visualized. **Isobars** (lines connecting points of equal pressure) are drawn at intervals of 4 mb, with 1000 mb being the base value. Note that the isobars do not pass through each point, but, rather, between many of them, with the exact values being interpolated from the data given on the chart. For example, follow the 1008-mb line from the top of the chart southward and observe that there is no plotted pressure of 1008 mb. The 1008-mb isobar, however, comes closer to the station with a sea level pressure of 1007 mb than it does to the station with a pressure of 1010 mb. With its isobars, the bottom chart (Fig. 6.7c) is now called a *sea level pressure chart* or simply a *surface map*.

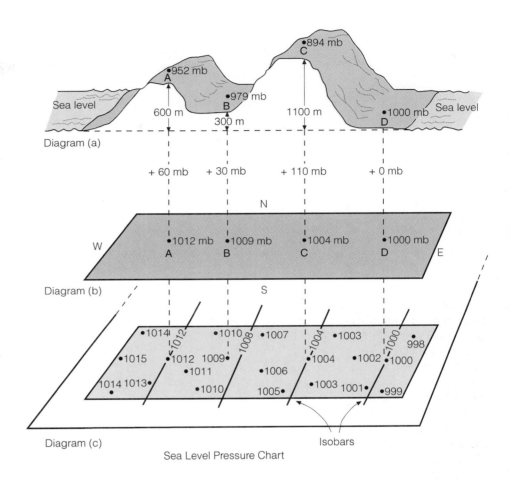

Figure 6.7
The top diagram (a) shows four cities (A, B, C, and D) at varying elevations above sea level, all with different station pressures. The middle diagram (b) represents sea level pressures of the four cities plotted on a sea level chart. The bottom diagram (c) shows isobars drawn on the chart (dark lines) at intervals of 4 millibars.

▲▼▲

Surface and Upper-Air Charts

Figure 6.8a is a simplified surface map that shows areas of high and low pressure and arrows that indicate wind direction—the direction from which the wind is blowing. The large H's on the map indicate the centers of high pressure, which are also called **anticyclones**. The large L's represent centers of low pressure, also known as *depressions* or **mid-latitude cyclones** because they form in the middle latitudes, outside of the tropics. The solid lines are isobars with units in millibars. Notice that the surface winds tend to blow across the isobars toward regions of lower pressure. In fact, as we briefly observed in Chapter 1, in the Northern Hemisphere the winds blow counterclockwise and inward toward the center of the lows and clockwise and outward from the center of the highs.

Figure 6.8b shows an upper-air chart for the same day as the surface map in Fig. 6.8a. The upper-air map is a constant pressure chart because it is constructed to show height variations along a constant pressure (isobaric) surface. That is why these maps are also known as **isobaric maps**. This particular isobaric map shows height variations at a pressure level of 500 millibars (which is about 5600 meters or 18,000 feet above sea level). Hence, this map is called a *500-millibar map*. The solid lines on the map are **contour lines**—lines that connect points of equal elevation above sea level. Although contour lines are height lines, they illustrate pressure much like isobars do. Consequently, *contour lines of low height represent a region of lower pressure*, and *contour lines of high height represent a region of higher pressure*.

Notice on the 500-millibar map (Fig. 6.8b) that the contour lines typically decrease in value from south to

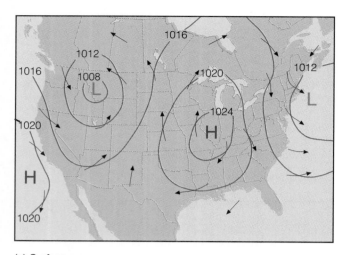

(a) Surface map

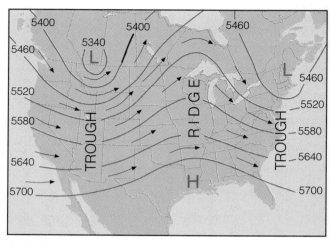

(b) Upper-air map (500 millibars)

Figure 6.8
(a) Surface map showing areas of high and low pressure. The solid lines are isobars and the arrows represent wind direction. (b) Upper-level map (for the same day as the surface map) at the 500-millibar level, which is about 5600 meters or 18,000 feet above sea level. Solid lines on the map are contour lines in meters above sea level. Arrows show wind direction. Length of arrows represent relative wind speed.

north. The reason for this fact is that, in the Northern Hemisphere, we normally find warmer air to the south and colder air to the north. (Recall from our earlier discussion that cold air aloft is associated with low pressure and warm air aloft with high pressure.) The lines, however, are not straight; they bend and turn, indicating **ridges** (*elongated highs*) where the air is warmer and indicating depressions, or **troughs** (*elongated lows*), where the air is colder.

The arrows on the 500-millibar map show the wind direction. Notice that, unlike the surface winds that cross the isobars in Fig. 6.8a, the winds on the 500-mb chart tend to flow parallel to the contour lines in a more or less west-to-east direction. Observe that where the contour lines are close together, the winds (as indicated by the length of the arrows) are stronger. And where the lines are farther apart, the winds are weaker. Later in this chapter, we learn why this is so.

Surface and upper-air charts are valuable tools for the meteorologist. Surface maps describe where the centers and high and low pressure are found, as well as the winds and weather associated with these systems. Upper-air charts, on the other hand, are extremely important in forecasting the weather. The upper-level winds not only determine the movement of

surface pressure systems but, as we will see in Chapter 8, they determine whether these surface systems will intensify or weaken.

At this point, however, our interest lies mainly in the movement of air. Consequently, now that we have looked at surface and upper-air maps, we will use them to study why the wind blows the way it does, at the surface and aloft.

▲▼▲

Why the Wind Blows

Our understanding of why the wind blows stretches back through several centuries, with many scientists contributing to our knowledge. When we think of the movement of air, however, one great scholar stands out—Isaac Newton (1642–1727), who formulated several fundamental laws of motion.

Newton's Laws of Motion   Newton's first law of motion states that *an object at rest will remain at rest and an object in motion will remain in motion (and travel at a constant velocity along a straight line) as long as no force is exerted on the object.* For example, a baseball in a pitcher's hand will remain there until a force (a push) acts upon the ball. Once the ball is pushed (thrown), it would continue to move in that direction forever if it were not for the force of air friction (which slows it down), the force of gravity (which pulls it toward the ground), and the catcher's mitt (which exerts an equal but opposite force to bring it to a halt). Similarly, to start air moving, to speed it up, to slow it down, or even to change its direction requires the action of an external force. This brings us to Newton's second law.

Newton's second law states that *the force exerted on an object equals its mass times the acceleration produced.* In symbolic form, this law is written as

$$F = ma.$$

From this relationship we can see that, when the mass of an object is constant, the force acting on the object is directly related to the acceleration that is produced. A force in its simplest form is a push or a pull. Acceleration is the speeding up, the slowing down, or the changing of direction of an object. (More precisely, acceleration is the change of velocity over a period of time.)

Because more than one force may act upon an object, Newton's second law always refers to the *net*, or total, force that results. An object will always accelerate in the direction of the total force acting on it. Therefore, to determine in which direction the wind will blow, we must identify and examine all of the forces that affect the horizontal movement of air. These forces include:

1. pressure gradient force
2. Coriolis force
3. friction

We will first study the forces that influence the flow of air aloft. Then we will see which forces modify winds near the ground.

Forces That Influence the Wind   We already know from an earlier discussion that horizontal differences in atmospheric pressure cause air to move and, hence, the wind to blow. Since air is an invisible gas, it may be easier to see how pressure differences cause motion if we examine a visible fluid, such as water.

In Fig. 6.9, the two large tanks are connected by a pipe. Tank A is two-thirds full and tank B is only one-half full. Since the water pressure at the bottom of each

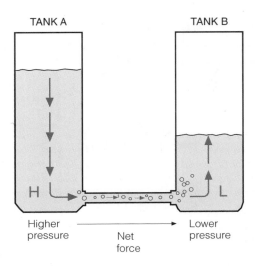

Figure 6.9
The higher water level creates higher fluid pressure at the bottom of tank A and a net force directed toward the lower fluid pressure at the bottom of tank B. This net force causes water to move from higher pressure toward lower pressure.

tank is proportional to the weight of water above, the pressure at the bottom of tank A is greater than the pressure at the bottom of tank B. Moreover, since fluid pressure is exerted equally in all directions, there is a greater pressure in the pipe directed from tank A toward tank B than from B toward A.

Since pressure is force per unit area, there must also be a net force directed from tank A toward tank B. This force causes the water to flow from left to right, from higher pressure toward lower pressure. The greater the pressure difference, the stronger the force, and the faster the water moves. In a similar way, horizontal differences in atmospheric pressure cause air to move.

Pressure Gradient Force   Figure 6.10 shows a region of higher pressure on the map's left side, lower pressure on the right. The isobars show how the horizontal pressure is changing. If we compute the amount of pressure change that occurs over a given distance, we have the **pressure gradient**:

$$\text{Pressure gradient} = \frac{\text{difference in pressure}}{\text{distance}}.$$

In Fig. 6.10, the pressure gradient between points 1 and 2 is 4 millibars per 100 kilometers.

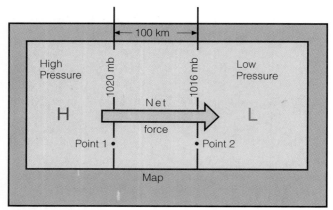

Figure 6.10
The pressure gradient between point 1 and point 2 is 4 millibars per 100 kilometers. The net force directed from higher toward lower pressure is the pressure gradient force.

Suppose the pressure in Fig. 6.10 were to change, and the isobars become closer together. This condition would produce a rapid change in pressure over a relatively short distance, or what is called a *steep* (or *strong*) *pressure gradient*. However, if the pressure were to change such that the isobars spread farther apart, then the difference in pressure would be small over a relatively large distance. This condition is called a *gentle* (or *weak*) *pressure gradient*.

Notice in Fig. 6.10 that when differences in horizontal air pressure exist there is a net force acting on

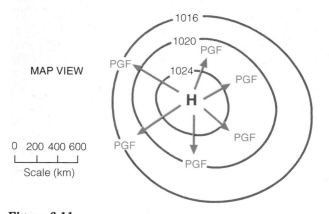

Figure 6.11
The closer the spacing of the isobars, the greater the pressure gradient. The greater the pressure gradient, the stronger the pressure gradient force (*PGF*). The arrows represent the relative magnitude of the force, which is always directed from higher toward lower pressure.

the air. This force, called the **pressure gradient force** (*PGF*), is *directed from higher toward lower pressure at right angles to the isobars.* The magnitude of the force is directly related to the pressure gradient. Steep pressure gradients correspond to strong pressure gradient forces and vice versa. Figure 6.11 shows the relationship between pressure gradient and pressure gradient force.

The *pressure gradient force* causes the wind to blow. Because of this, closely spaced isobars on a weather chart indicate steep pressure gradients, strong forces, and high winds. On the other hand, widely spaced isobars indicate gentle pressure gradients, weak forces, and light winds.

If the *PGF* were the only force acting upon air, we would always find winds blowing directly from higher toward lower pressure. However, the moment air starts to move, it is deflected in its path by the *Coriolis force*.

Did you know?
When you drive along a highway (at the speed limit), the Coriolis force would "pull" your car about 1500 feet to the right for every 100 miles you travel, if it were not for the friction between your tires and the road surface.

Coriolis Force   The Coriolis force describes an apparent force that is due to the rotation of the earth. To understand how it works, consider two people playing catch as they sit opposite one another on the rim of a merry-go-round (Fig. 6.12, platform A). If the merry-go-round is not moving, each time the ball is thrown, it moves in a straight line to the other person.

Suppose the merry-go-round starts turning counterclockwise—the same direction the earth spins as viewed from above the North Pole. If we watch the game of catch from above, we see that the ball moves in a straight-line path just as before. However, to the people playing catch on the merry-go-round, the ball seems to veer to its right each time it is thrown, always landing to the right of the point intended by the thrower (Fig. 6.12, platform B). This is due to the fact that, while the ball moves in a straight-line path, the merry-go-round rotates beneath it; by the time the ball reaches the opposite side, the catcher has moved. To anyone on the merry-go-round, it seems as if there is some force causing the ball to deflect to the right. This apparent force is called the **Coriolis force** after Gaspard Coriolis, a nineteenth-century French scientist who worked it out mathematically. (Because it is an *apparent force* due to the rotation of the earth, it is also called the *Coriolis effect*.) This effect occurs on the rotating earth, too. All free-moving objects, such as ocean currents, aircraft, artillery projectiles, and air molecules seem to deflect from a straight-line path because the earth rotates under them.

The Coriolis force *causes the wind to deflect to the right of its path in the Northern Hemisphere and to the left of its path in the Southern Hemisphere*. To illustrate this, consider a satellite in polar circular orbit. If the earth were not rotating, the path of the satellite would be observed to move directly north-south, parallel to the earth's meridian lines. However, the earth *does* rotate, carrying us and meridians eastward with it. Because of this rotation, in the Northern Hemisphere we see the satellite moving southwest instead of due south; it seems to veer off its path and move toward *its right*. In the Southern Hemisphere, the earth's direction of rotation is clockwise as viewed from above the South Pole. Consequently, a satellite moving northward from the South Pole would appear to move northwest and, hence, would veer to the *left* of its path.

As the wind speed increases, the Coriolis force increases; hence, *the stronger the wind, the greater the deflection*. Additionally, the Coriolis force increases for all wind speeds from a value of *zero at the equator to*

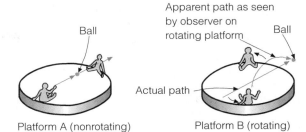

Figure 6.12
On nonrotating platform A, the thrown ball moves in a straight line. On platform B, which rotates counterclockwise, the ball continues to move in a straight line. However, platform B is rotating while the ball is in flight; thus, to anyone on platform B, the ball appears to deflect to the right of its intended path.

a maximum at the poles. This phenomenon is illustrated in Fig. 6.13 where three aircraft, each at a different latitude, are flying along a straight-line path, with no external forces acting on them. The destination of each aircraft is due east and is marked on the illustration in Fig. 6.13a. Each plane travels in a straight path relative to an observer positioned at a fixed spot in space. The earth rotates beneath the moving planes, causing the destination points at latitudes 30° and 60° to change direction slightly (to the observer in space) (Fig. 6.13b).

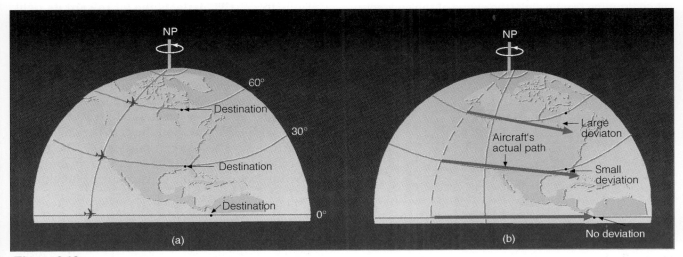

Figure 6.13
Except at the equator, a free-moving object heading either east or west (or any other direction) will appear from the earth to deviate from its path as the earth rotates beneath it. The deviation (Coriolis force) is greatest at the poles and decreases to zero at the equator.

To an observer standing on the earth, however, it is the plane that appears to deviate. The amount of deviation is greatest toward the pole and nonexistent at the equator. Therefore, the Coriolis force has a far greater effect on the plane at high latitudes (large deviation) than on the plane at low latitudes (small deviation). On the equator, it has no effect at all. The same is true of its effect on winds.

In summary, to an observer on the earth, objects moving in *any direction* (north, south, east, or west) are deflected to the *right* of their intended path in the Northern Hemisphere and to the *left* of their intended path in the Southern Hemisphere. The amount of deflection depends upon:

1. the rotation of the earth
2. the latitude
3. the object's speed

In addition, *the Coriolis force acts at right angles to the wind, only influencing wind direction and never wind speed.*

The Coriolis "force" behaves as a real force, constantly tending to "pull" the wind to its right in the Northern Hemisphere and to its left in the Southern Hemisphere. Moreover, this effect is present in all motions relative to the earth's surface. However, in most of our everyday experiences, the Coriolis force is so small that it is negligible and, contrary to popular belief, does not cause water to turn clockwise or counterclockwise when draining from a sink (the shape of the sink actu-

ally plays a much larger role). The Coriolis force is also minimal on small-scale winds, such as those that blow inland along coasts in summer. Only where winds blow over vast regions is the effect significant.

Wind Flow Aloft

We now know that the pressure gradient force is the driving force behind the wind and that the Coriolis force influences wind direction only. Let's examine these two factors to see how they produce the observed flow of air aloft. We will look at the wind flow above the *friction layer*—a level typically 1000 meters (3000 feet) above the ground.

Geostrophic Wind   Figure 6.14 shows a map of the Northern Hemisphere, with horizontal pressure variations at an altitude just above the friction layer. The evenly spaced isobars indicate a constant pressure gradient force (*PGF*) directed from south toward north as indicated by the red arrow at the left. Why, then, does the map show a west wind? We can answer this question by placing a parcel of air at position 1 in the diagram and watching its behavior.

At position 1, the *PGF* acts immediately upon the parcel, accelerating it northward toward lower pressure. However, the instant the air begins to move, the Coriolis force (CF) deflects the air toward its right, curving its path. As the parcel of air increases in speed (positions 2, 3, and 4), the magnitude of the Coriolis force increases (as shown by the longer blue arrows), bending the wind more and more to its right. Eventually, the wind speed increases to a point where the Coriolis force just balances the *PGF*. At this point (position 5), the wind no longer accelerates because the net force is zero. Here the wind flows in a straight path, parallel to the isobars at a constant speed.* This flow of air is called a **geostrophic** (*geo*: earth; *strophic*: turning) **wind**. Notice that the geostrophic wind blows in

---

*At first, it may seem odd that the wind blows at a constant speed with no net force acting on it. But when we remember that the net force is necessary only to accelerate ($F = ma$) the wind, it makes more sense. For example, it takes a considerable net force to push a car and get it rolling from rest. But once the car is moving, it only takes a force large enough to counterbalance friction to keep it going. There is no net force acting on the car, yet it rolls along at a constant speed.

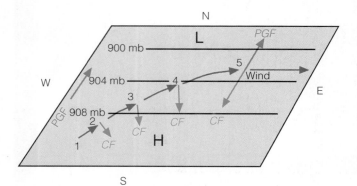

Figure 6.14
Above the level of friction, air initially at rest will accelerate until it flows parallel to the isobars at a steady speed with the pressure gradient force (*PGF*) balanced by the Coriolis force (*CF*). Wind blowing under these conditions is called *geostrophic*.

the Northern Hemisphere with lower pressure to its left and higher pressure to its right.

When the flow of air is purely geostrophic, the isobars are straight and evenly spaced, and the wind speed is constant. In the atmosphere, isobars are rarely straight or evenly spaced, and the wind normally changes speed as it flows along. So, the geostrophic wind is usually only an approximation of the real wind. However, the approximation is generally close enough to help us more clearly understand the behavior of the winds aloft.

The speed of the geostrophic wind is directly related to the pressure gradient. In Fig. 6.15, we can see that a wind flowing parallel to the isobars is similar to water in a stream flowing parallel to its banks. At position 1, the wind is blowing at a low speed; at position 2, the pressure gradient increases and the wind speed picks up.

On the upper-air chart (Fig. 6.16), notice that, as we would expect, the winds tend to parallel the isobars or contour lines. Moreover, where the lines are closer together, winds are stronger; where the lines are farther apart, the winds are weaker. Where the wind flows in large, looping meanders, following a more or less north-south trajectory (such as along the west coast of North America), the wind-flow pattern is called *meridional*. Where the winds are blowing in a west-to-

east direction (such as over the eastern third of the United States), the flow is termed *zonal*.

Because the winds aloft in middle latitudes generally blow from west to east, planes flying in this direction have a beneficial tail wind, which explains why a flight from San Francisco to New York City takes about thirty minutes less than the return flight. If the flow aloft is zonal, clouds, storms, and surface anticyclones tend to move more rapidly from west to east. However, where the flow aloft is meridional, as we will see in Chapter 8, surface storms tend to move more slowly, often intensifying into major storm systems.

As we can see from Fig. 6.16, if we know the contour or isobar patterns on an upper-level chart, we also know the direction and relative speed of the wind, even for regions where no direct wind measurements have been made. Similarly, if we know the wind direction and speed, we can estimate the orientation and spacing of the contours or isobars, even if we do not have a current upper-air chart. (It is also possible to estimate the wind flow and pressure pattern aloft by watching the

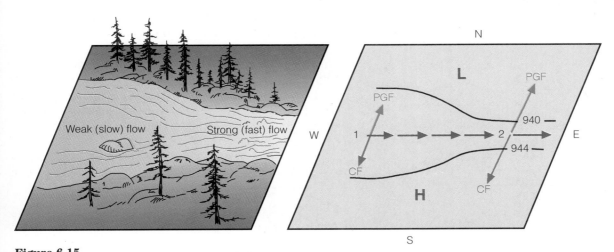

Figure 6.15
The isobars and contours on an upper-level chart are like the banks along a flowing stream. When they are widely spaced, the flow is weak; when they are narrowly spaced, the flow is stronger. The increase in winds on the chart results in a stronger Coriolis force (CF), which balances a larger pressure gradient force (PGF).

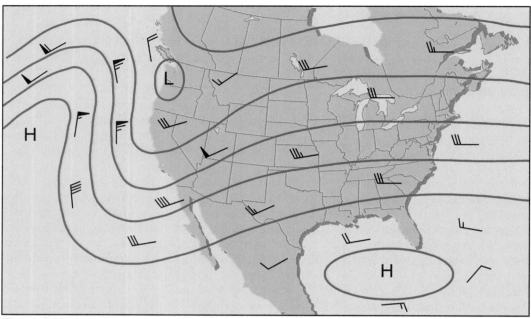

	Miles (statute) per hour	Knots
◎	Calm	Calm
	1–2	1–2
	3–8	3–7
	9–14	8–12
	15–20	13–17
	21–25	18–22
	26–31	23–27
	32–37	28–32
	38–43	33–37
	44–49	38–42
	50–54	43–47
	55–60	48–52
	61–66	53–57
	67–71	58–62
	72–77	63–67
	78–83	68–72
	84–89	73–77
	119–123	103–107

Figure 6.16
Upper-air chart showing wind direction and wind speed. Lines on the map represent contours (or isobars).

movement of clouds. The Focus section on p. 145 illustrates this further.)

Look again at the upper-air chart (Fig. 6.16) and observe that, in many places, the wind is not considered geostrophic because it bends into loops as it follows the pattern of curved contour lines or isobars. A wind that blows at a constant speed parallel to curved lines above the level of friction is termed a **gradient wind**. Figure 6.17 illustrates gradient winds blowing counterclockwise about an area of low pressure and clockwise about an area of high pressure.

Curved Winds Around Lows and Highs Because lows are also known as cyclones, the counterclockwise

CYCLONIC FLOW

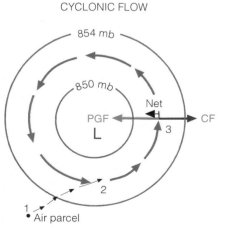

(a) Low pressure area (cyclone)

ANTICYCLONIC FLOW

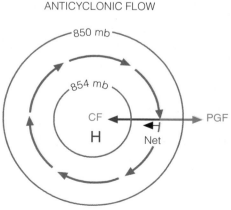

(b) High pressure area (anticyclone)

Figure 6.17
Winds and related forces around areas of low and high pressure above the friction level in the Northern Hemisphere.

Focus on an Observation
Estimating Wind Direction and Pressure Patterns Aloft

Both the wind direction and the orientation of the isobars aloft can be estimated by observing middle- and high-level clouds from the earth's surface. Suppose, for example, we are in the Northern Hemisphere watching clouds directly above us move from southwest to northeast at an elevation of about 3000 meters or 10,000 feet (see Fig. 1a). The cloud movement indicates that the wind at this level is southwesterly. Looking downwind, the wind blows parallel to the isobars with lower pressure on the left and higher pressure on the right. Thus, if we stand with our backs to the direction from which the clouds are moving, lower pressure aloft will always be to our left and higher pressure to our right.* From this observation, we can

*This statement for wind and pressure aloft in the Northern Hemisphere is often referred to as *Buys-Ballot's Law*, after the Dutch meteorologist Christoph Buys-Ballot (1817–1890), who formulated it.

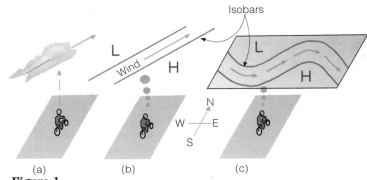

Figure 1
This drawing of a simplified upper-level chart is based on cloud observations. Upper-level clouds moving from the southwest (a) indicate isobars and (b) winds aloft. When extended horizontally, the upper-level chart appears as in (c), where lower pressure is to the northwest and higher pressure is to the southeast.

draw a rough upper-level chart (Fig. 1b), which shows isobars and wind direction for an elevation of approximately 10,000 feet.

The isobars aloft will not continue in a southwest-northeast direction indefinitely; rather, they will often bend into wavy patterns. We may carry our

observation one step further, then, by assuming a bending of the lines (Fig. 1c). Thus, with a southwesterly geostrophic wind aloft, a trough of low pressure will be found to our west and a ridge of high pressure to our east.

---

flow of air around them is often called *cyclonic flow.* Likewise, the clockwise flow of air around a high is called *anticyclonic flow.* Look at the cyclonic wind flow around the upper-level low (Northern Hemisphere) in Fig. 6.17a. At first, it appears as though the wind is defying the Coriolis force by bending to the left as it moves counterclockwise around the system. Let's see why the wind blows in this manner.

Suppose we consider a parcel of air initially at rest at position 1. The pressure gradient force accelerates the air inward toward the center of the low and the Coriolis force deflects the moving air to its right, until the air is moving parallel to the isobars at position 2. The gradient wind at position 3 is now blowing at a constant speed, but parallel to curved isobars.

Earlier in this chapter, we learned that an object accelerates when there is a change in its speed or

direction (or both). Therefore, the gradient wind blowing *around* the low-pressure center is constantly accelerating because it is constantly changing direction. This acceleration, called the *centripetal acceleration,* is directed at right angles to the wind, inward toward the low center.

Remember from Newton's second law that, if an object is accelerating, there must be a net force acting on it. In this case, the net force acting on the wind must be directed toward the center of the low, so that the air will keep moving in a counterclockwise, circular path. This inward-directed force is called the *centripetal force* (*centri*: center; *petal*: to push toward) and results from an imbalance between the Coriolis force and the pressure gradient force.

Again, look closely at position 3 (Fig. 6.17a) and observe that the inward-directed pressure gradient force (PGF) is greater than the outward-directed Coriolis force (CF). The difference between these forces—the net force—is the inward-directed centripetal force. In Fig. 6.17b, the wind blows clockwise around the center of the high. To keep the wind blowing in a circle, the inward-directed Coriolis force must be greater in magnitude than the outward-directed pressure gradient force, so that the centripetal force (again, the net force) is directed inward.

In the Southern Hemisphere, the pressure gradient force starts the air moving and the Coriolis force deflects it to the *left*, thereby causing the wind to blow clockwise around lows and counterclockwise around highs. Does this mean that the winds aloft in middle latitudes of the Southern Hemisphere blow from east to west? The answer is found in the Focus section on p. 148.

▲▼▲

Surface Winds

Winds on a surface weather map do not blow exactly parallel to the isobars; instead, they cross the isobars, moving from higher to lower pressure. The angle at which the wind crosses the isobars varies, but averages about 30°. The reason for this behavior is friction.

In Fig. 6.18a, the wind aloft is blowing at a level above the frictional influence of the ground. At this level (typically above about 1000 meters or 3000 feet),

the wind is approximately geostrophic and blows parallel to the isobars with the pressure gradient force (PGF) on its left balanced by the Coriolis force (CF) on its right. At the earth's surface, however, the same pressure gradient will not produce the same wind speed, and the wind will not blow in the same direction, as it does aloft.

Near the surface, *friction reduces the wind speed, which in turn reduces the Coriolis force*. Consequently, the weaker Coriolis force no longer balances the pressure gradient force, and the wind blows across the isobars toward lower pressure. The pressure gradient force is now balanced by the sum of the frictional force and the Coriolis force. Therefore, in the Northern Hemisphere, we find surface winds blowing counterclockwise and *into* a low; they flow clockwise and *out* of a high (Fig. 6.18b).

In the Southern Hemisphere, winds blow clockwise and inward around surface lows, counterclockwise and outward around surface highs. See the surface weather map and the general wind flow pattern for South America (Fig. 6.19).

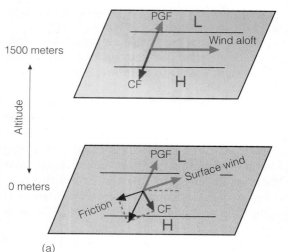

(a)

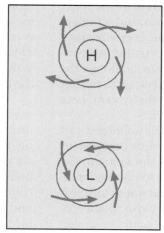

(b) Surface map
Northern Hemisphere

Figure 6.18
The effect of surface friction is to slow down the wind so that, near the ground, the wind crosses the isobars and blows toward lower pressure. This phenomenon produces an outflow of air around a high and an inflow around a low.

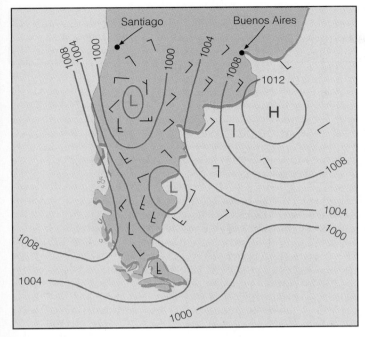

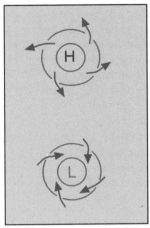

Southern Hemisphere

Figure 6.19
Surface weather map showing isobars and winds on December 20, 1971, in South America.
The boxed area shows the idealized flow around surface pressure systems in the Southern
Hemisphere.

Winds and Vertical Air Motions

Up to this point, we have seen that surface winds blow
in toward the center of low pressure and outward away
from the center of high pressure. As air moves inward
toward the center of a low pressure area (Fig. 6.20), it
must go somewhere. Since this converging air cannot
go into the ground, it slowly rises. Above the surface
low (at about 6000 meters or 20,000 feet), the air begins
to spread apart (diverge) to compensate for the converging surface air. As long as the upper-level diverging outflow of air balances the converging inflow of
surface air, the central pressure in the low does not
change. However, the surface pressure will change if
the upper-level outflow and the surface inflow are not
in balance. For example, if upper-level divergence exceeds surface convergence, the central pressure of the
low will decrease, and isobars around the low will become more tightly packed. This process increases the

pressure gradient (and, hence, the pressure gradient
force) which, in turn, increases the surface winds.

Surface winds move outward (diverge) away from
the center of a high pressure area. To replace this laterally spreading air, the air aloft converges and slowly
descends (Fig. 6.20). Again, as long as upper-level

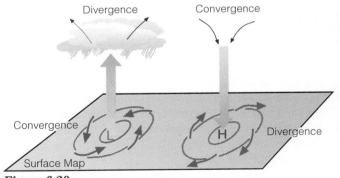

Figure 6.20
Winds and air motions associated with surface highs and lows in
the Northern Hemisphere.

Focus on a Special Topic
Winds Aloft in the Southern Hemisphere

In the Southern Hemisphere, just as in the Northern Hemisphere, the winds aloft blow because of horizontal differences in pressure. The pressure differences, in turn, are due to variations in temperature. Recall from an earlier discussion of pressure that warm air aloft is associated with high pressure and cold air aloft with low pressure. Look at Fig. 2. It shows an upper-level chart that extends from the Northern Hemisphere into the Southern Hemisphere. Over the equator where the air is warmer, the pressure is high. North and south of the equator, where the air is colder, the pressure is lower.

Let's assume, to begin with, that there is no wind on the chart. In the Northern Hemisphere, the pressure gradient force (PGF) directed northward starts the air moving toward lower pressure. Once the air is set in motion, the Coriolis force bends it to the right until it is a *west wind*, blowing parallel to the isobars. In the Southern Hemisphere, the pressure gradient force directed southward

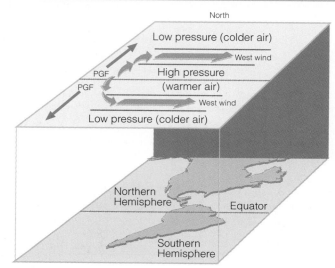

Figure 2
Upper-level chart that extends over the Northern and Southern Hemispheres. Solid gray lines on the chart are isobars.

starts the air moving south. But notice that the Coriolis force in the Southern Hemisphere bends the moving air to its *left*, until the wind is blowing parallel to the isobars *from the west*.

Hence, in the middle latitudes of both hemispheres, we generally find westerly winds aloft.

converging air balances surface diverging air, the central pressure in the high will not change. (Convergence and divergence of air are so important to the development or weakening of surface pressure systems that we will examine this topic again when we look closely at the vertical structure of pressure systems in Chapter 8.)

The rate at which air rises above a low or descends above a high is small compared to the horizontal winds that spiral about these systems. Generally, the vertical motions are usually only about an inch per second, or about a mile per day.

Earlier in this chapter we learned that air moves in response to pressure differences. Because air pressure decreases rapidly with increasing height above the sur-

face, there is always a strong pressure gradient force directed upward. Why, then, doesn't the air rush off into space?

Air does not rush off into space because the upward-directed pressure gradient force is nearly always exactly balanced by the downward force of gravity. When these two forces are in exact balance, the air is said to be in **hydrostatic equilibrium**. When air is in hydrostatic equilibrium, there is no net vertical force acting on it, and so there is no net vertical acceleration. Most of the time, the atmosphere approximates hydrostatic balance, even when air slowly rises or descends at a constant speed. However, this balance does not exist in violent thunderstorms and tornadoes,

Figure 6.21
An onshore wind blows from water to land, whereas an offshore wind blows from land to water.

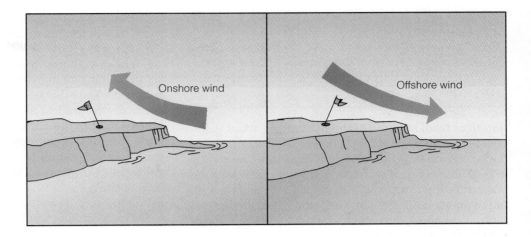

where the air shows appreciable vertical accelerations. But these occur over relatively small vertical distances, considering the total vertical extent of the atmosphere.

Measuring and Determining Winds

Wind is characterized by its direction, speed, and gustiness. Because air is invisible, we cannot really see it. Rather, we see things being moved by it. Therefore, we can determine wind direction by watching the movement of objects as air passes them. For example, the rustling of small leaves, smoke drifting near the ground, and flags waving on a pole all indicate wind direction. In a light breeze, a tried and true method of determining wind direction is to raise a wet finger into the air. The dampness quickly evaporates on the windward side, cooling the skin. Traffic sounds carried from nearby railroads or airports can be used to help figure out the direction of the wind. Even your nose can alert you to the wind direction as the smell of fried chicken or broiled hamburgers drifts with the wind from a local restaurant.

We already know that wind direction is given as the direction from which it is blowing—a north wind blows from the north toward the south. However, near large bodies of water and in hilly regions, wind direction may be expressed differently. For example, wind blowing from the water onto the land is referred to as an **onshore wind**, while wind blowing from land to water is called an **offshore wind** (Fig. 6.21). Air moving uphill is an *upslope wind*; air moving downhill is a

downslope wind. The wind direction may also be given as degrees about a 360° circle. These directions are expressed by the numbers shown in Fig. 6.22. For example: A wind direction of 360° is a north wind; an east wind is 90°; a south wind is 180°; and calm is expressed as zero. It is also common practice to express the wind direction in terms of compass points, such as N, NW, NE, and so on.

The Influence of Prevailing Winds   At many locations, the wind blows more frequently from one direc-

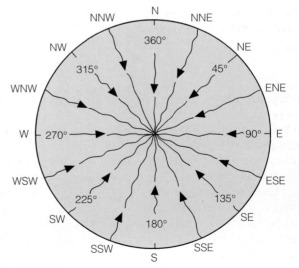

Figure 6.22
Wind direction can be expressed in degrees about a circle or as compass points.

Did you know?
During February, 1965, a "ghost train" rambled across the snow-swept prairies of North Dakota, as strong prevailing winds pushed five empty cars about 90 miles from Portal to Minot.

tion than from any other. The **prevailing wind** is the name given to the wind direction most often observed during a given time period. Prevailing winds can greatly affect the climate of a region. For example, where the prevailing winds are upslope, clouds, fog, and precipitation are more likely than where the winds are downslope. Prevailing onshore winds in summer carry moisture, cool air, and fog into coastal regions, while prevailing offshore breezes carry warmer and drier air into the same locations.

In city planning, the prevailing wind can help decide where industrial centers, factories, and city dumps should be built. All of these, of course, must be located so that the wind will not carry pollutants into populated areas. Sewage disposal plants must be situated downwind from large housing developments, and major runways at airports must be aligned with the prevailing wind to assist aircraft in taking off or landing. In the

high country, strong prevailing winds can bend and twist tree branches toward the downwind side, producing *wind-sculptured "flag" trees* (Fig. 6.23).

The prevailing wind can even be a significant factor in building an individual home. In the northeastern half of the United States, the prevailing wind in winter is northwest and in summer it is southwest. Thus, houses built in the northeastern United States should have windows facing southwest to provide summertime ventilation and few, if any, windows facing the cold winter winds from the northwest. The northwest side of the house should be thoroughly insulated and even protected by a windbreak.

The prevailing wind can be represented by a **wind rose**, which indicates the percentage of time the wind blows from different directions. Extensions from the center of a circle point to the wind direction, and the length of each extension indicates the percentage of time the wind blew from that direction. (See Fig. 6.24.)

Wind Instruments   A very old, yet reliable, weather instrument for determining wind direction is the **wind vane**. Most wind vanes consist of a long arrow with a tail, which is allowed to move freely about a vertical post (Fig. 6.25). The arrow always points into the wind and, hence, always gives the wind direction. Wind vanes can be made of almost any material. At airports, a cone-shaped bag opened at both ends so that it extends horizontally as the wind blows through it sits

Figure 6.23
The effect of a strong prevailing wind is to sculpture these unprotected trees into "flag" trees.

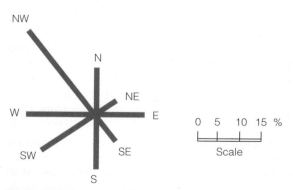

Figure 6.24
This wind rose represents the percent of time the wind blew from different directions at a given site during the month of January for the past ten years. The prevailing wind is NW and the wind direction of least frequency is NE.

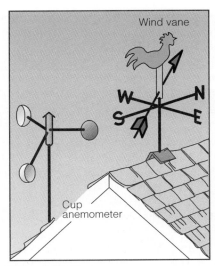

Figure 6.25
A wind vane and a cup anemometer.

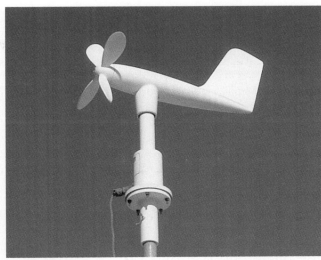

Figure 6.26
The aerovane (skyvane).

near the runway. This form of wind vane, called a *wind sock*, enables pilots to tell the surface wind direction when landing.

The instrument that measures wind speed is the **anemometer**. Most anemometers consist of three (or more) hemispherical cups (*cup anemometer*) mounted on a vertical shaft. The difference in wind pressure from one side of a cup to the other causes the cups to spin about the shaft. The rate at which they rotate is directly proportional to the speed of the wind. The spinning of the cups is usually translated into wind speed through a system of gears, and may be read from a dial or transmitted to a recorder.

The **aerovane** (*skyvane*) is an instrument commonly used to indicate both wind speed and direction. It consists of a three-bladed propeller that rotates at a rate proportional to the wind speed. Its streamlined shape and a vertical fin (Fig. 6.26) keep the blades facing into the wind. When attached to a recorder, a continuous record of both wind speed and direction is obtained.

The wind-measuring instruments described thus far are "ground-based" and only give wind speed or direction at a particular fixed location. But the wind is influenced by local conditions, such as buildings, trees, and so on. Also, wind speed normally increases rapidly with height above the ground. Thus, wind instruments

should be exposed to freely flowing air well above the roofs of buildings. In practice, unfortunately, anemometers are placed at various levels; the result, then, is often erratic wind observations.

Wind information can be obtained during a radiosonde observation. As a balloon carrying a *radiosonde* (an instrument package designed to measure the vertical profile of temperature, pressure, and humidity) rises from the surface, equipment located on the ground constantly tracks it, measuring its vertical and horizontal angles, as well as its distance from the observing station. From this information, a computer determines and prints the vertical profile of wind from the surface up to where the balloon normally pops, typically in the stratosphere near 30 kilometers or 100,000 feet. The observation of winds using a radiosonde balloon is called a *rawinsonde observation*.

In remote regions of the world where upper-air observations are lacking, satellites have been used to

Did you know?
The strongest wind ever measured at the ground with a cup anemometer occurred on New Hampshire's Mt. Washington, when on April 12, 1934, a momentary gust reached 231 miles per hour.

Focus on an Issue
Wind Power

Thousands of small windmills, their arms spinning in a stiff breeze, have pumped water, sawed wood, and even supplemented the electrical needs of small farms for many decades. It was not until the energy crisis of the early 1970s, however, that we seriously considered wind-driven turbines to run generators. In 1980, wind energy got a boost from Congress in the form of a bill—the Wind Energy Research, Demonstration, and Utilization Act—that authorized $900 million to define the nation's wind-energy potential and provide support for the research and development of wind generators called *wind turbines.*

Wind power seems an attractive way of producing energy—it is non-polluting and, unlike solar power, is not restricted to daytime use. It does, however, pose some problems. Large wind machines may interfere with television reception in their immediate vicinity. Also, the financial aspects are another problem, as the cost of a single wind turbine can range from as low as $6,000 to over $1 million. In addition, a region dotted with large wind machines is unaesthetic. (Probably, though, it would be no more of an eyesore than the parades of huge electrical towers marching across

Figure 3
A portion of a wind farm near the summit of Altamont Pass, California. With over 7000 wind turbines, this is the world's largest wind energy development project.

many open areas today.) Most important, if the wind turbine is to produce electricity, there must be wind, not just any wind, but a steady flow of air neither too weak nor too strong. A slight breeze will not turn the blades, and a powerful wind gust could severely damage the machine. Thus, regions with the greatest potential for wind-generated power would have moderate, steady winds.

With fossil fuels rapidly diminishing, the wind can provide a pollution-free alternative form of energy. For exam-

ple, a 1000 kilowatt wind turbine can supply power for approximately 400 average-size U.S. homes. In California alone, there are over 17,000 wind turbines, many of which are on *wind farms*—clusters of 50 or more wind turbines (Fig. 3). Present estimates are that wind power may be able to furnish up to a few percent of the nation's total energy needs by the early part of the twenty-first century.

obtain wind speed and direction. So far, the most reliable wind data have come from geostationary satellites positioned above a particular location. From this vantage point, the satellite shows the movement of clouds. The direction of cloud movement indicates wind direction, and the horizontal distance the cloud moves during a given time period indicates the wind speed.

Recently, Doppler radar has been employed to obtain a vertical profile of wind speed and direction up to an altitude of 10 kilometers or six miles above the ground. Such a profile is called a *wind sounding,* and the radar, a **wind profiler** (or simply a *profiler*). Doppler radar, like conventional radar, emits pulses of microwave radiation that are returned (backscattered) from a target, in this case the irregularities in moisture and temperature created by turbulent, twisting eddies that move with the wind. Doppler radar works on the principle that, as these eddies move toward or away from the receiving antenna, the returning radar pulse will change in frequency. The Doppler radar wind profilers are so sensitive that they can translate the back-scattered energy from these eddies into a vertical picture of wind speed and direction. At present, the National Oceanic and Atmospheric Administration (NOAA) is planning to deploy a network of thirty wind profilers across the central United States by the mid 1990s.

Summary

This chapter has given us a broad view of how and why the wind blows. Aloft where horizontal variations in temperature exist, there is a corresponding horizontal change in pressure. The difference in pressure establishes a force, the pressure gradient force (*PGF*), which starts the air moving from higher toward lower pressure.

Once the air is set in motion, the Coriolis force bends the moving air to the right of its intended path in the Northern Hemisphere and to the left in the Southern Hemisphere. Above the level of surface friction, the wind is bent enough so that it blows nearly parallel to the isobars, or contours. Where the wind blows in a straight-line path, and a balance exists between the *PGF* and the Coriolis force, the wind is termed geostrophic. Where the wind blows parallel to curved isobars (or contours), wind is called a gradient wind.

The interaction of the forces causes the winds in the Northern Hemisphere to blow clockwise around regions of high pressure and counterclockwise around areas of low pressure. In the Southern Hemisphere, the winds blow counterclockwise around highs and clockwise around lows. The effect of surface friction is to slow down the wind. This causes the surface air to blow across the isobars from higher pressure toward lower pressure. Consequently, in both hemispheres, surface winds blow outward, away from the center of a high, and inward, toward the center of a low.

At the end of the chapter, we looked at various methods and instruments used to determine and measure wind speed and direction.

Key Terms

The following terms are listed in the order they appear in the text. Define each.
Doing so will aid you in reviewing the material covered in this chapter.

air pressure	mid-latitude cyclone	hydrostatic equilibrium
millibar	isobaric map	onshore wind
standard atmospheric pressure	contour line	offshore wind
barometer	ridge	prevailing wind
mercury barometer	trough	wind rose
aneroid barometer	pressure gradient	wind vane
station pressure	pressure gradient force (PGF)	anemometer
sea level pressure	Coriolis force	aerovane
isobar	geostrophic wind	wind profiler
anticyclone	gradient wind	

Review Questions

1. (a) Explain why atmospheric pressure always decreases with increasing elevation.
   (b) Why does air pressure decrease more rapidly in a cold column of air?
2. What is the standard atmospheric pressure in millibars and in inches of mercury?
3. Would a sea level pressure of 1040 millibars be considered high or low pressure?
4. With the aid of a diagram, describe how a mercury barometer works.
5. How does an aneroid barometer differ from a mercury barometer?
6. Explain how sea level pressure differs from station pressure.
7. Why will Denver, Colorado, always have a lower station pressure than Chicago, Illinois?
8. What are isobars?
9. On an upper-level map, is cold air aloft generally associated with low or high pressure? What about warm air aloft?
10. What do Newton's first and second laws of motion tell us?
11. (a) Define the term *pressure gradient*.
    (b) What does a steep (or strong) pressure gradient mean?
12. What is the name of the force that initially sets the air in motion?

13. Explain why, on a map, closely spaced isobars (or contours) indicate strong winds, and widely spaced isobars (or contours) indicate weak winds.
14. What does the Coriolis force do to moving air
    (a) in the Northern Hemisphere?
    (b) in the Southern Hemisphere?
15. Explain how each of the following influences the Coriolis force: (a) wind speed; (b) latitude.
16. Why do upper-level winds in the middle latitude of both hemispheres generally blow from west to east?
17. What is a geostrophic wind?
18. What are the forces that affect the horizontal movement of air?
19. Describe how the wind blows around high pressure areas and low pressure areas aloft and near the surface (a) in the Northern Hemisphere; and (b) in the Southern Hemisphere.
20. If the clouds overhead are moving from north to south, would the upper-level center of low pressure be to the east or west of you?
21. On a surface map, why do surface winds tend to cross the isobars and flow from higher pressure toward lower pressure?
22. Since there is always an upward-directed pressure gradient force, why doesn't air rush off into space?

23. List as many ways as you can of determining wind direction and wind speed.

24. Below is a list of instruments. Describe how each one is able to measure wind speed, wind direction, or both.

    (a) wind vane
    (b) cup anemometer
    (c) aerovane (skyvane)
    (d) radiosonde
    (e) satellite
    (f) wind profiler

25. An upper wind direction is reported as 225°. From what compass direction is the wind blowing?

26. Explain how a wind rose can be used to show prevailing wind direction.

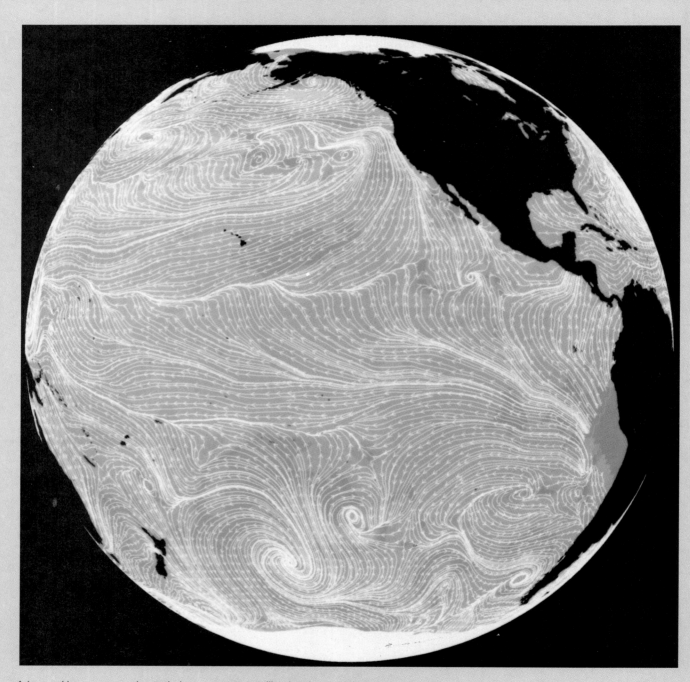

Advanced image processing techniques convert satellite data into a global picture of winds. Arrows show wind direction and colors represent wind speed, with yellow being the highest wind speed and blue the lowest. Notice that our atmosphere consists of whirling eddies of varying sizes. (Photo: NASA. The image was created by Peter Woiceshyn and Andrew Pursh of Jet Propulsion Lab, Professor Morton Wurtele of UCLA, and Dr. Stephen Peteherych of the Canadian Atmospheric Environment Service.)

Chapter 7

Atmospheric Circulations

Contents

On January 22, 1983, a DC-8 jetliner carrying 138 passengers and a crew of 6 was heading for San Francisco from Chicago. Some mild bumpiness over Colorado prompted the "fasten your seat belt" sign to go on. But other than that it was a routine, uneventful flight until suddenly over Utah a tremendous vibration ran through the plane. In an instant the aircraft nosed up, then dropped several thousand feet before stabilizing. Inside, screaming passengers were flung up against the walls and ceilings, then dropped. Those not fastened into their seats were sent hurling against overhead luggage compartments. Small bags and luggage, having slipped out from under the seats, were flying about inside the plane. Within seconds the entire ordeal was over. Twelve people suffered injuries and one passenger was knocked unconscious for about two minutes.

7 The aircraft in our opener encountered a turbulent eddy—an "air pocket"—in perfectly clear weather. Such eddies are not uncommon, especially in the vicinity of jet streams. In this chapter, we will examine a variety of eddy circulations. First, we will look at the formation of small-scale winds. Then, we will examine slightly larger circulations—local winds—such as the sea breeze and the chinook, describing how they form and the type of weather they generally bring. Finally, we will look at the general wind-flow pattern around the world.

▲▽▲
Scales of Atmospheric Motion

The air in motion—what we commonly call *wind*—is invisible, yet we see evidence of it nearly everywhere we look. It sculptures rocks, moves leaves, blows smoke, and lifts water vapor upward to where it can condense into clouds. The wind is with us wherever we go. On a hot day, it can cool us off; on a cold day, it can make us shiver. A breeze can sharpen our appetite when it blows the aroma from the local bakery in our direction. The wind is a powerful element. The workhorse of weather, it moves storms and large fair-weather systems around the globe. It transports heat, moisture, dust, insects, bacteria, and pollens from one area to another.

Circulations of all sizes exist within the atmosphere. Little whirls form inside bigger whirls, which encompass even larger whirls—one huge mass of turbulent, twisting eddies. For clarity, meteorologists arrange circulations according to their size. This hierarchy of motion from tiny gusts to giant storms is called the **scales of motion**.

Consider smoke rising into the otherwise clean air from a chimney in the industrial section of a large city (Fig. 7.1a). Within the smoke, small chaotic motions—tiny eddies—cause it to tumble and turn. These eddies constitute the smallest scale of motion—the **microscale**. At the microscale, eddies with diameters of a few meters or less not only disperse smoke, they also sway branches and swirl dust and papers into the air. They form by convection or by the wind blowing past obstructions and are usually short-lived, lasting only a few minutes at best.

In Fig. 7.1b observe that, as the smoke rises, it drifts toward the center of town. Here the smoke rises even higher and is carried back toward the industrial section. This circulation of city air constitutes the next larger scale—the **mesoscale** (meaning middle scale). Typical mesoscale winds range from a few kilometers to about a hundred kilometers in diameter. Generally, they last longer than microscale motions, often many minutes, hours, or in some cases as long as a day. Mesoscale circulations include local winds (which form along shorelines and mountains), as well as thunderstorms, tornadoes, and small tropical storms.

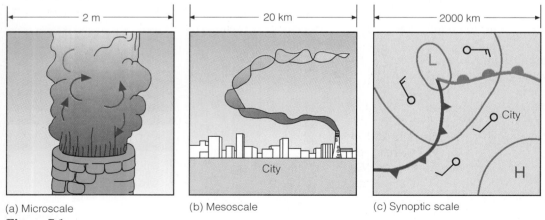

(a) Microscale   (b) Mesoscale   (c) Synoptic scale

Figure 7.1
Scales of atmospheric motion. The tiny microscale motions constitute a part of the larger mesoscale motions, which, in turn, are part of the much larger synoptic scale. Notice that as the scale becomes larger, motions observed at the smaller scale are no longer visible.

When we look at the smoke stack on a surface weather map (Fig. 7.1c), neither the smoke stack nor the circulation of city air shows up. All that we see are the circulations around high- and low-pressure areas—the cyclones and anticyclones of the middle latitude. We are now looking at the *synoptic scale*, or weather map scale. Circulations of this magnitude dominate regions of hundreds to even thousands of square kilometers and, although the life spans of these features vary, they typically last for days and sometimes weeks. The largest wind patterns are seen at the *global scale*. Here, we have wind patterns ranging over the entire earth. Sometimes, the synoptic and global scales are combined and referred to as the **macroscale**.

Eddies—Big and Small

When the wind encounters a solid object, a whirl of air— or *eddy*—forms on the object's downwind side. The size and shape of the eddy depend upon the size and shape of the obstacle and on the speed of the wind. Light winds produce small stationary eddies. Wind moving past trees, shrubs, and even your body produces small eddies. (You may have had the experience of dropping a piece of paper on a windy day only to have it carried away by a swirling eddy as you bend down to pick it up.) Air flowing over a building produces larger eddies that will, at best, be about the size of the building. Strong winds blowing past an open sports stadium can produce eddies that may rotate in such a way as to create surface winds on the playing field that move in a direction opposite to the wind flow above the stadium. Wind blowing over a fairly smooth surface produces few eddies, but when the surface is rough, many eddies form.

The eddies that form downwind from obstacles can produce a variety of interesting effects. For instance, wind moving over a mountain range in stable air with a speed greater than 40 knots usually produces waves and eddies, such as those shown in Fig. 7.2. We can see that eddies form both close to the mountain and beneath each wave crest. These so-called **rotors** have violent vertical motions that produce extreme turbulence and hazardous flying conditions. On a much smaller scale, the howling of wind on a blustery night is believed to be caused by eddies that are constantly be-

ing shed around obstructions, such as chimneys and roof corners.

Turbulent eddies form aloft as well as near the surface. Turbulence aloft can occur suddenly and unexpectedly, especially where the wind changes its speed or direction (or both) abruptly. Such a change is called a **wind shear**. The shearing creates forces that produce eddies along a mixing zone. If the eddies form in clear air, this form of turbulence is called **clear air turbulence**, or **CAT**. (Additional information on this topic is given in the Focus section on p. 160.)

Local Wind Systems

Every summer, millions of people flock to the New Jersey shore, hoping to escape the oppressive heat and humidity of the inland region. On hot, humid days, these travelers often encounter thunderstorms about 20 miles or so from the ocean, thunderstorms that invariably last for only a few minutes. In fact, by the time the vacationers arrive at the beach, skies are generally clear and air temperatures are much lower, as cool ocean breezes greet them. If the travelers return home in the afternoon, these "mysterious" showers occur at just about the same location as before.

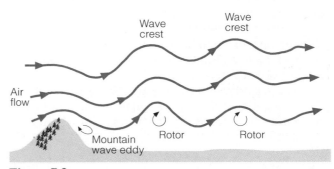

Figure 7.2
Under stable conditions, air flowing past a mountain range can create eddies many miles downwind from the mountain itself.

Focus on an Application
Eddies and "Air Pockets"

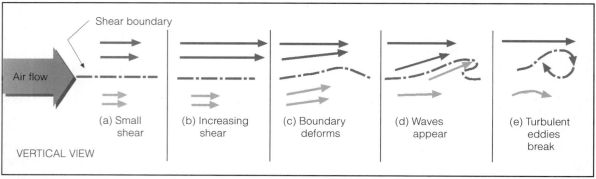

Figure 1
The formation of clear air turbulence (CAT) along a boundary of increasing wind speed shear.

To better understand how eddies form along a zone of wind shear, imagine that, high in the atmosphere, there is a stable layer of air having vertical wind speed shear as depicted in Fig. 1. The top half of the layer slowly slides over the bottom half, and the relative speed of both halves is low. As long as the wind shear between the top and bottom of the layer is small, few if any eddies form. How-

ever, if the shear and the corresponding relative speed of these layers increase, wavelike undulations may form. When the shearing exceeds a certain value, the waves break into large swirls, with significant vertical movement. (See Fig. 2.) Eddies such as these often form in the upper troposphere near jet streams, where large wind speed shears exist. They also occur in conjunction with moun-

tain waves, which may extend upward into the stratosphere. As we learned earlier, when these huge eddies develop in clear air, this form of turbulence is referred to as *clear air turbulence*, or *CAT*.

The eddies that form in clear air may have diameters ranging from a couple of meters to several hundred meters. An unsuspecting aircraft entering such a region may be in for

The showers are not really mysterious. Actually, they are caused by a local wind system—the sea breeze. As cooler ocean air pours inland, it forces the warmer, unstable humid air to rise and condense, producing majestic clouds and rainshowers along a line where the wind systems meet.

The sea breeze forms as part of a thermally driven circulation. Consequently, we will begin our study of local winds by examining the formation of thermal circulations.

Thermal Circulations   Consider the vertical distribution of pressure shown in Fig. 7.3a. The isobars all lie parallel to the earth's surface; thus, there is no horizontal variation in pressure (or temperature), and there

is no pressure gradient and no wind. Suppose the atmosphere is cooled to the north and warmed to the south (Fig. 7.3b). In the cold, dense air above the surface, the isobars bunch closer together, while in the warm, less dense air, they spread farther apart. This dipping of the isobars produces a horizontal pressure gradient force (PGF) aloft that causes the air to move from higher pressure toward lower pressure.

At the surface, the air pressure remains unchanged until the air aloft begins to move. As the air aloft moves from south to north, air leaves the southern area and "piles up" above the northern area. This redistribution of air reduces the surface air pressure to the south and raises it to the north. Consequently, a pressure gradient force is established at the earth's surface from

Focus on an Application
Eddies and "Air Pockets," *continued*

Figure 2
Turbulent eddies forming in a wind shear zone produce these clouds.

more than just a bumpy ride. If the aircraft flies into a zone of descending air, it may drop suddenly, producing the sensation that there is no air to support the wings. Consequently, these regions have come to be known as *air pockets*.

Commercial aircraft entering an air pocket have dropped hundreds of meters, injuring passengers and flight attendants not strapped into their

seats. For example, on April 4, 1981, a DC-10 jetliner flying at 37,000 feet over central Illinois encountered a region of severe clear air turbulence and reportedly plunged about 2000 feet toward the earth before stabilizing. Twenty-one of the 154 people aboard were injured; one person sustained a fractured hip and another person, after hitting the ceiling, jabbed himself in the nose with a fork,

then landed in the seat in front of him. Clear air turbulence has occasionally caused structural damage to aircraft by breaking off vertical stabilizers and tail structures. Fortunately, the effects are usually not this dramatic.

north to south and, hence, surface winds begin to blow from north to south.

We now have a distribution of pressure and temperature and a circulation of air, as shown in Fig. 7.3c. As the cool surface air flows southward, it warms and becomes less dense. In the region of surface low pressure, the warm air slowly rises, expands, cools, and flows out the top at an elevation of about 1 kilometer (3300 feet) above the surface. At this level, the air flows horizontally northward toward lower pressure, where it completes the circulation by slowly sinking and flowing out the bottom of the surface high. Circulations brought on by changes in air temperature, in which warmer air rises and colder air sinks, are termed **thermal circulations**.

The regions of surface high and low atmospheric pressure created as the atmosphere either cools or warms are called *thermal highs* and *thermal lows*. In general, they are shallow systems, usually extending no more than a few kilometers above the ground.

Sea and Land Breezes   The sea breeze is a type of thermal circulation. The uneven heating rates of land and water (described in Chapter 3) cause these mesoscale coastal winds. During the day, the land heats more quickly than the adjacent water, and the intensive heating of the air above produces a shallow thermal low. The air over the water remains cooler than the air over the land; hence, a shallow thermal high exists above the water. The overall effect of this pressure

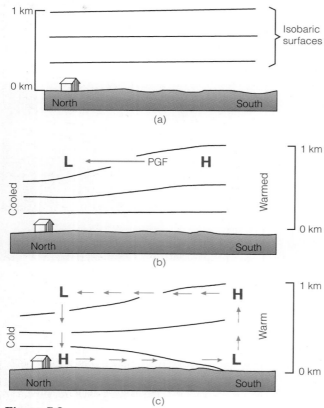

Figure 7.3
A thermal circulation produced by the heating and cooling of the atmosphere near the ground. (The Hs and Ls refer to atmospheric pressure.)

distribution is a **sea breeze** that blows from the sea toward the land (Fig. 7.4a). Since the strongest gradients of temperature and pressure occur near the land-water boundary, the strongest winds typically occur right near the beach and diminish inland. Further, since the greatest contrast in temperature between land and water usually occurs in the afternoon, sea breezes are strongest at this time. (The same type of breeze that develops along the shore of a large lake is called a *lake breeze*.)

At night, the land cools more quickly than the water. The air above the land becomes cooler than the air over the water, producing a distribution of pressure, such as the one shown in Fig. 7.4b. With higher surface pressure now over the land, the wind reverses itself and becomes a **land breeze**—a breeze that flows from the land toward the water. Temperature contrasts between land and water are generally much smaller at

night, hence land breezes are usually weaker than their daytime counterpart, the sea breeze. In regions where greater nighttime temperature contrasts exist, stronger land breezes occur over the water, off the coast. They are not usually noticed much on shore, but are frequently observed by ships in coastal waters.

Look at Fig. 7.4 again and observe that the rising air is over the land during the day and over the water during the night. Therefore, along the humid East Coast, daytime clouds tend to form over land and nighttime clouds over water. This explains why, at night, distant lightning flashes are sometimes seen over the ocean.

The leading edge of the sea breeze is called the *sea breeze front*. As the front moves inland, a rapid drop in temperature occurs just behind it. In some locations, this temperature change may be 5°C (9°F) or more during the first hours—a refreshing experience on a hot, sultry day. Since cities near the ocean usually experience the sea breeze by noon, their highest temperature usually occurs much earlier than in inland cities. Along the East Coast, the passage of the sea breeze front is marked by a wind shift, usually from west to east. In the cool ocean air, the relative humidity rises as the temperature drops. If the relative humidity increases to above 70 percent, water vapor begins to condense upon particles of sea salt or industrial smoke, producing haze. When the ocean air is highly concentrated with pollutants, the sea breeze front may meet relatively clear air and thus appear as a *smoke front*, or a *smog front*. If the ocean air becomes saturated, a mass of low clouds and fog will mark the leading edge of the marine air (Fig. 7.5).

When there is a sharp contrast in air temperature across the frontal boundary, the warmer, lighter air will converge and rise. In many regions, this makes for good sea breeze glider soaring. If this rising air is sufficiently moist, a line of cumulus clouds will form along the sea breeze front, and, if the air is also unstable, thunderstorms may form. As previously mentioned, on a hot, humid day one can drive toward the shore, encounter heavy showers several miles from the ocean, and arrive at the beach to find it sunny with a steady onshore breeze.

Sea breezes in Florida help produce that state's abundant summertime rainfall. On the Atlantic side of the state, the sea breeze blows in from the east; on the Gulf shore, it moves in from the west. The convergence of these two moist wind systems, coupled with daytime

Figure 7.4
Development of a sea breeze and a land breeze. (a) At the surface, a sea breeze blows from the water onto the land, while (b) the land breeze blows from the land out over the water.

(a) Sea breeze

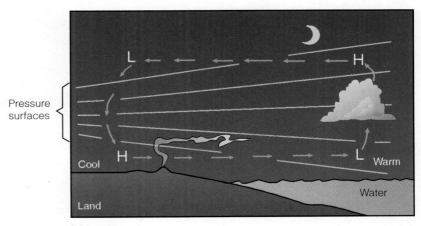

(b) Land breeze

convection, produces cloudy conditions and showery weather over the land (Fig. 7.6). Over the water (where cooler, more stable air lies close to the surface), skies often remain cloud-free.

Convergence of coastal breezes is not restricted to ocean areas. Both Lake Michigan and Lake Superior are capable of producing well-defined lake breezes. In upper Michigan, where these large bodies of water are separated by a narrow strip of land, two breezes push inland and converge near the center of the peninsula, creating afternoon clouds and showers, while the lakeshore area remains sunny, pleasantly cool, and dry.

Seasonally Changing Winds—the Monsoon The word *monsoon* derives from the Arabic *mausim*, which means seasons. A **monsoon wind system** is one

that *changes direction seasonally*, blowing from one direction in summer and from the opposite direction in winter. This seasonal reversal of winds is especially well developed in eastern and southern Asia.

In some ways, the monsoon is similar to a huge sea breeze. During the winter, the air over the continent becomes much colder than the air over the ocean. (See Fig. 3.7, p. 62.) A large, shallow high-pressure area develops over continental Siberia, producing a *clockwise* circulation of air that flows out over the Indian Ocean and South China Sea (Fig. 7.7). Subsiding air of the anticyclone and the downslope movement of northeasterly winds from the inland plateau provide eastern and southern Asia with generally fair weather and the dry season. Hence, the winter monsoon means clear skies, with winds that blow from land to sea.

Figure 7.5
The leading edge of a sea breeze, marked by a band of low clouds, moves into Northern
California.

Figure 7.6
Surface heating and lifting of air along a sea breeze combine to form thunderstorms almost
daily during the summer in southern Florida.

In summer, the wind flow pattern reverses itself as air over the continents becomes much warmer than air above the water. (See Fig. 3.8, p. 63.) A shallow thermal low develops over the continental interior. The heated air within the low rises, and the surrounding air responds by flowing *counterclockwise* into the low center. This results in moisture-bearing winds sweeping into the continent from the ocean. The humid air converges with a drier westerly flow, causing it to rise; further lifting is provided by hills and mountains. Lifting cools the air to its saturation point, resulting in heavy showers and thunderstorms. Thus, the summer monsoon of southeastern Asia means wet, rainy weather (wet season) with winds that blow from sea to land. (See Fig. 7.7.)

Summer monsoon rains over southern Asia can reach record amounts. Located inland on the southern slopes of the Khasi Hills in northeastern India, Cherrapunji receives an average of 425 inches of rainfall each year, most of it during the summer monsoon between April and October. The summer monsoon rains are essential to the agriculture of that part of the world. With a population of over 600 million people, India depends heavily on the summer rains so that food crops will grow. Unfortunately, the monsoon can be unreliable in both duration and intensity. Since the monsoon is vital to the survival of so many people, it is no wonder that meteorologists have investigated it extensively. They have tried to develop methods of accurately forecasting the intensity and duration of the monsoon. The outlook has been hopeful, but the results are not yet impressive.

Did you know?
Cherrapunji, India, received 90 feet of rain in 1861, most of which fell between April and October—the summer monsoon.

Monsoon wind systems exist in other regions of the world, where large contrasts in temperature develop between oceans and continents. (Usually, however, these systems are not as pronounced as in southeast Asia.) For example, a monsoon-like circulation exists in the southwestern United States, especially in Arizona and New Mexico, where spring and early summer are normally dry, as warm westerly winds sweep over the region. By mid-July, however, moist southerly winds are more common, and so are afternoon showers and thunderstorms.

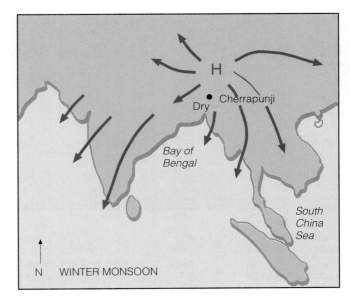

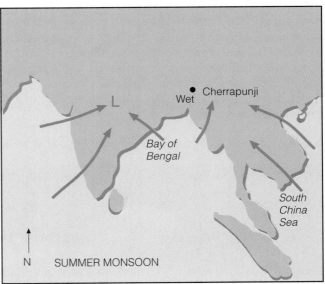

Figure 7.7
Changing annual wind flow patterns associated with the winter and summer Asian monsoon.

Mountain and Valley Breezes   Mountain and valley breezes develop along mountain slopes. Observe in Fig. 7.8 that, during the day, sunlight warms the valley walls, which in turn warm the air in contact with them. The heated air, being less dense than the air of the same altitude above the valley, rises as a gentle upslope wind known as a **valley breeze**. At night, the flow reverses. The mountain slopes cool quickly, chilling the air in

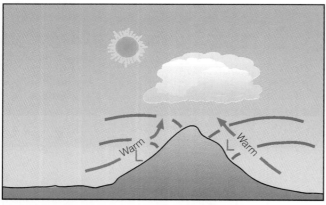 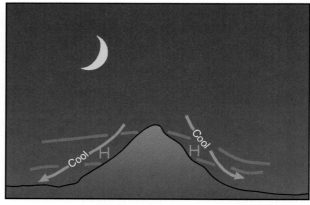

Valley Breeze   Mountain Breeze

Figure 7.8
Valley breezes blow uphill during the day; mountain breezes blow downhill at night. (The Ls
and Hs represent pressure, while the purple lines are pressure surfaces.)

contact with them. The cooler, more dense air glides downslope into the valley, providing a **mountain breeze**. (Because gravity is the force that directs these winds downhill, they are also referred to as *gravity winds*, or *drainage winds*.) This daily cycle of wind flow is best developed in clear, summer weather when prevailing winds are light.

When the upslope valley winds are well developed and have sufficient moisture, they can reveal themselves as building cumulus clouds above mountain summits (Fig. 7.9). Since valley breezes usually reach their maximum strength in the early afternoon, cloudiness, showers, and even thunderstorms are common over mountains during the warmest part of the day—a

Figure 7.9
As mountain slopes warm during the day, air rises and often condenses into cumuliform
clouds, such as these.

fact well known to climbers, hikers, and seasoned mountain picnickers.

Katabatic Winds   Although any downslope wind is technically a **katabatic wind**, the name is usually reserved for downslope winds that are much stronger than mountain breezes. Katabatic (or *fall*) winds can rush down elevated slopes at hurricane speeds, but most are not that intense and many are on the order of 10 knots or less.

The ideal setting for a katabatic wind is an elevated plateau surrounded by mountains, with an opening that slopes rapidly downhill. When winter snows accumulate on the plateau, the overlying air grows extremely cold. Along the edge of the plateau the cold, dense air begins to descend through gaps and saddles in the hills, usually as a gentle or moderate cold breeze. If the breeze, however, is confined to a narrow canyon or channel, the flow of air can increase, often destructively, as cold air rushes downslope like water flowing over a fall.

Katabatic winds are observed in various regions of the world. For example, along the northern Adriatic coast in Yugoslavia, a polar invasion of cold air from Russia descends the slopes from a high plateau and reaches the lowlands as the *bora*—a cold, gusty, north-easterly wind with speeds sometimes in excess of 100 knots. A similar, but often less violent, cold wind known as the *mistral* descends the western mountains into the Rhone Valley of France, and then out over the Mediterranean Sea. It frequently causes frost damage to exposed vineyards and makes people bundle up in the otherwise mild climate along the Riviera. Strong, cold katabatic winds also blow downslope off the icecaps in Greenland and Antarctica occasionally, with speeds greater than 100 knots.

In North America, when cold air accumulates over the Columbia plateau, it may flow westward through the Columbia River Gorge as a strong, gusty, and sometimes violent wind. Even though the sinking air warms by compression, it is so cold to begin with that it reaches the ocean side of the Cascade Mountains much colder than the marine air it replaces. The *Columbia Gorge wind* is often the harbinger of a prolonged cold spell.

Strong downslope katabatic-type winds funneled through a mountain canyon can do extensive damage. For example, during January, 1984, a ferocious downslope wind blew through Yosemite National Park at

Did you know?
Dressing for the weather in Granville, North Dakota, on February 21, 1918, would have been a difficult task, as a chinook wind caused the air temperature to jump 83°F—from –33°F in the morning to 50°F in the afternoon.

speeds estimated at 100 knots. The wind toppled trees and, unfortunately, caused a fatality when a tree fell on a park employee sleeping in a tent.

Chinook (Foehn) Winds   The **chinook wind** is a warm, dry wind that descends the eastern slope of the Rocky Mountains. The region of the chinook is rather narrow and extends from northeastern New Mexico northward into Canada. Similar winds occur along the leeward slopes of mountains in other regions of the world. In the Alps, for example, such a wind is called a *foehn*. When these winds move through an area, the temperature rises sharply, sometimes over 20°C (36°F) in one hour, and a corresponding sharp drop in the relative humidity occurs, occasionally to less than 5 percent. (More information on temperature changes associated with chinooks is given in the Focus section on p. 169.)

Chinooks occur when strong westerly winds aloft flow over a north-south-trending mountain range, such as the Rockies and Cascades. Such conditions can produce a trough of low pressure on the mountain's eastern side, a trough that tends to force the air downslope. As the air descends, it is compressed and warms. So the main source of warmth for a chinook is *compressional heating*, as potentially warmer (and drier) air is brought down from aloft.

When clouds and precipitation occur on the mountain's windward side, they can enhance the chinook. For example, as the cloud forms on the windward side of the mountain in Fig. 7.10, the conversion of latent heat to sensible heat supplements the compressional heating on the leeward side. This phenomenon makes the descending air at the base of the mountain on the leeward side warmer than it was before it started its upward journey on the windward side. The air is also drier, since much of its moisture was removed as precipitation on the windward side.

Along the front range of the Rockies, a bank of clouds forming over the mountains is a telltale sign of an impending chinook. This *chinook wall cloud*

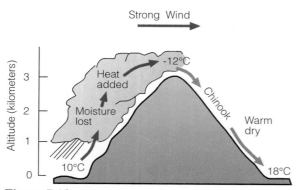

Figure 7.10
Conditions that may enhance a chinook.

usually remains stationary as air rises, condenses, and then rapidly descends the leeward slopes, often causing strong winds in foothill communities. Figure 7.11 shows how a chinook wall cloud appears as one looks west toward the Rockies from the Colorado plains. The photograph was taken on a winter afternoon with the air temperature about −7°C (20°F). That evening, the chinook moved downslope at high speeds through foothill valleys, picking up sand and pebbles (which dented cars and cracked windshields). The chinook

spread out over the plains like a warm blanket, raising the air temperature the following day to a mild 15°C (59°F). The chinook and its wall of clouds remained for several days, bringing with it a welcomed break from the cold grasp of winter.

Santa Ana Winds   A warm, dry wind that blows from the east or northeast into Southern California is the **Santa Ana wind**. As the air descends from the elevated desert plateau, it funnels through mountain canyons in the San Gabriel and San Bernardino Mountains, finally spreading over the Los Angeles Basin and San Fernando Valley. The wind often blows with exceptional speed in the Santa Ana Canyon (the canyon from which it derives its name).

These warm, dry winds develop as a region of high pressure builds over the Great Basin. The clockwise circulation around the anticyclone forces air downslope from the high plateau. Thus, *compressional heating* provides the primary source of warming. The air is dry, since it originated in the desert, and it dries out even more as it is heated. Figure 7.12 shows a typical wintertime Santa Ana situation.

As the wind rushes through canyon passes, it lifts dust and sand and dries out vegetation. This sets the stage for serious brush fires, especially in autumn,

Figure 7.11
A chinook wall cloud forming over the Colorado Rockies (viewed from the plains).

Focus on a Special Topic
Snow Eaters and Rapid Temperature Changes

Chinooks are thirsty winds. As they move over a heavy snow cover, they can melt and evaporate a foot of snow in less than a day. This has led to some tall tales about these so-called "snow eaters." Canadian folklore has it that a sled-driving traveler once tried to outrun a chinook. During the entire ordeal his front runners were in snow while his back runners were on bare soil.

Actually, the chinook is important economically. It not only brings relief from the winter cold, but it uncovers prairie grass, so that livestock can graze on the open range. Also, these warm winds have kept railroad tracks clear of snow, so that trains can keep running. On the other hand, the drying effect of a chinook can create an extreme fire hazard. And when a chinook follows spring planting, the seeds may die in the parched soil. Along with the dry air comes a buildup of static electricity, making a simple handshake a shocking experience. These warm, dry winds have sometimes adversely affected human behavior. During periods of chinook winds some people feel irritable and

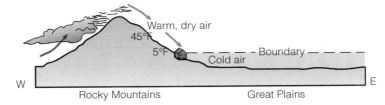

Figure 3
Cities near the warm air–cold air boundary can experience sharp temperature changes, if cold air should rock up and down like water in a bowl.

depressed and others become ill. The exact reason for this phenomenon is not clearly understood.

Chinook winds have been associated with rapid temperature changes. Figure 3 shows a shallow layer of extremely cold air that has moved southward out of Canada and is now resting against the Rocky Mountains. In the cold air, temperatures are near 5°F, while just a short distance up the mountain a warm chinook wind raises the air temperature to 45°F. The cold air behaves just as any fluid, and, in some cases, atmospheric conditions may cause the air to move up and down much like water does when a bowl is rocked back and forth. This can cause extreme temperature varia-

tions for cities located at the base of the hills along the periphery of the cold air–warm air boundary, as they are alternately in and then out of the cold air. Such a situation is held to be responsible for the unbelievable two-minute temperature change of 49°F recorded at Spearfish, South Dakota, during the morning of January 22, 1943. On the same morning, in nearby Rapid City, the temperature fluctuated from –4°F at 5:30 A.M. to 54°F at 9:40 A.M., then down to 11°F at 10:30 A.M. and up to 55°F just 15 minutes later. At nearby cities, the undulating cold air produced similar temperature variations that lasted for several hours.

when chaparral-covered hills are already parched from the dry summer. One such fire in November of 1961—the infamous Bel Air fire—burned for three days, destroying 484 homes and causing over $25 million in damage. During October, 1977, Santa Ana-driven flames scorched 25,000 acres along a 10-mile-wide swath across the Santa Monica Mountains, destroying 91 homes collectively valued at several million dollars. Four hundred miles to the north in Oakland, California, a ferocious Santa Ana-type wind was responsible for the disastrous Oakland hills fire during October, 1991,

that damaged or destroyed over 3000 dwellings and took 25 lives. With the protective vegetation cover removed, the land is ripe for erosion, as winter rains wash away topsoil and, in some areas, create serious mudslides. The adverse effects of a wind-driven Santa Ana fire may be felt throughout the year.

Desert Winds   Local winds form in deserts, too. Dust storms form in dry regions, where strong winds are able to lift and fill the air with particles of fine dust. In

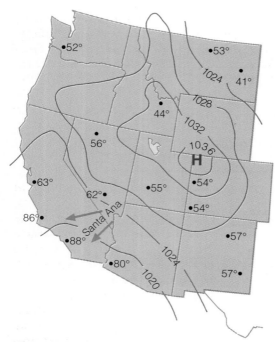

Figure 7.12
Surface weather map showing Santa Ana conditions for January 17, 1976. Maximum temperatures for this day are given in °F. Observe that the downslope winds blowing into Southern California raised temperatures into the upper 80s, while elsewhere temperature readings were much lower.

Figure 7.13
A dust devil forming on a clear, hot summer day just south of Phoenix, Arizona.

desert areas where loose sand is more prevalent, *sandstorms* develop, as high winds enhanced by surface heating rapidly carry sand particles close to the ground. A spectacular example of a storm composed of dust or sand is the **haboob** (from Arabic *hebbe*: blown). The haboob forms as cold downdrafts along the leading edge of a thunderstorm lift dust or sand into a huge, tumbling dark cloud that may extend horizontally for over a hundred miles and rise vertically to the base of the thunderstorm. Spinning whirlwinds of dust frequently form along the turbulent cold air boundary, giving rise to sightings of huge *dust devils* and even tornadoes. Haboobs are most common in the African Sudan (where about twenty-four occur each year) and

in the desert southwest of the United States, especially in southern Arizona.

The spinning vortices so commonly seen on hot days in dry areas are called **whirlwinds**, or **dust devils**. (In Australia, the Aboriginal word *willy-willy* is used to refer to a dust devil.) Generally, dust devils form on clear, hot days, as warm air rises above a heated surface. Wind, often deflected by small topographic barriers, flows into this region, rotating the rising air. Depending on the nature of the topographic feature, the spin of a dust devil around its central eye may be cyclonic or anticyclonic, and both directions occur with about equal frequency.

Having diameters of only a few meters and heights of less than 100 meters (300 feet), most dust devils are small and last only a short time. (See Fig. 7.13.) There are, however, some dust devils of sizable dimension, extending upward from the surface for many hundreds of meters. Such whirlwinds are capable of considerable damage; winds exceeding 75 knots may overturn mobile homes and tear the roofs off buildings. Fortunately, the majority of dust devils are small.

Did you know?
A raging dust storm on November 29, 1991, near Coalinga, California, triggered a horrific 164-vehicle pileup along Interstate 5 that injured 150 people and killed 17.

Global Winds

Up to now, we have seen that local winds vary considerably from day to day and from season to season. As you may suspect, these winds are part of a much larger circulation—the little whirls within larger whirls that we spoke of earlier in this chapter. Indeed, if the rotating high- and low-pressure areas are like spinning eddies in a huge river, then the flow of air around the globe is like the meandering river itself. When winds throughout the world are averaged over a long period, the local wind patterns vanish, and what we see is a picture of the winds on a global scale—what is commonly called the **general circulation of the atmosphere**.

General Circulation of the Atmosphere   Before we study the general circulation, we must remember that it only represents the *average* air flow around the world. Actual winds at any one place and at any given time may vary considerably from this average. Nevertheless, the average can answer why and how the winds blow around the world the way they do—why, for example, prevailing surface winds are northeasterly in Honolulu and westerly in New York City. The average can also give a picture of the driving mechanism behind these winds, as well as a model of how heat and momentum are transported from equatorial regions poleward, keeping the climate in middle latitudes tolerable.

The underlying cause of the general circulation is the unequal heating of the earth's surface. We learned in Chapter 2 that, averaged over the entire earth, incoming solar radiation is roughly equal to outgoing earth radiation. However, we also know that this energy balance is not maintained for each latitude, since the tropics experience a net gain in energy, while polar regions suffer a net loss. To balance these inequities, the atmosphere transports warm air poleward and cool air equatorward. Although seemingly simple, the actual flow of air is complex; certainly not everything is known about it. In order to better understand it, we will first look at some models (that is, artificially constructed analogies) that eliminate some of the complexities of the general circulation.

Single-Cell Model   The first model is the single-cell model, in which we assume that the earth's surface is uniformly covered with water, so that differential heat-

ing between land and water does not come into play. We will further assume that the sun is always directly over the equator, so that the winds will not shift seasonally. Finally, we assume that the earth does not rotate, so that the only force we need deal with is the pressure gradient force. With these assumptions, the general circulation of the atmosphere would look much like Fig. 7.14, a huge thermally driven convection cell in each hemisphere.

This is the **Hadley cell** (named after the eighteenth-century English meteorologist George Hadley, who first proposed the idea). It is driven by energy from the sun. Excessive heating of the equatorial area produces a broad region of surface low pressure, while at the poles excessive cooling creates a region of surface high pressure. In response to the horizontal pressure gradient, cold surface polar air flows equatorward, while at higher levels air flows toward the poles. The entire circulation consists of rising air near the equator, sinking air over the poles, an equatorward flow of air near the surface, with a return flow aloft. In this manner, some of the excess energy of the tropics is transported as sensible and latent heat to the regions of energy deficit at the poles.

Such a simple cellular circulation as this does not actually exist on the earth. For one thing, the earth rotates, so the Coriolis force would deflect the southward-moving surface air in the Northern Hemisphere to the right, producing easterly surface winds at practically all latitudes. These winds would be moving in a direction opposite to that of the earth's rotation and, due to friction with the surface, would slow down the earth's spin. We know that this does not happen and that prevailing winds in middle latitudes actually blow from the west. Therefore, observations alone tell us that a closed circulation of air between the equator and the poles is not the proper model for a rotating earth. (Models that simulate air flow around the globe have also verified this.) How, then, does the wind blow on a rotating planet? To answer, we will keep our model simple by retaining our first two assumptions—that is, that the earth is covered with water and that the sun is always directly above the equator.

Three-Cell Model   If we allow the earth to spin, the simple convection system breaks into a series of rotating cells as shown in Fig. 7.15a. Although this model is considerably more complex than the single-cell model, there are some similarities. The tropical regions still

Figure 7.14
The general circulation of air on a nonrotating earth uniformly covered with water (with the sun directly above the equator).

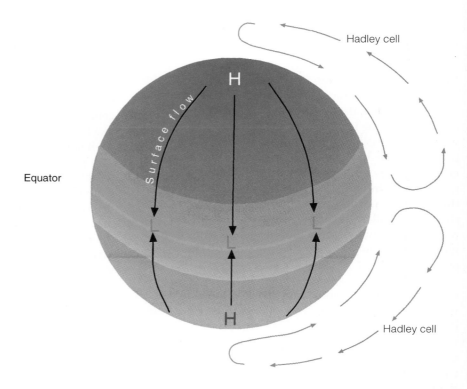

receive an excess of heat and the poles a deficit. In each hemisphere, three cells instead of one have the task of energy redistribution. A surface high-pressure area is located at the poles, and a broad trough of surface low pressure still exists at the equator. From the equator to latitude 30°, the circulation closely resembles that of a Hadley cell. Let's look at this model more closely by examining what happens to the air above the equator. (Refer to Fig. 7.15 as you read the following section.)

Over equatorial waters, the air is warm, horizontal pressure gradients are weak, and winds are light. This region is referred to as the **doldrums**. (The monotony of the weather in this area has given rise to the expression "down in the doldrums.") Here, warm air rises, often condensing into huge cumulus clouds and thunderstorms called *convective "hot" towers* because of the enormous amount of latent heat they liberate. This heat makes the air more buoyant and provides energy to drive the Hadley cell. The rising air reaches the tropopause, which acts like a barrier, causing the air to move laterally toward the poles. The Coriolis force deflects this poleward flow toward the right in the Northern Hemisphere and to the left in the Southern Hemisphere, providing westerly winds aloft in both hemispheres. (We will see later that these westerly

winds reach maximum velocity and produce jet streams near 30° and 60° latitudes.)

Air moving poleward from the tropics constantly cools by radiation, and at the same time it also begins to converge, especially as it approaches the middle latitudes.* This convergence (piling up) of air aloft increases the mass of air above the surface, which in turn causes the air pressure at the surface to increase. Hence, at latitudes near 30°, the convergence of air aloft produces belts of high pressure called **subtropical highs** (or anticyclones). As the converging, relatively dry air above the highs slowly descends, it warms by compression. This subsiding air produces generally clear skies and warm surface temperatures; hence, it is here that we find the major deserts of the world. Over the ocean, the weak pressure gradients in the center of the high produce only weak winds. According to legend, sailing ships traveling to the New World were frequently becalmed in this region; and, as food and supplies dwindled, horses were either thrown over-

---

*You can see why the air converges if you have a globe of the world. Put your fingers on meridian lines at the equator and then follow the meridians poleward. Notice how the lines and your fingers bunch together in the middle latitudes.

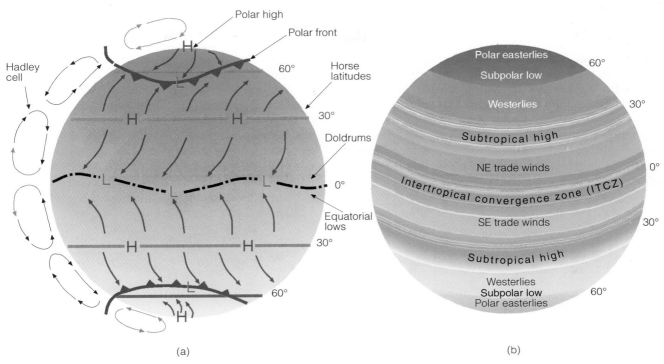

Figure 7.15
Diagram (a) shows the idealized wind and surface pressure distribution over a uniformly water-covered rotating earth. Diagram (b) gives the names of surface winds and pressure systems over a uniformly water-covered rotating earth.

board or eaten. As a consequence, this region is sometimes called the *horse latitudes*.

From the horse latitudes, some of the surface air moves back toward the equator. It does not flow straight back, however, because the Coriolis force deflects the air, causing it to blow from the northeast in the Northern Hemisphere and from the southeast in the Southern Hemisphere. These steady winds provided sailing ships with an ocean route to the New World; hence, these winds are called the **trade winds**. Near the equator, the *northeast trades* converge with the *southeast trades* along a boundary called the **intertropical convergence zone** (ITCZ). In this region of surface convergence, air rises and continues its cellular journey.

Meanwhile, at latitude 30°, not all of the surface air moves equatorward. Some air moves toward the poles and deflects toward the east, resulting in a more or less westerly air flow—called the *prevailing westerlies*, or, simply, **westerlies**—in both hemispheres. Consequently, from Texas northward into Canada, it is much

more common to experience winds blowing out of the west than from the east. The westerly flow is not constant; migrating areas of high and low pressure break up the surface flow pattern from time to time.

As this mild air travels poleward, it encounters cold air moving down from the poles. These two air masses of contrasting temperature do not readily mix. They are separated by a boundary called the **polar front**, a zone of low pressure—the **subpolar low**—where surface air converges and rises and storms develop. Some of the rising air returns at high levels to the horse latitudes, where it sinks back to the surface in the vicinity of the subtropical high. This middle cell is

Did you know?
Strong easterly winds during August, 1983, swept several species of butterflies over 600 miles of open water, from India to the Arabian Peninsula.

completed when surface air from the horse latitudes flows poleward toward the polar front.

Behind the polar front, the cold air from the poles is deflected by the Coriolis force, so that the general flow of air is northeasterly. Hence, this is the region of the **polar easterlies**. In winter, the polar front with its cold air can move into middle and subtropical latitudes, producing a cold polar outbreak. Along the front, a portion of the rising air moves poleward, and the Coriolis force deflects the air into a westerly wind at high levels. Air aloft eventually reaches the poles, slowly sinks to the surface, and flows back toward the polar front, completing the weak polar cell.

We can summarize all of this by referring back to Fig. 7.15 and noting that, at the surface, there are two major areas of high pressure and two major areas of low pressure. Areas of high pressure exist near latitude 30° and the poles; areas of low pressure exist over the equator and near 60° latitude in the vicinity of the polar front. By knowing the way the winds blow around these systems, we have a generalized picture of surface winds throughout the world. The trade winds extend from the subtropical high to the equator, the westerlies from the subtropical high to the polar front, and the polar easterlies from the poles to the polar front.

How does this three-cell model compare with ac-

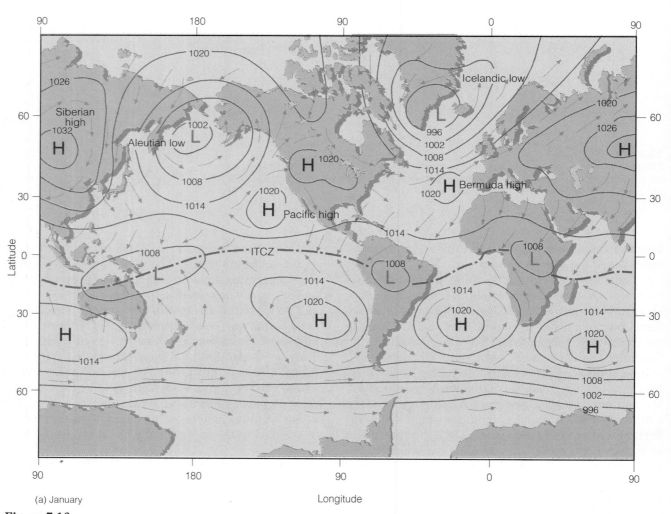

(a) January

Longitude

Figure 7.16
Average sea level pressure distribution and surface wind-flow patterns for January (a) and for July (b). The heavy dashed line represents the position of the ITCZ.

tual observations of winds and pressure? We know, for example, that upper-level winds at middle latitudes generally blow from the west. The middle cell, however, suggests an east wind aloft as air flows equatorward. Hence, discrepancies exist between this model and atmospheric observations. This model does, however, agree closely with the winds and pressure distribution at the *surface*, and so we will examine this next.

Average Surface Winds and Pressure: The Real World   When we examine the real world with its continents and oceans, mountains and ice fields, we obtain an average distribution of sea level pressure and winds

for January and July, as shown in Figs. 7.16a and b. Even though these data are based on sparse observations, especially in unpopulated areas, we can see that there are regions where pressure systems appear to persist throughout the year. These systems are referred to as *semipermanent highs and lows* because they move only slightly during the course of a year.

In Fig. 7.16a, we can see that there are four semipermanent pressure systems in the Northern Hemisphere during January. In the eastern Atlantic, between latitudes 25° and 35°N is the **Bermuda–Azores high**, and, in the Pacific Ocean, its counterpart, the **Pacific high**. These are the subtropical anticyclones that

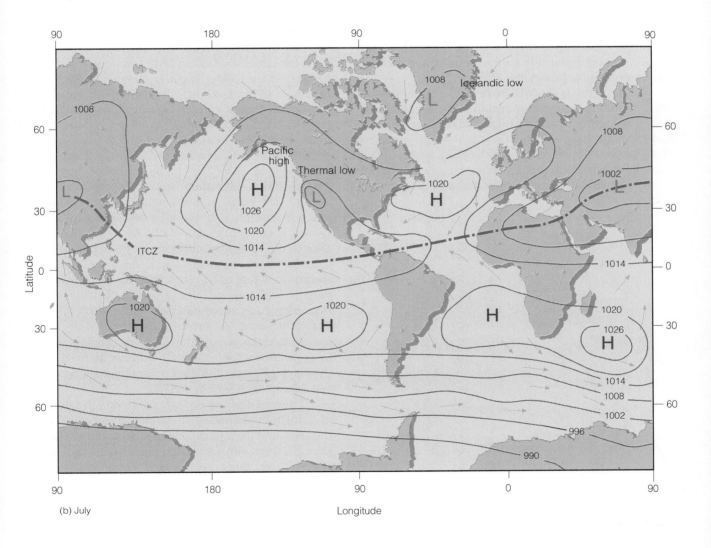

(b) July   Longitude

develop in response to the convergence of air aloft. Since surface winds blow clockwise around these systems, we find the trade winds to the south and the prevailing westerlies to the north. In the Southern Hemisphere, where there is relatively less land area, there is less contrast between land and water, and the subtropical highs show up as well-developed systems with a clearly defined circulation.

Where we would expect to observe the polar front (between latitudes 40° and 65°), there are two semipermanent subpolar lows. In the North Atlantic, there is the **Icelandic low**, which covers Iceland and southern Greenland, while the **Aleutian low** sits over the Aleutian Islands in the North Pacific. These zones of cyclonic activity actually represent regions where numerous storms, having traveled eastward, tend to converge, especially in winter. In the Southern Hemisphere, the subpolar low forms a continuous trough that completely encircles the globe.

On the January map, there are other pressure systems, which are not semipermanent in nature. Over Asia, for example, there is a huge (but shallow) thermal anticyclone called the **Siberian high**, which forms because of the intense cooling of the land. South of this system, the winter monsoon shows up clearly, as air flows away from the high across Asia and out over the

ocean. A similar (but less intense) anticyclone is evident over North America.

As summer approaches, the land warms and the cold, shallow highs disappear. In some regions, areas of surface low pressure replace areas of high pressure. The lows that form over the warm land are *thermal lows*. On the July map (Fig. 7.16b), warm thermal lows are found over the desert southwest of the United States, over the plateau of Iran and over India. As the thermal low over India intensifies, warm, moist air from the ocean is drawn into it, producing the wet summer monsoon so characteristic of India and Southeast Asia.

When we compare the January and July maps, we can see several changes in the semipermanent pressure systems. The strong subpolar lows so well developed in January over the Northern Hemisphere are hardly discernible on the July map. The subtropical highs, however, remain dominant in both seasons. Because the sun is overhead in the Northern Hemisphere in July and overhead in the Southern Hemisphere in January, the zone of maximum surface heating shifts seasonally. In response to this, the major pressure systems, wind belts, and ITCZ (heavy dashed line) shift toward the north in July and toward the south in January.

The General Circulation and Precipitation Patterns The position of the major features of the general circulation and their latitudinal displacement (which annually averages about 10° to 15°) strongly influence the climate of many areas. For example, on the global scale, we would expect abundant rainfall where the air rises and very little where the air sinks. Consequently, areas of high rainfall exist in the tropics, where humid air rises in conjunction with the ITCZ, and between 40° and 55° latitude, where middle latitude storms and the polar front force air upward. Areas of low rainfall are found near 30° latitude in the vicinity of the subtropical highs and in polar regions where the air is cold and dry. (See Fig. 7.17.)

Poleward of the equator, between the doldrums and the horse latitudes, the area is influenced by both the ITCZ and the subtropical high. In summer (high sun period), the subtropical high moves poleward and the ITCZ invades this area, bringing with it ample rainfall. In winter (low sun period), the subtropical high moves equatorward, bringing with it clear, dry weather.

During the summer, the Pacific high drifts northward to a position off the California coast (Fig. 7.18). Sinking air on its eastern side produces a strong

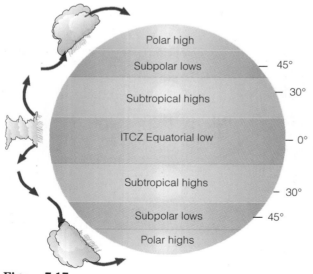

Figure 7.17
Major pressure systems and idealized air motions and precipitation patterns of the general circulation. (Areas shaded blue represent abundant rainfall.)

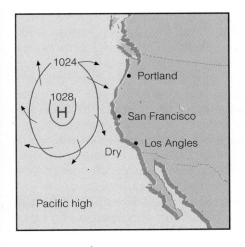

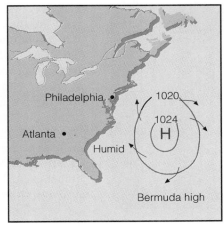

Figure 7.18
During the summer, the Pacific high moves northward. Sinking air along its eastern margin produces a strong subsidence inversion, which causes relatively dry weather to prevail. Along the western margin of the Bermuda high, southerly winds bring in humid air, which rises, condenses, and produces abundant rainfall.

upper-level subsidence inversion. This tends to keep summer weather along the West Coast relatively dry. The rainy season typically occurs in winter when the high moves south and storms are able to penetrate the region. Along the East Coast, the clockwise circulation of winds around the Bermuda high (Fig. 7.18) brings warm, tropical air northward into the United States and southern Canada from the Gulf of Mexico. Because subsiding air is not as well developed on this side of the high, the humid air can rise and condense into towering cumulus clouds and thunderstorms. So, in part, it is

the air motions associated with the subtropical highs that keep summer weather dry in California and moist in Georgia. (Compare the rainfall patterns for Los Angeles, California, and Atlanta, Georgia—Fig. 7.19.)

Westerly Winds and the Jet Stream   In Chapter 6, we learned that the winds above the middle latitudes in both hemispheres blow in a more or less west-to-east direction. The reason for these westerly winds is that, aloft, we generally find higher pressure over equatorial regions and lower pressures over polar regions. Where

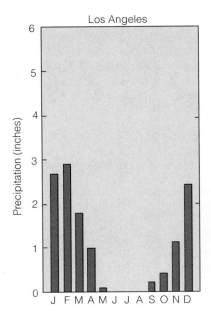

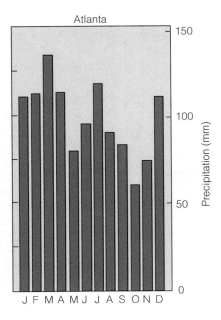

Figure 7.19
Average annual precipitation for Los Angeles, California, and Atlanta, Georgia.

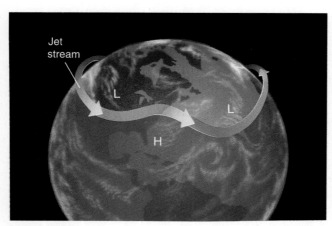

Figure 7.20
A jet stream is a swiftly flowing current of air that moves in a wavy west-to-east direction. It forms along a boundary where colder air lies to the north and warmer air to the south.

these upper-level winds tend to concentrate into narrow bands, we find rivers of fast-flowing air—what we call **jet streams**.

Atmospheric jet streams are swiftly flowing air currents hundreds of miles long, normally less than several hundred miles wide, and typically less than a mile thick. (See Fig. 7.20.) Wind speeds in the central core of a jet stream often exceed 100 knots and occasionally 250 knots. Jet streams are usually found at the tropopause at elevations between 10 and 14 km (33,000 and 46,000 ft) although they may occur at both higher and lower altitudes.

Since jet streams are bands of strong winds, they must form in the same manner as all winds—due to horizontal differences in pressure. In Fig. 7.20, notice that the jet stream is situated along the boundary where

cold, polar air lies to the north and milder, subtropical air lies to the south. Recall from our earlier discussion that this boundary is marked by the polar front. (See Fig. 7.15.) Aloft, sharp contrast in temperatures along the front produce rapid horizontal pressure changes, which sets up a steep pressure gradient. This condition intensifies the wind speed along the front and causes the jet stream. Because the north-south temperature contrast along the front is strongest in winter and weakest in summer, the polar jet shows seasonal variations. In winter, the winds blow stronger and the jet moves farther south, as the leading edge of the cold air may extend into Southern California, south Texas, and even Florida. In summer, the polar jet is weaker and is usually found farther north, such as over Canada.

Jet streams were first encountered by high-flying military aircraft during World War II, but their existence was suspected before that time. Ground-based observations of fast-moving cirrus clouds had revealed that westerly winds aloft must be moving rapidly indeed.

Figure 7.21 illustrates the average position of the jet streams, tropopause, and general air flow for the Northern Hemisphere in winter. From this diagram, we can see that there are two jet streams, both located in the tropopause gaps, where mixing between tropospheric and stratospheric air takes place. The jet stream situated at nearly 13 km (43,000 ft) above the subtropical high is the **subtropical jet**. The jet stream situated at about 10 km (33,000 ft) near the polar front is known as the **polar jet stream**, or, simply, the *polar jet*.

In Fig. 7.21, the wind in the jet core would be flowing as a westerly wind away from the viewer. This direction, of course, is only an average, as jet streams often meander into broad loops that sweep north and

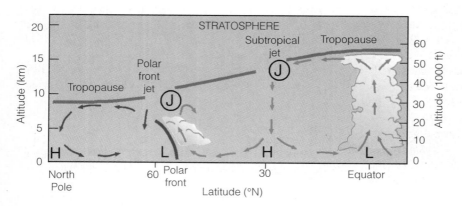

Figure 7.21
Average position of the polar jet stream and the subtropical jet stream, with respect to a model of the general circulation in winter. Both jet streams are flowing into the page, away from the viewer, which would be from west to east.

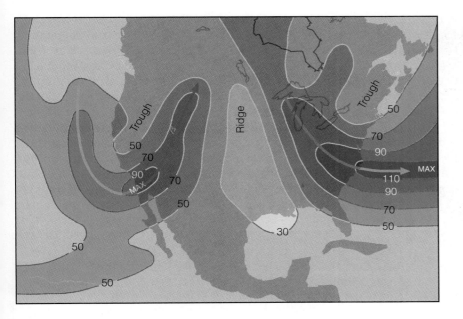

Figure 7.22
Position of the polar jet stream about 9 kilometers (30,000 feet) above the surface on April 18, 1979. Solid lines are lines of equal wind speed (isotachs) in knots. Heavy line is the jet stream axis.

south. The polar jet may split into two jet streams, and it may even merge with the subtropical jet.

We can see the looping pattern of the jet by studying Fig. 7.22. This diagram shows the position of the jet on April 18, 1979. The airflow pattern and jet core are given by the heavy dark arrow; the solid lines represent lines of equal wind speed (*isotachs*). Since the wind flow at this level nearly parallels the contour lines, troughs of low pressure exist over the western states and along the East Coast, while a ridge of high pressure covers the Central Plains. Note that the strongest winds are located in the troughs. This region of strong winds is called a *jet maximum* or *jet streak*. In Chapter 8, we will see how this region of high winds is an important factor in developing and intensifying surface storm systems.

Look at Figure 7.22 again and notice that air moving in the jet core over Los Angeles would move northward into Canada, then loop around and head southeastward, eventually moving off the coast of Virginia. This wavy pattern illustrates an important function of jet streams. On the eastern side of the trough, swiftly moving air carries warmer air poleward, while, on the western side, the more northerly flow brings cold air equatorward. Jet streams are, therefore, significant in the global transfer of heat. Since jet streams tend to meander around the world, we can also see how a radioactive cloud from the nuclear accident at Chernobyl, Russia, during April, 1986, could end up over the United States and elsewhere.

Although the polar and subtropical jets are the two most frequently in the news, there are other jet streams that deserve mentioning. For example, there is a *low-level jet stream* that forms just above the central plains of the United States. During the summer, this jet (which usually has peak winds of less than 60 knots) often contributes to the formation of nighttime thunderstorms. Higher up in the atmosphere, over the subtropics, a summertime easterly jet called the *tropical easterly jet* forms at the base of the tropopause. And during the dark polar winter, a *stratospheric polar jet* forms near the top of the stratosphere.

▲▼▲

Global Wind Patterns and the Oceans

Although scientific understanding of all the interactions between the oceans and the atmosphere is far from complete, there are some relationships that deserve mentioning here.

As the wind blows over the oceans, it causes the surface water to drift along with it. The moving water

Did you know?
Because of the jet stream, a plane flight from New York City to Europe takes about an hour less time than the return flight.

gradually piles up, creating pressure differences within the water itself. This leads to further motion several hundreds of feet down into the water. In this manner, the general wind flow around the globe starts the major surface ocean currents moving. The relationship between the general circulation and ocean currents can be seen by comparing Figs. 7.16 (p. 174) and Fig. 7.23 below.

Because of the larger frictional drag in water, ocean currents move more slowly than the prevailing wind. Typically, they range in speed from several miles per day to several miles per hour. In Fig. 7.23, we can see that ocean currents tend to spiral in semiclosed whirls. In the North Atlantic, flowing northward along the east coast of the United States, is a tremendous warm water current called the *Gulf Stream*, which carries vast quantities of warm, tropical water into higher latitudes. Off the coast of North Carolina, the Gulf Stream provides warmth and moisture for developing mid-latitude storms.

Notice in Fig. 7.23 that as the Gulf Stream moves northward, the prevailing westerlies steer it away from the coast of North America and eastward toward Europe. Generally, it widens and slows as it merges into the broader *North Atlantic Drift*. As this current approaches Europe, part of it flows northward along the

Table 7.1 **Major Ocean Currents**		
1. Gulf Stream	9. South Equatorial Current	17. Peru or Humbolt Current
2. North Atlantic Drift	10. South Equatorial Countercurrent	18. Brazil Current
3. Labrador Current	11. Equatorial Countercurrent	19. Falkland Current
4. West Greenland Drift	12. Kuroshio Current	20. Benguela Current
5. East Greenland Drift	13. North Pacific Drift	21. Agulhas Current
6. Canary Current	14. Alaska Current	22. West Wind Drift
7. North Equatorial Current	15. Oyashio Current	
8. North Equatorial Countercurrent	16. California Current	

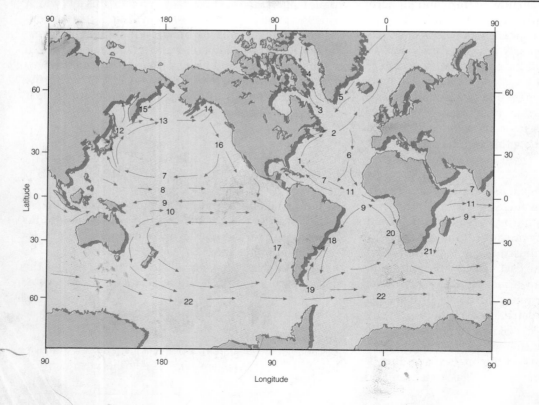

Figure 7.23
Average position and extent of the major surface ocean currents. Cold currents are shown in blue; warm currents are shown in red. Names of the ocean currents are given in Table 7.1.

coasts of Great Britain and Norway, bringing with it warm water (which helps keep winter temperatures much warmer than one would expect this far north). The other part flows southward as the *Canary Current*, which transports cool, northern water equatorward. In the Pacific Ocean, the counterpart to the Canary Current is the *California Current* that carries cool water southward along the coastline of the western United States.

Up to now, we have seen that atmospheric circulations and ocean circulations are closely linked; wind blowing over the oceans produces surface ocean currents. The currents, along with the wind, transfer heat from tropical areas, where there is a surplus of energy, to polar regions, where there is a deficit. This helps to equalize the latitudinal energy imbalance with about 40 percent of the total heat transport in the Northern Hemisphere coming from surface ocean currents. The environmental implications of this heat transfer are tremendous. If the energy imbalance were to go unchecked, yearly temperature differences between low and high latitudes would increase greatly, and the climate would gradually change.*

Winds and Upwelling   Earlier, we saw that the cool California Current flows roughly parallel to the west coast of North America. From this, we might conclude that summer surface water temperatures would be cool along the coast of Washington and gradually warm as we move south. A quick glance at the water temperatures along the west coast of the United States during August (Fig. 7.24) quickly alters that notion. The coldest water is observed along the Northern California coast near Cape Mendocino. The reason for the cold, coastal water is **upwelling**—the rising of cold water from below.

For upwelling to occur, the wind must flow more or less parallel to the coastline. Notice in Fig. 7.25 that summer winds tend to parallel the coastline of California. As the wind blows over the ocean, the surface water beneath it is set in motion. As the surface water moves, it bends slightly to its right due to the Coriolis effect. (Remember, it would bend to the left in the Southern Hemisphere.) The water beneath the surface also moves, and it too bends slightly to its right. The net effect of this phenomenon is that a rather shallow layer

*In an attempt to better understand the circulation of the ocean and its role on climate, the World Ocean Circulation Experiment (WOCE) began during the 1990s.

of surface water moves at right angles to the wind and heads seaward. As the surface water drifts away from the coast, cold, nutrient-rich water from below rises (upwells) to replace it. Upwelling is strongest and surface water is coolest where the wind parallels the coast, such as it does in summer along the coast of Northern California.

Because of the cold coastal water, summertime weather along the West Coast often consists of low clouds and fog, as the air over the water is chilled to its saturation point. On the brighter side, upwelling produces good fishing, as higher concentrations of nutrients are brought to the surface. But swimming is only for the hardiest of souls, since the average surface water temperature in summer is nearly 10°C (18°F) colder than the average coastal water temperature found at the same latitude along the Atlantic Coast.

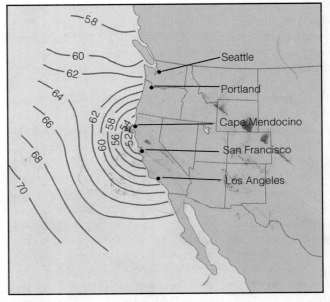

Figure 7.24
Average sea surface temperatures (°F) along the west coast of the United States during August.

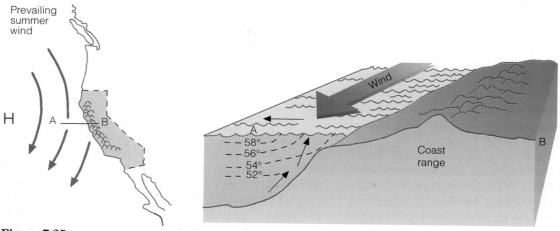

Figure 7.25
As winds blow parallel to the west coast of North America, surface water is transported to the right (out to sea). Cold water moves up from below (upwells) to replace the surface water.

Between the ocean surface and the atmosphere, there is an exchange of heat and moisture that depends, in part, on temperature differences between water and air. In winter, when air-water temperature contrasts are greatest, there is a substantial transfer of sensible and latent heat from the ocean surface into the atmosphere. This energy helps to maintain the global air flow. Consequently, even a relatively small change in surface ocean temperatures could modify atmospheric circulations and have far-reaching effects on global weather patterns. The next section describes how weather events can be linked to surface ocean temperature changes in the tropical Pacific.

El Niño and the Southern Oscillation   Along the west coast of South America, where the cool Peru Current sweeps northward (see Fig. 7.23), southerly winds promote upwelling of cold, nutrient-rich water that gives rise to large fish populations, especially anchovies. The abundance of fish supports a large population of sea birds whose droppings (called *guano*) produce huge phosphate-rich deposits, which support the fertilizer industry. Near the end of each calendar year, a warm current of nutrient-poor tropical water moves southward, replacing the cold, nutrient-rich surface water. Because this condition frequently occurs around Christmas, local residents called it **El Niño** (Spanish for boy child), referring to the Christ child.

In most years, the warming lasts for only a few weeks to a month or more, after which weather pat-

terns usually return to normal and fishing improves. However, when El Niño conditions last for many months, and a more extensive ocean warming occurs, the economic results can be catastrophic. This extremely warm episode, which occurs at irregular intervals of three to seven years, is now referred to as a *major El Niño event.*

During a major El Niño event, large numbers of fish and marine plants may die. Dead fish and birds may litter the water and beaches of Peru; their decomposing carcasses deplete the water's oxygen supply, which leads to the bacterial production of huge amounts of smelly hydrogen sulfide. The El Niño of 1972–1973 drastically reduced the annual Peruvian anchovy catch. Since much of the harvest of this fish is converted into fishmeal and exported for use in feeding livestock and poultry, the world's fishmeal production in 1972 was greatly reduced. Countries such as the United States that rely on fishmeal for animal feed had to use soybeans as an alternative. This raised poultry prices in the United States by more than 40 percent. A less severe El Niño occurred in 1976–1977. But an extremely strong El Niño—one of the strongest ever seen—occurred during 1982–1983.

Normally, in the tropical Pacific Ocean, the trades are persistent winds that blow westward from a region of higher pressure over the eastern Pacific toward a region of lower pressure centered over Indonesia. (See Fig. 7.16a.) The westward-moving trades drag some of the cool water located along the South American coast with them. As this water moves westward, it is heated by sunlight and the atmosphere. Consequently, in the Pacific Ocean, surface water along the equator is cool in the east and warm in the west. In addition, the dragging of surface water raises the sea level in the western Pacific and lowers it in the eastern Pacific. This produces a thick layer of warm water over the tropical western Pacific Ocean and a weak ocean current (called the countercurrent) that flows eastward toward South America.

Every few years, the surface atmospheric pressure patterns break down, as air pressure rises over the region of the western Pacific and falls over the eastern Pacific. This change in pressure weakens the trades, and, during strong pressure reversals, east winds are replaced by west winds. The west winds strengthen the countercurrent, causing warm water to head eastward toward South America over broad areas of the tropical Pacific. Near the end of the warming period, which may last between one and two years, atmospheric pressure over the eastern Pacific reverses and begins to rise, whereas, over the western Pacific, it falls. This seesaw pattern of reversing of surface air pressure at opposite ends of the Pacific Ocean is called the **Southern Oscillation**. Because the pressure reversals and ocean warming are more or less simultaneous, scientists call this phenomenon the *El Niño/Southern Oscillation* or *ENSO* for short. Although most ENSO episodes follow a similar evolution, each event has its own personality, differing in both strength and behavior.

Did you know?
The total damage worldwide (due to flooding, winds, and drought) attributed to the 1982–83 El Niño exceeded $8 billion.

During the 1982–1983 ENSO event, the west winds near the equator were stronger than during any previous episode. As these winds pushed eastward, they dragged surface water with them. This raised the sea level in the east and lowered it in the west. The eastward moving water gradually warmed under the tropical sun, becoming as much as 6°C (11°F) warmer than normal in the eastern equatorial Pacific. Gradually, a thick layer of warm water pushed into coastal areas of Ecuador and Peru, choking off the upwelling that supplies cold, nutrient-rich water to this region. The unusually warm water extended from South America's coastal region for many thousands of miles westward along the equator. (See Fig. 7.26.) The warm tropical water also spread northward along the west coast of North America.

Such a large area of abnormally warm water can have an effect on global wind patterns. The warm tropical water fuels the atmosphere with additional warmth and moisture, which the atmosphere turns into additional storminess and rainfall. The added warmth from the oceans and the release of latent heat during condensation apparently influence the westerly winds aloft

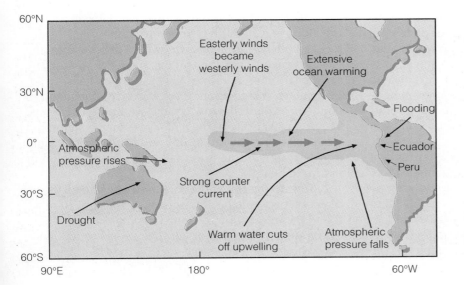

Figure 7.26
Some of the conditions that occurred during the major El Niño event of 1982–1983.

in such a way that certain regions of the world experience too much rainfall, while others have too little.

Although the actual mechanism by which changes in surface ocean temperatures influence global wind patterns is not fully understood, the by-products are plain to see. For example, during the El Niño of 1982–1983, severe drought was felt in Indonesia, southern Africa, and Australia, where the 1982 production of wheat, oats, and barley was half that of the previous year. Meanwhile, record rains and flooding occurred over Ecuador and Peru, where the 1982 commercial fish catch was 50 percent of the 1981 total. In the Northern Hemisphere, an unusually strong subtropical westerly jet stream (that frequently merged with the polar jet) brought storms from California into the Gulf Coast states.

A much weaker ENSO episode during 1986–1987 caused heavy rains and flooding over coastal Ecuador and northwestern Peru. At the same time, the subtropical jet stream (being fueled by warm tropical waters and huge thunderstorms) curved its way over the southeastern United States, where it brought abundant rainfall to a region that, during the previous summer, had suffered through a devastating drought. During the ENSO event of 1991–1992, the subtropical jet stream once again swung over North America. This time it caused extensive flooding in Texas, but it did manage to bring substantial rains to parched Southern California, which had been in a five-year drought.

Following an ENSO event, the trade winds usually return to normal. However, if the trades are exceptionally strong, unusually cold surface water moves over the eastern and central Pacific, and the warm water and rainy weather is confined mainly to the western tropical Pacific. This cold-water episode, which is the opposite of El Niño conditions, has been termed *La Niña* (the girl child). Some scientists speculate that the exceptionally cold La Niña of 1988 may have contributed to the summer drought over North America that year.

Are ENSO episodes predictable? Recent models, which simulate atmospheric and oceanic conditions, did a good job of predicting the 1991–92 event. At present, however, the models are much better at predicting weather trends over a large region than they are in predicting specific weather events for an individual area. Since El Niño and the Southern Oscillation are part of a large scale ocean-atmosphere interaction that can take several years to run its course, the hope is that a better understanding of this phenomenon will provide improved long-range forecasts of weather and climate.

Summary

In this chapter, we examined a variety of atmospheric circulations. We looked at small scale winds and found that eddies can form in a region of strong wind shear, especially in the vicinity of a jet stream. On a slightly larger scale, land and sea breezes blow in response to local pressure differences created by the uneven heating and cooling rates of land and water. Monsoon winds change direction seasonally, while mountain and valley winds change direction daily.

A warm, dry wind that descends the eastern side of the Rocky Mountains is the chinook. The same type of wind in the Alps is the foehn. A warm, dry downslope wind that blows into Southern California is the Santa Ana wind. Local intense heating of the surface can produce small rotating winds, such as the dust devil, while downdrafts in a thunderstorm are responsible for the desert haboob.

The largest pattern of winds that persists around the globe is called the general circulation. At the surface in both hemispheres, winds tend to blow from the east or northeast in the tropics, from the west in the middle latitudes, and from the east or northeast in polar regions. Where upper-level westerly winds tend to concentrate into narrow bands, we find jet streams. The annual shifting of the major pressure systems and wind belts—northward in July and southward in January—strongly influences the annual precipitation of many regions.

Toward the end of the chapter we examined the interaction between the atmosphere and oceans. Here we found the interaction to be an ongoing process where everything, in one way or another, seems to influence everything else. On a large scale, winds blowing over the surface of the water drive the major ocean currents;

the oceans in turn, release energy to the atmosphere, which helps to maintain the general circulation. When warm water extends over a broad area of the tropical Pacific during a condition known as a major El Niño event, the large-scale interactions between the atmosphere and the ocean can have a dramatic effect on the weather and climate of many areas of the world.

Key Terms

The following terms are listed in the order they appear in the text. Define each. Doing so will aid you in reviewing the material covered in this chapter.

scales of motion	monsoon wind system	doldrums	Icelandic low
microscale	valley breeze	subtropical highs	Aleutian low
mesoscale	mountain breeze	trade winds	Siberian high
macroscale	katabatic wind	intertropical convergence	jet stream
rotor	chinook wind	zone (ITCZ)	subtropical jet stream
wind shear	Santa Ana wind	westerlies	polar jet stream
clear air turbulence	haboob	polar front	upwelling
(CAT)	dust devils (whirlwinds)	subpolar low	El Niño
thermal circulation	general circulation	Polar easterlies	Southern Oscillation
sea breeze	of the atmosphere	Bermuda-Azores high	
land breeze	Hadley cell	Pacific high	

Review Questions

1. Describe the various scales of motion and give an example of each.
2. Define the term *wind shear*.
3. Using a diagram, explain how a thermal circulation develops.
4. Why does a sea breeze blow from sea to land and a land breeze from land to sea?
5. (a) Briefly explain how the monsoon wind system develops over eastern and southern Asia.
   (b) Why in India is the summer monsoon wet and the winter monsoon dry?
6. You are fly fishing in a mountain stream during the early morning. Would you expect the wind to be blowing upstream or downstream? Explain.
7. Which wind will produce clouds: a valley breeze or a mountain breeze? Why?
8. How do katabatic winds form?
9. Explain why chinook winds are warm and dry.
10. (a) What is the primary source of warmth for a Santa Ana wind?
    (b) What atmospheric conditions contribute to the development of a strong Santa Ana?

11. Why do dust devils mainly form on sunny, hot days?
12. Draw a large circle. Now, place the major surface pressure and wind belts of the world at their appropriate latitudes.
13. According to Fig. 7.15 (p. 173) most of the United States is located in what wind belt?
14. Explain how and why the average surface pressure features shift from summer to winter.
15. Explain the relationship between the general circulation of air and the circulation of ocean currents.
16. Describe how the winds along the west coast of North America produce upwelling.
17. (a) What is a major El Niño event?
    (b) What happens to the surface pressure at opposite ends of the Pacific Ocean during the Southern Oscillation?
    (c) Describe how the Southern Oscillation influences a major El Niño event.

A dissipating middle latitude cyclonic storm, photographed by *Apollo 9* astronauts, spins counterclockwise over the North Pacific Ocean. (Photo: NASA)

Chapter 8

Air Masses, Fronts, and Middle Latitude Storms

Contents

About two o'clock in the afternoon it began to grow dark from a heavy, black cloud which was seen in the northwest. Almost instantly the strong wind, traveling at the rate of 70 miles an hour, accompanied by a deep bellowing sound, with its icy blast, swept over the land, and everything was frozen hard. The water in the little ponds in the roads froze in waves, sharp edged and pointed, as the gale had blown it. The chickens, pigs and other small animals were frozen in their tracks. Wagon wheels ceased to roll, froze to the ground. Men, going from their barns or fields a short distance from their homes, in slush and water, returned a few minutes later walking on the ice. Those caught out on horseback were frozen to their saddles, and had to be lifted off and carried to the fire to be thawed apart. Two young men were frozen to death near Rushville. One of them was found with his back against a tree, with his horse's bridle over his arm and his horse frozen in front of him. The other was partly in a kneeling position, with a tinder box in one hand and a flint in the other, with both eyes wide open as if intent on trying to strike a light. Many other casualties were reported. As to the exact temperature, however, no instrument has left any record; but the ice was frozen in the stream, as variously reported, from six inches to a foot in thickness in a few hours.

John Moses, *Illinois: Historical and Statistical*

8 The opening details the passage of a spectacular cold front as it moved through Illinois on December 21, 1836. Although no reliable temperature records are available, estimates are that, as the front swept through, air temperatures dropped almost instantly from the balmy 40s (°F) to 0 degrees. Fortunately, temperature changes of this magnitude are quite rare with cold fronts.

In this chapter, we will examine the more typical weather associated with cold fronts and warm fronts. We will address questions such as: Why are cold fronts usually associated with showery weather? How can warm fronts cause freezing rain and sleet to form over a vast area during the winter? And how can one read the story of an approaching warm front by observing its clouds? We will also see how weather fronts are an integral part of a mid-latitude cyclonic storm. But, first,

so that we may better understand fronts and storms, we will examine air masses. We will look at where and how they form and the type of weather usually associated with them.

Air Masses

An **air mass** is an extremely large body of air whose properties of temperature and moisture are fairly similar in any horizontal direction at any given altitude. Air masses may cover many thousands of square miles. In Fig. 8.1, a large winter air mass, associated with a high-pressure area, covers over half of the United States on January 9, 1976. Note that, although the surface air temperature and dew point vary somewhat, everywhere

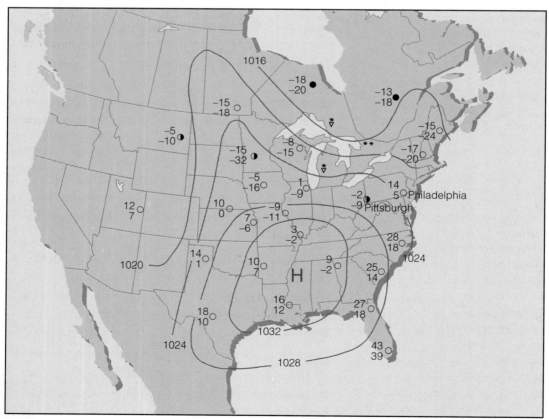

Figure 8.1
A large, extremely cold winter air mass is dominating the weather over much of the United States on the morning of January 9, 1976. At almost all cities, the air is cold and dry. Upper number is air temperature (°F); bottom number is dew point (°F).

the air is cold and dry, with the exception of the zone of snow showers on the eastern shores of the Great Lakes. This cold, shallow anticyclone will drift eastward, carrying with it the temperature and moisture characteristic of the region where the air mass formed; hence, in a day or two, cold air will be located over the central Atlantic Ocean. Part of weather forecasting is, then, a matter of determining air mass characteristics, predicting how and why they change, and in what direction the systems will move.

Source Regions   Regions where air masses originate are known as **source regions**. In order for a huge mass of air to develop uniform characteristics, its source region should be generally flat and of uniform composition, with light surface winds. The longer the air remains stagnant over its source region, the more likely it will acquire properties of the surface below. Consequently, ideal source regions are usually those areas dominated by high pressure. They include the ice- and snow-covered arctic plains in winter and subtropical oceans and desert regions in summer. The middle latitudes, where surface temperatures and moisture characteristics vary considerably, are not good source regions. Instead, this region is a transition zone where air masses with different physical properties move in, clash, and produce an exciting array of weather activity.

Classification   Air masses are grouped into four general categories according to their source region. Air masses that originate in polar latitudes are designated by the capital letter "P" (for *Polar*); those that form in warm tropical regions are designated by the capital letter "T" (for *Tropical*). If the source region is land, the air mass will be dry and the lowercase letter "c" (for *continental*) precedes the P or T. If the air mass originates over water, it will be moist—at least in the lower layers—and the lowercase letter "m" (for *maritime*) precedes the P or T. We can now see that polar air originating over land will be classified cP on a surface weather chart, while tropical air originating over water will be marked as mT. In winter, an extremely cold cP air mass is designated as cA, *continental arctic*. Often, however, it is difficult to distinguish between arctic and polar air masses, especially when the arctic air mass has traveled over warmer terrain. By the same token, an extremely hot, humid air mass originating over equatorial waters is sometimes designated as mE,

for *maritime equatorial*. Distinguishing between equatorial and tropical air masses is usually difficult. Table 8.1 lists the four basic air masses.

When the air mass is colder than the underlying surface, it is warmed from below, which makes the air unstable at low levels. In this case, increased convection and turbulent mixing near the surface usually produce good visibility, cumuliform clouds, and showers of rain or snow. On the other hand, when the air mass is warmer than the surface below, the lower layers are chilled by contact with the cold earth. Warm air above cooler air produces stable air with little vertical mixing. This situation causes the accumulation of dust, smoke, and pollutants, which restricts surface visibilities. In moist air, stratiform clouds accompanied by drizzle or fog may form.

Air Masses of North America   The principal air masses (with their source regions) that invade the United States are shown in Fig. 8.2. We are now in a position to study the formation and modification of each of these air masses and the variety of weather that accompanies them.

cP (Continental Polar) and cA (Continental Arctic) Air Masses   The bitterly cold weather that enters the United States in winter is associated with **continental polar** and **continental arctic** air masses. These originate over the ice- and snow-covered regions of northern Canada and Alaska where long, clear nights allow for strong radiational cooling of the surface. Air in contact with the surface becomes quite cold and stable. Since little moisture is added to the air, it is also quite dry. Eventually a portion of this cold air breaks away and, under the influence of the air flow aloft, moves southward as an enormous shallow high-pressure area.

Table 8.1
Air Mass Classification and Characteristics

Source Region	Polar (P)	Tropical (T)
Land continental (c)	cP cold, dry, stable	cT hot, dry, stable air aloft; unstable surface air
Water maritime (m)	mP cool, moist, unstable	mT warm, moist; usually unstable

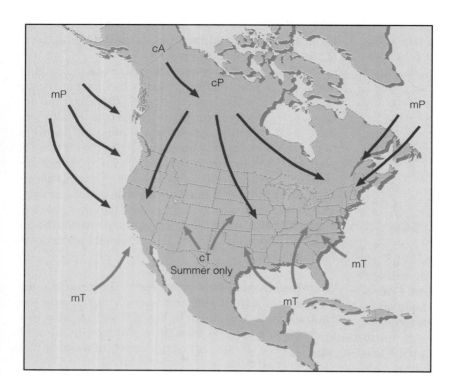

Figure 8.2
Air mass source regions and their paths.

As the cold air moves into the interior plains, there are no topographic barriers to restrain it, so it continues southward, bringing with it cold wave warnings and frigid temperatures. As the air mass moves over warmer land to the south, the air temperature moderates slightly. However, even during the afternoon, when the surface air is most unstable, cumulus clouds are rare because of the extreme dryness of the air mass. At night, when the winds die down, rapid surface cooling and clear skies combine to produce low minimum temperatures. If the cold air moves as far south as central or southern Florida, the winter vegetable crop may be severely damaged. When the cold, dry air mass moves over a relatively warm body of water, such as the Great Lakes, heavy snow showers—called **lake-effect snows**—often form on the eastern shores. (More information on lake-effect snows is provided in the Focus section on p. 191.)

In winter, the generally fair weather accompanying cP air is due to the stable nature of the atmosphere aloft. Sinking air develops above the large dome of high pressure. The subsiding air warms by compression and creates warmer air, which lies above colder surface air. Therefore, a strong upper-level subsidence inversion

often forms. Should the anticyclone stagnate over a region for several days, the visibility gradually drops as pollutants become trapped in the cold air near the ground. Usually, however, winds aloft move the cold air mass either eastward or southeastward.

The Rockies, Sierra Nevada, and Cascades normally protect the Pacific Northwest from the onslaught of cP air, but, occasionally, cP air masses do invade these regions. When the upper-level winds over Washington and Oregon blow from the north or northeast on a trajectory beginning over northern Canada or Alaska, cold cP (and occasionally cA) air can slip over the mountains and extend its icy fingers all the way to the Pacific Ocean. As the air moves off the high plateau, over the mountains, and on into the lower valleys, compressional heating of the sinking air causes its temperature to rise, so that by the time it reaches the lowlands, it is considerably warmer than it was originally. However, in no way would this air be considered warm. In some cases, the subfreezing temperatures slip over the Cascades and extend southward into the coastal areas of southern California.

A similar but less dramatic warming of cP and cA air occurs along the eastern coast of the United States.

Focus on a Special Topic
Lake-effect Snows

During the winter, when the weather in the Midwest is dominated by clear, brisk cP air, people living on the eastern shores of the Great Lakes brace themselves for heavy snow showers. Snowstorms that form on the downwind side of one of these lakes are known as *lake-effect snows*. These storms are highly localized, extending from just a few miles to more than 30 miles inland. The snow usually falls as a heavy shower or squall in a concentrated zone. So centralized is the region of snowfall, that one part of a city may accumulate many inches of snow, while, in another part, the ground is bare.

Lake-effect snows are most numerous from November to January. During these months, cP air moves over the lakes when they are relatively warm and not quite frozen. The contrast in temperature between water and air can be as much as 25°C (45°F). Studies show that the greater the contrast in temperature, the greater the potential for snow showers. In Fig. 1, we can see that, as the cold air moves over the warmer water, the air mass is quickly warmed from below, making it more buoyant and less stable. Rapidly, the air sweeps up moisture, soon becoming saturated. Out over the water, the vapor condenses into steam fog. As the air continues to warm, it rises and forms billowing cumuliform clouds, which continue to grow as the air becomes more unstable. Eventually, these clouds produce heavy showers of snow, which make the lake seem like a snow factory. Once the air and

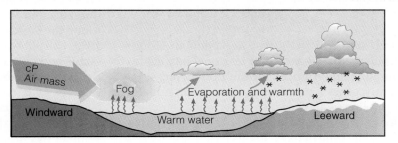

Figure 1
The formation of lake-effect snows. Cold, dry air crossing the lake gains moisture and warmth from the water. The more buoyant air now rises, forming clouds that deposit large quantities of snow on the lake's leeward shores.

clouds reach the downwind side of the lake, additional lifting is provided by low hills and the convergence of air as it slows down over the rougher terrain. In late winter, the frequency and intensity of lake-effect snows taper off as the temperature contrast between water and air diminishes and larger portions of the lakes freeze.

Generally, the longer the stretch of water over which the air mass travels (the longer the fetch), the greater the amount of warmth and moisture derived from the lake, and the greater the potential for heavy snow showers. Consequently, forecasting lake-effect snowfalls depends to a large degree on determining the trajectory of the air as it flows over the lake. Regions that experience heavy lake-effect snowfalls are shown in Fig. 2.

As the cP air moves farther east, the heavy snow showers usually taper off; however, the western slope of the Appalachian Mountains produces further lifting, enhancing the possibility of more and heavier showers. The heat given off during condensation

Figure 2
Shaded areas show regions that experience heavy lake-effect snows.

warms the air and, as the air descends the eastern slope, compressional heating warms it even more. Snowfall ceases, and by the time the air arrives at Philadelphia, New York, or Boston, the only remaining trace of the snow showers occurring on the other side of the mountains are the puffy cumulus clouds drifting overhead.

Did you know?
An arctic air mass from northern Canada on January 19, 1977, brought frigid weather to Florida and even snow to Miami and the northern Keys.

Air rides up and over the lower Appalachian Mountains. Turbulent mixing and compressional heating increase the air temperatures on the downwind side. Consequently, cities located to the east of the Appalachian Mountains usually do not experience temperatures as low as those on the west side. In Fig. 8.1, notice that for the same time of day—in this case 7 A.M. EST—Philadelphia, with an air temperature of 14°F, is 16°F warmer than Pittsburgh, at –2°F.

Figure 8.3 shows two upper-air patterns that led to extremely cold outbreaks of arctic air during December 1989 and 1990. Upper-level winds typically blow from west to east, but, in both of these cases, the flow, as given by the heavy, dark arrows, had a strong north-south (meridional) trajectory. The H represents the positions of the cold surface anticyclones. Numbers

on the map represent minimum temperatures (°F) recorded during the cold spells. East of the Rocky Mountains, over 350 record low temperatures were set between December 21 and 24, 1989, with the arctic outbreak causing an estimated $480 million in damage to the fruit and vegetable crops in Texas and Florida. Along the West Coast, the frigid air during December, 1990, caused over $300 million in damage to the vegetable and citrus crops, as temperatures over parts of California plummeted to their lowest readings in more than fifty years. Notice in both cases how the upper-level wind directs the paths of the air masses.

The cP air that moves into the United States in summer has properties much different from its winter counterpart. The source region remains the same but is now characterized by long summer days that melt snow and warm the land. The air is only moderately cool, and surface evaporation adds water vapor to the air. A summertime cP air mass usually brings relief from the oppressive heat in the central and eastern states, as cooler air lowers the air temperature to more comfortable levels. Daytime heating warms the lower layers, producing surface instability. With its added

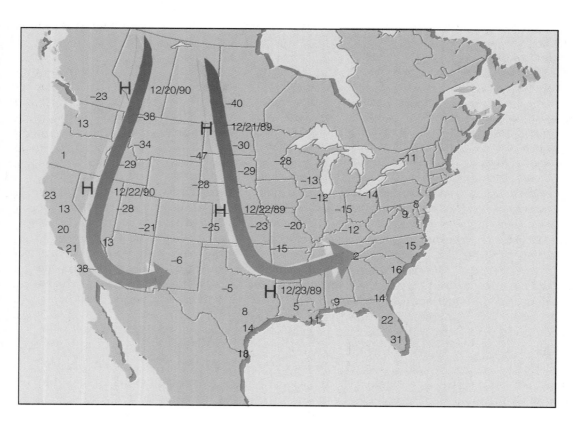

Figure 8.3
Average upper-level wind flow (heavy arrows) and surface position of anticyclones (H) associated with two extremely cold outbreaks of arctic air during December. Numbers on the map represent minimum temperatures (°F) measured during each cold snap.

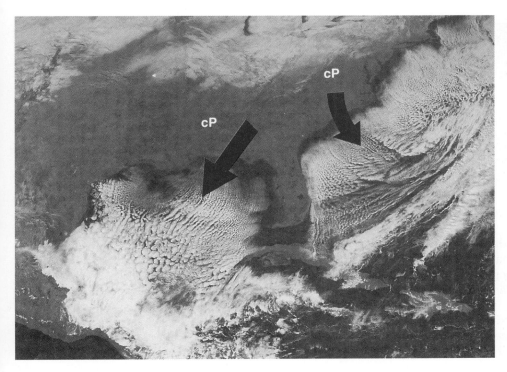

Figure 8.4
Visible satellite picture showing the modification of cP air as it moves over the warmer Gulf of Mexico and the Atlantic Ocean.

moisture, the rising air may condense and create a sky dotted with fair weather cumulus clouds.

When an air mass moves over a large body of water, its original properties may change considerably. For instance, cold, dry cP air moving over the Gulf of Mexico warms rapidly and gains moisture. The air quickly assumes the qualities of a maritime air mass. Notice in Fig. 8.4 that rows of cumulus clouds are forming over the Gulf parallel to northerly surface winds as cP air is being warmed by the water beneath it. As the air continues its journey southward into Mexico and Central America, strong, moist northerly winds build into heavy clouds (bright area) and showers along the northern coast. Hence, a once cold, dry, and stable air mass can be modified to such an extent that its original characteristics are no longer discernible. When this happens, the air mass is given a new designation.

In summary, polar and arctic air masses are responsible for the bitter cold winter weather that can cover wide sections of North America. When the air mass originates over the Canadian Northwest territories, frigid air can bring record-breaking low temperatures. Such was the case on Christmas Eve, 1983, when arctic air covered most of North America. (A detailed look at this air mass and its accompanying record-

setting low temperatures is given in the Focus section beginning on p. 194.)

mP (Maritime Polar) Air Masses During the winter, cP air originating over Asia and frozen polar regions is carried eastward and southward over the Pacific Ocean by the circulation around the Aleutian low. The ocean water modifies the cP air by adding warmth and moisture to it. Since this air has to travel over water many hundreds or even thousands of miles, it gradually changes into a **maritime polar** air mass.

By the time this air mass reaches the Pacific Coast it is cool, moist, and unstable. The ocean's effect is to keep air near the surface warmer than the air aloft. Temperature readings in the 40s and 50s (°F) are common near the surface, while air several thousand feet above the surface may be at the freezing point. Within this colder air, characteristics of the original cP air mass may still prevail. As the air moves inland, coastal mountains force it to rise, and much of its water vapor condenses into rain-producing clouds. In the colder air aloft, the rain changes to snow, with heavy amounts accumulating in mountain regions. A typical upper-level wind-flow pattern that brings mP air onto the west coast of North America is shown in Fig. 8.5.

Focus on a Special Topic
The Return of the Siberian Express

The winter of 1983–1984 was one of the coldest on record across North America. Unseasonably cold weather arrived in December, which, for much of the country, was one of the coldest Decembers since records have been kept. During the first part of the month, continental polar air covered most of the northern and central plains. As the cold air moderated slightly, far to the north a huge mass of bitter cold arctic air was forming over the frozen reaches of the Canadian Northwest territories.

By midmonth, the frigid air, associated with a massive high-pressure area, covered all of northwest Canada. Meanwhile, aloft, strong northerly winds directed the leading edge of the frigid air southward over the prairie provinces of Canada and southward into the United States. Because the extraordinarily cold air was accompanied in some regions by winds gusting to 45 knots, at least one news reporter dubbed the onslaught of this arctic blast, "the Siberian Express."

The Express dropped temperatures to some of the lowest readings ever recorded during the month of December. On December 22, Elk Park, Montana, recorded an unofficial low of –53°C (–64°F), only 4°C higher than

the all-time low of –57°C (–70°F) for the nation (excluding Alaska) recorded at Rogers Pass, Montana, on January 20, 1954.

The center of the massive anticyclone gradually pushed southward out of Canada. By December 24, its center was over eastern Montana (Fig. 3), where the sea level pressure at Miles City reached an incredible 1064 mb (31.40 in.)—a new United States record that topped the old mark of 1063 mb set in Helena, Montana, on January 10, 1962. An enormous ridge of high pressure stretched from the Canadian arctic coast to the Gulf of Mexico. On the east side of the ridge, cold westerly winds brought lake-effect snows to the eastern shores of the Great Lakes. To the south of the high-pressure center, cold easterly winds, rising along the elevated plains, brought light amounts of *upslope snow* to sections of the Rocky Mountain states. Notice in Fig. 3 that, on Christmas Eve, arctic air covered almost 90 percent of the United States. As the cold air swept eastward and southward, a hard freeze caused hundreds of millions of dollars in damage to the fruit and vegetable crops in Texas, Louisiana, and Florida. On Christmas Day, 125 record low temperature readings were set in

twenty-four states. That afternoon, at 1:00 P.M., it was actually colder in Atlanta, Georgia, than it was in Fairbanks, Alaska. One of the worst cold waves to occur in December during this century continued through the week, as many new record lows were established in the Deep South from Texas to Louisiana.

By January 1, the extreme cold had moderated, as the upper-level winds became more westerly. These winds brought milder mP Pacific air eastward into the Great Plains. The warmer pattern continued until about January 10, when the Siberian Express decided to make a return visit. Driven by strong upper-level northerly winds, impulse after impulse of arctic air from Canada swept across the United States. On January 18, an all-time record low of –54°C (–65°F) was recorded for the state of Utah at Middle Sinks. On January 19, temperatures plummeted to a new low of –22°C (–7°F) for the airports in Philadelphia and Baltimore. Toward the end of the month, the upper-level winds once again became more westerly. Over much of the nation, the cold air moderated. But the express was to return at least one more time.

The beginning of February saw relatively warm air covering much of the

---

When the mP air moves inland, it loses much of its moisture as it crosses a series of mountain ranges. Beyond these mountains, it travels over a cold, elevated plateau that chills the surface air and slowly transforms the lower level into dry, stable cP air. East of the Rockies this air mass is referred to as *Pacific air*. Here it often brings fair weather and temperatures not nearly as cold as the cP air, which invades this region from northern Canada. In fact, when mP air from the west replaces retreating cP air from the north, chinook winds often develop. Furthermore, when the modified mP air replaces moist tropical air, storms can form along the boundary separating the two air masses.

Along the East Coast, mP air originates in the North Atlantic as cP air moves southward some distance off the Atlantic Coast. It then swings southwest-

Focus on a Special Topic
The Return of the Siberian Express, *continued*

Figure 3
Surface weather map
for 7 A.M., EST, De-
cember 24, 1983.
Solid lines are isobars.
Shaded area repre-
sents precipitation.
An extremely cold arc-
tic air mass covers
nearly 90 percent of
the United States.
(Weather symbols for
the surface map are
given in Appendix C.)

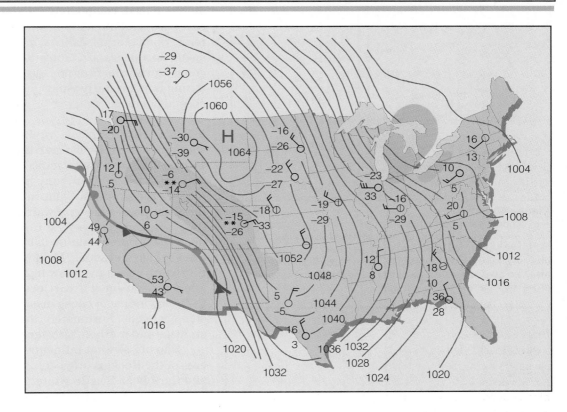

nation from California to the Atlantic coast. On February 4, an arctic outbreak of cA air spread southward and eastward across the nation. Although freezing air extended southward into central Florida, the express ran out of steam, and a February heat wave soon engulfed most of the United States east of the Rocky Mountains. Maritime tropical air from the Gulf of Mexico brought record warmth to much of the eastern two-thirds of the nation. Near the middle of the month, Louisville, Kentucky, reported 23°C (73°F) and Columbus, Ohio, 21°C (69°F). Even though February was one of the warmest months on record over parts of the United States, the winter of 1983–1984 (December, January, and February) will go down in the record books as one of the coldest winters for the United States as a whole since reliable record keeping began in 1931.

ward toward the northeastern states. Because the water of the North Atlantic is very cold and the air mass travels only a short distance over water, wintertime Atlantic mP air masses are usually much colder than their Pacific counterparts. Because the prevailing winds aloft are westerly, Atlantic mP air masses are also much less common.

Figure 8.6 illustrates a typical late winter or early spring surface weather pattern that carries mP air from the Atlantic into the New England and middle Atlantic states. A slow-moving, cold anticyclone drifting to the east (north of New England) causes a northeasterly flow of mP air to the south. The boundary separating this invading colder air from warmer air even farther south is marked by a stationary front. North of this front, northeasterly winds provide generally undesir-

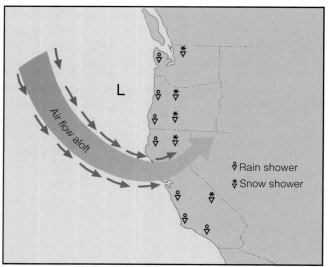

Figure 8.5
A winter upper-air pattern that brings mP air into the west coast of
North America. The large arrow represents the upper-level flow.
Note the trough of low pressure along the coast. The small arrows
show the trajectory of the mP air at the surface. Regions that normally experience precipitation under these conditions are also
shown on the map.

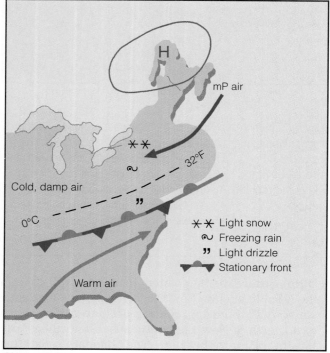

Figure 8.6
Winter and early spring surface weather patterns that usually pre-
vail during the invasion of mP air into the mid-Atlantic and New
England states. (Green-shaded areas represent precipitation.)

able weather, consisting of damp air and low, thick
clouds from which light precipitation falls in the form
of rain, drizzle, or snow. As we will see later in this
chapter, when upper atmospheric conditions are right,
storms may develop along the stationary front, move
eastward, and intensify near the shores of Cape Hatteras. Such storms, called *Hatteras lows*, sometimes
swing northeastward along the coast, where they become *northeasters* (or nor'easters) bringing with them
strong northeasterly winds, heavy rain or snow, and
coastal flooding.

mT (Maritime Tropical) Air Masses   The wintertime
source region for Pacific **maritime tropical** air masses
is the subtropical east Pacific Ocean. Air from this region must travel over many miles of water before it
reaches the southern California coast. Consequently,
these air masses are very warm and moist by the time
they arrive along the West Coast. The warm air produces heavy precipitation usually in the form of rain,
even at high elevations. Melting snow and rain quickly
fill rivers, which overflow into the low-lying valleys.
The rapid snowmelt leaves local ski slopes barren, and
the heavy rain can cause disastrous mudslides in the
steep canyons. Fortunately, the invasion of a real mT
air mass into northern California is rare.

The mT air that influences much of the weather
east of the Rockies originates over the Gulf of Mexico
and Caribbean Sea. In winter, cold polar air tends to
dominate the continental weather scene, so mT air is
usually confined to the Gulf and extreme southern
states. Occasionally, a slow-moving storm system over
the Central Plains draws mT air northward. Gentle,
moist south or southwesterly winds blow into the central and eastern parts of the nation in advance of the
system. Since the land is still extremely cold, air near
the surface is chilled to its dew point. Fog and low
clouds form in the early morning, dissipate by midday,
and reform in the evening. This mild winter weather in
the Mississippi and Ohio valleys lasts, at best, only a
few days. Soon cold polar air will move down from the
north behind the eastward-moving storm system. Along
the boundary between the two air masses, the mT air
is lifted above the more dense cP air, which often leads
to heavy and widespread precipitation and storminess.

When a storm system stalls over the Central
Plains, a constant supply of mT air from the Gulf of
Mexico can bring record-breaking maximum temperatures to the eastern half of the country. Sometimes the
air temperatures are higher in the mid-Atlantic states

than they are in the Deep South, as compressional heating warms the air even more as it moves downslope after crossing the Appalachian Mountains. Figure 8.7 shows a surface weather map and the associated upper air flow (heavy arrow) that brought unseasonably warm mT air into the central and eastern states during April, 1976. A large high centered off the southeast coast coupled with a strong southwesterly flow aloft carried warm, moist air into the Midwest and East, causing a record-breaking April heat wave. The flow aloft prevented the surface low and the cP air behind it from making much eastward progress, so that the warm spell lasted for five days. Note that, on the west side of the surface low, the winds aloft funneled cold cP air from the north into the western states, creating unseasonably cold weather from California to the Rockies. Hence, while people in the Southwest were huddled around heaters, others several thousand miles

away in the Northeast were turning on air conditioners. We can see that it is the upper-level flow, directing cP air southward and mT air northward, which makes these contrasts in temperature possible.

As maritime air moves inland over the hot continent, it warms, rises, and frequently causes cumuliform clouds, which produce afternoon showers and thunderstorms. You can almost count on thunderstorms developing along the Gulf Coast each afternoon in summer. As evening approaches, thunderstorm activity typically dies off. Nighttime cooling lowers the air temperature and, if the air becomes saturated, fog or low clouds form. These, of course, dissipate by late morning as surface heating warms the air again.

Normally, mT air originating in the tropical eastern Pacific remains far south of California. Occasionally, a weak upper-level southerly flow will spread this humid air northward into the southwestern United States,

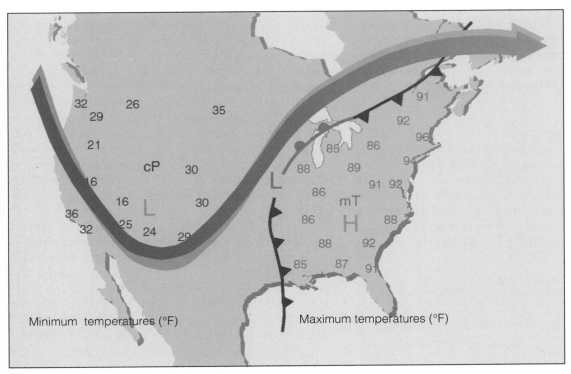

Figure 8.7
Weather conditions during an unseasonally hot spell in the East that occurred between the 15th and 20th of April, 1976. The surface low-pressure area and fronts are shown for April 17. Numbers to the east of the low (in red) are maximum temperatures recorded during the hot spell, while those to the west of the low (in blue) are minimums reached during the same time period. The heavy arrow is the average upper-level flow during the period. The faint L and H show average positions of the upper-level trough and ridge.

Figure 8.8
Humid maritime tropical air produces towering cumulus clouds and thunderstorms as it is lifted up along the Sierra Nevada.

most often Arizona, Nevada, and southern California. In many places, the moist, unstable air aloft only shows up as middle and high cloudiness. However, where the moist flow meets a mountain barrier, it usually rises and condenses into towering shower-producing clouds. (See Fig. 8.8.)

cT (Continental Tropical) Air Masses   The only real source region for hot, dry **continental tropical** air masses in North America is found during the summer in northern Mexico and the adjacent arid southwestern United States. Here, the air mass is hot, dry, and unstable at low levels, with frequent dust devils forming during the day. Because of the low relative humidity (typically less than 10 percent during the afternoon),

air must rise to great heights before condensation begins. Furthermore, an upper-level ridge usually produces weak subsidence over the region, tending to make the air aloft rather stable and the surface air even warmer. Consequently, skies are generally clear, the weather is hot, and rainfall is practically nonexistent where cT air masses prevail. If this air mass moves outside its source region and into the Great Plains and stagnates over that region for any length of time, a severe drought may result. Figure 8.9 shows a weather map situation where cT air covers a large portion of the western United States and produces hot, dry weather northward to Canada.

So far, we have examined the various air masses that enter North America annually. The characteristics of each depend upon the air mass source region and the type of surface over which the air mass moves. The winds aloft determine the trajectories of these air masses. Occasionally, an air mass will control the weather in a region for some time. These persistent weather conditions are known as **air mass weather**.

Air mass weather is especially common in the southeastern United States during summer as, day after day, mT air from the Gulf brings sultry conditions and afternoon thunderstorms. It is also common in the Pacific Northwest in winter when unstable, cool mP air accompanied by widely scattered showers dominates the weather for several days or more. The real weather

---

Did you know?
A continental tropical air mass, stretching from Southern California to the heart of Texas, brought record warmth to the desert southwest during the last week of June, 1990. The temperature reached a sweltering peak of 122°F in Phoenix, Arizona, on June 26. Officials suspended aircraft takeoffs at Sky Harbor Airport, as the reduced lift brought on by the extreme heat had caused exceptionally low air densities.

Figure 8.9

During August 6th and 7th, 1983, cT air covered a large area of the western United States. Numbers on the map represent maximum temperatures (°F) during this period. The faint H with the isobar shows the upper-level position of a subtropical high. Sinking air associated with the high contributed to the hot weather. Winds aloft were weak with the main flow (heavy line) positioned over Canada.

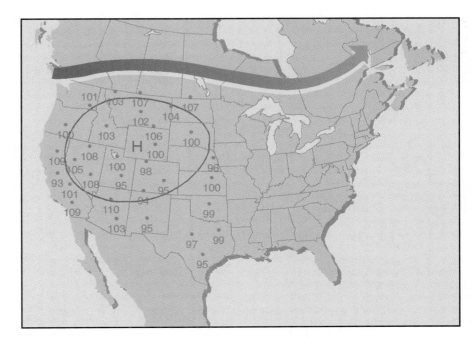

action, however, usually occurs not within air masses but at their margins, where air masses with sharply contrasting properties meet—in the zone marked by weather fronts.*

Fronts

Although we briefly looked at fronts in Chapter 1, we are now in a position to study them in depth, which will aid us in forecasting the weather. We will now learn about the general nature of fronts—how they move and what weather patterns are associated with them.

A **front** is the transition zone between two air masses of different densities. Since density differences are most often caused by temperature differences, fronts usually separate air masses with contrasting temperatures. Often, they separate air masses with different humidities as well. Remember that air masses have both horizontal and vertical extent; consequently, the upward extension of a front is referred to as a *frontal surface*, or a *frontal zone*.

---

*The word "front" is used to denote the clashing of two air masses, probably because it resembles the fighting in Western Europe during World War I, when the term originated.

Figure 8.10 shows a simplified weather map illustrating four different fronts. As we move from west to east across the map, the fronts appear in the following order: a stationary front between points A and B; a cold front between points B and C; a warm front between points C and D; and an occluded front between points C and L. Let's examine the properties of each of these fronts.

Stationary Fronts   A **stationary front** has essentially no movement. On a colored weather map, it is drawn as an alternating red and blue line. Semicircles face toward colder air on the red line and triangles point toward warmer air on the blue line. The stationary front between points A and B in Fig. 8.10 marks the boundary where cold, dense cP air from Canada butts up against the north-south trending Rocky Mountains. Unable to cross the barrier, the cold air shows little or no westward movement. The stationary front is drawn along a line separating the cP from the milder mP air to the west. Notice that the surface winds tend to blow parallel to the front, but in opposite directions on either side of it.

The weather along the front is clear to partly cloudy, with much colder air lying on its eastern side. Because both air masses are dry, there is no precipitation. This is not, however, always the case. When warm, moist air rides up and over the cold air, widespread

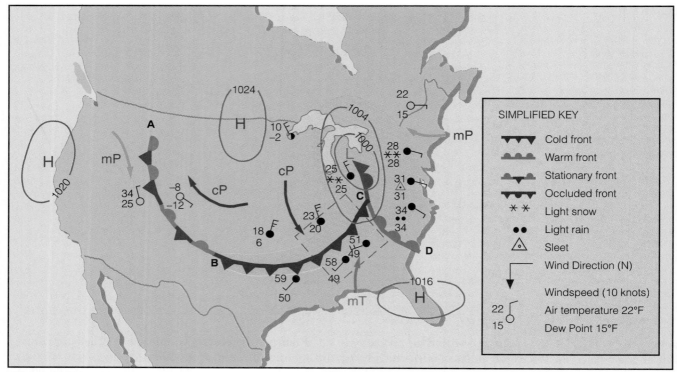

Figure 8.10
A simplified weather map showing surface pressure systems, air masses, and fronts. (Green-shaded area represents precipitation.)

cloudiness with light precipitation can cover a vast area. These are the conditions that prevail north of the east-west running stationary front depicted in Fig. 8.6.

If the warmer air to the west begins to move and replace the colder air to the east, the front in Fig. 8.10 will no longer remain stationary; it will become a warm front. If, on the other hand, the colder air slides up over the mountain and replaces the warmer air on the other side, the front will become a cold front.

Cold Fronts   The **cold front** between points B and C on the surface weather map (Fig. 8.10) represents a zone where cold, dry, stable polar air is replacing warm, moist unstable subtropical air. The front is drawn as a solid blue line with the triangles along the front showing its direction of movement. How did the meteorologist know to draw the front at that location? A closer look at the situation will give us the answer.

The weather in the immediate vicinity of this cold front in the southeastern United States is shown in Fig. 8.11. The data plotted on the map represent the current weather at selected cities. The station model used to represent the data at each reporting station is a sim-

plified one that shows temperature, dew point, present weather, cloud cover, sea level pressure, wind direction and speed. The little line in the lower right-hand corner of each station shows the pressure change—the pressure tendency, whether rising (/) or falling (\)—during the last three hours. With all of this information, the front can be properly located. (Appendix C explains the weather symbols and the station model more completely.)

The following criteria are used to locate a front on a surface weather map:

1. sharp temperature changes over a relatively short distance
2. changes in the air's moisture content (as shown by marked changes in the dew point)
3. shifts in wind direction
4. pressure and pressure changes
5. clouds and precipitation patterns

In Fig. 8.11, we can see a large contrast in air temperature and dew point on either side of the front. There is also a wind shift from southwesterly ahead of the front, to northwesterly behind it. Notice that each

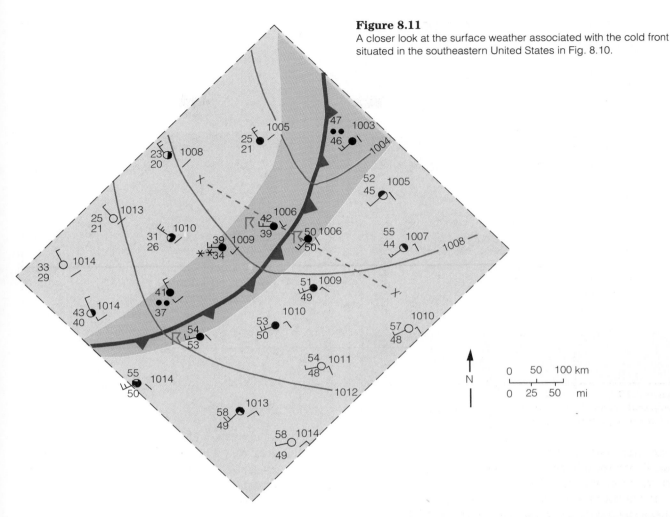

Figure 8.11
A closer look at the surface weather associated with the cold front situated in the southeastern United States in Fig. 8.10.

isobar kinks as it crosses the front, forming an elongated area of low pressure—a *trough*—which accounts for the wind shift. Since surface winds normally blow across the isobars toward lower pressure, we find winds with a southerly component ahead of the front and winds with a northerly component behind it.

Since the cold front is a trough of low pressure, sharp changes in pressure can be significant in locating the front's position. One important fact to remember is that the lowest pressure usually occurs just as the front passes a station. Notice that, as you move toward the front, the pressure drops, and, as you move away from it, the pressure rises.

The cloud and precipitation patterns are better seen in a side view of the front along the line *X–X'* (Fig. 8.12). We can see from Fig. 8.12 that, at the front, the cold, dense air wedges under the warm air, forcing it upward. As the moist, unstable air rises, it condenses

into a series of cumuliform clouds. Strong, upper-level westerly winds blow the delicate ice crystals (which form near the top of the cumulonimbus) into cirrostratus (Cs) and cirrus (Ci). These clouds usually appear far in advance of the approaching front. At the front itself, a relatively narrow band of thunderstorms (Cb) produces heavy showers with gusty winds. Behind the front, the air cools quickly. (Notice how the freezing level dips as it crosses the front.) The winds shift from

Did you know?
The temperature change during the passage of a strong cold front can be quite dramatic. On the evening of January 19, 1810, Portsmouth, New Hampshire, had a relatively mild air temperature of 41°F. Within a few hours, after the passage of a cold front, the temperature plummeted to −13°F.

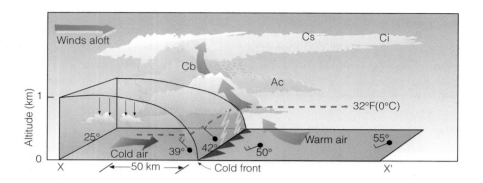

Figure 8.12
A vertical view of the weather across the cold front in Fig. 8.11 along the line *X–X'*.

southwesterly to northwesterly, pressure rises, and precipitation ends. As the air dries out, the skies clear, except for a few lingering fair weather cumulus clouds.

Observe that the leading edge of the front is steep. The steepness is due to friction, which slows the air

Figure 8.13
Clouds developing along an approaching cold front. The front is moving from west to east and the photograph was taken east of the advancing front.

flow near the ground. The air aloft pushes forward, blunting the frontal surface. If we could walk from where the front touches the surface back into the cold air, a distance of 50 kilometers, the front would be about 1 kilometer above us. Thus, the slope of the front—the ratio of vertical rise to horizontal distance—is 1:50. This is typical for a fast-moving cold front—those that move about 25 knots. In a slower-moving cold front the slope is much more gentle.

With slow-moving cold fronts, clouds and precipitation usually cover a broad area behind the front. When the ascending warm air is stable, stratiform clouds, such as nimbostratus, become the predominate cloud type and even fog may develop in the rainy area. Occasionally, out ahead of a fast-moving front, a line of active showers and thunderstorms, called a *squall line*, develops parallel to and ahead of the advancing front. Figure 8.13 shows clouds developing along an advancing cold front.

So far, we have considered the general weather patterns of "typical" cold fronts. There are, of course, exceptions. For example, if the rising warm air is dry and stable, scattered clouds are all that form, and there is no precipitation. In extremely dry weather, a marked change in the dew point, accompanied by a slight wind shift, may be the only clue to a passing front. During the winter, a series of cold polar outbreaks may travel across the United States so quickly that warm air is unable to develop ahead of the front. In this case, frigid arctic air usually replaces cold polar air, and a drop in temperature is the only indication that a cold front has moved through your area. Along the West Coast, the Pacific Ocean modifies the air so much that cold fronts, such as those described in the previous section, are never seen. In fact, as a cold front moves inland from the Pacific Ocean, the surface temperature contrast across the front may be quite small. Topographic features usually distort the wind pattern so much that locating the position of the front and the time of its passage are exceedingly difficult. In this case, the pressure tendency is the most reliable indication of a frontal passage.

Most cold fronts move toward the south, southeast, or east. But sometimes they will move southwestward out of Canada into New England. Cold fronts that move in this manner are called *back door cold fronts*.

Even though cold-front weather patterns have many exceptions, learning these patterns can be to your advantage if you live in the eastern two-thirds of the United States, where well-defined cold fronts are experienced. Knowing them improves your own ability to make short-range weather forecasts. For your reference, Table 8.2 summarizes idealized cold-frontal weather.

Warm Fronts   In Fig. 8.10, a **warm front** is drawn along the solid red line running from points C to D. Here, the leading edge of advancing warm, moist subtropical air from the Gulf of Mexico replaces the retreating cold maritime polar air from the North Atlantic. The direction of frontal movement is given by the half circles, which point into the cold air; this front is heading toward the northeast. The average speed of a warm front is about 10 knots, or about half that of an average cold front. During the day, as mixing occurs on both sides of the front, its movement may be much faster. Warm fronts often move in a series of rapid jumps, which show up on successive weather maps.

At night, however, radiational cooling creates cool, dense surface air behind the front. This inhibits both lifting and the front's forward progress. When the forward surface edge of the warm front passes a station, the wind shifts, the temperature rises, and the overall weather conditions improve. To see why, we will examine the weather commonly associated with the warm front both at the surface and aloft.

Look at Figs. 8.14 and 8.15 closely and observe that the warmer, less-dense air rides up and over the colder, more-dense surface air. This rising of warm air over cold, called **overrunning**, produces clouds and precipitation well in advance of the front's surface boundary. The warm front that separates the two air masses has an average slope of about 1:150, although it can be as gentle as 1:300—a much more inclined shape than that of a typical cold front.

Suppose we are standing at the position marked P' in Figs. 8.14 and 8.15. Note that we are over 1200 km (750 mi) ahead of the surface front. Here, the surface winds are light and variable. The air is cold and about the only indication of an approaching warm front is the

Table 8.2 Typical Weather Conditions Associated with a Cold Front			
Weather Element	**Before Passing**	**While Passing**	**After Passing**
Winds	south-southwest	gusty, shifting	west-northwest
Temperature	warm	sudden drop	steadily dropping
Pressure	falling steadily	minimum, then sharp rise	rising steadily
Clouds	increasing Ci, Cs, then either Tcu or Cb	Tcu or Cb*	often Cu
Precipitation	short period of showers	heavy showers of rain or snow, sometimes with hail, thunder, and lightning	decreasing intensity of showers, then clearing
Visibility	fair to poor in haze	poor, followed by improving	good except in showers
Dew point	high; remains steady	sharp drop	lowering

*Tcu stands for towering cumulus, such as cumulus congestus; whereas Cb stands for cumulonimbus.

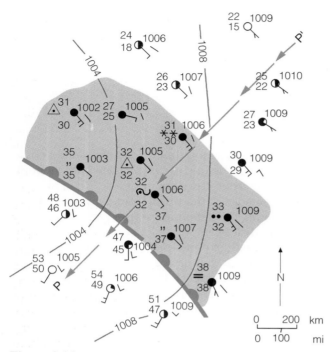

Figure 8.14
Surface weather associated with a typical warm front. (Green-shaded area represents precipitation.)

high cirrus clouds overhead. We know the front is moving slowly toward us and that within a day or so it will pass our area. Suppose that, instead of waiting for the front to pass us, we drive toward it, observing the weather as we go.

Heading toward the front, we notice that the cirrus (Ci) clouds gradually thicken into a thin, white veil of cirrostratus (Cs) whose ice crystals cast a halo around

the sun.* Almost imperceptibly, the clouds thicken and lower, becoming altocumulus (Ac) and altostratus (As) through which the sun shows only as a faint spot against an overcast gray sky. Snowflakes begin to fall, and we are still over 500 km (300 mi) from the surface front. The snow increases, and the clouds thicken into a sheetlike covering of nimbostratus (Ns). The winds become brisk and out of the southeast, while the barometer slowly falls. Within 300 km (180 mi) of the front, the cold surface air mass is now quite shallow. The surface air temperature moderates and, as we approach the front, the light snow changes first into sleet. It then becomes freezing rain and finally rain and drizzle as the air temperature climbs above freezing. Overall, the precipitation remains light or moderate but covers a broad area. Moving still closer to the front, warm, moist air mixes with cold, moist air producing ragged wind-blown stratus (St) and fog. (Thus, flying in the vicinity of a warm front is quite hazardous.)

Finally, after a trip of over 1200 kilometers, we reach the warm front's surface boundary. As we cross the front, the weather changes are noticeable, but much less pronounced than those experienced with the cold front; they show up more as a gradual transition rather than a sharp change. On the warm side of the front, the air temperature and dew point rise, the wind shifts from southeast to south or southwest, and the barometer stops falling. The light rain ends and, except for a few stratocumulus, the fog and low clouds vanish.

This scenario of an approaching warm front represents average warm-front weather. In some in-

---

*If the warm air is relatively unstable, ripples or waves of cirrocumulus clouds will appear as a "mackerel sky."

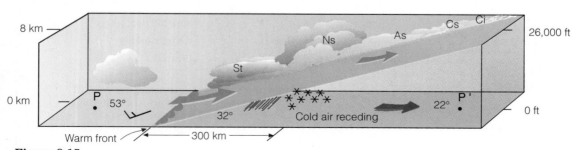

Figure 8.15
Vertical view of clouds, precipitation, and winds across the warm front in Fig. 8.14 along the line *P–P'*.

stances, the weather can differ from this dramatically. For example, if the overrunning warm air is relatively dry and stable, only high and middle clouds will form, and no precipitation will occur. On the other hand, if the warm air is relatively moist and unstable (as is often the case during the summer), heavy showers can develop as thunderstorms become embedded in the cloud mass.

Along the West Coast, the Pacific Ocean significantly modifies the surface air so that warm fronts are difficult to locate on a surface weather map. Also, not all warm fronts move northward or northeastward. On rare occasions, a front will move into the eastern seaboard from the Atlantic Ocean as the front spins all the way around a deep storm positioned off the coast. Cold northeasterly winds ahead of the front usually become warm northeasterly winds behind it. Even with these exceptions, knowing the normal sequence of warm-front weather will be useful, especially if you live east of the Rockies, where warm fronts become well developed. You can look for certain cloud and weather patterns and make reasonably accurate short-range forecasts of your own. Table 8.3 summarizes typical warm-front weather.

Occluded Fronts When a cold front catches up to and overtakes a warm front, the frontal boundary created between the two air masses is called an **occluded front**, or, simply, an **occlusion**. On the surface weather map, it is represented as alternating cold-front triangles and warm-front half circles; both symbols point in the direction toward which the front is moving. In Fig. 8.10,

the occluded front is shown by a solid purple line. Notice that the air behind the front is colder than the air ahead of it. This is known as a *cold-type occluded front*, or *cold occlusion*. Let's see how this front develops.

The development of a cold occlusion is shown in Fig. 8.16. Along line *A–A'*, the cold front is rapidly approaching the slower-moving warm front. Along line *B–B'*, the cold front overtakes the warm front, and, as we can see in the vertical view across *C–C'*, lifts both the warm front and the warm air mass off the ground. As a cold-occluded front approaches, the weather sequence is similar to that of a warm front with high clouds lowering and thickening into middle and low clouds, with precipitation forming well in advance of the surface front. Since the front represents a trough of low pressure, southeasterly winds and falling barometers occur ahead of it. The frontal passage, however, brings weather similar to that of a cold front: heavy, often showery precipitation with winds shifting to west or northwest. After a period of wet weather, the sky begins to clear, barometers rise, and the air turns colder. The most violent weather usually occurs where the cold front is just overtaking the warm front, at the point of occlusion, where the greatest contrast in temperature occurs. Cold occlusions are the most prevalent type of front that moves into the Pacific coastal states. Occluded fronts frequently form over the North Pacific and North Atlantic, as well as in the vicinity of the Great Lakes.

Continental polar air over eastern Washington and Oregon may be much colder than milder maritime polar air moving inland from the Pacific Ocean. Figure

Table 8.3 Typical Weather Conditions Associated with a Warm Front			
Weather Element	**Before Passing**	**While Passing**	**After Passing**
Winds	south-southeast	variable	south-southwest
Temperature	cool–cold, slow warming	steady rise	warmer, then steady
Pressure	usually falling	leveling off	slight rise, followed by fall
Clouds	in this order: Ci, Cs, As, Ns St and fog; occasionally Cb in summer	stratus-type	clearing with scattered Sc; occasionally Cb in summer
Precipitation	light to moderate rain, snow, sleet, or drizzle	drizzle or none	usually none; sometimes light rain or showers
Visibility	poor	poor, but improving	fair in haze
Dew point	steady rise	steady	rise, then steady

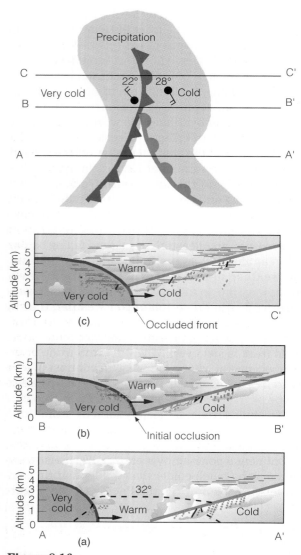

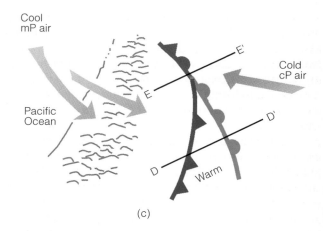

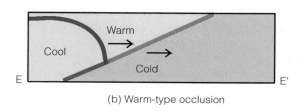

(b) Warm-type occlusion

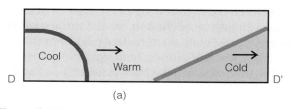

(a)

Figure 8.16
The formation of a cold-occluded front. The faster-moving cold front (a) catches up to the slower-moving warm front (b) and forces it to rise off the ground (c).

Figure 8.17
The formation of a warm-type occluded front. The faster-moving cold front in (a) overtakes the slower-moving warm front in (b). The lighter air behind the cold front rises up and over the denser air ahead of the warm front. Diagram (c) shows a surface map of the situation.

8.17 illustrates this situation. Observe that the air ahead of the warm front is colder than the air behind the cold front. Consequently, when the cold front catches up to and overtakes the warm front, the milder, lighter air behind the cold front is unable to lift the colder, heavier air off the ground. As a result, the cold front rides "piggyback" along the sloping warm front. This produces a *warm-type occluded front*, or a *warm occlusion*. The surface weather associated with a warm occlusion is similar to that of a warm front.

Contrast Fig. 8.16 and Fig. 8.17. Note that the primary difference between the warm- and cold-type occluded front is the location of the upper-level front. In a warm occlusion, the upper-level cold front *precedes* the surface occluded front, whereas in a cold occlusion the upper warm front *follows* the surface occluded front.

In the world of weather fronts, occluded fronts are the mavericks. In our discussion, we treated occluded fronts as forming when a cold front overtakes a warm front. Some do form in this manner, but others apparently form as new fronts, which develop when a surface storm intensifies in a region of cold air after its trailing cold and warm fronts have broken away and moved eastward. The new occluded front shows up on a surface chart as a trough of low pressure separating two cold air masses. Because of this, locating and defining occluded fronts at the surface is often difficult for the meteorologist. Similarly, you too may find it hard to recognize an occlusion. In spite of this, we will assume that the weather associated with occluded fronts behaves in a similar way to that shown in Table 8.4.

The frontal systems described so far are actually part of a much larger storm system—the middle latitude cyclone. The following section details these storms, explaining where, why, and how they form.

▲▼▲
Middle Latitude Cyclones

Early weather forecasters were aware that precipitation generally accompanied falling barometers and areas of low pressure. However, it was not until after the turn of this century that scientists began to piece together the information that yielded the ideas of modern meteorology and cyclonic storm development.

Working largely from surface observations, a group of scientists in Bergen, Norway, developed a model explaining the life cycle of an extratropical storm; that is, a storm that forms at middle and high latitudes outside of the tropics. This extraordinary group of meteorologists included Vilhelm Bjerknes, his son Jakob, Halvor Solberg, and Tor Bergeron. They published their theory shortly after World War I. It was widely acclaimed and became known as the "polar front theory of a developing wave cyclone" or, simply, the **polar front theory**. What these meteorologists gave to the world was a working model of how a mid-latitude cyclone progresses through the stages of birth, growth, and decay. An important part of the model involved the development of weather along the polar front. As new information became available, the original work was modified, so that, today, it serves as a convenient way to describe the structure and weather associated with a migratory storm system.

Polar Front Theory   The development of a wave cyclone, according to the Norwegian model, begins along the polar front. You will remember (from our discus-

Table 8.4   Summary of Typical Weather Associated with Occluded Fronts

Weather Element	Before Passing	While Passing	After Passing
Winds	southeast-south	variable	west to northwest
Temperature			
Cold type	cold–cool	dropping	colder
Warm type	cold	rising	milder
Pressure	usually falling	low point	usually rising
Clouds	in this order: Ci, Cs, As, Ns	Ns, sometimes Tcu and Cb	Ns, As, or scattered Cu
Precipitation	light, moderate, or heavy precipitation	light, moderate, or heavy continuous precipitation or showers	light-to-moderate precipitation followed by general clearing
Visibility	poor in precipitation	poor in precipitation	improving
Dew point	steady	usually slight drop, especially if cold-occluded	slight drop, although may rise a bit if warm-occluded

sion of the general circulation in Chapter 7) that the polar front is a semicontinuous global boundary separating cold polar air from warm subtropical air. The stages of a developing wave cyclone are illustrated in the sequence of surface weather maps shown in Fig. 8.18.

Figure 8.18a shows a segment of the polar front as a stationary front. It represents a trough of lower pressure with higher pressure on both sides. Cold air to the north and warm air to the south flow parallel to the front, but in opposite directions. This type of flow sets up a cyclonic wind shear. You can conceptualize the shear more clearly if you place a pen between the palms of your hands and move your left hand toward your body; the pen turns counterclockwise, cyclonically.

Under the right conditions (described later), a wavelike kink forms on the front, as shown in Fig. 8.18b. The wave that forms is known as a **frontal wave**. Watching the formation of a frontal wave on a weather map is like watching a water wave from its side as it approaches a beach: It first builds, then breaks, and finally dissipates. This is why a cyclonic storm system is known as a **wave cyclone**. Figure 8.18b shows the newly formed wave with a cold front

pushing southward and a warm front moving northward. The region of lowest pressure is at the junction of the two fronts. As the cold air displaces the warm air upward along the cold front, and as *overrunning* occurs ahead of the warm front, a narrow band of precipitation forms (shaded area). Steered by the winds aloft, the system typically moves east or northeastward and gradually becomes a fully developed **open wave** in 12 to 24 hours (Fig. 8.18c). The central pressure is now much lower, and several isobars encircle the wave's apex. These more tightly packed isobars create a stronger cyclonic flow, as the winds swirl counterclockwise and inward toward the low center. Precipitation forms in a wide band *ahead* of the warm front and along a narrow band of the cold front. The region of warm air between the cold and warm fronts is known as the *warm sector*. Here, the weather tends to be partly cloudy, although scattered showers may develop if the air is unstable.

Energy for the storm is derived from several sources. As the air masses try to attain equilibrium, warm air rises and cold air sinks, transforming potential energy into kinetic energy (that is, energy of motion). Condensation supplies energy to the system in the form of latent heat. And, as the surface air con-

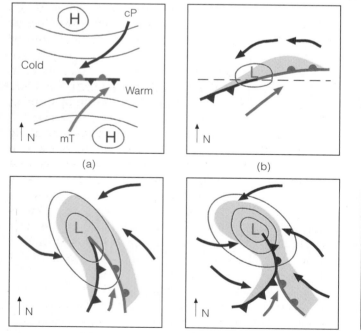

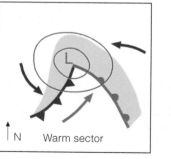

Figure 8.18
A series of consecutive weather maps (a through f) that shows the idealized life cycle of a wave cyclone (Northern Hemisphere) based on the polar front theory. As the life cycle progresses, the system moves eastward in a dynamic fashion.

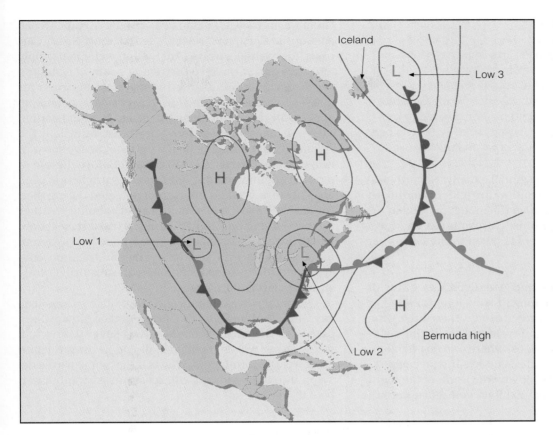

Figure 8.19
A series of wave cyclones ("family" of cyclones) forming along the polar front.

verges toward the low center, wind speeds may increase, producing an increase in kinetic energy.

As the open wave moves eastward, central pressures continue to decrease, and the winds blow more vigorously. The faster-moving cold front constantly inches closer to the warm front, squeezing the warm sector into a smaller area, as shown in Fig. 8.18d. Eventually, the cold front overtakes the warm front and the system becomes occluded. At this point, the storm is usually most intense, with clouds and precipitation covering a large area. The intense storm system shown in Fig. 8.18e gradually dissipates, because cold air now lies on both sides of the occluded front. Without the supply of energy provided by the rising warm, moist air, the old storm system dies out and gradually disappears (Fig. 8.18f). Occasionally, however, a new wave will form on the westward end of the trailing cold front. We can think of the sequence of a developing wave cyclone as a whirling eddy in a stream of water that forms behind an obstacle, moves with the flow, and gradually vanishes downstream. The entire life cycle of a wave cyclone can last from a few days to over a week.

Figure 8.19 shows a series of wave cyclones at various stages of development along the polar front in winter. Such a succession of storms is known as a *"family" of cyclones*. Observe that to the north of the front are cold anticyclones; to the south over the Atlantic Ocean is the warm, semipermanent Bermuda high. The polar front itself has developed into a series of loops, and at the apex of each loop is a cyclone. The cyclone over the northern plains (Low 1) is just forming; the one along the East Coast (Low 2) is an open wave; and the system near Iceland (Low 3) is dying out. If the average rate of movement of a wave cyclone from birth to decay is 25 knots, then it is entirely possible for a storm to develop over the central part of the United States, intensify into a large storm over New England, become occluded over the ocean, and reach the coast of England in its dissipating stage less than a week after it formed.

Up to now, we have considered the polar front model of a developing wave cyclone, which represents a rather simplified version of the stages that a midlatitude storm system must go through. In fact, few (if

Did you know?
Mid-latitude storms have been known to change the course of history. One such storm, in March of 1776, forced the British to cancel a massive amphibious assault at Dorchester Heights, south of Boston, where General Washington held a well-fortified position. The storm kept Dorchester Heights from becoming a bloody, historic battlefield, such as Lexington, Concord, and Bunker Hill.

any) storms adhere to the model exactly. Nevertheless, it serves as a good foundation for understanding the structure of storms. So keep the model in mind as you read the following sections. (Information on storms that form on other planets is given in the Focus section on p. 212.)

Developing Cyclones and Anticyclones   Any development or strengthening of a cyclone is called **cyclogenesis**. Within the United States, there are regions that show a propensity for cyclogenesis, including the eastern slopes of the Rockies, where a strengthening or developing storm is called a **lee-side low** because it is forming on the leeward (downwind) side of the mountain. Additional areas that exhibit cyclogenesis are the

Great Basin, the Gulf of Mexico, and the Atlantic Ocean east of the Carolinas. Near Cape Hatteras, North Carolina, for example, warm Gulf Stream water can supply moisture and warmth to the region south of a stationary front, thus increasing the contrast between air masses to a point where storms may suddenly and unexpectedly spring up along the front. As noted earlier, these cyclones normally move northeastward along the Atlantic Coast, bringing high winds and heavy snow or rain to coastal areas. (See Fig. 8.20.) Before the age of modern communications and weather prediction such coastal storms would often go undetected during their formative stages; and sometimes an evening weather forecast of "fair and colder" along the eastern seaboard would have to be changed to "heavy snowfall" by morning. Fortunately, with today's weather information gathering and forecasting techniques, these storms rarely strike by surprise.

Frontal waves that develop into huge storms usually form suddenly, grow in size, and then slowly dissipate with the entire process taking several days to a week. Other waves remain small and never grow into giant weather producers. Why is it that some waves develop into huge storms, whereas others simply dissipate in a day or so?

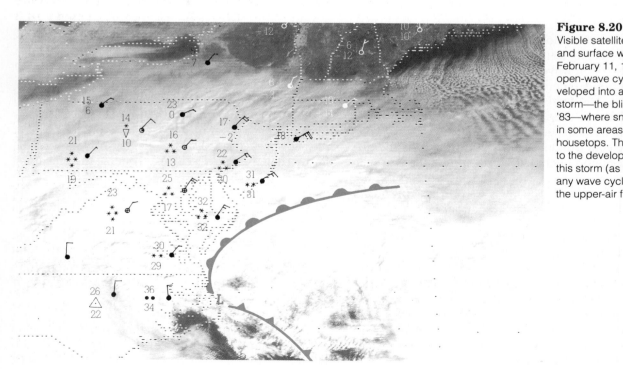

Figure 8.20
Visible satellite picture and surface weather for February 11, 1983. This open-wave cyclone developed into a ferocious storm—the blizzard of '83—where snow drifts in some areas rose to housetops. The real key to the development of this storm (as well as any wave cyclone) is in the upper-air flow.

This question poses one of the real challenges in weather forecasting. The answer is complex. Indeed, there are many surface conditions that do influence the formation of a wave, including mountain ranges and land-ocean temperature contrasts. However, the real key to the development of a wave cyclone is found in the *upper-wind flow*, in the region of the high-level westerlies. Therefore, before we can arrive at a reasonable answer to our question, we need to see how the winds aloft influence surface pressure systems.

In Chapter 7, we learned that thermal pressure systems are shallow and weaken with increasing elevation. On the other hand, developing surface storm systems are deep lows that usually intensify with height. This means that a surface low pressure area will appear on an upper-level chart as either a closed low or a trough.

Suppose the upper-level low is directly above the surface low (see Fig. 8.21). Notice that only at the surface do the winds blow inward toward the low's center. As these winds converge (flow together) and the air "piles up," air density increases directly above the surface low. This increase in mass causes surface pressures to rise; gradually, the low fills and the surface low dissipates. The same reasoning can be applied to surface anticyclones. Winds blow outward away from the center of a surface high. If a closed high or ridge lies directly over the surface anticyclone, divergence (the spreading out of air) at the surface will remove air from the column directly above the high. Surface pressures fall and the system weakens. Consequently, it appears that, if upper-level pressure systems were always located directly above those at the surface, cyclones and anticyclones would die out soon after they form (if they could form at all). What, then, is it that allows these systems to develop and intensify?

Waves in the Westerlies Recall from Chapter 6 that the flow aloft above the middle latitudes usually consists of a series of troughs and ridges. Look at Fig. 8.22a and observe that the distance between two troughs, called a **longwave**, is normally thousands of miles in length. Typically, at any given time, there are between four and six of these waves looping around the earth. These longwaves are also known as *Rossby waves*, after C. G. Rossby, a famous meteorologist who carefully studied their motion. In Fig. 8.22a we can see that embedded in longwaves are **shortwaves** which are small disturbances or ripples.

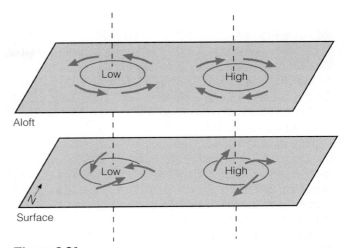

Figure 8.21
If lows and highs aloft were always directly above lows and highs at the surface, the surface systems would quickly dissipate.

Notice in Fig. 8.22b that, while the longwaves move eastward very slowly, the shortwaves move fairly quickly around the longwaves. Generally, shortwaves deepen when they approach a longwave trough and weaken when they approach a ridge. Also notice in Fig. 8.22b that, when a shortwave moves into a longwave trough, the trough tends to deepen. The upper flow is now capable of providing the proper ingredients for the development or intensification of a surface cyclonic storm.

Convergence and Divergence of Air The piling up of air above a region is called **convergence**, whereas the spreading out of air above a region is called **divergence**. For mid-latitude cyclones and anticyclones to maintain themselves or intensify, the winds aloft must blow in such a way that zones of converging and diverging air form. (Additional information on convergence and divergence is given in the Focus section on p. 215.)

In Fig. 8.23, notice that the surface winds are converging about the center of the low while aloft, directly above the low, the winds are diverging. For the surface low to develop into a major storm system, upper-level divergence of air must be greater than surface convergence of air. (More air must be removed above the storm than is brought in at the surface.) When this event happens, surface air pressure decreases, and we say that the storm system is *intensifying* or *deepening*.

Focus on an Observation
Storms on Other Planets

Earth is not the only planet to have storms. Probably the best known and, certainly, the largest storm in our solar system is Jupiter's *Great Red Spot*—a giant reddish, swirling eddy that has persisted in Jupiter's atmosphere for more than 300 years.

The Great Red Spot (which is about three times larger than the earth) spins counterclockwise in Jupiter's southern hemisphere (Fig. 4). The winds that circulate around the spot take about six days to make one complete rotation. Unlike the earth's weather machine, which is driven by the sun, Jupiter's massive swirling clouds appear to be driven by a collapsing core of hot hydrogen. Energy from this lower region rises toward the surface, where it (along with Jupiter's rapid rotation) stirs the cloud layers into more or less horizontal bands. The Great Red Spot, as well as the smaller white ones, apparently form as two flowing cloud layers move rapidly past one another, but in opposite directions.

Although the Great Red Spot is similar to a storm on earth, it is also different. On earth, a mid-latitude storm or a hurricane in the Southern Hemisphere spins *clockwise* (because of the Coriolis effect) about a center of low pressure. The Great Red Spot, however, spins *counterclockwise* about its center, which means

Figure 4
A portion of Jupiter extending from the equator to the southern polar latitudes. The Great Red Spot, as well as the smaller ones, are spinning eddies similar to storms that exist in the Earth's atmosphere.

that the Red Spot must be a storm spinning about a center of high pressure.

In 1989, when *Voyager 2* flew by Neptune, it photographed (in Neptune's southern hemisphere) a large dark oval known as the Great Dark Spot (Fig. 5). It, like Jupiter's Great Red Spot, appears to be a storm that spins counterclockwise about a center of high pressure. Further pictures revealed smaller dark and bright spots that tend to move around the planet at various speeds. Also visible were white, wispy, cirrus-like clouds,

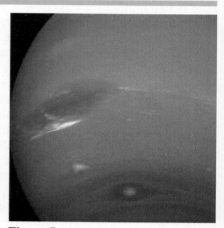

Figure 5
The Great Dark Spot on Neptune. The white, wispy clouds are similar to cirrus clouds on Earth. However, on Neptune, they are probably composed of methane ice crystals.

probably composed of methane ice crystals.

Similar swirling storms have also been detected on the planet Saturn. One huge cloud system measured more than twice the diameter of earth. Saturn appears to be dotted with storms, most of which are too small to be seen from earth. As we learn more about the mechanisms that produce storms on other planets, it is hoped that we will obtain a better understanding of the workings of our own atmosphere.

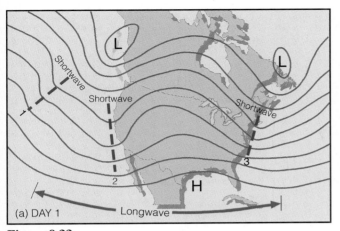
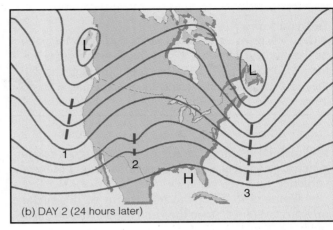

(a) DAY 1

(b) DAY 2 (24 hours later)

Figure 8.22
(a) Upper-air chart showing a longwave with three shortwaves embedded in the flow.
(b) Twenty-four hours later the shortwaves have moved rapidly around the longwave. Notice
that the shortwaves labeled 1 and 3 tend to deepen the longwave trough, while shortwave 2
has weakened as it moves into a ridge.

If the reverse should occur (more air flows in at the surface than is removed at the top), surface pressures will rise, and the storm system will weaken and gradually dissipate.

Notice also in Fig. 8.23 that surface winds are diverging about the center of the high, while aloft, directly above the anticyclone, they are converging. In order for the surface high to strengthen, upper-level convergence of air must exceed low-level divergence of air (more air must be brought in above the anticyclone than is removed at the surface). When this occurs, surface air pressure increases, and we say that the high-pressure area is *building*.

We can see in Fig. 8.23 that the convergence of air aloft causes an accumulation of air above the surface high, which allows the air to sink slowly and replace the diverging surface air. Above the surface low, divergence allows the converging surface air to rise and flow out the top of the column.

In the atmosphere, when a shortwave disturbs the flow aloft, as shown on the upper-air chart in Fig. 8.24, a region of converging air usually forms on the west side of the trough and a region of diverging air forms on its east side. Notice that the area of divergence is directly above the surface low and the area of convergence is directly above the surface high. This configuration means that for the surface storm to intensify, the upper-level trough must be located behind (or to the *west*) of the surface low. When the upper-level trough is in this position, the atmosphere is able to

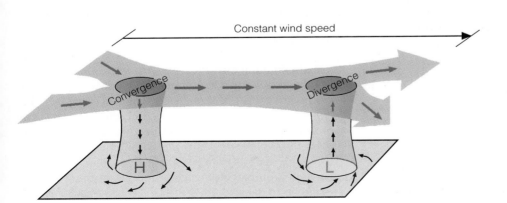

Figure 8.23
Horizontal convergence (flowing together) and divergence (spreading out) of air created by upper-level winds allows surface pressure systems to maintain themselves or intensify.

redistribute its mass, as regions of low-level convergence are compensated for by regions of upper-level divergence, and vice versa.

Winds aloft steer the movement of the surface pressure systems. Since the winds above the storm are blowing from the southwest, the surface low should move northeastward. The northwesterly winds above the surface high should direct it toward the southeast. These paths are typical of the average movement of surface pressure systems in the eastern two-thirds of the United States. (See Fig. 8.25.)

The Role of the Jet Stream Jet streams play an additional part in the formation of surface cyclones and anticyclones. When the jet stream flows in a wavy west-

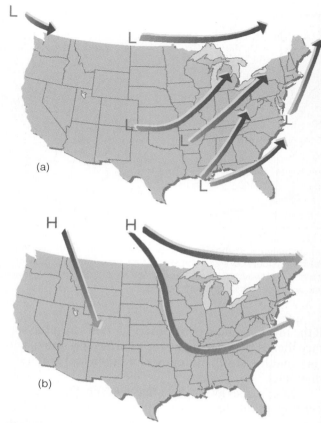

(a)

(b)

Figure 8.25
Typical paths of (a) winter mid-latitude cyclones and (b) winter anticyclones.

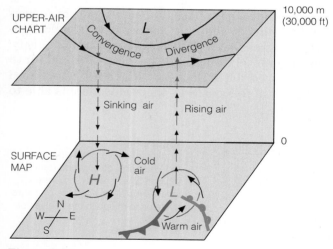

Figure 8.24
Convergence, divergence, and vertical motions associated with surface pressure systems. Notice that for the surface storm to intensify, the upper trough of low pressure must be located to the left (or west) of the surface low.

to-east pattern (as illustrated in Fig. 8.26a), deep troughs and ridges exist in the flow aloft. Notice that in the trough a strong core of winds—a *jet maximum*—forms. The curving of the jet stream coupled with the changing wind speeds around the jet maximum produces regions of strong convergence and divergence along the flanks of the jet. The region of diverging air above the storm draws warm surface air upward to the jet, which quickly sweeps the air downstream, causing surface pressure in the low to drop rapidly. Above the anticyclone, cold, dense air brought southward by the jet stream slowly sinks to replace the diverging surface air. Hence, we find the jet stream *removing air above the surface cyclone and supplying air to the surface anticyclone*. Additionally, the sinking of cold air and the rising of warm air provide energy for the developing cyclones as potential energy is transformed into energy of motion (kinetic energy).

Focus on a Special Topic
A Closer Look at Convergence and Divergence

Convergence and divergence of air may result from changes in wind direction. For example, convergence occurs when moving air is funneled into an area, much in the way cars converge when they enter a crowded freeway. Divergence occurs when moving air spreads apart, much as cars spread out when a congested two-lane freeway becomes three lanes. On an upper-level chart (Fig. 6) this type of convergence (also called *confluence*) occurs when contour lines move closer together, as a steady wind flows parallel to them. On the same chart, this type of divergence (also called *diffluence*) occurs when the contour lines move apart as a steady wind flows parallel to them.

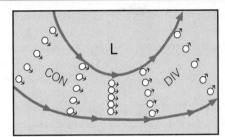

Figure 6
The formation of convergence (CON) and divergence (DIV) of air with a constant wind speed.

Convergence and divergence may also result from changes in wind speed. Convergence occurs when the wind slows down as it moves

along, while divergence occurs when the wind speeds up. We can grasp these relationships more clearly if we imagine air molecules to be marching in a band. When the marchers in front slow down, the rest of the band members squeeze together, causing convergence; when the marchers in front start to run, the band members spread apart, or diverge. In summary, *speed convergence* takes place when the wind speed decreases downwind, and *speed divergence* takes place when the wind speed increases downwind.

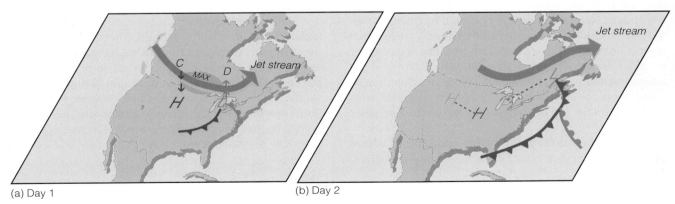

(a) Day 1   (b) Day 2

Figure 8.26
(a) As the polar jet stream swings over a developing mid-latitude cyclone, an area of divergence (D) draws warm, surface air upward and an area of convergence (C) allows cold air to sink. The jet stream, removing air above the surface storm and supplying air to the surface anticyclone, allows the two surface pressure systems to develop. When the surface storm moves northeastward and occludes (b), it no longer has the upper-level support of the jet stream, and the surface storm gradually dies out.

As the jet stream steers the storm along—toward the northeast, in this case—the surface cold front catches up to the warm front and the storm system occludes (Fig. 8.26b). With the supply of warm surface air cut off (as cold air now lies on both sides of the occluded front), and the region of diverging air no longer above the surface low, the storm system gradually dies out.

Since the polar jet stream is strongest and moves farther south in winter, we can see why mid-latitude cyclonic storms are better developed and move more quickly during the coldest months. During the summer when the polar jet shifts northward, developing mid-latitude storm activity shifts northward and occurs principally over the Canadian provinces of Alberta and the Northwest Territories.

In general, we now have a fairly good picture as to why some surface lows intensify into huge storms while others do not. For a storm to intensify, there must be an upper-level counterpart—a trough of low pressure—that lies to the *west* of the surface low. At the same time, the polar jet forms into waves and swings slightly south of the developing storm. When these conditions exist, zones of converging and diverging air, along with rising and sinking air, provide energy conversions for the storm's growth. When these conditions do not exist, we say that the surface storm does not have the proper *upper-air support* for its development.

Summary

In this chapter, we considered the different types of air masses and the various weather each brings to a particular region. Continental arctic air masses are responsible for the extremely cold (arctic) outbreaks of winter, while continental polar air masses are responsible for cold, dry weather in winter and cool, pleasant weather in summer. Maritime polar air, having traveled over an ocean for a considerable distance, brings to a region cool, moist weather. The hot, dry weather of summer is associated with continental tropical air masses, while warm, humid conditions are due to maritime tropical air masses. Where air masses with sharply contrasting properties meet, we find weather fronts.

Along the leading edge of a cold front, where colder air replaces warmer air, showers are prevalent, especially if the warmer air is moist and unstable. Along a warm front, warmer air rides up and over colder surface air, producing widespread cloudiness and precipitation. Occluded fronts, which are often difficult to locate and define on a surface weather map, may have characteristics of both cold and warm fronts.

We learned that fronts are actually part of the mid-latitude cyclone. We examined where, why, and how these storms form and found that an important influence on their development is the upper-air flow, including the jet stream. When an upper-level low lies to the west of the surface low, and the jet stream bends and then dips south of the surface storm, the ingredients are there for the mid-latitude storm to develop into a deep low pressure area.

Key Terms

The following terms are listed in the order they appear in the text. Define each.
Doing so will aid you in reviewing the material covered in this chapter.

air mass
source region (for
  air masses)
continental polar
  (air mass)
continental arctic
  (air mass)
lake-effect snows

maritime polar
  (air mass)
maritime tropical
  (air mass)
continental tropical
  (air mass)
air mass weather
front

stationary front
cold front
warm front
overrunning
occluded front
  (occlusion)
polar front theory
frontal wave

wave cyclone
open wave
cyclogenesis
lee-side low
longwave
shortwave
convergence
divergence

Review Questions

1. (a) What is an air mass?
   (b) If an area is described as a "good air mass source region," what information can you give about it?
2. How does a cA air mass differ from a cP air mass?
3. Why is cP air not welcome to the Central Plains in winter and yet very welcome in summer?
4. What are lake-effect snows and how do they form?
5. List the temperature and moisture characteristics of each of the major air mass types.
6. Which air mass only forms in summer over the United States?
7. Why are mP air masses along the East Coast of the United States usually colder than those along the nation's West Coast? Why are they also *less* prevalent?
8. Explain how the air flow aloft regulates the movement of air masses.
9. Why are the boundaries between air masses more strongly developed in winter than in summer?
10. What type of air mass would be responsible for the weather conditions listed below?
   (a) hot, muggy summer weather in the Midwest and the East
   (b) refreshing, cool, dry breezes after a long summer hot spell on the Central Plains
   (c) persistent cold, damp weather with drizzle along the East Coast
   (d) drought with high temperatures over the Great Plains

   (e) record-breaking low temperatures over a large portion of North America
   (f) cool weather with showers over the Pacific Northwest
11. Describe the characteristics of:
   (a) a warm front
   (b) a cold front
   (c) an occluded front
12. Draw side views of a typical cold front, warm front, and cold-occluded front. Include in each diagram cloud types and patterns, areas of precipitation, and relative temperature on each side of the front.
13. Describe the states of a developing wave cyclone using the polar front theory.
14. List four regions in North America where wave cyclones tend to develop.
15. How are longwaves in the upper-level westerlies different from shortwaves?
16. Why is it important that for a surface low to develop or intensify, its upper-level counterpart must be to the left (or west) of the surface storm?
17. Describe some of the necessary ingredients for a wave cyclone to develop into a huge mid-latitude storm system. Explain the role that upper-level divergence plays in the storm's development.
18. How does the polar jet stream influence the formation of a wave cyclone?

A watchful eye and a little weather information are important when making a local short-range weather forecast. These altocumulus clouds, northwesterly surface winds, and a rising barometer all suggest that fair weather will remain into the next day. (Photo by author)

Chapter 9

Weather Forecasting

Contents

On the eve of John Fitzgerald Kennedy's inauguration in 1961, it began to snow in the nation's capital. At first, only a few flakes glistened the pavements. But by afternoon a heavy snow began to fall, accompanied by strong and gusty northeasterly winds. By nightfall, streets and highways were impassable. Thousands of cars littered the roadways, and stranded motorists abandoned their stalled autos and set off on foot through the blinding storm. An eight-inch snowfall, driven by blustery winds, had paralyzed the city. Would the snow end by morning? Should the inauguration be postponed? These were the questions that ran through the minds of the inauguration committee and the forecasters at the National Weather Service office.

Milling over their various charts, the forecasters projected that the upper-level winds would steer the storm out of the area by daybreak. By then, they concluded, the snow should end and skies should begin to clear. However, as a large anticyclone pushed eastward toward the city, the forecasters warned that cold, northerly winds would persist throughout the day.

Their forecast was right on target. On inauguration day, the snow ended by sunrise. By noon, when the ceremonies began, the estimated one million people who watched the parade were greeted by air temperatures hovering in the low 20s (°F), northwesterly winds gusting to 20 miles per hour, and an icy wind chill of −10°F. At dusk, when the parade finally ended, the bitter cold had driven most of the onlookers away, and on the reviewing stand about the only ones left were a partially frozen president, his brother Robert, and Robert's wife, Ethel.

9 Weather forecasts are issued to save lives, to save property and crops, and to tell us what to expect in our atmospheric environment. In addition, knowing what the weather will be like in the future is vital to many human activities. For example, a summer forecast of extended heavy rain and cool weather would have construction supervisors planning work under protective cover, department stores advertising umbrellas instead of bathing suits, and ice cream vendors vacationing as their business dropped off. The forecast would alert farmers to harvest their crops before their fields became too soggy to support the heavy machinery needed for the job. And the commuter? Well, the commuter knows that prolonged rain could mean clogged gutters, flooded highways, stalled traffic, blocked railway lines and late dinners.

On the other side of the coin, a forecast calling for extended high temperatures with low humidity has an entirely different effect. As ice cream vendors prepare for record sales, the dairy farmer anticipates a decrease in milk and egg production. The forest ranger prepares warnings of fire danger in parched timber and grasslands. The construction worker is on the job outside once again, but the workday begins in the early morning and ends by early afternoon to avoid the oppressive heat. And the commuter prepares for increased traffic stalls due to overheated car engines.

Put yourself in the shoes of a weather forecaster: It is your responsibility to predict the weather accurately so that thousands (possibly millions) of people will know whether to carry an umbrella, wear an overcoat, or prepare for a winter storm. Since weather forecasting is not an exact science, your predictions will occasionally be incorrect. If your erroneous forecast misleads many people, you may become the target of jokes, insults, and even anger. There are even people who expect you to be able to predict the unpredictable. For example, on Monday you may be asked whether next Saturday will be a nice day for a picnic. And, of course, what about next winter? Will it be bitterly cold? Unfortunately, accurate answers to such questions are beyond meteorology's present technical capabili-

ties. Will forecasters ever be able to answer such questions confidently? If so, what steps are being taken to improve the forecasting art? How are forecasts made, and why do they sometimes go wrong? These are just a few of the questions we will address in this chapter.

▲▼▲

Acquisition of Weather Information

Weather forecasting basically entails predicting how the present state of the atmosphere will change. Consequently, if we wish to make a weather forecast, present weather conditions over a large area must be known. To obtain this information, a network of observing stations is located throughout the world. Over 10,000 land-based stations and hundreds of ships provide surface weather information four times a day. Most airports observe conditions hourly. Additional information, especially upper-air data, is supplied by radiosondes, aircraft, and satellites.

A United Nations agency—the World Meteorological Organization (WMO)—consists of over 130 nations. The WMO is responsible for the international exchange of weather data and certifies that the observation procedures do not vary among nations, an extremely important task, since the observations must be comparable.

Weather information from all over the world is transmitted electronically to the *National Meteorological Center* (NMC) located in Camp Springs, Maryland, just outside Washington, D.C. Here, the massive job of analyzing the data, preparing weather maps and charts, and predicting the weather on a global and national scale begins. From NMC, weather information is transmitted to private and public agencies, such as forecast offices that use the information to issue local and regional weather forecasts.

The public hears weather forecasts over radio or television. Many stations hire professional meteorologists to make their own forecasts aided by NMC material or to modify a weather service forecast. Other stations hire meteorologically untrained announcers who usually read the forecasts of the National Weather Service word for word, sometimes without properly accrediting the originating agency.

Today, the forecaster has access to over 200 maps and charts, as well as vertical profiles (called *soundings*) of temperature, dew point, and winds. Also avail-

Did you know?
Incredibly poor weather warnings and forecasts in 1868 resulted in the sinking or damaging of 1164 ships on the Great Lakes.

able is radar information that can detect and monitor the severity of precipitation and thunderstorms.

When hazardous weather is likely, the National Weather Service issues advisories in the form of weather watches and warnings. A **watch** indicates that atmospheric conditions favor hazardous weather occurring over a particular region during a specified time period. A **warning**, on the other hand, indicates that hazardous weather is either imminent or actually occurring within the specified forecast area. *Advisories* are issued to inform the public of hazardous driving conditions caused by wind, dust, fog, snow, sleet, or freezing rain. (Additional information on watches, warnings, and advisories is given in the Focus section on p. 222.)

▲▼▲ Weather Forecasting Tools

As late as the mid-1950s, all weather maps and charts were plotted by hand and analyzed by individuals. Meteorologists predicted the weather using certain rules that related to the particular weather system in question. For short-range forecasts of six hours or less, surface weather systems were moved along at a steady rate. Upper-air charts were used to predict where surface storms would develop and where pressure systems aloft would intensify or weaken. The predicted positions of these systems were extrapolated into the future using current maps. In many cases, these forecasts turned out to be amazingly accurate. They were good but, with the advent of the modern computers, today's forecasts are even better.

The Computer and Weather Forecasting Modern electronic computers can analyze large quantities of data extremely fast. Each day the many thousands of observations transmitted to NMC are fed into a high-speed computer, which plots and draws lines that interpret the weather patterns. Human intervention then corrects any errors that may be present. The final chart is referred to as an **analysis**.

The computer not only plots and analyzes data, it also predicts the weather. The routine daily forecasting of weather by the computer has come to be known as **numerical weather prediction**.

Because the many weather variables are constantly changing, meteorologists have devised **atmos-** **pheric models** that describe the present state of the atmosphere. These are not physical models that paint a picture of a developing storm; they are, rather, mathematical models that describe how atmospheric temperature, pressure, and moisture will change with time. Actually, the models do not fully represent the real atmosphere but are approximations formulated to retain the most important aspects of the atmosphere's behavior.

The models are programmed into the computer, and surface and upper-air observations of temperature, pressure, moisture, winds, and air density are fed into the equations. To determine how each of these variables will change, each equation is solved for a small increment of future time—say five minutes—for a large number of locations called *grid points*, each situated a given distance apart. In addition, each equation is solved for as many as eighteen levels in the atmosphere. The results of these computations are then fed back into the original equations. The computer again solves the equations with the new "data," thus predicting weather over the following five minutes. This procedure is done repeatedly until it reaches some desired time in the future, usually 12, 24, 36, or 48 hours. The computer than prints this information, analyzes it, and draws the projected positions of pressure systems with their isobars or contour lines. The final forecast chart representing the atmosphere at a specified future time is called a **prognostic chart**, or, simply, a **prog**. Computer-drawn progs have come to be known as "machine-made" forecasts.

The forecaster uses the progs as a guide to predicting the weather. At present, there are a variety of models (and, hence, progs) from which to choose, each producing a slightly different interpretation of the weather for the same projected time and atmospheric level. Some models predict some features better than others: One model works best in predicting the position

Did you know?

Lewis F. Richardson was the first person to make a mathematical weather prediction without the use of a computer or even a calculator. In the early 1920s, Richardson trudged six long weeks through the myriad of computations that went into making a single forecast, for a single station, only six hours into the future. Unfortunately, due to an error, his forecast turned out to be way off.

Focus on a Special Topic
Watches and Warnings

As we have seen, where severe or hazardous weather is either occurring or possible, the National Weather Service issues a forecast in the form of a watch or warning. The public, however, is not always certain as to what this forecast actually means. For example, a *high wind warning* indicates that there will be high winds—but how high and for how long? The following describes a few of the various watches, warnings, and advisories issued by the National Weather Service and the necessary precautions that should be taken during the event.

Wind advisories
Issued when sustained winds reach 25 to 39 miles per hour or wind gusts are up to 57 miles per hour.

High wind warning
Issued when sustained winds are at least 40 miles per hour, or wind gusts exceed 57 miles per hour. Caution should be taken when driving high-profile vehicles, such as trucks, trailers, and motor homes.

Wind-chill advisory
Issued for wind-chill temperatures of –30° to –35°F or below.

Flash-flood watch
Heavy rains may result in flash flooding in the specified area. Be alert and prepared for the possibility of a flood emergency that will require immediate action.

Flash-flood warning
Flash flooding is occurring or is imminent in the specified area. Move to safe ground immediately.

Severe thunderstorm watch
Thunderstorms (with winds exceeding

| Small craft advisory | Gale warning | Storm warning | Hurricane warning |

Figure 1
Flags indicating advisories and warnings in maritime areas.

57 miles per hour and/or hail three-fourths of an inch or more in diameter) are possible.

Severe thunderstorm warning
Severe thunderstorms have been visually sighted or indicated by radar. Be prepared for lightning, heavy rains, strong winds, and large hail. (Tornadoes can form with severe thunderstorms.)

Tornado watch
Issued to alert people that tornadoes may develop within a specified area during a certain time period.

Tornado warning
Issued to alert people that a tornado has been spotted either visually or by radar. Take shelter immediately.

Snow advisory
In non-mountainous areas, expect a snowfall of two inches or more in twelve hours, or three inches or more in twenty-four hours.

Winter storm warning
(formerly heavy snow warning) In non-mountainous areas, expect a snowfall of four inches or more in twelve hours or six inches or more in twenty-four hours. (Where heavy snow is infrequent, a snow fall of several inches may justify a warning.)

Blizzard warning
Issued when falling or blowing snow and winds of at least 35 miles per hour frequently restrict visibility to less than one-quarter mile for several hours.

Dense fog advisory
Issued when fog limits visibility to less than one-quarter mile, or in some parts of the country to less than one-eighth mile.

Warnings Over the Water

Small craft advisories
Issued to alert mariners that weather or sea conditions might be hazardous to small boats. Expect winds of 18 to 34 knots (21 to 39 miles per hour). Figure 1 displays the posted advisory and warning flags.

Gale warnings
Winds will range between 34 and 47 knots (39 to 54 miles per hour) in the forecast area.

Storm warnings
Winds in excess of 47 knots (54 miles per hour) are to be expected in the forecast area.

Hurricane watch
Issued when a tropical storm or hurricane becomes a threat to a coastal area. Be prepared to take precautionary action in case hurricane warnings are issued.

Hurricane warning
Issued when it appears that the storm will strike an area within twenty-four hours. Expect wind speeds in excess of 64 knots (74 miles per hour).

of troughs on upper-level charts, while another forecasts the position of surface lows quite well. Some models even forecast the state of the atmosphere 132 hours (5.5 days) into the future. (See Fig. 9.1.)

A good forecaster knows the idiosyncrasies of each model and carefully scrutinizes all the progs. The forecaster then makes a prediction based on the *guidance* of the computer, a personalized practical interpretation of the weather situation and any local geographic features that influence the weather within the specific forecast area.

There are flaws inherent in the computer models that limit the accuracy of weather forecasts. For example, computer-forecast models idealize the real atmosphere, meaning that each model makes certain assumptions about the atmosphere. These assumptions may be on target for some weather situations and be way off for others. Consequently, the computer may produce a prog that on one day comes quite close to describing the actual state of the atmosphere (such as the example in Fig. 9.1) and not so close on another. A

forecaster who bases a prediction on an "off day" computer prog may find a forecast of "rain and windy" turning out to be a day of "clear and colder."

Even though many thousands of weather observations are taken worldwide each day, there are still regions where observations are sparse, particularly in the upper air over the oceans. Temperature data recorded by satellites have helped to alleviate this problem, although not completely because correlating temperatures measured by satellites with those measured by radiosondes has not been an easy task. Since the computer's forecast is only as good as the data fed into it, and since observing stations may be so far apart that they miss certain weather features, a denser network of observations is needed, especially in remote areas, to ensure better forecasts in the future.

Another forecasting problem is that computer models cannot adequately interpret many of the factors that influence surface weather, such as the interactions of water, ice, and local terrain on weather systems. Some models do take large geographic features (such

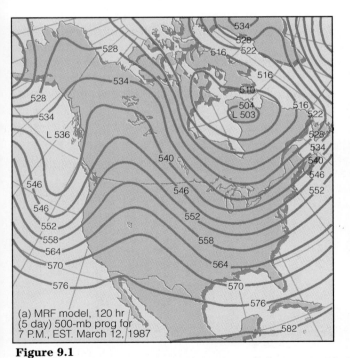
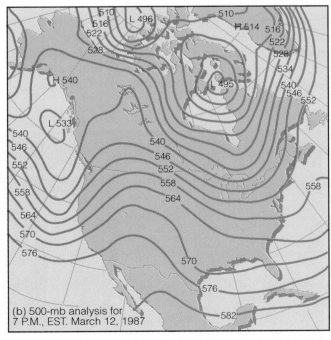

Figure 9.1

(a) An upper-level forecast chart (prog) for 7 P.M., EST, March 12, 1987—5 days (120 hr) into the future. The prediction was made on March 7. Solid lines on the map are height contours, where 552 equals 5520 meters. (The name of this model, MRF, means *medium-range forecast*.). (b) The upper-level analysis for 7 P.M., EST, March 12, 1987. This chart shows that the forecast model did an excellent job of predicting the positions of troughs and ridges across North America.

Did you know?
Nightly news weather presentations have come a long way since the early days of television. The "weather girl," which became popular during the 1960s, was a crazy fad that searched for gimmicks to attract viewers. Females gave weather forecasts in various attire (from bathing suits to bunny outfits), sometimes with the aid of hand puppets that resembled the mutant ninja turtles. One West Coast television station actually hired a woman to do the nightly news weather segment with the requirement that she be able to write backwards on a clear, plexiglass screen.

as mountain chains and oceans) into account, while ignoring smaller features (such as hills and lakes). These smaller features can have a marked influence on the local weather. Given the effect of local terrain, as well as the impact of some of the other problems previously mentioned, computer forecasts presently do an inadequate job of predicting local weather conditions, such as surface temperatures, winds, and precipitation.

Even with a denser network of observing stations and near perfect computer models, there are countless small, unpredictable atmospheric fluctuations that fall under the heading of *chaos*. These small disturbances, as well as small errors (uncertainties) in the data, generally amplify with time as the computer tries to project

the weather farther and farther into the future. After a number of days, these initial imperfections (errors) tend to dominate, and the forecast shows little or no skill in predicting the behavior of the real atmosphere.

In summary, imperfect numerical weather predictions may result from flaws in the computer models, from a sparseness of data, and/or from inadequate representation of the many pertinent processes, interactions, and inherently chaotic behavior that occurs within the atmosphere.

The Use of Minicomputers   To help forecasters handle all the available charts and maps, high-speed data modeling systems using minicomputers are employed. This particular communication system, known as **AFOS** (*A*utomation of *F*ield *O*perations and *S*ervices), provides weather information from any desired region on a television screen, as well as graphic overlays of satellite pictures and charts that enable forecasters to view and analyze dynamic weather events over a broad area. The AFOS system is shown in Fig. 9.2.

At present, more advanced computerized systems are being developed and tested for future use. During the 1990s, the National Weather Service is planning a major modernization and restructuring program to replace the current AFOS system with a superior *A*dvanced *W*eather *I*nteractive *P*rocessing *S*ystem,

Figure 9.2
The AFOS all-electronic computerized system displays weather information on television-type consoles at the push of a button.

Focus on an Observation
TV Weathercasters—How Do They Do It?

Figure 2
On your home television, weather forecaster Tom Loffman of KOVR-TV Sacramento, appears to be pointing to weather information directly behind him.

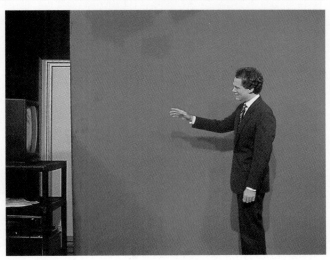

Figure 3
In the studio, however, Tom is actually standing in front of a blank board.

As you watch the TV weathercaster, you typically see a person describing and pointing to specific weather information, such as weather maps, satellite photos, and radar images (Fig. 2). What you may not know is that the weathercaster is actually pointing to a blank board (usually blue or green) on which there is nothing (Fig. 3). This process of electronically superimposing weather information in the TV camera against a blank wall is called color-separation overlay or *chroma key*.

The chroma key process works because the studio camera is constructed to pick up all colors except (in this case) blue. The various maps, charts, satellite photos, and other graphics are electronically inserted from a computer into this blue area of the color spectrum. The person in the TV studio should not wear blue clothes because such clothing would not be picked up by the camera—what you would see on your home screen would be a head and hands moving about the weather graphics!

How, then, does a TV weathercaster know where to point on the blank wall? Positioned on each side of the blue wall are TV monitors (look carefully at Fig. 3) that weathercasters watch so that they know where to point.

AWIPS. The AWIPS system will have new data communications, storage, processing, and display capabilities to better help the individual forecaster extract and assimilate information from the mass of available data. In addition, AWIPS will integrate and process information received from NEXRAD (the new Doppler radar system being installed at weather forecast offices) and the new Automated Surface Observing Systems (ASOS) that are planned for airports and other sites throughout the United States. The ASOS system is designed to provide nearly continuous information about wind, temperature, pressure, cloud base height, and runway visibility at various airports. Meteorologists believe that information from all of these sources will improve

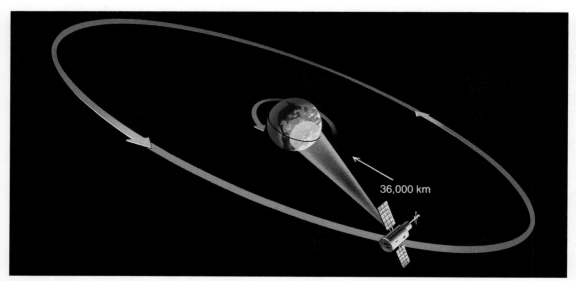

36,000 km

Figure 9.3
The geostationary satellite moves through space at the same rate that the earth rotates, so it remains above a fixed spot on the equator and monitors one area constantly.

the accuracy of weather forecasts by providing previously unobtainable data for integration into numerical forecast models.

Satellites and Weather Forecasting   The weather satellite is a cloud-observing platform in earth orbit. It provides extremely valuable cloud photographs of areas where there are no ground-based observations. Because water covers over 70 percent of the earth's surface, there are vast regions where few (if any) surface cloud observations are made. Before weather satellites were used, severe storms, such as hurricanes and typhoons, went undetected until they moved dangerously near inhabited areas. Residents of the regions affected had little advance warning. Today, satellites spot these storms while they are still far out in the ocean and track them accurately.

There are two primary types of weather satellites used in weather forecasting. The first are called **geostationary satellites*** (or *geosynchronous satellites*) because they remain at nearly 36,000 kilometers (27,300 miles) above a fixed spot on the earth's surface

---

*Geostationary satellites are also known as GOES (Fig. 9.5), after their more formal name: Geostationary Operational Environmental Satellite.

(Fig. 9.3). This positioning allows continuous monitoring of a specific region.

Geostationary satellites are also important because they use a "real time" data system, meaning that the satellites transmit photographs to the receiving system on the ground as soon as the camera takes the picture. Successive cloud photographs from these satellites can be put into a time-lapse movie sequence to show the cloud movement, dissipation, or development associated with weather fronts and storms. This is a great help in forecasting the progress of large weather systems. Wind directions and speeds at various levels may also be approximated by monitoring cloud movement with the geostationary satellite.

To complement the geostationary satellites, there are **polar orbiting satellites**, which closely parallel the earth's meridian lines. They pass over the north and south polar regions on each revolution. As the earth rotates to the east beneath the satellite, each pass monitors an area to the west of the previous pass (Fig. 9.4). Eventually, the satellite covers the entire earth.

Polar orbiting satellites have the advantage of photographing clouds directly beneath them. Thus, they provide sharp pictures in polar regions, where photographs from a geostationary satellite are distorted because of the low angle at which the satellite "sees" this

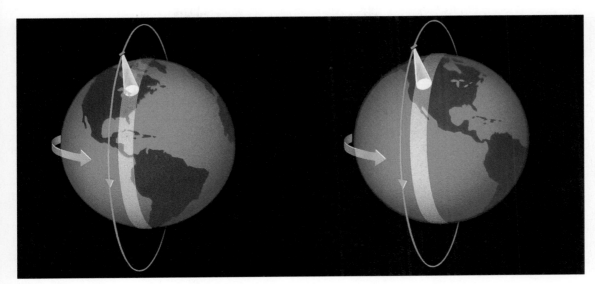

Figure 9.4
Polar orbiting satellites scan from north to south, and on each successive orbit the satellite scans an area further to the west.

region. Polar orbiters also circle the earth at a much lower altitude (about 850 kilometers or 530 miles) than geostationary satellites and provide detailed photographic information about objects, such as violent storms and cloud systems.

Continuously improved detection devices make our satellites more versatile than ever. Early satellites, such as *TIROS I*, launched on April 1, 1960, used television cameras to photograph clouds. Contemporary satellites use radiometers, which can observe clouds during both day and night by detecting radiation coming from their tops.

The forecaster can obtain information on cloud thickness and height from satellite photographs. Visible photographs show the sunlight reflected from a cloud's upper surface. Because thick clouds have a higher reflectivity than thin clouds, they appear brighter on a visible satellite photograph. However, high, middle, and low clouds have just about the same reflectivity, so it is difficult to distinguish among them simply by using visible light photographs. To make this distinction, *infrared cloud pictures* are used. Such pictures produce a better image of the actual radiating surface because they do not show the strong visible reflected light. Since warm objects radiate more energy than cold objects, high temperature regions can be artificially made

to appear darker on an infrared photograph. Because the tops of low clouds are warmer than those of high clouds, cloud observations made in the infrared can distinguish between warm, low clouds (dark) and cold, high clouds (light) (Fig. 9.6).

Figure 9.5
Artist's view of the newest *GOES* meteorological satellite.

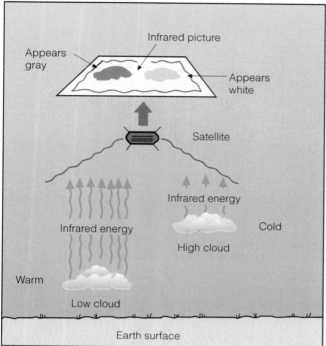

Figure 9.6
Generally, the lower the cloud, the warmer its top. Warm objects emit more infrared energy than do cold objects. Thus, an infrared satellite picture can distinguish warm, low (gray) clouds from cold, high (white) clouds.

Figure 9.7a shows a visible satellite picture (from a geostationary satellite) of an occluded storm system in the eastern Pacific. Notice that all of the clouds in the photo appear white. However, in the infrared photograph (Fig. 9.7b), taken on the same day (and just about the same time), the clouds appear to have many shades of gray. In the visible photograph, the clouds covering part of Oregon and Northern California appear relatively thin compared to the thicker, bright clouds to the west. Furthermore, these thin clouds must be high because they also appear bright in the infrared picture.

Along the elongated band of clouds associated with the occluded front, the clouds appear white and bright in both pictures, indicating a zone of thick, heavy clouds. Behind the front, the forecaster knows that the lumpy clouds are probably cumulus because they appear gray in the infrared photo, suggesting that their tops are low and relatively warm.

When temperature differences are small, it is difficult to directly identify significant cloud and surface features on an infrared picture. Some way must be found to increase the contrast between features and their backgrounds. This can be done by a process called *computer enhancement*. Certain temperature ranges in the infrared photograph are assigned specific shades of gray—grading from black to white. These shades of gray are then color-contoured to make specific features, such as deep cloud layers and the freezing level, more obvious. Often, dark blue or red is assigned to clouds with the coldest (highest) tops. Figure 9.8 is an infrared-enhanced picture for the same date and area as shown in Fig. 9.7. Note the dark and light contouring in the picture. Clouds with cold tops, and those with tops near freezing, are assigned the darkest color. Hence, the dark gray areas embedded along the front represent the region where the coldest and, therefore, highest and thickest clouds are found. It is here where the stormiest weather is probably occurring. Also notice that, near the southern tip of the picture, the dark gray blotches surrounded by areas of white are thunderstorms that have developed over warm tropical waters. They show up clearly as white, thick clouds in both the visible and infrared photographs. By examining the movement of these clouds on successive satellite photographs, the forecaster can predict the arrival of clouds and storms, and the passage of weather fronts.

In regions where there are no clouds, it is difficult for the forecaster to observe the movement of the air aloft. To help with this situation, certain geostationary satellites are equipped with a sensor that detects atmospheric water vapor. Notice in Fig. 9.9 that the swirling pattern of moisture clearly reveals the large-scale atmospheric features.

Methods of Weather Forecasting

Probably the easiest weather forecast to make is a **persistence forecast**, which is simply a prediction that future weather will be the same as present weather. If it is snowing today, a persistence forecast would call for snow through tomorrow. Such forecasts are most accurate for time periods of several hours and become less and less accurate after that.

Another method of forecasting is the **steady-state**, or **trend method**. The principle involved here is that surface weather systems tend to move in the same

(a)   (b)

Figure 9.7
A visible picture (a) and an infrared picture (b) of the eastern Pacific taken on the same day (April 23, 1978) at just about the same time.

direction and at approximately the same speed as they have been moving, providing no evidence exists to indicate otherwise. Suppose, for example, that a cold front is moving eastward at an average speed of thirty miles per hour and it is ninety miles west of your home. Using the steady-state method, we might extrapolate and predict that the front should pass through your area in three hours.

In recent years, the trend method has been employed in the making of forecasts from minutes for up to a few hours. Such short-term forecasting has come to be called **nowcasting**.

The **analogue method** is yet another form of weather forecasting. Basically, this method relies on the fact that existing features on a weather chart (or a series of charts) may strongly resemble features that produced certain weather conditions sometime in the past. Prior weather events can then be utilized as a guide to the future. The problem here is that, even though weather situations may appear similar, they are never *exactly* the same. There are always sufficient differences in the variables to make applying this method a challenge.

The analogue method can be used to predict a number of weather elements, such as maximum temperature. Suppose that in New York City the average maximum temperature on a particular date for the past thirty years is 10°C (50°F). By statistically relating the maximum temperatures on this date to other weather elements—such as the wind, cloud cover, and humidity—a relationship between these variables and maximum temperature can be drawn. By comparing these

Figure 9.8
An enhanced infrared picture of the eastern Pacific taken on April 23, 1978.

relationships with current weather information, the forecaster can predict the maximum temperature for the day.

Predicting the weather by **weather types** employs the analogue method. In general, weather patterns are categorized into similar groups or "types," using such criteria as the position of the subtropical highs, the upper-level flow, and the prevailing storm track. As an example, when the Pacific high is weak or depressed southward and the flow aloft is zonal (west-to-east), surface storms tend to travel rapidly eastward across the Pacific Ocean and into the United States without developing into deep systems. But when the Pacific

high is to the north of its normal position and the upper-air flow is meridional (north-south), looping waves form in the flow with surface lows usually developing into huge storms. Since upper-level longwaves move slowly, usually remaining almost stationary for perhaps a few days to a week or more, the particular surface weather at different positions around the wave is likely to persist for some time. (See Fig. 9.10.)

Weather types can be used as an approach to *long-range* (a month or more in advance) *weather forecasting*. Typically, the upper-air circulation changes gradually from zonal to meridional over 4 to 6 weeks. As this slow change occurs in the upper air, the surface weather may repeat itself at specific intervals. For instance, winter cold fronts may sweep into New England every four days or so, bringing showers and below-normal temperatures. By projecting trends such as these, and assuming that the atmosphere's behavior will not change radically (an assumption not always valid), *extended weather forecasts* can be made. At best, these forecasts only show the broad-scale weather features. They do not adequately predict specific weather elements.

Despite the apparently unsuccessful nature of long-range forecasts, the National Weather Service currently issues extended forecasts of six to ten days, as well as a thirty-day outlook for the coming month, and a ninety-day seasonal outlook. These are not forecasts in the strict sense, but rather an overview of how average precipitation and temperature patterns may compare with normal conditions. Figure 9.11 gives a typical thirty-day outlook.

To improve medium- and extended-range forecasts, meteorologists are turning to a technique called **ensemble forecasting**. This approach is based on running several forecast models—or different versions (simulations) of a single model—each beginning with slightly different weather information to reflect the errors inherent in the measurements. If at the end of a specified time the models match each other fairly well, then the forecaster can issue a prediction with a high degree of confidence. If the models disagree, the fore-

Did you know?
Long-range weather forecasts were first made by Benjamin Franklin in his *Poor Richard's Almanac*, beginning in 1732. Ironically, the fact that storms move was not known until about 1743.

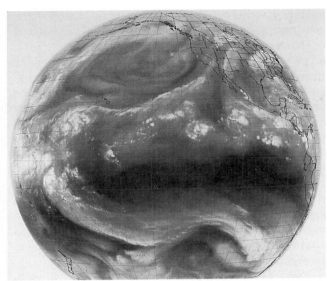

Figure 9.9
Infrared water-vapor image taken by a geostationary satellite on August 30, 1982. The cyclonic swirl of water vapor off the California coast defines a low pressure area. The white band of clouds (blobs) north of the equator is the intertropical convergence zone. The stretched-out band of moisture in the Southern Hemisphere is the jet stream.

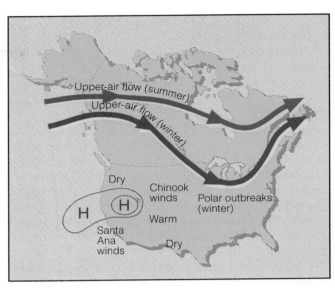

Figure 9.10
Weather type showing upper-air flow (heavy arrows), surface position of Pacific high, and general weather conditions that should prevail.

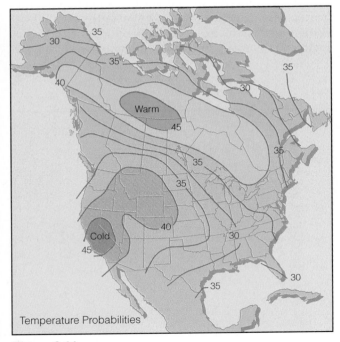

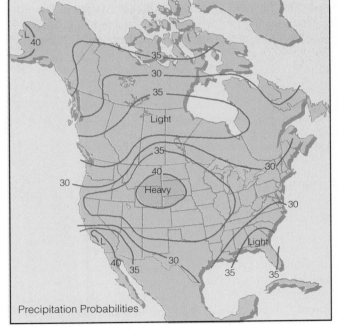

Figure 9.11
The 30-day outlook for October, 1985. Notice that the greatest probability for a warm month (45 percent or greater) occurs over Canada, while the greatest probability for a wet month (40 percent or greater) occurs over the central United States.

Did you know?
Due to extremely limited availability of accurate weather reports and forecasts, the average life expectancy for an airmail pilot between 1918 and 1925 was about four years.

caster, with little faith in the computer model prediction, issues a forecast with limited confidence, perhaps by giving a number ranging from 0 (no confidence) to 5 (great confidence). In essence, *the less agreement among the models, the less predictable the weather.* Consequently, it would not be wise to make outdoor plans for Saturday when on Monday the weekend forecast calls for "sunny and warm" with a low degree of confidence.

A forecast based on the climatology (average weather) of a particular region is known as a **climatological forecast**. Anyone who has lived in Los Angeles for a while knows that July and August are practically rain-free. In fact, rainfall data for the summer months taken over many years reveal that rainfall amounts of more than a trace occur in Los Angeles about one day in every ninety, or only about 1 percent of the time. Therefore, if we predict that it will not rain on some day next year during July or August in Los Angeles, our chances are nearly 99 percent that the forecast will be correct based on past records. Since it is

unlikely that this pattern will significantly change in the near future, we can confidently make the same forecast for the year 2020.

When the Weather Service issues a forecast calling for rain, it is usually followed by a probability. For example: "The chance of rain is 60 percent." Does this mean (a) that it will rain on 60 percent of the forecast area or (b) that there is a 60 percent chance that it will rain within the forecast area? Neither one! The expression means that there is a 60 percent chance that any random place in the forecast area, such as your home, will receive measurable rainfall. Looking at the forecast in another way, if the forecast on ten days calls for a 60 percent chance of rain, it should rain where you live on six of those days. The verification of the forecast (as to whether it actually rained or not) is usually made at the Weather Service office.

An example of a **probability forecast** using climatological data is given in Fig. 9.12. The map shows the probability of a "White Christmas"—one inch or more of snow on the ground—across the United States. The map is based on the average of thirty years of data and gives the likelihood of snow in terms of a probability. For instance, the chances are 90 percent (nine Christmases out of ten) that portions of northern Minnesota, Michigan, and Maine will experience a White Christmas. In Chicago, it is 50 percent; and in Washington, D.C., about 20 percent. Many places in the far

Figure 9.12
Probability of a "White Christmas"—one inch or more of snow on the ground—based on a thirty-year average.

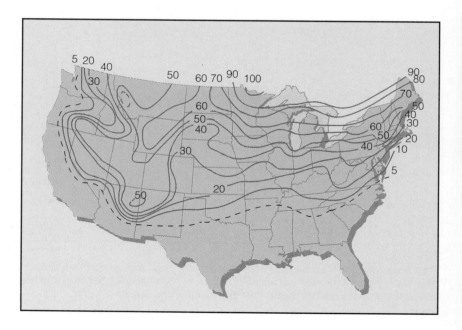

Focus on an Issue
A Word about Accuracy and Skill

In spite of the complexity and ever-changing nature of the atmosphere, forecasts made for between twelve and twenty-four hours are usually quite accurate. Those made for between one and three days are fairly good. Beyond about five days, however, forecast accuracy falls off rapidly. Although weather predictions made for up to three days are by no means perfect, they are far better than simply flipping a coin. But how accurate are they?

One problem with determining forecast accuracy is deciding what constitutes a right or wrong forecast. Suppose tomorrow's forecast calls for a minimum temperature of 5°C. If the official minimum turns out to be 6°C, is the forecast incorrect? Is it as incorrect as one ten degrees off? By the same token, what about a forecast for snow over a large city, and the snow line cuts the city in half with the southern portion receiving heavy amounts and the northern portion none? Is the forecast right or wrong? At present, there is no clear-cut answer to the question of determining forecast accuracy.

How does forecast accuracy compare with forecast skill? Suppose you are forecasting the daily summertime weather in Los Angeles. It is not raining today and your forecast for tomorrow calls for "no rain." Suppose that tomorrow it doesn't rain. You made an accurate forecast, but did you show any skill in so doing? In a previous section, we saw that the chance of measurable rain in Los Angeles on any summer day is very small indeed; chances are good that day after day it will not rain. For a forecast to show skill, it should be more than one based solely on the current weather (*persistence*) or on the "normal" weather (*climatology*) for a given region. Therefore, during the summer in Los Angeles, a forecaster will have many accurate forecasts calling for "no measurable rain," but will need skill to predict correctly on which summer days it will rain.

Meteorological forecasts, then, show skill when they are more accurate than a forecast utilizing only persistence or climatology. Persistence forecasts are usually difficult to improve upon for a period of time of several hours or less. Weather forecasts ranging from twelve hours to a few days generally show much more skill than those of persistence. However, as the range of the forecast period increases, the skill drops quickly. The three- to five-day and six- to ten-day mean outlooks both show some skill in predicting temperature and precipitation. In fact, five-day forecasts now show about as much skill as three-day forecasts did a decade ago. Beyond ten days, specific forecasts are generally about as good as those based on climatology. The greatest improvement in forecasting skill during the past thirty years has been made in the area of severe storm warnings for hurricanes and tornadoes. Despite large population increases in areas generally threatened by these storms, there has been a decrease over the years in the number of lives lost because of them.

---

west and south have probabilities less than 5 percent, but nowhere is the probability exactly 0, for there is always some chance (no matter how small) that a mantle of white will cover the ground on Christmas day.

In most locations throughout North America, the weather is fair more often than rainy. Consequently, there is a forecasting bias toward fair weather, which means that, if you made a forecast of no-rain where you live for each day of the year, your forecast would be correct more than 50 percent of the time. But did you show any *skill* in making your correct forecast? What constitutes skill, anyway? And how accurate are the forecasts issued by the National Weather Service? (These questions are addressed in the Focus section above.)

Predicting the Weather from Local Signs

Because the weather affects every aspect of our daily lives, attempts to predict it accurately have been made for centuries. One of the earliest attempts was undertaken by Theophrastus, a pupil of Aristotle, who in 300 B.C. compiled all sorts of weather indicators in his *Book of Signs*. A dominant influence in the field of weather forecasting for 2000 years, this work consists of ways to foretell the weather by examining natural signs, such as the color and shape of clouds, and the intensity at which a fly bites. Some of these signs have validity and are a part of our own weather folklore—"a halo around the moon portends rain" is one of these. Today, we realize that the halo is caused by the bending of light as it passes through ice crystals and that ice crystal-type clouds (cirrostratus) are often the forerunners of an approaching storm.

Weather predictions can be made by observing the sky and using a little weather wisdom. If you keep your eyes open and your senses keenly tuned to your environment, you should, with a little practice, be able to make fairly good short-range local weather forecasts by interpreting the messages written in the weather elements. Table 9.1 is designed to help you with this endeavor.

▲▼▲

Weather Forecasting Using Surface Charts

We are now in a position to forecast the weather, utilizing more sophisticated techniques. Suppose, for ex-

ample, that we wish to make a short-range weather prediction and the only information available is a surface weather map. Can we make a forecast from such a chart? Most definitely. And our chances of that forecast being correct improve markedly if we have maps available from several days back. We can use these past maps to locate the previous position of surface features and predict their movement.

A simplified surface weather map is shown in Fig. 9.13. The map portrays early winter weather conditions on Tuesday morning at 6:00 A.M. A single isobar is drawn around the pressure centers to show their positions without cluttering the map. Note that an open wave cyclone is developing over the Central Plains with showers forming along a cold front and light rain and snow ahead of a warm front. The dashed lines on the map represent the position of the weather systems six hours ago. Our first question is: How will these systems move?

Determining the Movement of Weather Systems
There are several methods we can use in forecasting

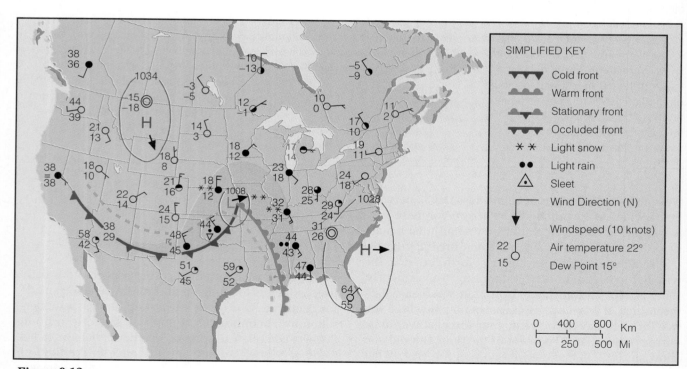

Figure 9.13
Surface weather map for 6:00 A.M. Tuesday. Dashed lines indicate positions of weather features six hours ago. Green-shaded areas are receiving precipitation.

Table 9.1   Forecast at a Glance. Forecasting the Weather from Local Weather Signs.
Listed below are a few forecasting rules that may be applied when making a short-range local weather forecast.

Observation	Indication	Local Weather Forecast
Surface winds from the S or from the SW; clouds building to the west; warm (hot) and humid	Possible cool front and thunderstorms approaching from the west	Possible showers; possibly turning cooler; windy
Surface winds from the E or from the SE, cool or cold; high clouds thickening and lowering; halo around the sun or moon	Possible approach of a warm front	Possibility of precipitation within 12–24 hours; windy (rain with possible thunderstorms during the summer; snow changing to sleet or rain in winter)
Winter night		
(a) If clear, relatively calm with low humidity (low dew-point temperature)	(a) Rapid radiational cooling will occur	(a) A very cold night
(b) If clear, relatively calm with low humidity and snow covering the ground	(b) Rapid radiational cooling will occur	(b) A very cold night with minimum temperatures lower than in (a)
(c) If cloudy, relatively calm with low humidity	(c) Clouds will absorb and reradiate infrared (IR) energy back to surface	(c) Minimum temperature will not be as low as in (a) or (b)
Summer night		
(a) Clear, hot, humid (high dew points)	(a) Strong absorption and re-emission of IR energy back to surface by water vapor	(a) High minimum temperatures
(b) Clear and relatively dry	(b) More rapid radiational cooling	(b) Lower minimum temperatures
If surface winds are from the N and they become NE, then E, then SE (winds that change direction in this clockwise sense are known as *veering winds*)	A surface high-pressure area may be moving to your E, and a surface low pressure area may be approaching from the W	Increasing clouds, warmer with the possibility of precipitation within 24 hours
If surface winds are from the NE and they become N, then NW (winds that change direction in this counterclockwise sense are known as *backing winds*)	A surface low-pressure area is moving to your E, and a surface high-pressure area may be approaching from the W	Clearing and colder (cooler in summer)
Scattered cumulus clouds that show extensive vertical growth by mid morning	Atmosphere is relatively unstable	Possible showers or thunderstorms by afternoon with gusty winds
Afternoon cumulus clouds with flat bases, and tops at just about the same level	Stable layer above clouds (region dominated by high pressure)	Continued partly cloudy with no precipitation; probably clearing by nightfall

the movement of surface pressure systems and fronts. The following are a few of these forecasting rules of thumb:

1. For short-time intervals, storms and fronts tend to move in the same direction and at approximately the same speed as they did during the previous six hours (providing, of course, there is no evidence to indicate otherwise).
2. Lows tend to move in a direction that parallels the isobars in the warm air ahead of the cold front.
3. Lows tend to move toward the region of greatest surface pressure drop, whereas highs tend to move toward the region of greatest surface rise.
4. Surface pressure systems tend to move in the same direction as the wind at 5600 meters (18,000 feet)—the 500-millibar level. The speed at which surface systems move is about half the speed of the winds at this level.

When the surface map (Fig. 9.13) is examined carefully and when rules of thumb 1 and 2 are applied, it appears that—based on present trends—the storm center over the Central Plains should move northeast. When we observe the 500-millibar upper-air chart (Fig.

9.14), it too suggests that the surface low should move northeast at a speed of about 25 knots.

A Forecast for Six Cities   We are now in a position to make a weather forecast for six cities. To do this, we will project the pressure systems, fronts, and current weather into the future by assuming steady-state conditions. Figure 9.15 gives the 12- and 24-hour projected positions of these features.

A word of caution before we make our forecasts. We are assuming that the pressure systems and fronts are moving at a constant rate, which may or may not occur. Storm systems, for example, tend to accelerate until they occlude, after which their rate of movement slows. Furthermore, the direction of moving systems may change due to "blocking" highs and lows that exist in their path or because of shifting upper-level wind patterns. We will assume a constant rate of movement and forecast accordingly, always keeping in mind that the longer our forecasts extend into the future, the more susceptible they are to error.

Using Fig. 9.15 to follow the storm center eastward, we can make a basic forecast. The cold front moving into north Texas on Tuesday morning is pro-

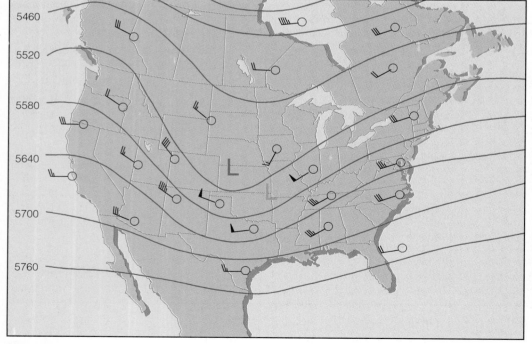

	Miles (statute) per hour	Knots
◎	Calm	Calm
—	1–2	1–2
﹀	3–8	3–7
﹂	9–14	8–12
﹄	15–20	13–17
﹅	21–25	18–22
﹆	26–31	23–27
﹇	32–37	28–32
﹈	38–43	33–37
﹉	44–49	38–42
﹊	50–54	43–47
﹋	55–60	48–52
﹌	61–66	53–57
﹍	67–71	58–62
﹎	72–77	63–67
﹏	78–83	68–72
﹐	84–89	73–77
﹑	119–123	103–107

Figure 9.14
Upper-air chart (500-millibar map) for 6:00 A.M. Tuesday, showing wind flow. The light red L represents the position of the surface low. (Solid lines on the map are height contours.)

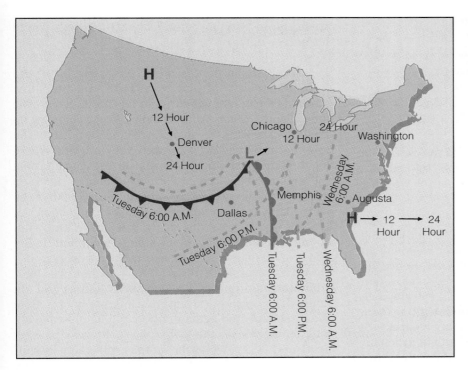

Figure 9.15
Projected 12- and 24-hour movement of fronts, pressure systems, and precipitation (green-shaded area) from 6:00 A.M. Tuesday until 6:00 A.M. Wednesday.

jected to pass Dallas by that evening, so a forecast for the Dallas area would be "warm with showers, then turning colder." But we can do much better than this. Knowing the weather conditions that accompany advancing pressure areas and fronts, we can make more detailed weather forecasts that will take into account changes in temperature, pressure, humidity, cloud cover, precipitation, and winds. Our forecast will include the 24-hour period from Tuesday morning to Wednesday morning for the cities of Augusta, Georgia; Washington, D.C.; Chicago, Illinois; Memphis, Tennessee; Dallas, Texas; and Denver, Colorado. We will begin with Augusta.

Weather Forecast for Augusta, Georgia   On Tuesday morning, continental polar air associated with a high-pressure center brought freezing temperatures and fair weather to the Augusta area (Fig. 9.13). Clear skies, light winds, and low humidities allowed rapid nighttime cooling so that, by morning, temperatures were in the low thirties. Now look closely at Fig. 9.15 and observe that the anticyclone is moving slowly eastward. Southerly winds on the western side of this system will bring warmer and more moist air to the region. Therefore, afternoon temperatures will be warmer than those

of the day before. As the warm front approaches from the west, clouds will increase, appearing first as cirrus, then thickening and lowering into the normal sequence of warm-front clouds. Barometric pressure should fall. Clouds and high humidity should keep minimum temperatures well above freezing on Tuesday night. Note that the projected area of precipitation (green-shaded region) does not quite reach Augusta. With all of this in mind, our forecast might sound something like this:

> Clear and cold this morning with moderating temperatures by afternoon. Increasing high clouds with skies becoming overcast by evening. Cloudy and not nearly as cold tonight and tomorrow morning. Winds will be light and out of the south or southeast. Barometric pressure will fall slowly.

Wednesday morning we discover that the weather in Augusta is foggy with temperatures in the upper 40s (°F). But fog was not in the forecast. What went wrong? We forgot to consider that the ground was still cold from the recent cold snap. The warm, moist air moving over the cold surface was chilled below its dew point, resulting in fog. Above the fog were the low clouds we predicted. The minimum temperatures remained higher than anticipated because of the release of latent heat during fog formation and the absorption of infrared energy by the fog droplets. Not bad for a start. Now we will forecast the weather for Washington, D.C.

Rain or Snow for Washington, D.C.?   Look at Fig. 9.15 and observe that the storm center is slowly approaching Washington, D.C., from the west. Hence, the clear weather, light southwesterly winds, and low temperatures on Tuesday morning (Fig. 9.13) will gradually give way to increasing cloudiness, winds shifting to the southeast, and slightly higher temperatures. By Wednesday morning, the projected band of precipitation will be over the city. Will it be in the form of rain or snow? Without a vertical profile of temperature, this question is difficult to answer. We can see in Fig. 9.13, however, that cities south of Washington, D.C.'s latitude are receiving snow. So a reasonable forecast would call for snow, possibly changing to rain as warm air moves in aloft in advance of the approaching fronts. A 24-hour forecast for Washington, D.C., might sound like this:

> Increasing clouds today and continued cold. Snow beginning by early Wednesday morning, possibly changing to rain. Winds will be out of the southeast. Pressures will fall.

Wednesday morning a friend in Washington, D.C., calls to tell us that the sleet began to fall but has since changed to rain. Sleet? Another fractured forecast! Well, almost. What we forgot to account for this time was the intensification of the storm. As the storm moved eastward, it deepened; central pressure lowered, pressure gradients tightened, and southeasterly winds blew stronger than anticipated. As air moved inland off the warmer Atlantic, it rode up and over the colder surface air. Snow falling into this warm layer at least partially melted; it then refroze as it entered the colder air near ground level. The influx of warmer air from the ocean slowly raised the surface temperatures, and the sleet soon became rain. Although we did not see this possibility when we made our forecast, a fore-

caster more familiar with local surroundings would have. Let's move on to Chicago.

Big Snow Storm for Chicago   From Figs. 9.13 and 9.15, it appears that Chicago is in for a major snow storm. Overrunning of warm air has produced a wide area of snow which, from all indications, is heading directly for the Chicago area. Since cold air north of the low center will be over Chicago, precipitation reaching the ground should be frozen. On Tuesday morning the leading edge of precipitation is less than six hours away from Chicago. Based on the projected path of the storm, light snow should begin to fall around noon.

By evening, as the storm intensifies, snowfall should become heavy. It should taper off and finally end around midnight as the storm moves on east. If it snows for a total of twelve hours—six hours as light snow (around one inch every three hours) and six hours as heavy snow (around one inch per hour)—then the total expected accumulation will be between six and ten inches. As the low moves eastward, passing south of Chicago, winds on Tuesday will gradually shift from southeasterly to easterly, then northeasterly by evening. Since the system is intensifying, it should produce strong winds that will swirl the snow into huge drifts, which may bring traffic to a crawl.

The winds will continue to shift to the north and finally become northwesterly by Wednesday morning. By then the storm center will probably be far enough east so that skies should begin to clear. Cold air moving in from the northwest behind the storm will cause temperatures to drop further. Barometer readings during the storm will fall as the low center approaches and reach a low value sometime Tuesday night, after which they will begin to rise. A weather forecast for Chicago might be:

> Cloudy and cold with light snow beginning by noon, becoming heavy by evening and ending by Wednesday morning. Total accumulations will range between six and ten inches. Winds will be strong and gusty out of the east or northeast today becoming northerly tonight and northwesterly by Wednesday morning. Barometric pressure will fall sharply today and rise tomorrow.

A call Wednesday morning to a friend in Chicago reveals that our forecast was correct except that the total snow accumulation so far is thirteen inches. We were off in our forecast because the storm system slowed as it became occluded. We did not consider this because we moved the system by the steady-state method. At this time of year (early winter), Lake Michigan is not quite frozen over and the added moisture picked up from the lake by the strong easterly winds also helped to produce a heavier-than-predicted snowfall. Again, a knowledge of the local surroundings would have helped make a more accurate forecast. The weather about five hundred miles south of Chicago should be much different from this.

Mixed Bag of Weather for Memphis   Observe in Fig. 9.15 that, within twenty-four hours, both a warm and a cold front should move past Memphis. The light rain that began Tuesday morning should saturate the cool air, creating a blanket of low clouds and fog by midday. The warm front, as it moves through sometime Tuesday afternoon, should cause temperatures to rise slightly as winds shift to the south or southwest. At night, clear to partly cloudy skies should allow the ground and air above to cool, offsetting any tendency for a rapid rise in temperature. Falling pressures should level off in the warm air, then fall once again as the cold front approaches. According to the projection in Fig. 9.15, the cold front should arrive sometime before midnight on Tuesday, bringing with it gusty northwesterly winds, showers, the possibility of thunderstorms, rising pressures, and colder air. Taking all of this into account, our weather forecast for Memphis will be:

> Cloudy and cool with light rain, low clouds, and fog early today, becoming partly cloudy and warmer by late this afternoon. Clouds increasing with possible showers and thunderstorms later tonight or early Wednesday morning and turning colder. Winds southeasterly this morning, becoming southerly or southwesterly this evening and shifting to northwesterly by Wednesday morning. Pressures falling this morning, leveling off this evening, then falling again tonight and rising by Wednesday morning.

A friend who lives near Memphis calls Wednesday to inform us that our forecast was correct except that the thunderstorms did not materialize and that Tuesday night dense fog formed in low-lying valleys, but by Wednesday morning it had dissipated. Apparently, in the warm air, winds were not strong enough to mix the cold, moist air that had settled in the valleys with the warm air above. It's on to Dallas.

Cold Wave for Dallas   From Fig. 9.15, it appears that our weather forecast for Dallas should be straightforward, since a cold front is expected to pass the area

around noon. Weather along the front is showery with a few thunderstorms developing; behind the front the air is clear but cold. By Wednesday morning it looks as if the cold front will be far to the east and south of Dallas and an anticyclone will be centered over Colorado. North or northwesterly winds on the east side of the high will bring cold continental polar air into Texas, dropping temperatures as much as 40°F within a 24-hour period. With minimum temperatures well below freezing, Dallas will be in the grip of a cold wave. Our weather forecast should therefore sound something like this:

> Increasing cloudiness and mild this morning with the possibility of showers and thunderstorms this afternoon. Clearing and turning much colder tonight and tomorrow. Winds will be southwesterly today, becoming gusty north or northwesterly this afternoon and tonight. Pressures falling this morning, then rising later today.

How did our forecast turn out? A quick call to Dallas on Wednesday morning reveals that the weather there is cold but not as cold as expected, and the sky is overcast. Cloudy weather? How can this be?

The cold front moved through on schedule Tuesday afternoon, bringing showers, gusty winds, and cold weather with it. Moving southward, the front gradually slowed and became stationary along a line stretching from the Gulf of Mexico westward through southern Texas and northern Mexico. (Without an upper-level forecast chart, we had no way of knowing this would happen.) Along the stationary front a wave formed, which caused warm, moist Gulf air to slide northward up and over the cold surface air. Clouds formed, minimum temperatures did not go as low as expected, and we are left with a fractured forecast. Let's try Denver.

Clear but Cold for Denver   In Fig. 9.15, we can see that, based on our projections, the cold anticyclone will be almost directly over Denver by Wednesday morning. Sinking air aloft should keep the sky relatively free of clouds. Weak pressure gradients will produce only weak winds and this, coupled with dry air, will allow for intense radiational cooling. Minimum temperatures will probably drop to well below 0°F. Our forecast should therefore read:

> Clear and cold through tomorrow. Northerly winds today becoming light and variable by tonight. Low temperatures tomorrow morning will be below zero. Barometric pressure will continue to rise.

Almost reluctantly Wednesday morning, we inquire about the weather conditions at Denver. "Clear and very cold" is the reply. A successful forecast at last! We are told, however, that the minimum temperature did not go below zero; in fact, 13°F was as cold as it got. A downslope wind coming off the mountains to the west of Denver kept the air mixed and the minimum temperature higher than expected. Again, a forecaster familiar with the local topography of the Denver area would have foreseen the conditions that lead to such downslope winds and would have taken this into account when making the forecast.

A complete picture of the surface weather systems for 6:00 A.M. Wednesday morning is given in Fig. 9.16.

By comparing this chart with Fig. 9.15, we can summarize why our forecasts did not turn out exactly as we had predicted. For one thing, the storm center over the Central Plains moved slower than expected. This slow movement allowed a southeasterly flow of mild Atlantic air to overrun cooler surface air ahead of the storm while, behind the low, cities remained in the snow area for a longer time. The weak wave that developed along the trailing cold front brought cloudiness and precipitation to Texas and prevented the really cold air from penetrating deep into the south. Further west, the high originally over Montana moved more southerly than southeasterly, which set up a pressure gradient that brought westerly downslope winds to eastern Colorado.

In summary, the forecasting techniques discussed in this section are those you can use when making a short-range weather forecast. Keep in mind, however, that this chapter was not intended to make you an expert weather forecaster, nor was it designed to show you all the methods of weather prediction. It is hoped that you now have a better understanding of some of the problems confronting anyone who attempts to predict the behavior of this churning mass of air we call our atmosphere.

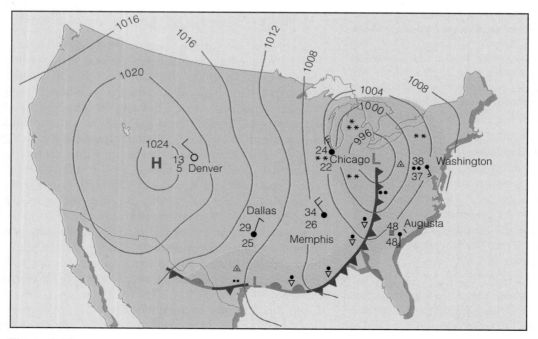

Figure 9.16
Surface weather map for 6:00 A.M. Wednesday.

Summary

Forecasting tomorrow's weather entails a variety of techniques and methods. Persistence and steady-state forecasts are useful when making a short-range (0–6 hour) prediction. For a longer range forecast, the current analysis, satellite data, weather typing, intuition, and experience, along with guidance from the many computer progs supplied by the National Weather Service, all go into making a prediction.

Different computer progs are based upon different atmospheric models that describe the state of the atmosphere and how it will change with time. Currently, flaws in the models—as well as tiny errors (uncertainties) in the data—generally amplify as the computer tries to project weather farther and farther into the future. At present, computer progs are better at forecasting the position of mid-latitude highs and lows than local showers and thunderstorms.

Satellites aid the forecaster by providing a bird's-eye view of clouds and storms. Polar orbiting satellites obtain data covering the earth from pole to pole, whereas geostationary satellites situated above the equator supply the forecaster with dynamic photographs of cloud and storm development and movement. To show where the highest and thickest clouds are located in a particular storm, infrared pictures are often enhanced by computer.

In the latter part of this chapter we learned how a person, by observing the weather around them, and by watching the weather systems on surface weather maps, can make fairly good short-range weather predictions.

Key Terms

The following terms are listed in the order they appear in the text. Define each. Doing so will aid you in reviewing the material covered in this chapter.

weather watch	prognostic chart (prog)	steady-state (trend)	weather type
weather warning	AFOS	forecast	forecasting
analysis	AWIPS	nowcasting	ensemble forecasting
numerical weather	geostationary satellites	analogue forecasting	climatological forecast
prediction	polar orbiting satellites	method	probability forecast
atmospheric models	persistence forecast		

Review Questions

1. What is the function of NMC?
2. How does a *weather watch* differ from a *weather warning*?
3. How does a prog differ from an analysis?
4. In what ways have high-speed computers assisted the meteorologist in making weather forecasts?
5. Briefly explain how the computer goes about making a weather forecast.
6. What are some of the problems associated with computer model forecasts?
7. (a) Explain how satellites aid in forecasting the weather.
   (b) Using infrared satellite information, how can a forecaster distinguish high clouds from low clouds?
   (c) Why are infrared images often enhanced?
8. Describe four methods of forecasting the weather and give an example for each one.

9. Suppose that where you live, the middle of January is typically several degrees warmer than the rest of the month. If you forecast this "January thaw" for the middle of next January, what type of weather forecast will you have made?
10. (a) Look out the window and make a persistence forecast for tomorrow at this time.
    (b) Did you use any skill in making this prediction?
11. Do extended weather forecasts make specific predictions of rain or snow? Explain.
12. Describe the technique of ensemble forecasting.
13. Suppose that the weather forecast for your area today calls for a "70 percent chance of rain." What exactly does that mean?
14. List three methods that you would use to predict the movement of a surface mid-latitude storm system.

A tornado with winds exceeding 100 knots swirls across the plains of North Dakota.
(Photo: Edi Ann Otto)

Chapter 10

Thunderstorms and Tornadoes

Contents

Wednesday, March 18, 1925, was a day that began un-eventfully, but within hours turned into a day that changed the lives of thousands of people and made meteorological history. Shortly after 1:00 P.M. the sky turned a dark green-ish-black and the wind began whipping around the small town of Murphysboro, Illinois. Arthur and Ella Flatt lived on the outskirts of town with their only son, Art, who would be four years old in two weeks. Arthur was working in the garage when he heard the roar of the wind and saw the threatening dark clouds whirling overhead. Instantly con-cerned for the safety of his family, he ran toward the house as the tornado began its deadly pass over the area. With debris from the house flying in his path and the deafening thunder of destruction all around him, Arthur reached the front door. As he struggled in vain to get to his family, whose screams he could hear inside, the porch and its massive support pillars caved in on him. Inside the house, Ella had scooped up young Art in her arms and was mak-ing a panicked dash down the front hallway towards the door when the walls collapsed, knocking her to the floor, with Art cradled beneath her. Within seconds, the rest of the house fell down upon them. Both Arthur and Ella were killed instantly, but Art was spared, nestled safely under his mother's body.

As the dead and survivors were pulled from the devasta-tion that remained, the death toll mounted. Few families es-caped the grief of lost loved ones, as 40 percent of the town of Murphysboro was leveled and 234 people were killed.*

---

*This introduction is written in memory of the author's wife's great aunt and great uncle, Arthur and Ella Flatt, who lost their lives in the tri-state tornado outbreak on March 18, 1925.

10 The devastating tornado described in our opening cut a mile-wide path for a distance of more than 200 miles through the states of Missouri, Illinois, and Indiana. It totally obliterated 4 towns, killed an estimated 695 persons, and left over 2000 injured. Tornadoes such as these, as well as much smaller ones, are associated with severe thunderstorms. Consequently, we will first examine the different types of thunderstorms. Later, we will focus on tornadoes, examining how and where they form, and why they are so destructive.

What Are Thunderstorms?

It probably comes as no surprise that a thunderstorm is merely a storm containing lightning and thunder. Sometimes it produces gusty surface winds with heavy rain and hail. The storm itself may be a single cumulonimbus cloud, a cluster of them, or even a line of clouds that in some cases extends for more than 100 kilometers (62 miles).

The birth of a thunderstorm occurs when warm, humid air rises in an unstable environment. The trigger needed to start air moving upward may be the unequal heating of the surface, the effect of terrain, or the lifting of warm air along a frontal zone. Diverging upper-level winds also provide a favorable region for thunderstorm development, as they tend to draw air upward beneath them. Usually, several of these mechanisms work together to generate severe thunderstorms.

Scattered thunderstorms that form in summer are often referred to as **air-mass thunderstorms** (or *ordinary thunderstorms*) because they tend to develop in warm, humid air masses away from weather fronts. These storms are usually short-lived and rarely produce strong winds and large hail. On the other hand, the *severe thunderstorms* may produce high winds, flash floods, damaging hail, and even tornadoes. Let's examine the air-mass thunderstorms first.

Air-mass Thunderstorms

Extensive studies indicate that thunderstorms go through a cycle of development from birth, to maturity, to decay. The first stage is known as the **cumulus stage**. As humid air rises, it cools and condenses into a single cumulus cloud or a cluster of clouds (Fig. 10.1). If you have ever watched a thunderstorm develop, you may have noticed that at first the cumulus clouds grow upward only a short distance, then they dissipate. This is because the cloud droplets evaporate as the drier air surrounding the cloud mixes with it. However, after the water drops evaporate, the air is more moist than before. So, the rising air is now able to condense at successively higher levels, and the cumulus cloud grows taller, often appearing as a rising dome or tower.

As the cloud builds, the transformation of water vapor into liquid or solid cloud particles releases large quantities of latent heat. This keeps the air inside the cloud warmer than the air surrounding it. The cloud continues to grow in the unstable air as long as it is constantly fed by rising air from below. In this manner, a cumulus cloud may show extensive vertical development in just a few minutes. During the cumulus stage, there is insufficient time for precipitation to form, and the updrafts keep water droplets and ice crystals suspended within the cloud. Also, there is no lightning or thunder during this stage.

As the cloud builds well above the freezing level, the cloud particles grow larger. They also become heavier; eventually, the rising air is no longer able to keep them suspended, and they begin to fall. While this phenomenon is taking place, drier air from around the cloud is being drawn into it in a process called *entrainment*. The entrainment of drier air causes some of

Figure 10.1
Simplified model depicting the life cycle of an air-mass thunderstorm that is nearly stationary. (Arrows show vertical air currents.)

Cumulus   Mature   Dissipating

the raindrops to evaporate, which chills the air. The air, now being colder and heavier than the air around it, begins to descend as a *downdraft*. The downdraft may be enhanced as falling precipitation drags some of the air along with it.

The downdraft and updraft within the cloud constitute a *cell*. In most storms, there are several cells, each of which may last for half an hour or so. The appearance of the downdraft and the storm cell marks the beginning of the **mature thunderstorm**.

During its mature stage, the thunderstorm is most intense. The top of the cloud, having reached the stable stratosphere, begins to take on the familiar anvil shape, as strong upper-level winds spread the cloud's ice crystals horizontally. The cloud itself may extend upward to an altitude of over 12 kilometers (40,000 feet) and be several kilometers in diameter near its base. Updrafts and downdrafts reach their greatest strength in the middle of the cloud, creating severe turbulence. Lightning and thunder are also present. Heavy rain (and occasionally small hail) falls from the cloud. The rainfall may or may not reach the surface, depending on the relative humidity beneath the storm. In the dry air of the desert southwest, for example, a mature air-mass thun-

derstorm may look ominous and contain all of the ingredients of any other storm, except that the raindrops evaporate before reaching the ground.

About fifteen minutes to half an hour after the storm enters the mature stage, it begins to dissipate. The **dissipating stage** occurs when the updrafts weaken and downdrafts tend to dominate throughout much of the cloud. Deprived of its rich supply of warm, humid air, cloud droplets no longer form. Light precipitation now falls from the cloud, accompanied by only weak downdrafts. As the storm dies, the lower-level cloud particles evaporate rapidly, sometimes leaving only the cirrus anvil as the reminder of the once mighty presence (Fig. 10.2).

A single air-mass thunderstorm may go through these three stages in an hour or less. The reason it does not last very long is that the storm's own precipitation

Figure 10.2
A dissipating thunderstorm. Most of the cloud particles in the lower half of the storm have evaporated. Only the cirrus anvil stands out as a distinguishing feature.

enhances the downdrafts that cut off the storm's fuel supply by destroying the humid updrafts. Hence, the storm tends to collapse on itself.

Not only do these storms produce summer rainfall for a large portion of the United States but they also bring with them momentary cooling after an oppressively hot day. The cooling comes during the mature stage, as the downdraft reaches the surface in the form of a blast of welcome relief. Sometimes, the air temperature may lower as much as 10°C (18°F) in just a few minutes. Unfortunately, the cooling effect is short-lived, as the downdraft diminishes or the thunderstorm moves on. In fact, after the storm has ended, the air temperature usually rises; and as the moisture from the rainfall evaporates into the air, the humidity increases, sometimes to a level where it actually feels more oppressive after the storm than it did before.

Upon reaching the surface, the cold downdraft has another effect. It may force warm, moist surface air upward. This rising air then condenses and gradually builds into a new thunderstorm. Thus, it is entirely possible for a series of thunderstorms to grow in a line, one next to the other, each in a different stage of development. (See Fig. 10.3.) Thunderstorms that form in this manner are termed **multicell storms**. Most air-mass thunderstorms are multicell storms, as are most severe thunderstorms.

There are many thunderstorms that form in response to forced lifting, such as a frontal boundary (*frontal thunderstorms*), a mountain range (*orographic thunderstorms*), or along the leading edge of a sea breeze. Most of these thunderstorms are not severe, and their life cycle can be described in much the same way as the air-mass storms.

Severe Thunderstorms

Severe thunderstorms are capable of producing large hail, strong, gusty surface winds, flash floods, and tornadoes.* Just as the air-mass thunderstorm, they form

---

*The National Weather Service defines a severe thunderstorm as having three-quarter inch hail and/or wind gusts of 50 knots (57 miles per hour).

Figure 10.3
A multicell storm. This storm is composed of a series of cells in successive stages of growth. The thunderstorm in the middle is in its mature stage, with a well-defined anvil. Heavy rain is falling from its base. To the right of this cell, a thunderstorm is in its cumulus stage. To the left, a well-developed cumulus congestus cloud is about ready to become a mature thunderstorm.

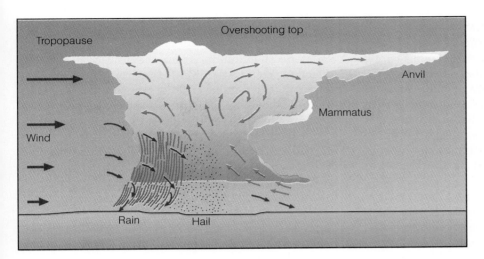

Figure 10.4
A simplified model describing air motions and other features associated with a severe thunderstorm. The severity depends on the intensity of the storm's circulation pattern.

as moist air is forced to rise into unstable air. But, severe thunderstorms also form in areas with a strong vertical wind shear, which causes the updraft to tilt in the mature stage.

Figure 10.4 is a model depicting the air motions within a severe thunderstorm. The storm is moving from left to right. Strong winds aloft cause the system to tilt so that the updrafts move up and over the downdrafts. This phenomenon is very important to the development, continued existence, and propagation of the system. When the precipitation becomes too heavy for the updrafts to support, it falls into the downdrafts rather than into the updrafts, as in the air-mass thunderstorm. Hence, the updrafts remain strong and do not dissipate. Because the updrafts flow unabated, they are capable of obtaining speeds of more than 50 knots.

The updrafts may be so strong that the cloud top is able to intrude well into the stable stratosphere, a condition known as *overshooting*. In some cases, the top of the cloud may extend to more than 18 kilometers (60,000 feet) above the surface. The violent updrafts keep hailstones suspended in the cloud long enough for them to grow to considerable size. Once they are large enough, they either fall out the bottom of the cloud with the downdraft or a strong updraft may toss them out the side of the cloud, or even from the base of the anvil. Aircraft have actually encountered hail in clear air several kilometers from a storm. Also, strong downdrafts within the upper part of the cloud may produce beautiful mammatus clouds.

When we look more closely at the lower half of a severe thunderstorm (Fig. 10.5), we see that the down-

draft is fed by surrounding drier air being entrained into the system. As some of the falling precipitation evaporates, it cools the air and actually produces the downdraft. The cool air that reaches the ground acts like a wedge, forcing warm, moist surface air up into the system. Thus, the downdraft helps to maintain the updraft and vice versa, so that the severe thunderstorm is able to maintain itself—for many hours in some cases.

The Gust Front and Microburst   Look at Fig. 10.5 again and notice that the downdraft spreads laterally after striking the ground. The boundary separating this cold downdraft from the warm surface air is known as

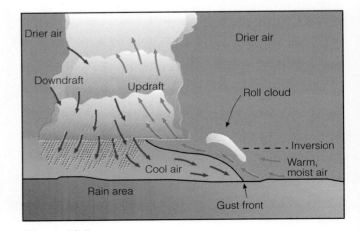

Figure 10.5
The lower half of a severe squall-line-type thunderstorm and some of the features associated with it.

a **gust front**. To an observer on the ground, the passage of the gust front resembles that of a cold front. During its passage, the wind shifts and becomes strong and gusty, with speeds occasionally exceeding 55 knots; temperatures drop sharply and, in the cold heavy air of the downdraft, the surface pressure rises. Sometimes it may jump several millibars, producing a small area of high pressure called a *mesohigh* (meaning mesoscale high). The cold air may linger close to the ground for several hours, well after the thunderstorm activity has ceased.

Along the leading edge of the gust front, the air is quite turbulent. Here, strong winds can pick up loose dust and soil and lift them into a huge tumbling cloud—the haboob that we described in Chapter 7. As warm, moist air rises along the forward edge of the gust front, a **roll cloud** (also called an *arcus cloud*) may form, such as the one shown in Fig. 10.6. These clouds are especially prevalent when an inversion exists near the base of the thunderstorm. Sometimes the gust front forces warm, moist air upward, producing new thunderstorms.

Beneath a thunderstorm, the downdraft may become localized so that it hits the ground and spreads horizontally in a radial burst of wind, much like water pouring from a tap and striking the sink below. Such downdrafts are called **downbursts**. A downburst with winds extending only 4 kilometers or less is termed a **microburst**. In spite of its small size, an intense microburst can induce damaging winds as high as 146 knots. Figure 10.7 shows the dust clouds generated from a microburst north of Denver, Colorado.

When the microburst reaches the ground and continues as an expanding outflow, a gust front may form. Hence, a microburst can evolve into a gust front.

Microbursts are capable of blowing down trees and inflicting heavy damage upon poorly built structures. In fact, microbursts may be responsible for some damage once attributed to tornadoes. Moreover, microbursts and their accompanying *wind shear*—rapid

Figure 10.6
A dramatic example of a roll cloud associated with a severe thunderstorm. The photograph was taken in the Philippines as the thunderstorm approached from the northwest.

Figure 10.7
Dust clouds rising in response to the outburst winds of a microburst north of Denver, Colorado.

changes in wind speed or wind direction—appear to be responsible for several airline crashes. When an aircraft flies through a microburst, it first encounters a headwind that generates extra lift. However, in a matter of seconds, the headwind is replaced by a tailwind that causes a sudden loss of lift and a subsequent decrease in the performance of the aircraft. One accident attributed to a microburst occurred north of Dallas–Fort Worth Regional Airport during April, 1985. Just as the aircraft was making its final approach, it encountered severe wind shear beneath a small, but intense thunderstorm. The aircraft then dropped to the ground and crashed in a spectacular fireball, killing over 100 passengers.

To better predict the hazardous wind shear associated with microbursts, a new and improved ground-based wind-shear detection system called LLWAS (for *Low-Level Wind Shear Alert System*) is being installed at selected airports. The system utilizes surface wind speed and direction sensors to measure the presence of wind shear.

It should be noted that although microbursts can be associated with severe thunderstorms, producing strong, damaging winds, studies show that they can also occur with clouds and thunderstorms that produce only isolated showers—clouds that may or may not contain thunder and lightning.

Supercell and Squall-line Thunderstorms   There are two main types of severe thunderstorms: the **supercell storm** and the **squall line**. The *supercell storm* is an enormous thunderstorm whose updrafts and downdrafts are so sufficiently in balance that it is able to maintain itself as a single entity for hours on end. Figure 10.8 shows a supercell storm in eastern Colorado. Storms of this type are capable of producing updrafts that may exceed 90 miles per hour, hail the size of grapefruit, damaging surface winds, and large, long-lasting tornadoes.

The *squall line* forms as a line of severe thunderstorms. Sometimes they are right along a cold front, but more often they are 100 to 300 kilometers out ahead of

Figure 10.8
A supercell thunderstorm. This awesome storm dumped hail the size of golf balls and more than 5 centimeters (2 in.) of rain over portions of eastern Colorado.

it. These prefrontal squall-line thunderstorms may be caused by air aloft flowing over the cold front and developing into waves, much like the waves that form downwind of mountain chains. Notice in Fig. 10.9 that the downward-moving part of the wave inhibits cloud formation, while the rising part, about 100 kilometers from the cold front, favors uplift. It is here that clouds and thunderstorms form in unstable air.

Another possible cause of the prefrontal squall-line thunderstorm is shown in Fig. 10.10. The map shows conditions that lead to severe thunderstorms over the Central Plains, especially during the spring. A developing mid-latitude cyclone forms as an open wave with a cold front, a warm front, and three distinct air masses. Behind the cold front, cold, dry air pushes in from the north. In the warm air, ahead of the cold front, warm, dry air moves in from the southwest. Further

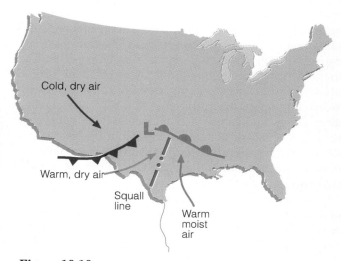

Figure 10.9
Squall-line thunderstorms may form ahead of an advancing cold front as the upper-air flow develops waves downwind from the cold front.

Figure 10.10
Surface weather conditions favorable for the generation of prefrontal severe thunderstorms.

east, warm but very humid air moves northward from the Gulf of Mexico. Along the cold front—where cold, dry air replaces warm, dry air—there is insufficient moisture for thunderstorms to form. However, in the warm air, where the more-dense dry air encounters the less-dense humid air, convergence and lifting occur. It is along this boundary (called a **dry line**) that squall-line thunderstorms form. Once these thunderstorms develop, the outflow of cold air along the ground initiates the uplift necessary for generating new (possibly more severe) thunderstorms.

Severe thunderstorms are often associated with **flash floods**—floods that rise rapidly with little or no advance warning. Such flooding often results when severe thunderstorms stall or move very slowly, causing intense rainfall over a relatively small area. In recent years, flash floods have claimed an average of more than 100 lives a year and have accounted for untold property and crop damage. One such storm raged through eastern Ohio during June, 1990, taking the lives of dozens of people and causing damage in the millions

of dollars. (An example of a terrible flash flood that took the lives of more than 135 people is given in the Focus section on p. 252.)

Mesoscale Convective Complexes   Where conditions are favorable for convection, a number of individual thunderstorms will occasionally grow in size and organize into a large convective weather system. These convectively driven systems, called **Mesoscale Convective Complexes (MCCs)**, are quite large—they can be as much as 1000 times larger than an individual air-mass thunderstorm. In fact, they are often large enough to cover an entire state, an area in excess of 100,000 square kilometers (Fig. 10.11).

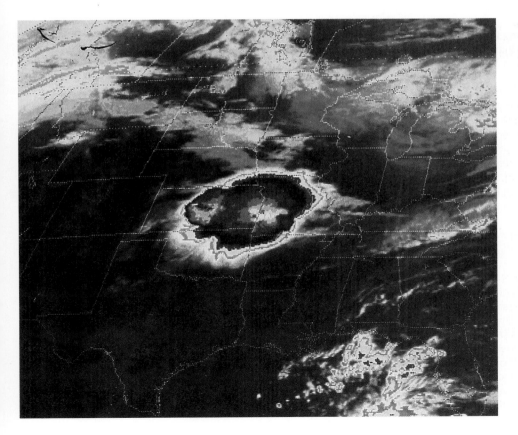

Figure 10.11
An enhanced infrared satellite picture for June 22, 1981, which shows a Mesoscale Convective Complex extending from central Kansas across western Missouri. This organized mass of thunderstorms brought hail, heavy rain, and flooding to this area.

Focus on a Special Topic
Terrifying Flash Flood in the Big Thompson Canyon

July 31, 1976, was like any other summer day in the Colorado Rockies, as small cumulus clouds with flat bases and dome-shaped tops began to develop over the eastern slopes near the Big Thompson and Cache La Poudre rivers. At first glance, there was nothing unusual about these clouds, as almost every summer afternoon they form along the warm mountain slopes. Normally, strong upper-level winds push them over the plains, causing rainshowers of short duration. But the cumulus clouds on this day were different. For one thing, they were much lower than usual, indicating that the southeasterly surface winds were bringing in a great deal of moisture. Also, their tops were somewhat flattened, suggesting that an inversion aloft was stunting their growth. But these harmless-looking clouds gave no clue that later that evening in the Big Thompson Canyon more than 135 people would lose their lives in a terrible flash flood.

By late afternoon, a few of the cumulus clouds were able to puncture the inversion. Fed by moist southeasterly winds, these clouds soon developed into spectacular supercell thunderstorms with tops exceeding 18 kilometers (60,000 feet). By early evening, these same clouds were producing incredible downpours in the mountains.

In the narrow canyon of the Big Thompson River, some places received as much as 30.5 centimeters (12 inches) of rain in the four hours between 6:30 P.M. and 10:30 P.M. local time. This is an incredible amount of precipitation, considering that the area normally receives about 40.5 centimeters (16 inches) for an entire year. The heavy downpours turned small creeks into raging torrents, and the Big Thompson River was quickly filled to capacity. Where the canyon narrowed, the river overflowed its banks and water covered

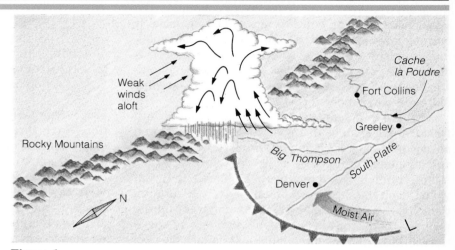

Figure 1
Weather conditions that led to the development of severe thunderstorms that remained nearly stationary over the Big Thompson Canyon in the Colorado Rockies. The arrows within the thunderstorm represent air motions.

the road. The relentless pounding of water caused the road to give way.

Soon cars, tents, mobile homes, resort homes, and campgrounds were being claimed by the river. Where the debris entered a narrow constriction, it became a dam. Water backed up behind it, then broke through, causing a wall of water to rush downstream. Of the approximately 2000 people in the canyon that evening, over 135 lost their lives. Property damage exceeded $35.5 million.

Figure 1 shows the weather conditions during the evening of July 31, 1976. A cool front moved through earlier in the day and is south of Denver. The weak inversion layer associated with the front kept the cumulus clouds from building to great heights earlier in the afternoon. However, the strong southeasterly flow behind the cool front pushed unusually moist air upslope along the mountain range. Heated from below, the conditionally unstable air eventually punctured the inversion and developed into severe supercell thun-

derstorms. These remained nearly stationary for several hours due to the weak southerly winds aloft. The deluge may have deposited 19 centimeters (7.5 inches) of rain on the main fork of the Big Thompson River in about one hour.

The flash floods associated with such severe thunderstorms occur with unpleasant frequency. On June 9, 1972, a destructive flood occurred at Rapid City, South Dakota. The weather conditions were very similar to those that caused the Big Thompson flood. A slowly moving cold front was situated to the south of Rapid City, with moist easterly winds behind it. Winds aloft were weak, so the severe thunderstorms moved slowly. Although there are other similarities between the two storms, we will not examine them here. However, a meteorologist, recognizing these conditions, would be able to alert a community to the probability of a flash flood and, hopefully, save many lives.

Within the MCC, the individual thunderstorms apparently work together to generate a long-lasting weather system that moves slowly (normally less than 20 knots) and often exists for periods exceeding twelve hours. The circulation of the MCC supports the growth of new thunderstorms as well as a region of widespread precipitation. These systems are beneficial, as they provide a significant portion of the growing season rainfall over much of the corn and wheat belts of the United States. However, MCCs can also produce a wide variety of severe weather, including hail, high winds, destructive flash floods, and tornadoes.

Incidentally, there are a variety of Mesoscale Convective Systems (MCS) that have many of the characteristics of an MCC, but they do not last as long or grow as large.

Distribution of Thunderstorms

It is estimated that more than 40,000 thunderstorms occur each day throughout the world. Hence, over 14 million occur annually. The combination of warmth and moisture make equatorial land masses especially conducive to thunderstorm formation. Here, thunderstorms occur on about one out of every three days. Thunderstorms are also prevalent over water along the intertropical convergence zone, where the low-level convergence of air helps to initiate uplift. The heat energy liberated in these storms helps the earth maintain its heat balance by distributing heat poleward (see Chapter 7). Thunderstorms rarely occur in dry climates, such as the polar regions and the desert areas of the subtropical highs.

Figure 10.12 shows the average number of days each year having thunderstorms in various parts of the United States. Notice that they occur most frequently in the southeastern states along the Gulf Coast with a maximum in Florida. A secondary maximum exists over the central Rockies. The region with the fewest thunderstorms is the Pacific coastal and interior valleys.

In many areas, thunderstorms form primarily in spring and summer during the warmest part of the day when the surface air is most unstable. Over the Central Plains, thunderstorms tend to form more frequently at night. These storms may be caused by a low-level southerly jet stream that forms at night, and not only carries humid air northward but also produces areas of converging surface air, which helps to trigger uplift and, hence, thunderstorms.

At this point it is interesting to compare Fig. 10.12 and Fig. 10.13. Notice that, even though the greatest frequency of thunderstorms is near the Gulf Coast, the greatest frequency of hailstorms is over the western

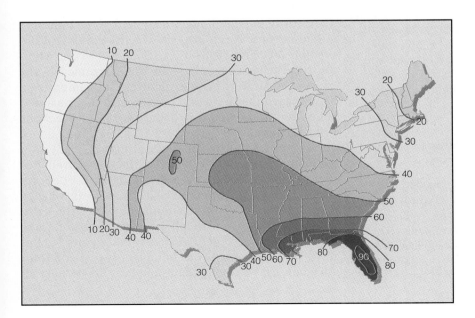

Figure 10.12
The average number of days each year on which thunderstorms are observed throughout the United States. (Due to the scarcity of data, the number of thunderstorms is underestimated in the mountainous west.)

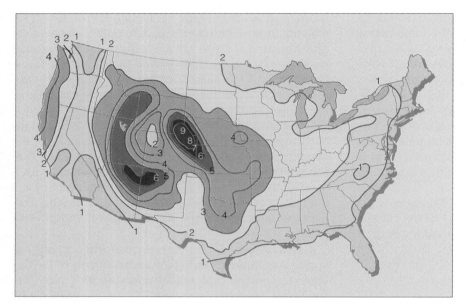

Figure 10.13
The average number of days each year on which hail is observed throughout the United States.

Great Plains. One reason for this is that conditions over the Great Plains are more favorable for the development of severe thunderstorms. We also find that, in summer along the Gulf Coast, a thick layer of warm, moist air extends upward from the surface. Most hailstones falling into this layer will melt before reaching the surface. Over the plains, the warm surface layer is much shallower and drier. Falling hailstones do begin to melt, but the water around their periphery quickly evaporates in the dry air. This cools the hailstones and slows the melting rate so that many survive as ice all the way to the surface.

Now that we have looked at the development and distribution of thunderstorms, we are ready to examine an interesting, though yet not fully understood, aspect of all thunderstorms—lightning.

Did you know?
The costliest hailstorm in the United States pounded parts of Colorado with hail the size of golfballs and baseballs, causing over $600 million in damage on July 11, 1990. Many thousands of vehicles, street lights, and windows were battered by hail that, in some places, piled into drifts several feet deep.

▲▼▲
Lightning and Thunder

Lightning is simply a discharge of electricity, a giant spark, which occurs in mature thunderstorms. Lightning may take place within a cloud, from one cloud to another, from a cloud to the surrounding air, or from a cloud to the ground. (The majority of lightning strikes occur within the cloud, while only about 20 percent or so occur between cloud and ground.) The lightning stroke can heat the air through which it travels to an incredible 30,000°C (54,000°F), which is five times hotter than the surface of the sun. This extreme heating causes the air to expand explosively, thus initiating a shock wave that becomes a booming sound wave—called **thunder**—that travels outward in all directions from the flash.

Light travels so fast that we see light instantly after a lightning flash. But the sound of thunder, traveling at only about 330 meters per second (1100 feet per second), takes much longer to reach the ear. If we start counting seconds from the moment we see the lightning until we hear the thunder, we can determine how far away the stroke is. Because it takes sound about three seconds to travel one kilometer (five seconds for each mile), if we see lightning and hear the thunder fif-

Focus on an Application
Don't Sit Under the Apple Tree

Because a single lightning stroke may involve a current as great as 100,000 amperes, animals and humans can be electrocuted when struck by lightning. The average yearly death toll in the United States attributed to lightning is about 80, with Florida accounting for the most fatalities. Many victims are struck in open places, riding on farm equipment, playing golf, or sailing in a small boat. Some live to tell about it, as did the champion golfer Lee Trevino and the former Shenandoah National Park ranger Roy "Dooms" Sullivan, who was struck seven times and dubbed "the lightning conductor of Virginia" by the *Guinness Book of World Records*. Others are less fortunate. When you see someone struck by lightning, immediately give CPR (cardiopulmonary resuscitation), as lightning normally leaves its victims unconscious and stops their breathing.

The largest single location of lightning fatalities is in the vicinity of relatively isolated trees. Because a positive charge tends to concentrate in upward projecting objects, the upward return stroke that meets the stepped leader is most likely to originate from such objects. Clearly, sitting under a tree during an electrical storm is not wise. What *should* you do?

Figure 2
Lightning tends to strike elevated objects, such as towers, buildings, trees, and even flag sticks on golf courses.

When caught in a thunderstorm, the best protection, of course, is to get inside a building. Automobiles and trucks (but not golf carts) may also provide protection. If no such shelter exists, be sure to avoid elevated places and isolated trees (Fig. 2). If you are on level ground, try to keep your head as low as possible, but do not lie down. Because lightning channels usually emanate outward through the ground at the point of a lightning strike, a surface current may travel through your body and injure or kill you. Therefore, crouch down as low as possible and minimize the contact area you have with the ground. There are some warning signs to alert you to a strike. If your hair begins to stand on end, or your skin begins to tingle and you hear clicking sounds, beware—lightning may be about to strike. And if you are standing upright, you may be acting as a lightning rod.

teen seconds later, the lightning stroke occurred 5 kilometers (3 miles) away.

When the lightning stroke is very close—several hundred feet or less—thunder sounds like a clap or a crack followed immediately by a loud bang. When it is farther away, it often rumbles. The rumbling can be due to the sound emanating from different areas of the stroke. Moreover, the rumbling is accentuated when the sound wave reaches an observer after having bounced off obstructions, such as hills and buildings.

In some instances, lightning is seen but no thunder is heard. Does this mean that thunder was not produced by the lightning? Actually, there is thunder, but the atmosphere refracts (bends) and attenuates the sound waves, making the thunder inaudible. Sound travels faster in warm air than in cold air. Because thunderstorms form in unstable air, where the temperature normally drops rapidly with height, sound waves usually travel faster in the warm air near the surface and bend upward, away from an observer at the surface. Consequently, an observer closer than about 5 kilometers (3 miles) to a lightning stroke will usually hear thunder, while an observer 15 kilometers (about 9 miles) away will not.

What causes lightning? The normal fair weather electric field of the atmosphere is characterized by a negatively charged surface and a positively charged upper atmosphere. For lightning to occur, separate regions containing opposite electrical charges must exist within a cumulonimbus cloud. Exactly how this charge separation comes about is not totally comprehended; however, there are many theories to account for it.

Figure 10.14
The generalized charge distribution in a mature thunderstorm.

Electrification of Clouds One popular theory proposes that clouds become electrified as ice particles such as graupel and hail fall through a region of supercooled (liquid) droplets and ice crystals. As liquid droplets collide with an ice particle, they freeze on contact and release latent heat. This keeps the surface of the ice particle warmer than that of the surrounding ice crystals. When the warmer ice particle comes in contact with a colder ice crystal, an important phenomenon occurs: *There is a net transfer of positive ions from the warmer object to the colder object.* Hence, the ice particle becomes negatively charged and the ice crystal positively charged, as the positive ions are incorporated into the ice. The same effect occurs when colder supercooled (liquid) droplets freeze on contact with a warmer ice particle and tiny splinters of positively charged ice break off. These lighter, positively charged particles are then carried to the upper part of the cloud by updrafts. The larger ice particles, left with a negative charge, fall toward the bottom of the cloud. By this mechanism, the cold upper part of the cloud becomes positively charged, while the middle of the cloud becomes negatively charged. The lower part of the cloud is generally of negative and mixed charge except for an occasional positive region located in the falling precipitation near the melting level. (See Fig. 10.14.)

The Lightning Stroke Because unlike charges attract one another, the negative charge at the bottom of the cloud causes a region of the ground beneath it to become positively charged. As the thunderstorm moves along, this region of positive charge follows the cloud like a shadow. The positive charge is most dense on protruding objects, such as trees, poles, and buildings. The difference in charges causes an electric field between the cloud and ground. In dry air, however, a flow of current does not occur because the air is a good electrical insulator. Gradually, the electrical field builds, and when it becomes sufficiently large (on the order of one million volts per meter), the insulating properties of the air break down, a current flows, and lightning occurs.

Cloud-to-ground lightning begins within the cloud when the localized electrical field exceeds a critical value along a path perhaps 50 meters long. This situation causes a surge of electrons to rush toward the cloud base and then toward the ground in a series of steps. As the electrons flow out of the cloud, they collide with air molecules, which ionizes them and produces a conducting channel that makes it easier for other electrons to travel along.

Each surge of electrons covers about 50 to 100 meters, then stops for about 50-millionths of a second, then occurs again over another 50 meters or so. This **stepped leader** is very faint and is usually invisible to the human eye. As the tip of the stepped leader approaches the ground, a current of positive charge starts upward from the ground (usually along elevated objects) to meet it. After they meet, large numbers of electrons flow to the ground and a much larger, more luminous **return stroke** several centimeters in diameter flows upward to the cloud along the path followed by the stepped leader (Fig. 10.15). Hence, the downward flow of electrons establishes the bright channel of upward propagating current. Even though the bright return stroke travels from the ground up to the cloud, it happens so quickly—in one ten-thousandth of a second—that our eyes cannot resolve the motion, and we see what appears to be a continuous bright flash of light. (See Fig. 10.16.)

Sometimes there is only one lightning stroke, but more often the leader-and-stroke process is repeated in the same ionized channel at intervals of about a tenth of a millionth of a second. The subsequent leader, called a **dart leader**, proceeds from the cloud along the same channel as the original stepped leader; however, it proceeds downward more quickly because the electrical resistance of the path is now lower. As the leader approaches the ground, normally a less energetic return stroke than the first one travels from the ground to the cloud. Typically, a *lightning flash* will have three or four leaders, each followed by a return

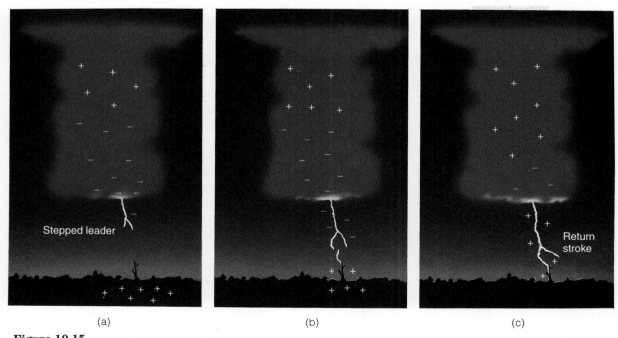

(a)   (b)   (c)

Figure 10.15
The development of a lightning stroke. (a) When the negative charge near the bottom of the cloud becomes large enough to overcome the air's resistance, a flow of electrons—the stepped leader—rushes toward the earth. (b) As the electrons approach the ground, a region of positive charge moves up into the air through any conducting object such as trees, buildings and even humans. (c) When the downward flow of electrons meets the upward surge of positive charge, a strong electric current—a bright return stroke—carries positive charge upwards into the cloud.

Figure 10.16
Time exposure of lightning in Tulsa, Oklahoma, during May, 1987. The bright flashes are return strokes. The lighter forked flashes probably are dart leaders propagating down new channels.

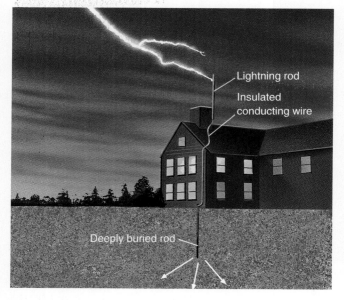

Lightning rod

Insulated conducting wire

Deeply buried rod

Figure 10.17
The lightning rod extends above the building, increasing the likelihood that lightning will strike the rod rather than some other part of the structure. After lightning strikes the metal rod, it follows an insulated conducting wire harmlessly into the ground.

stroke. A lightning flash consisting of many strokes (one photographed flash had twenty-six strokes) usually lasts less than a second. During this short period of time, our eyes may barely be able to perceive the individual strokes, and the flash appears to flicker. (For additional information on the various forms of lightning, read the Focus section on p. 259.)

Lightning rods are placed on buildings to protect them from lightning damage. The rod is made of metal and has a pointed tip, which extends well above the structure. (See Fig. 10.17.) The positive charge concentration will be maximum on the tip of the rod, thus increasing the probability that the lightning will strike the tip and follow the metal rod harmlessly down into the ground, where the other end is deeply buried.

Each year, approximately 10,000 fires are started by lightning in the United States alone and millions of dollars worth of timber is destroyed. For this reason,

Did you know?
A blind man from Falmouth, Maine, had his eyesight restored when struck by lightning during June, 1980.

Focus on an Observation
The Various Forms of Lightning

Lightning may take on a variety of shapes and forms. (See Fig. 3.) When a dart leader moving toward the ground deviates from the original path taken by the stepped leader, the lightning appears crooked or forked, and it is called *forked lightning. Ribbon lightning* forms when the wind moves the ionized channel between each return stroke, causing the lightning to appear as a ribbon hanging from the cloud. If the lightning channel breaks up, or appears to break up, the lightning looks like a series of beads tied to a string. The actual cause of this *bead lightning* is not known. However, one theory proposes that it occurs when the lightning stroke is partially obscured by clouds or falling rain. *Ball lightning* looks like a luminous sphere that appears to float in the air or slowly dart about for several seconds. Although many theories have been proposed, the actual cause of ball lightning remains an enigma. *Sheet lightning* forms when either the lightning flash occurs inside a cloud or intervening clouds obscure the flash, such that a portion of the cloud (or clouds) appears as a luminous white sheet. Distant lightning from thunderstorms that is seen but not heard is commonly called *heat lightning* because it frequently occurs on hot summer nights when the overhead sky is clear. As the light from

Figure 3
A time exposure of lightning during an intense thunderstorm. The cloud-to-ground discharge is forked lightning. Bright, in-cloud discharges are producing sheet lightning. A distant observer, unable to hear thunder, might call the illuminated sky heat lightning.

distant electrical storms is refracted through the atmosphere, air molecules and fine dust scatter the shorter wavelengths of visible light, often causing heat lightning to appear orange to a distant observer (Fig. 3).

As the electric field near the ground increases, a current of positive charge moves up pointed objects, such as antennas and masts of ships. However, instead of a lightning stroke, a luminous greenish or bluish halo may appear above them, as a continuous supply of sparks—a *corona dis-*charge—is sent into the air. This electric discharge, which can cause the top of a ship's mast to glow, is known as *St. Elmo's Fire*, named after the patron saint of sailors. St. Elmo's Fire is also seen around power lines and the wings of aircraft. When St. Elmo's Fire is visible and a thunderstorm is nearby, a lightning flash may occur in the near future, especially if the electric potential gradient of the atmosphere is increasing.

tests have been conducted to see whether the number of cloud-to-ground lightning discharges can be reduced. One technique that has shown some success in suppressing lightning involves seeding a cumulonimbus cloud with hair-thin pieces of aluminum about ten centimeters long. The idea is that these pieces of metal will produce many tiny sparks, called *corona discharges*, and prevent the electrical potential in the cloud from building to a point where lightning occurs. While the results of this experiment are inconclusive, many forestry specialists point out that nature itself may use a similar mechanism to prevent excessive

Did you know?
Tied to a tree, Norma Jean died from a lightning bolt during July, 1974. Lightning apparently struck the tree, traveled down the metal chain, and electrocuted the 6500-pound circus elephant.

lightning damage. The long, pointed needles of pine trees may act as tiny lightning rods, diffusing the concentration of electric charges and preventing massive lightning strokes.

Up to now, we have examined thunderstorms and their associated thunder and lightning. We are now ready to explore a product of a thunderstorm that is one of nature's most awesome phenomena: the tornado, a rapidly spiraling column of air that can strike sporadically and violently.

Tornadoes

Tornadoes are rapidly rotating winds that blow around a small area of intense low pressure. Sometimes called *twisters* or *cyclones*, tornadoes nearly always begin as a funnel-shaped cloud that looks like an elephant's trunk hanging from a large cumulonimbus cloud (Fig. 10.18). The **funnel cloud** is called a tornado only after it touches the ground. When viewed from above, the majority of tornadoes rotate counterclockwise. A few have been seen rotating clockwise, but these are rare.

The diameter of most tornadoes is between 100 and 600 meters (about 300 to 2000 feet), although some are just a few meters wide and others have diameters exceeding 1600 meters (1 mile). Tornadoes that form ahead of an advancing cold front are often steered by southwesterly winds and therefore tend to move from

Figure 10.18
Tornado photographed near Tracy, Minnesota.

the southwest toward the northeast at speeds usually between 20 and 40 knots. However, some have been clocked at speeds greater than 70 knots. Most tornadoes last only a few minutes and have an average path length of about 7 kilometers (4 miles). There are cases where they have traveled for hundreds of kilometers and have existed for many hours, such as the one that lasted over 7 hours and cut a path 470 kilometers (292 miles) long through portions of Illinois and Indiana on May 26, 1917.

Each year, tornadoes take the lives of many people. The yearly average is less than 100, although over 100 may die in a single day. In recent years, an alarming statistic is that 45 percent of all fatalities occurred in mobile homes. The deadliest tornadoes are those that occur in *families*; that is, different tornadoes spawned by the same thunderstorm. (Some thunderstorms produce a sequence of several tornadoes over two or more hours and over distances of 100 kilometers or more.) Tornado families usually form along squall lines and often constitute what is termed a **tornado outbreak**. One of the most violent outbreaks ever recorded occurred on April 3 and 4, 1974. During a 16-hour period, 148 tornadoes cut through parts of 13 states, killing 307 people, injuring more than 6000, and causing an estimated $600 million in damage. Some of these tornadoes were among the most powerful ever witnessed. The combined path of all the tornadoes during this *super outbreak* amounted to 4181 kilometers (2598 miles), well over half of the total path for an average year. The greatest loss of life attributed to tornadoes occurred during the tri-state outbreak of March 18, 1925, when an estimated 695 people died as 7 tornadoes traveled a total of 703 kilometers (437 miles) across portions of Missouri, Illinois, and Indiana.

Tornado Occurrence   Tornadoes occur in many parts of the world, but no country experiences more tornadoes than the United States, which averages more than 800 annually and experienced a record 1293 tornadoes during 1992. Although tornadoes have occurred in every state, including Alaska and Hawaii, the greatest number occur in the "tornado belt" of the Central Plains, which stretches from central Texas to Nebraska.* (See Fig. 10.19.)

---

*Many of the tornadoes that form along the Gulf Coast are generated by thunderstorms embedded within the circulation of hurricanes.

The Central Plains region is most susceptible to tornadoes because it provides the proper atmospheric setting for the development of the severe thunderstorms that spawn tornadoes. Here (especially in spring) warm, humid surface air is overlain by cooler, dryer air aloft, producing an unstable atmosphere. When a strong vertical wind shear exists and the surface air is forced upward, large thunderstorms capable of spawning tornadoes may form. Therefore, tornado frequency is highest during the spring and lowest during the winter when the warm surface air is absent.

About three-fourths of all tornadoes in the United States develop from March to July. The month of May normally has the greatest number of tornadoes (the average is about five per day) while the most violent tornadoes seem to occur in April, when horizontal and vertical temperature and moisture contrasts are greatest. Although tornadoes have occurred at all times of the day and night, they are most frequent in the late afternoon (between 4:00 P.M. and 6:00 P.M.), when the surface air is most unstable; they are least frequent in the early morning before sunrise, when the air is most stable.

Tornado Winds   The strong winds of a tornado can destroy buildings, uproot trees, and hurl all sorts of lethal missiles into the air. People, animals, and home appliances all have been picked up, carried several kilometers, then deposited. Tornadoes have accomplished some astonishing feats, such as lifting a railroad coach with its 117 passengers and dumping it in a ditch about 90 feet away. Showers of toads and frogs have poured out of a cloud after tornadic winds sucked them up from a nearby pond. Other oddities include chickens losing all of their feathers and pieces of straw being driven into metal pipes. Miraculous events have occurred, too. In one instance, a schoolhouse was demolished and the 85 students inside were carried over 300 feet without one of them being killed.

Our early knowledge of the furious winds of a tornado comes mainly from observations of the damage

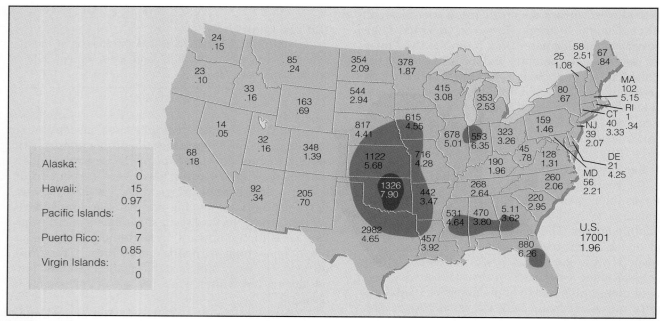

Figure 10.19
Tornado incidence by state. The upper figure shows the number of tornadoes reported by each state during a 25-year period. The lower figure is the average annual number of tornadoes per 10,000 square miles. The darker the shading, the greater the frequency of tornadoes.

done and the analysis of motion pictures. Because of the destructive nature of the tornado, it was once thought that it packed winds greater than 500 knots. However, more recent studies reveal that even the most powerful twisters seldom have winds exceeding 220 knots, and most tornadoes probably have winds of less than 125 knots. Nevertheless, being confronted with even a small tornado can be terrifying.

Few people have ever witnessed the inside of a tornado and lived to tell about it. However, Will Keller, a Kansas farmer, has. (His account, as told to Alonzo

Did you know?
In 1991, a Kansas tornado sucked a mother and her seven-year-old son out of their house in a bathtub. The tub with its occupants hit the ground hard, rose into the air, then hit the ground again, tossing its passengers into the neighbor's backyard. Battered and bruised with scratches and lumps, the mother and son survived. The bathtub, which disappeared in the tornado, has not been found.

A. Justice of the Weather Service office in Dodge City, is given in the Focus section on p. 263.)

The high winds of the tornado cause the most damage as walls of buildings buckle and collapse when blasted by the extreme wind force. Also, as high winds blow over a roof, lower air pressure forms above the roof. The greater air pressure inside the building then lifts the roof just high enough for the strong winds to carry it away. A similar effect occurs when the tornado's intense low-pressure center passes overhead. Because the pressure in the center of a tornado may be more than 100 mb (3 in.) lower than that of its surroundings, there is a momentary drop in outside pressure when the tornado is above the structure. It was once thought that opening windows and allowing inside and outside pressures to equalize would minimize the chances of the building exploding. However, it now appears that opening windows during a tornado actually increases the pressure on the opposite wall and *increases* the chances that the building will collapse. Damage from tornadoes may also be inflicted on people and structures by flying debris. Hence, the wisest course to take when confronted with an approaching tornado is to *seek shelter immediately*.

Focus on an Observation
Viewing the Inside of a Tornado

On the afternoon of June 22, 1928, I was out in my field with my family looking over the ruins of our wheat crop which had just been completely destroyed by a hailstorm. I noticed an umbrella-shaped cloud in the west and southwest and from its appearance suspected that there was a tornado in it. The air had that peculiar oppressiveness which nearly always precedes a tornado.

I saw at once that my suspicions were correct, for hanging from the greenish-black base of the cloud were three tornadoes. One was perilously near and apparently heading for my place. I lost no time hurrying my family to our cyclone cellar.

The family had entered the cellar and I was in the doorway just about to enter and close the door when I decided to take a last look at the approaching twister. I have seen a number of these, so I did not lose my head, although the approaching tornado was an impressive sight.

The surrounding country is level and there was nothing to obstruct the view. There was little or no rain falling from the cloud. Two of the tornadoes were some distance away and looked like great ropes dangling from the parent cloud, but the one nearest was shaped more like a funnel, with ragged clouds surrounding it. It appeared much larger and more energetic than the others and occupied the central position of the cloud, with a massive cumulus dome being directly over it.

Steadily the tornado came on, the end gradually rising above the ground. I probably stood there only a few seconds, but was so impressed with the sight that it seemed like a long time. At last the great shaggy end of the funnel hung directly overhead. Everything was as still as death. There was a strong, gassy odor and it seemed as though I could not breathe. There was a screaming, hissing sound coming directly from the end of the funnel. I looked up and, to my astonishment I saw right into the heart of the tornado. There was a circular spinning in the center of the tornado, about 50 to 100 feet in diameter, which extended straight up for a distance of at least one-half mile, as best I could judge under the circumstances. The walls of this opening were rotating clouds and the whole was brilliantly lighted with constant flashes of lightning which zig-zagged from side to side. Had it not been for the lightning I could not have seen the opening, or any distance into it.

Around the lower rim of the great vortex, small tornadoes were constantly forming and breaking away. These looked like tails as they writhed their way around the end of the funnel. It was these that made the hissing sound. I noticed that the direction of rotation of the great whirl was counterclockwise, but some of the smaller tornadoes rotated clockwise. The opening was entirely hollow, except for something I could not exactly make out—perhaps a detached wind cloud—that kept moving up and down. The tornado was not traveling at a great speed so I had plenty of time to get a good view of the whole thing, inside and out.

Will Keller
Kansas farmer

At home, take shelter in a basement. In a large building without a basement, the safest place is usually in a small room, such as a bathroom, closet, or interior hallway, preferably on the lowest floor and near the middle of the edifice. At school, move to the hallway and lie flat with your head covered. In a mobile home, leave immediately and seek substantial shelter. If none exists, lie flat on the ground in a depression or ravine. Don't try to outrun an oncoming tornado in a car or truck, as tornadoes often cover erratic paths with speeds sometimes exceeding 70 knots. If caught outdoors in an open field, look for a ditch, stream bed, or ravine, and lie flat with your head covered. Also, when a tornado is approaching from the southwest, its strongest winds are on its southeast side. (See Fig. 10.20.)

It now appears that the most violent tornadoes (with winds exceeding 180 knots) contain smaller whirls that rotate within them. Such tornadoes are called *multi-vortex tornadoes* and the smaller whirls are called **suction vortices** (Fig. 10.21). Suction vortices are only about 10 meters (30 feet) in diameter,

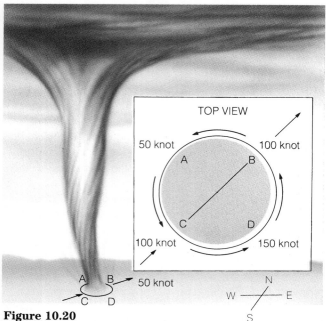

Figure 10.20
The total wind speed of a tornado is greater on one side than on the other. When facing an on-rushing tornado, the strongest winds will be on your left side. If the tornado is heading northeast at 50 knots and its rotational speed is 100 knots, then its forward speed will add 50 knots to its southeastern side (position D) and subtract 50 knots from its northwestern side (position A).

but they rotate very fast and apparently do a great deal of damage.

In the late 1960s, Dr. T. Theodore Fujita, a noted authority on tornadoes at the University of Chicago, proposed a scale (called the **Fujita scale**) for classifying tornadoes according to their rotational wind speed and the damage done by the storm. Table 10.1 presents this scale.

Statistics compiled by the staff of the National Severe Storms Forecast Center in Kansas City, Missouri, reveal that the majority of tornadoes are F0 and F1 (weak tornadoes) and only a few percent are above the F3 classification (violent), with about one F5 tornado reported annually. However, it is the violent tornadoes that account for the majority of tornado-related deaths. As an example, a powerful F5 tornado marched through the southern part of Andover, Kansas, on the evening of April 26, 1991. The tornado, which stayed on the ground for nearly 70 miles, destroyed more than 100 homes and businesses, injured several hundred people, and out of the 39 tornado fatalities in 1991, this

F5 tornado alone took the lives of 17. A powerful F5 tornado is shown in Fig. 10.22.

Tornado Formation Although everything is not known about the formation of a tornado, we do know that tornadoes tend to form with severe thunderstorms that rotate and that unstable air is essential for their development. One atmospheric situation that frequently leads to severe thunderstorms with tornadoes in the spring is shown in Fig. 10.23.

At the surface, we find an open wave middle-latitude cyclone with cold, dry air moving in behind a cold front, and warm, humid air pushing northward from the Gulf of Mexico behind a warm front. Above the warm, surface air a wedge of warm, moist air is streaming northward. Directly above the moist layer is a wedge of colder, drier air moving in from the southwest. Higher up, a trough of low pressure exists to the west of the

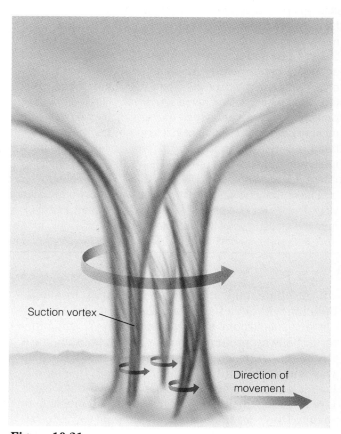

Figure 10.21
A powerful multi-vortex tornado with three suction vortices.

Figure 10.22
A devastating F5 tornado about 200 meters wide plows through Hesston, Kansas, on March 13, 1990, leaving almost 300 people homeless and 13 injured.

surface low and the polar jet stream swings over the region. At this level, the jet stream takes air away so quickly that air from below is drawn up to replace it. The stage is now set for the development of severe storms.

The boxed-off area on the surface map (Fig. 10.23) shows where tornadoes are most likely to form. They tend to form in this region because the position of cold air above warm air produces an unstable atmosphere. However, there is more to it than this. Above the warm, humid surface air, there often exists a shallow temperature inversion that acts like a lid on the moist air below. During the morning, the inversion caps the moist

air, and only small cumulus clouds form. As the day progresses, and the surface becomes warmer, rising blobs of air are able to break through the inversion at isolated places, and clouds build rapidly, sometimes explosively, as the humid air is vented upward through the opening. (The inversion is important because it prevents many small thunderstorms from forming.) The polar jet stream then draws this humid air upward into the cold unstable air aloft, and a huge thunderstorm quickly develops.

While it is difficult to tell which thunderstorm will spawn a tornado, it is easier to predict where tornado-generating storms are most likely to form. Notice in

Table 10.1	Fujita Scale for Damaging Wind			
Scale	**Category**	**Mi/Hr**	**Knots**	**Expected Damage**
F0	Weak	40–72	35–62	light: tree branches broken, sign boards damaged
F1		73–112	63–97	moderate: trees snapped, windows broken
F2	Strong	113–157	98–136	considerable: large trees uprooted, weak structures destroyed
F3		158–206	137–179	severe: trees leveled, cars overturned, walls removed from buildings
F4	Violent	207–260	180–226	devastating: frame houses destroyed
F5		261–318	227–276	incredible: structures the size of autos moved over 100 meters, steel-reinforced structures highly damaged

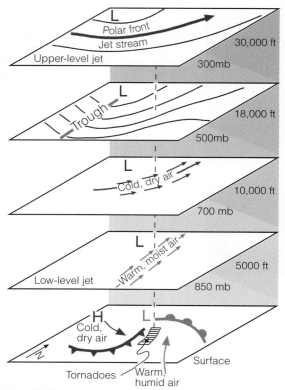

Figure 10.23
Conditions leading to the formation of severe thunderstorms that can spawn tornadoes.

Fig. 10.23 that this area (the boxed-off area on the surface map) is situated where the polar front jet stream and the cold tongue of air cross the warm, humid surface air. Knowing this helps to explain why the region of greatest tornado activity shifts northward from winter to summer. During the winter, tornadoes are most likely to form over the southern Gulf states when the polar front jet is above this region, and the contrast between warm and cold air masses is greatest. In spring, humid Gulf air surges northward; contrasting air masses and the jet stream also move northward and tornadoes become more prevalent from the southern Atlantic states westward into the southern Great

Did you know?
During the last fifteen years, the Los Angeles area has experienced more tornadoes than the Kansas City area.

Plains. In summer, the contrast between air masses lessens, and the jet stream is normally near the Canadian border; hence, tornado activity tends to be concentrated from the northern plains eastward to New York State.

In order for a thunderstorm to spawn a tornado, the updraft (and the thunderstorm itself) must rotate. Remember from our earlier discussion that severe thunderstorms form in a region of strong vertical wind shear. In Fig. 10.23, the rapidly increasing wind speed with height (vertical wind speed shear) and the changing wind direction with height—from southerly at low levels to westerly at high levels (vertical wind direction shear)—cause the updraft inside the storm to rotate cyclonically. This rising, spinning column of air, perhaps 5 to 10 kilometers across, is called a **mesocyclone**.

As the mesocyclone stretches vertically (and shrinks horizontally) the spinning air increases in speed and rises rapidly. Inside the mesocyclone a spinning vortex of increasing wind speed (a tornado) may—for reasons not fully understood—appear near the mid-level of the cloud and gradually extend downward to the cloud base (Fig. 10.24).

As air rushes into the low-pressure vortex from all directions, the air expands, cools, and, if sufficiently moist, condenses into a visible cloud—the *funnel cloud*. As the air beneath the funnel is drawn into the core, it cools rapidly and condenses, and the funnel cloud descends toward the surface. Upon reaching the ground it is called a *tornado*. Here, it usually picks up dirt and debris, making it appear both dark and ominous.

Observing Tornadoes   Observations reveal that the most strong and violent tornadoes develop near the right rear sector of a severe thunderstorm (on the southwestern side of an eastward moving storm, Fig. 10.24). However, weaker tornadoes may not only develop in the main updraft, but along the gust front, where the cool downdraft forces warm inflowing air upward. Although it appears that most strong and violent tornadoes form within the mesocyclone, not all mesocyclones produce tornadoes.

The first sign that the thunderstorm is about to give birth to a tornado is the sight of *rotating* clouds at the base of the storm. If the area of rotating clouds lowers, it becomes a **wall cloud** (compare Fig. 10.24 with Fig. 10.25). Usually within the wall cloud, a smaller, rapidly rotating funnel extends toward the sur-

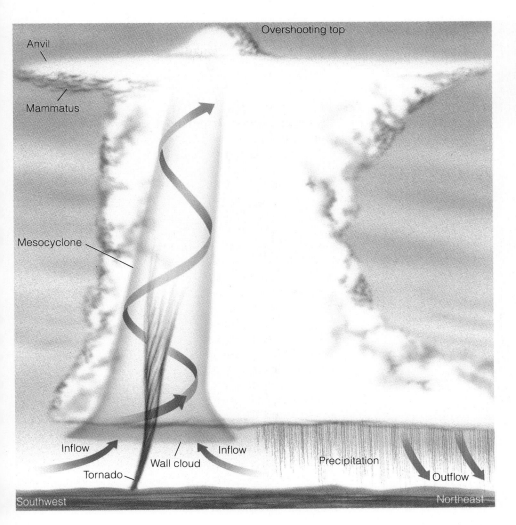

Anvil

Mammatus

Mesocyclone

Overshooting top

Inflow

Tornado

Wall cloud

Inflow

Precipitation

Outflow

Southwest

Northeast

Figure 10.24
Some of the features associated with a tornado-breeding thunderstorm as viewed from the southeast. The thunderstorm is moving to the northeast. The tornado forms in the southwest part of the thunderstorm.

face. Sometimes the air is so dry that the swirling wind remains invisible until it reaches the ground and begins to pick up dust. Unfortunately, people have mistaken these "invisible tornadoes" for dust devils, only to find out (often too late) that they were not. Occasionally, the funnel cannot be seen due to falling rain, clouds of dust, or darkness. Even when not clearly visible, many tornadoes have a distinctive roar that can be heard for several miles. This sound, which has been described as "a roar like a thousand freight trains," appears to be loudest when the tornado is touching the surface. However, not all tornadoes make this sound and, when these storms strike, they become silent killers.

When tornadoes are likely to form during the next few hours, a **tornado watch** is issued by the National Severe Storms Forecast Center in Kansas City, Missouri, to alert the public that tornadoes may develop within a specific area during a certain time period. Many communities have trained volunteer spotters, who look for tornadoes after the watch is issued. Once a tornado is spotted—either visually or on a radar screen—a **tornado warning** is issued by the local National Weather Service Office. In some communities, sirens are sounded to alert people of the approaching storm. Radio and television stations interrupt regular programming to broadcast the warning. Although not completely effective, this warning system is apparently saving many lives. Despite the large increase in population in the tornado belt during the past thirty years, tornado-related deaths have actually shown a decrease.

Figure 10.25
A wall cloud photographed just southwest of Canyon, Texas.

In an attempt to unravel some of the mysteries of the tornado, several studies are under way. In one—conducted jointly by the University of Oklahoma and the National Severe Storms Laboratory in Norman, Oklahoma—still and motion pictures of tornadoes are correlated with radar data in an attempt to better estimate wind speeds. On the more theoretical side, numerical cloud modeling studies are offering new insights into the formation and development of tornado-breeding thunderstorms.

Doppler Radar

Most of our knowledge about what goes on inside a tornado-generating thunderstorm has been gathered through the use of Doppler radar. Before we investigate this remote-sensing device, we will examine how pre-cipitation inside a severe thunderstorm appears on the scope of conventional radar.

Remember from Chapter 5 that a conventional radar transmitter sends out microwave pulses and that, when this energy strikes an object, a small fraction is scattered back to the antenna. Because precipitation particles are large enough to bounce microwaves back to the antenna, the white area on the radar scope in Fig. 10.26 represents precipitation inside a severe thunderstorm. Notice that the pattern is in the shape of a hook. A **hook-shape echo** such as this indicates the possible presence of a tornado. The dark area within the hook echo represents the region inside the severe thunderstorm where strong updrafts carry cloud particles upward so rapidly that they are unable to grow large enough to reflect microwaves. When tornadoes form, they do so near the tip of the hook. However, many severe thunderstorms (as well as smaller ones) do not show a hook echo, but still spawn tornadoes. Sometimes, when the hook echo does appear, the tornado is already touching the ground. Consequently, a better technique was needed in detecting tornado-producing storms. To address this need, Doppler radar was developed.

Doppler radar is like a conventional radar in that it can detect areas of precipitation and measure the speed of falling precipitation. But a Doppler radar can do more—it can actually measure the speed at which precipitation is moving horizontally toward or away from the radar antenna. Because precipitation particles are carried by the wind, Doppler radar can peer into a severe storm and unveil its winds.

Doppler radar works on the principle that, as precipitation moves toward or away from the antenna, the returning radar pulse will change in frequency. A similar change occurs when the high-pitched sound (high frequency) of an approaching noise source, such as a siren or train whistle, becomes lower in pitch (lower frequency) after it passes by the person hearing it. This change in frequency is called the *Doppler shift* and this, of course, is where the Doppler radar gets its name.

A single Doppler radar cannot detect winds that blow parallel to the antenna. Consequently, two or more units probing the same thunderstorm are able to give a three-dimensional picture of the winds within the storm. To help distinguish the storm's air motions, wind velocities can be displayed in color. Color contouring the wind field gives a good picture of the storm. (See Fig. 10.27.)

Unfortunately, the resolution of the Doppler radar is not high enough to measure actual wind speeds of most tornadoes, whose diameters are only a few hundred meters or less. However, a new and experimental Doppler system—called *Doppler lidar*—uses a light beam (instead of microwaves) to measure the change in frequency of falling precipitation, cloud particles, and dust. Because it uses a shorter wavelength of radiation, it has a narrower beam and a higher resolution than does Doppler radar.

Today, in an attempt to obtain tornado wind information at fairly close range (less than 10 km), smaller portable Doppler radar units are peering into tornado-generating storms. (See Fig. 10.28.) During the 1990s, the National Weather Service plans to install a network of Doppler radar units at selected weather stations. This planned radar network, called **NEXRAD** (an acronym for *NEX*t Generation Weather *RAD*ar) will replace the aging conventional radar units. (By the way,

the new Doppler radar's name is WSR 88D.) Detecting the images of mesocyclones and tornadoes with Doppler radar will assist forecasters in determining which severe thunderstorms will likely spawn tornadoes. In addition, they should give advanced and improved warning of an approaching tornado. More reliable warnings, of course, should cut down on the number of false alarms.

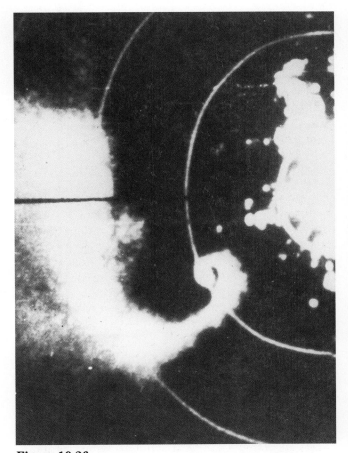

Figure 10.26
A hook-shaped echo often appears on the conventional radar screen with tornado-spawning thunderstorms.

Figure 10.27
Doppler radar display. The color contouring shows wind velocity in a severe thunderstorm with a black dot (in the middle left of the picture) depicting the point where a tornado is forming.

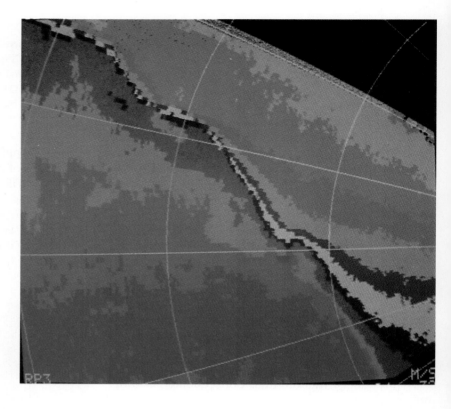

Figure 10.28
Graduate students of the University of Oklahoma use a portable Doppler radar to probe a tornado near Hodges, Oklahoma.

Because the Doppler radar shows air motions within a storm, it can help to identify the magnitude of other severe weather phenomena, such as gust fronts, microbursts, and wind shears that are dangerous to aircraft. Certainly, as Doppler radar becomes part of the major radar network, our understanding of the processes that generate severe thunderstorms will be enhanced, and hopefully there will be an even better tornado and severe storm-warning system, resulting in fewer deaths and injuries.

Waterspouts

A **waterspout** is a rotating column of air over a large body of water. The waterspout may be a tornado that formed over land and then traveled over water. In such a case, the waterspout is called a *tornadic waterspout*. Waterspouts that form over water, especially above the warm, shallow coastal waters of the Florida Keys, where almost 100 occur each month during the summer, are referred to as *"fair weather" waterspouts*. These waterspouts are generally much smaller than an average tornado, as they have diameters usually between 3 and 100 meters (10 and 300 feet). Fair weather waterspouts are also less intense, as their rotating winds are typically less than 45 knots. In addition, they tend to move more slowly than tornadoes and they only last for about ten to fifteen minutes, although some have existed for up to one hour.

Fair weather waterspouts tend to form when the air is unstable and clouds are developing. Unlike the tornado, they do not need a severe thunderstorm to generate them. Some form with small thunderstorms, but most form with developing cumulus congestus clouds whose tops are frequently no higher than 3600 meters (12,000 feet) and do not extend to the freezing level. Apparently, the warm, humid air near the water helps to create atmospheric instability, and the updraft beneath the resulting cloud helps initiate uplift of the surface air. Recent studies even suggest that gust fronts and converging sea breezes may play a role in the formation of some of the waterspouts that form over the Florida Keys.

The waterspout funnel is similar to the tornado funnel in that both are clouds of condensed water vapor with converging winds that rise about a central core. Contrary to popular belief, the waterspout does not draw water up into its core; however, swirling spray may be lifted several meters when the waterspout funnel touches the water. Apparently, the most destructive waterspouts are those that begin as tornadoes over land, then move over water. A photograph of a particularly well-developed and intense fair weather waterspout near the Florida Keys is shown in Fig. 10.29.

Figure 10.29
A waterspout over the warm waters of the Florida Keys.

Summary

In this chapter, we examined thunderstorms and the atmospheric conditions that produce them. The ingredients for an isolated air-mass thunderstorm are humid surface air, plenty of sunlight to warm the ground, and an unstable atmosphere. When these conditions prevail, small cumulus clouds may grow into towering clouds and thunderstorms within 20 minutes.

When conditions are ripe for thunderstorm development and a strong vertical wind shear exists, the stage is set for the generation of severe thunderstorms. Supercell thunderstorms may exist for many hours, as their updrafts and downdrafts are nearly in balance. Thunderstorms that form in a line, especially ahead of an advancing cold front, are called a squall line.

Lightning is a discharge of electricity that occurs in mature thunderstorms. The lightning stroke momentarily heats the air to an incredibly high temperature. The rapidly expanding air produces a sound called thunder. Along with lightning and thunder, severe thunderstorms produce violent weather, such as destructive hail, strong downdrafts, and the most feared of all atmospheric storms—the tornado.

Tornadoes are rapidly rotating columns of air that extend downward from the base of a thunderstorm. Most tornadoes are less than a few hundred meters wide with wind speeds less than 100 knots, although violent tornadoes may have wind speeds that exceed 250 knots. A violent tornado may actually have smaller whirls (suction vortices) rotating within it. With the aid of Doppler radar, scientists are probing tornado-spawning thunderstorms, hoping to better predict tornadoes and to better understand where, when, and how they form.

A normally small and less destructive cousin of the tornado is the "fair weather" waterspout that commonly forms above the warm waters of the Florida Keys.

Key Terms

The following terms are listed in the order they appear in the text. Define each. Doing so will aid you in reviewing the material covered in this chapter.

air-mass thunderstorm
cumulus stage
mature thunderstorm
dissipating stage
multicell storms
severe thunderstorms
gust front
roll cloud
downburst

microburst
supercell storm
squall line
dry line
flash floods
Mesoscale Convective Complexes (MCCs)
lightning
thunder

stepped leader
return stroke
dart leader
tornadoes
funnel cloud
tornado outbreak
suction vortices
Fujita scale

mesocyclone
wall cloud
tornado watch
tornado warning
hook-shape echo
Doppler radar
NEXRAD
waterspout

Review Questions

1. What is a thunderstorm?
2. How would you be able to differentiate an air-mass thunderstorm (ordinary thunderstorm) from a severe thunderstorm?
3. Describe the stages of development of an air-mass (ordinary) thunderstorm.
4. How do downdrafts form in thunderstorms?
5. Why do air-mass thunderstorms most frequently form in the afternoon?
6. Explain why air-mass thunderstorms tend to dissipate much sooner than severe thunderstorms.
7. (a) How do gust fronts form?
(b) What type of weather does a gust front bring when it passes?

8. (a) Describe how a microburst forms.
   (b) Why is the term *wind shear* often used in conjunction with a microburst?
9. Why are severe thunderstorms not very common in polar regions?
10. Give a possible explanation for the generation of prefrontal squall-line thunderstorms.
11. What do severe thunderstorms tend to do when they produce devastating flash floods?
12. What is a Mesoscale Convective Complex (MCC)?
13. Where does the highest frequency of thunderstorms occur in the United States? Why there?
14. Why is large hail more common in Kansas than in Florida?
15. Explain how a cloud-to-ground lightning stroke develops.
16. How is thunder produced?
17. If you see lightning and ten seconds later you hear thunder, how far away is the lightning stroke?
18. Why is it dangerous to seek shelter under a tree during a thunderstorm?
19. (a) What are tornadoes?
    (b) Give some average statistics about tornado size, winds, and general direction of movement.

20. How does a tornado watch differ from a tornado warning?
21. Why is it suggested that one *not* open windows when a tornado is approaching?
22. Explain why the central part of the United States is more susceptible to tornadoes than any other region of the world.
23. Describe the atmospheric conditions at the surface and aloft that are responsible for the development of the majority of tornado-spawning thunderstorms.
24. What must the updraft (and the thunderstorm itself) do in order to produce a tornado?
25. (a) Describe how Doppler radar measures the winds inside a severe thunderstorm.
    (b) Why is Doppler radar superior to conventional radar for this purpose?
26. How does a "fair weather" waterspout differ from a tornadic waterspout?

Hurricane Elena over the Gulf of Mexico, about 130 kilometers southwest of Apalachicola, Florida, as photographed from the space shuttle *Discovery* during September, 1985. With surface winds of 105 knots and a central pressure near 955 millibars, Elena rates as a category 3 on the Saffir-Simpson hurricane scale. (Photo: NASA)

Chapter 11

Hurricanes

Contents

▲▼▲

On September 18, 1926, as a hurricane approached Miami, Florida, everyone braced themselves for the devastating high winds and storm surge. Just before dawn the hurricane struck with full force—torrential rains, flooding, and easterly winds that gusted to over 100 miles per hour. Then, all of a sudden, it grew calm and a beautiful sunrise appeared. People wandered outside to inspect their property for damage. Some headed for work, and scores of adventurous young people crossed the long causeway to Miami Beach for the thrill of swimming in the huge surf. But the lull lasted for less than an hour. And from the south, ominous black clouds quickly moved overhead. In what seemed like an instant, hurricane force winds from the west were pounding the area and pushing water from Biscayne Bay over the causeway. Many astonished bathers, unable to swim against the great surge of water, were swept to their deaths. Hundreds more drowned as Miami Beach virtually disappeared under the rising wind-driven tide.

▲

11 Born over warm tropical waters and nurtured by a rich supply of water vapor, the *hurricane* can indeed grow into a ferocious storm that generates enormous waves, heavy rains, and winds that may exceed 150 knots. What exactly are hurricanes? How do they form? And why do they strike the east coast of the United States more frequently than the west coast? These are some of the questions we will consider in this chapter.

▲▼▲
Tropical Weather

In the broad belt around the earth known as the tropics—the region 23½° north and south of the equator—the weather is much different from that of the middle latitudes. In the tropics, the noon sun is always high in the sky, and so diurnal and seasonal changes in temperature are small. The daily heating of the surface and high humidity favor the development of cumulus clouds and afternoon thunderstorms. Most of these are individual thunderstorms that are not severe. However, if a jet stream should exist above the developing storm, a strong vertical wind shear can cause the updraft to tilt, producing a more violent *squall-line-type thunderstorm* similar to those that form in middle latitudes.

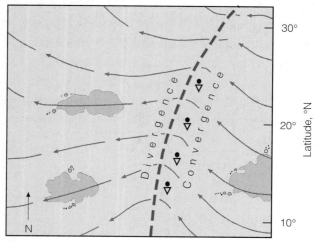

Figure 11.1
An easterly wave as shown by the bending of streamlines. (The heavy dashed line is the axis of the trough.) The wave moves slowly westward, bringing fair weather on its western side and showers on its eastern side.

As it is warm all year long in the tropics, the weather is not characterized by four seasons which, for the most part, are determined by temperature variations. Rather, most of the tropics are marked by seasonal differences in precipitation. The greatest cloudiness and precipitation occur during the high-sun period, when the intertropical convergence zone moves into the region. Even during the dry season, precipitation can be irregular, as periods of heavy rain, lasting for several days, may follow an extremely dry spell.

The winds in the tropics generally blow from the east, northeast, or southeast. Because the variation of sea level pressure is normally quite small, drawing isobars on a weather map provides little useful information. Instead of isobars, **streamlines** that depict wind flow are drawn. Streamlines are useful because they show where surface air converges and diverges. Occasionally, the streamlines will be disturbed by a weak trough of low pressure called an **easterly wave** (Fig. 11.1).

Easterly waves travel from east to west at speeds between 10 and 20 knots. Look at Fig. 11.1 and observe that, on the western side of the trough (heavy dashed line), where easterly and northeasterly surface winds diverge, sinking air produces generally fair weather. On its eastern side, where the southeasterly winds converge, rising air generates showers and thunderstorms. Consequently, the main area of showers form *behind* the trough. Occasionally, an easterly wave will intensify and grow into a hurricane.

▲▼▲
Anatomy of a Hurricane

A **hurricane** is an intense storm of tropical origin, with sustained winds exceeding 64 knots (74 miles per hour), which forms over the warm northern Atlantic and eastern North Pacific oceans. This same type of storm is given different names in different regions of the world. In the western North Pacific, it is called a **typhoon**, in the Philippines a *baguio* (or a typhoon), and in India and Australia a *cyclone*. For simplicity, we will refer to all of these storms as hurricanes.

Figure 11.2 is a satellite picture of hurricane John situated in the eastern Pacific on August 25, 1978. The storm is approximately 550 kilometers (340 miles) in diameter, which is about average for hurricanes. The area of broken clouds at the center is its **eye**. John's eye is

Figure 11.2
Hurricane John. (Visible satellite picture.)

almost 60 kilometers (37 miles) wide, somewhat larger than the average 20- to 50-kilometer eye diameter. Within the eye, winds are light and clouds are mainly broken. The surface air pressure is very low, nearly 965 millibars (28.50 inches).* The dark blotches in the eye are regions where the sky is clear. Notice that the clouds align themselves in bands (called *rain bands*) that spiral in toward the storm's center, where they wrap themselves around the eye. Surface winds in-

crease in speed as they blow counterclockwise and inward toward this center. Adjacent to the eye is the **eye wall**, a ring of intense thunderstorms that whirl around the storm's center and extend upward to almost 15 kilometers above sea level. Within the eye wall, we find the

---

*An extreme low pressure of 870 millibars (25.70 inches) was recorded in typhoon Tip during October, 1979, and hurricane Gilbert had a pressure reading of 888 millibars (26.22 inches) during September, 1988.

Did you know?
A Pacific typhoon, named Cobra, off the coast of Luzon in the Philippines, inflicted great damage on the United States Navy during the final year of World War II. In just a matter of hours during December, 1944, it claimed 3 destroyers, 146 aircraft, and the lives of 790 sailors.

heaviest precipitation and the strongest winds, which, in this storm, are 100 knots, with peak gusts of 115 knots.

If we were to venture from west to east (left to right) through the storm in Fig. 11.2, what might we experience? As we approach the hurricane, the sky becomes overcast with cirrostratus clouds; barometric pressure drops slowly at first, then more rapidly as we move closer to the center. Winds blow from the north and northwest with ever-increasing speed as we near the eye. The high winds, which generate huge waves over 10 meters (33 feet) high, are accompanied by heavy rainshowers. As we move into the eye, the air temperature rises, winds slacken, rainfall ceases, and the sky brightens, as middle and high clouds appear overhead. The barometer is now at its lowest point (965 millibars), some 50 millibars lower than the pressure measured on the outskirts of the storm. The brief respite ends as we enter the eastern region of the eye wall. Here, we are greeted by heavy rain and strong southerly winds. As we move away from the eye wall, the pressure rises, the winds diminish, the heavy rain lets up, and eventually the sky begins to clear.

This brief, imaginary venture raises many unanswered questions. Why, for example, is the surface pressure lowest at the center of the storm? And why is the weather clear almost immediately outside the storm area? To help us answer such questions, we need to look at a vertical view, a profile of the hurricane along a slice that runs directly through its center. Of course, this view would be almost impossible to obtain in reality. However, a model that describes such a profile is given in Fig. 11.3.

The model shows that the hurricane is composed of an organized mass of thunderstorms that are an integral part of the storm's circulation. Near the surface, moist tropical air flows in toward the hurricane's center. Adjacent to the eye, this air rises and condenses into huge thunderstorms that produce heavy rainfall, as much as 25 centimeters (10 inches) per hour. Near the top of the thunderstorms, the relatively dry air, having lost much of its moisture, begins to flow outward away from the center. This diverging air aloft actually produces a clockwise (anticyclone) flow of air several hundred kilometers from the eye. As this outflow reaches the storm's periphery, it begins to sink and warm, inducing clear skies. In the vigorous thunderstorms of the eye wall, the air warms due to the release of large quantities of latent heat. This produces slightly higher pressures aloft, which initiate downward air motion within the eye. As the air subsides, it warms by compression. This helps to account for the warm air and the absence of thunderstorms in the center of the storm.

Figure 11.3
A model that shows a vertical view of air motions, clouds, and precipitation in a typical hurricane.

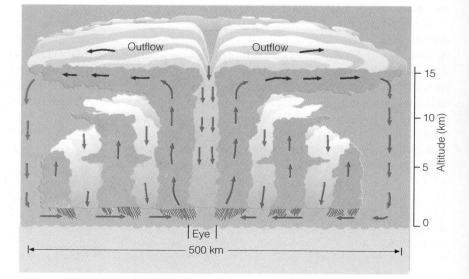

As surface air rushes in toward the region of much lower surface pressure, it should expand and cool, and we might expect to observe cooler air around the eye, with warmer air further away. But, apparently, so much heat is added to the air from the warm ocean surface that the surface air temperature remains fairly uniform throughout the hurricane.

We are now left with an important question: Where and how do hurricanes form? Although not everything is known about their formation, it is known that certain necessary ingredients are required before a weak tropical disturbance will develop into a full-fledged hurricane.

Hurricane Formation and Dissipation   Hurricanes form over tropical waters where the winds are light, the humidity is high in a deep layer, and the surface water temperature is warm over a vast area, typically 26°C (79°F) or greater. Over the tropical and subtropical North Atlantic and North Pacific oceans these conditions prevail in summer and early fall; hence, the hurricane season normally runs from June through November.

For a mass of unorganized thunderstorms to develop into a hurricane, the surface winds must converge. In the Northern Hemisphere, converging air spins counterclockwise. Because this type of rotation will not develop on the equator where the Coriolis force is zero (see Chapter 6), hurricanes form in subtropical regions, usually between 5° and 20° latitude. Convergence may occur along a front that has moved into the tropics from middle latitudes. Although the temperature contrast between the air on both sides of the front is gone, developing thunderstorms and converging surface winds may form, especially when the front is accompanied by a cold upper-level trough, where the air is spreading or diverging.

We know from Chapter 7 that the surface winds converge along the intertropical convergence zone (ITCZ). Occasionally, when the ITCZ is displaced away from the equator, a wave in the ITCZ forms into an area of low pressure, convection becomes organized, and the system grows into a hurricane. Weak convergence also occurs on the eastern side of an easterly wave, where hurricanes have been known to form. In fact, many if not most Atlantic hurricanes can be traced to easterly waves that form over Africa. However, only a small fraction of all of the tropical disturbances that form over the course of a year ever grow into hurricanes.

Figure 11.4
Hurricanes form over warm, tropical waters where the winds are light and the humidity, in a deep layer, is high.

Even when all of the surface conditions appear near perfect for the formation of a hurricane (e.g., warm water, converging surface winds, and so forth), the storm may not develop if the weather conditions aloft are not just right. For example, in the region of the trade winds and especially near latitude 20°, the air is often sinking due to the subtropical high. The sinking air warms and creates an inversion known as the **trade wind inversion**. When the inversion is strong it can inhibit the formation of intense thunderstorms and hurricanes. Also, hurricanes do not form where the upper-level winds are strong. Strong winds tend to disrupt the organized pattern of convection and disperse the heat, which is necessary for the growth of the storm.

For hurricanes to form, the thunderstorms must become organized so that the latent heat that drives the system can be confined to a limited area. If thunderstorms start to organize along the ITCZ or along an easterly wave and if the trade wind inversion is weak, the stage may be set for the birth of a hurricane. The

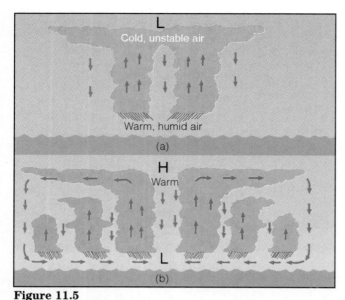

Figure 11.5
Development of a hurricane. (a) Cold air above an organized mass of tropical thunderstorms generates unstable air and large cumulonimbus clouds. (b) The release of latent heat warms the upper troposphere, creating an area of high pressure. Upper-level winds move outward away from the high. This, coupled with the warming of the air layer, causes surface pressures to drop. As air near the surface moves toward the lower pressure, it converges, rises, and fuels more thunderstorms. Soon a chain reaction develops, and a hurricane forms.

likelihood of hurricane development is enhanced if the air aloft is unstable. Such instability can be brought on when a cold upper-level trough from middle latitudes moves over the storm area. When this happens, the cumulonimbus clouds are able to build rapidly and grow into enormous thunderstorms. (See Fig. 11.5.)

Although the upper air is initially cold, it warms rapidly due to the huge amount of latent heat released during condensation. As this cold air is transformed into much warmer air, the air pressure in the upper troposphere above the developing storm rises, producing an area of high pressure (see Chapter 6). Now the air aloft begins to move outward away from the region of developing thunderstorms. This diverging air aloft, coupled with warming of the air layer, causes surface pressure to drop, and a small area of surface low pressure forms. The surface air begins to spin counterclockwise (Northern Hemisphere) and in toward the region of low pressure. Just as ice skaters spin faster as their arms are brought in close to their bodies, so the air increases in speed as it moves in toward the low-pressure cen-

ter. The winds then generate rough seas, which increase the friction on the moving air. This causes the winds to converge and ascend about the center of the storm.

We now have a chain reaction in progress, or what meteorologists call a *feedback mechanism.* The rising air, having picked up added moisture and warmth from the choppy sea, fuels more thunderstorms and releases more heat, which causes the surface pressure to lower even more. The lower pressure near the center creates a greater friction, more convergence, more rising air, more thunderstorms, more heat, lower surface pressure, stronger winds, and so on until a full-blown hurricane is born.

As long as the upper-level outflow of air is greater than the surface inflow, the storm will intensify and the surface pressure will drop. Because the air pressure within the system is controlled to a large extent by the warmth of the air, the storm will intensify only up to a point. The controlling factors are the temperature of the water and the release of latent heat.* Consequently, when the storm is literally full of thunderstorms, it will use up just about all of the available energy, so that air temperature will no longer rise and pressure will level off. Because there is a limit to how intense the storm can become, peak wind gusts seldom exceed 200 knots. When the converging surface air near the center exceeds the outflow at the top, surface pressure begins to increase, and the storm dies out.

If the hurricane remains over warm water, it may survive for several weeks. However, most hurricanes last for less than a week; they weaken rapidly when they travel over colder water and lose their heat source. They also dissipate rapidly over land. Here, not only is their energy source removed, but their winds decrease in strength (due to the added friction) and blow more directly into the center, causing the central pressure to rise.

Hurricane Stages of Development   Hurricanes go through a set of stages from birth to death. Initially, the mass of thunderstorms with only a slight wind circulation is known as a **tropical disturbance**. The tropical disturbance becomes a **tropical depression** when the winds increase to between 20 and 34 knots and several closed isobars appear about its center on a surface

---

*The maximum strength that a storm can achieve is determined by the temperature difference between the sea surface and the top of the storm.

weather map. When the isobars are packed together and the winds are between 35 and 64 knots, the tropical depression becomes a **tropical storm**. The tropical storm is classified as a *hurricane* only when its winds exceed 64 knots (74 miles per hour).

Figure 11.6 shows four tropical systems in various stages of development. Moving from east to west, we see a weak tropical disturbance (an easterly wave) crossing over Panama. Further west, a tropical depression is organizing around a developing center with winds less than 25 knots. In a few days, this system will develop into hurricane Gilma. Further west is a full-fledged hurricane with peak winds in excess of 110 knots. The swirling band of clouds to the north is Emilia; once a hurricane (but now with winds less than 40 knots), it is rapidly weakening over colder water.

Up to now, we have seen that hurricanes are storms with winds in excess of 64 knots (74 miles per hour)—storms that form and intensify over warm tropical waters. Hurricanes are similar to middle latitude

cyclones in that, at the surface, both have central cores of low pressure and winds that spiral counterclockwise about their respective centers (Northern Hemisphere). (The many differences between the two systems are described in detail in the Focus section on p. 282.)

Hurricane Movement   Figure 11.7 shows where most hurricanes are born and the general direction in which they move. Notice that they form over tropical oceans, except in the South Atlantic and in the eastern South Pacific. Presumably, the surface water temperatures are too cold in these areas for their development.

Figure 11.6
Visible satellite picture showing four tropical systems in different stages of development.

Focus on a Special Topic

How Do Hurricanes Compare with Middle Latitude Storms?

By now, it should be apparent that a hurricane is much different from the mid-latitude cyclone that we discussed in Chapter 8. A hurricane derives its energy from the warm water and the latent heat of condensation, whereas the mid-latitude storm derives its energy from horizontal temperature contrasts. The vertical structure of a hurricane is such that its central column of air is warm from the surface upward; consequently, hurricanes are called *warm core lows*. A hurricane weakens with height, and the area of low pressure at the surface may actually become an area of high pressure above 12 kilometers (40,000 feet). Mid-latitude cyclones, on the other hand, usually intensify with increasing height, and a cold upper-level low or trough exists to the west of the surface system. A hurricane usually contains an eye where the air is sinking, while mid-latitude cyclones are characterized by centers of rising air. Hurricane winds are strongest near the surface, whereas the strongest winds of the mid-latitude storm are found aloft in the jet stream.

Further contrasts can be seen on a surface weather map. Figure 1 shows hurricane Allen over the Gulf of Mexico and a mid-latitude storm north of New England. Around the hurricane, the isobars are more circular, the pressure gradient is much steeper, and the winds are stronger. The hurricane has no fronts and is smaller (although Allen is larger than most hurricanes). There are similarities between the two systems: Both are areas of surface low pressure, with winds moving more or less counter-

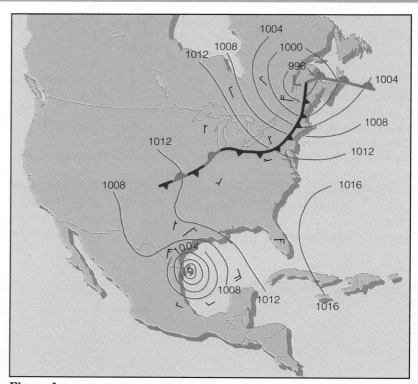

Figure 1
Surface weather map for the morning of August 9, 1980, showing hurricane Allen over the Gulf of Mexico and a middle latitude storm system north of New England.

clockwise about their respective centers.

It is interesting to note that some northeasters (winter storms that move northeastward along the coastline of North America, bringing with them heavy precipitation, high surf, and strong winds) may actually possess some of the characteristics of a hurricane. For example, a particularly powerful northeaster during January, 1989, was observed to have a cloud-free eye, with surface winds in excess

of 85 knots spinning about a warm inner core.

Even though hurricanes weaken rapidly as they move inland, their counterclockwise circulation may draw in air with contrasting properties. If the hurricane links with an upper-level trough, it may actually become a mid-latitude cyclone. This is what happened to hurricane Hazel in 1954, to Agnes in 1972, and to Hugo in 1989.

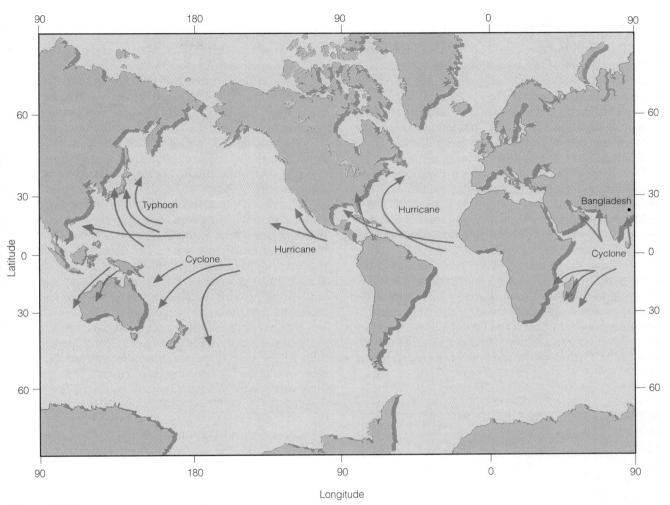

Figure 11.7
Regions where tropical storms form, the names given to storms, and the typical paths they take.

It is also possible that the unfavorable location of the ITCZ during the Southern Hemisphere's warm season discourages their development.

Hurricanes that form over the North Pacific and North Atlantic are steered by easterly winds and move west or northwestward at about 10 knots for a week or so. Gradually, they swing poleward around the subtropical high, and when they move far enough north, they become caught in the westerly flow, which curves them to the north or northeast. In the middle latitudes, the hurricane's forward speed normally increases, sometimes to more than 50 knots. The actual path of a hurricane may vary considerably. Some take erratic paths and make odd turns that occasionally catch

weather forecasters by surprise. There have been many instances where a storm heading directly for land suddenly veered away and spared the region from almost certain disaster. As an example, hurricane Elena with peak winds of 90 knots moved northwestward into the Gulf of Mexico on August 29, 1985. It then veered eastward toward the west coast of Florida. After stalling offshore, it headed northwest. After weakening, it then moved onshore near Biloxi, Mississippi, on the morning of September 2.

As we saw in an earlier section, many hurricanes form off the coast of Mexico over the North Pacific. In fact, this area usually spawns about eight hurricanes each year, which is slightly more than the yearly average of six storms born over the tropical North Atlantic. Eastern North Pacific hurricanes normally move westward, away from the coast, and so little is heard about them. When one does move northwestward, it normally weakens rapidly over the cool water of the North Pacific. Occasionally, however, one will curve northward or even northeastward and slam into Mexico, causing destructive flooding. Hurricane Tico left 25,000 people homeless and caused an estimated $66 million in property damage after passing over Mazatlán, Mexico, in October, 1983. The remains of Tico even produced record rains and flooding in Texas and Oklahoma. Even less frequently, a hurricane will stray far enough north to bring summer rains to southern California and Arizona, as did the remains of hurricane Boris during June, 1990 and hurricane Lester during August, 1992.

The Hawaiian Islands, which are situated in the central North Pacific between about 20° and 23°N, appear to be in the direct path of many eastern Pacific hurricanes and tropical storms. By the time most of these storms have reached the islands, however, they have weakened considerably, and pass harmlessly to the south or northeast. The exceptions were hurricane

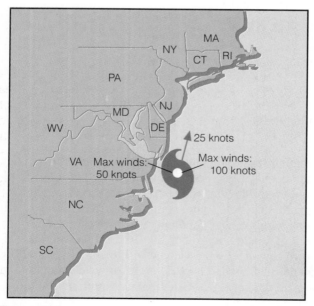

Figure 11.8
Hurricane Gloria on the morning of September 27, 1985. Moving northward at 25 knots, Gloria has sustained winds of 100 knots on its right side and 50 knots on its left side. (The central pressure of the storm is about 945 millibars.)

Iwo during November, 1982, and hurricane Iniki during September, 1992. Iwo lashed part of Hawaii with 100-knot winds and huge surf, causing an estimated $312 million in damages. Iniki, the worst hurricane to hit Hawaii this century, battered the island of Kawai with torrential rain, sustained winds of 114 knots that gusted to 140 knots, and twenty-foot waves that crashed over coastal highways. Major damage was sustained by most of the hotels and about 50 percent of the homes on the island. Iniki flattened sugar cane fields, destroyed the macadamia nut crop and caused at least three deaths.

Hurricanes that form over the tropical North Atlantic also move westward or northwestward on a collision course with Central or North America. Most hurricanes, however, swing away from land and move northward, parallel to the coastline of the United States. A few storms, perhaps three per year, move inland, bringing with them high winds, huge waves, and torrential rain that may last for days.*

A hurricane moving northward over the Atlantic will normally survive as a hurricane for a longer time than will its counterpart at the same latitude over the eastern Pacific. The reason is, of course, that the surface water of the Atlantic is much warmer. (See Chapter 7.)

Destruction and Warning   When a hurricane is approaching from the east, its highest winds are usually on its north (poleward) side. The reason for this phenomenon is that the winds that push the storm along add to the winds on the north side and subtract from the winds on the south (equator) side. Hence, a hurricane with 110-knot winds moving westward at 10 knots will have 120-knot winds on its north side and only 100-knot winds on its south side.

The same type of reasoning can be applied to a northward-moving hurricane. For example, as hurricane Gloria moved northward along the coast of Virginia on the morning of September 27, 1985 (see Fig. 11.8), winds of 75 knots were swirling counterclockwise about its center. Because the storm was moving northward at about 25 knots, sustained winds on its eastern (right) side were about 100 knots, while on its western (left) side—on the coast—the winds were only about 50 knots. Even so, these winds were strong

---

*Six hurricanes struck the United States mainland in 1985, a seventy-year record.

Figure 11.9
Although the high winds of a hurricane can inflict a great deal of damage, the majority of destruction is usually brought about by huge waves and flooding. Here hurricane Kate slams into Key West, Florida, on November 20, 1985. The hurricane was responsible for seven deaths in the United States and over one billion dollars in damages.

enough to cause significant beach erosion along the coasts of Maryland, Delaware, and New Jersey.

The high winds of a hurricane also generate large waves, sometimes 10 to 15 meters (30 to 50 feet) high. These waves move outward, away from the storm, in the form of *swells* that carry the storm's energy to distant beaches. Consequently, the effects of the storm may be felt days before the hurricane arrives.

Although the hurricane's high winds inflict a great deal of damage, it is the huge waves, high seas, and *flooding* that cause most of the destruction.* The flooding is due, in part, to winds pushing water onto the shore and to the heavy rains, which may exceed 63 centimeters (25 inches) in 24 hours. Flooding is also aided by the low pressure of the storm. The region of low pressure allows the ocean level to rise (perhaps half a

meter), much like a soft drink rises up a straw as air is withdrawn. (A 1-millibar drop in air pressure produces a 1-centimeter rise in ocean level.) The combined effect of high water (which is usually well above the high-tide level) and high winds produces the **storm surge**—an abnormal rise of several meters in the ocean level—that inundates low-lying areas and turns beachfront homes into piles of splinters (Fig. 11.10). The storm surge is particularly damaging when it coincides with normal high tides.

Considerable damage may also occur from hurricane-spawned tornadoes. The exact mechanism by

*Hurricanes are not without their beneficial aspect, as they can provide much needed rainfall in drought-stricken areas.

Did you know?
Hurricane Agnes, during June 1972, dumped an estimated 28 trillion gallons of rainfall over the eastern United States. The deluge caused extensive flooding and over $2 billion in damage.

Figure 11.10
When a storm surge moves in at high tide it can inundate and destroy a wide swath of coastal lowlands.

which these tornadoes form is not yet known; however, studies suggest that surface topography may play a role by initiating the convergence (and, hence, rising) of surface air. Recent studies suggest that swathlike areas of extreme damage once attributed to tornadoes may actually be due to downbursts associated with the large thunderstorms around the eye wall. In addition, scientists studying hurricane Debby with Doppler radar in 1982 observed a mesocyclone (Chapter 10) in the southern part of the developing eye wall.

With the aid of ship reports, satellites, radar, buoys, and reconnaissance aircraft, the location and intensity of hurricanes are pinpointed and their movements carefully monitored. When a hurricane poses a direct threat to an area, a **hurricane watch** is issued, if possible several days before the storm arrives. When it appears that the storm will strike an area within 24 hours, a **hurricane warning** is issued by the National Hurricane Center in Coral Gables, Florida, or by the Pacific Hurricane Center in Honolulu, Hawaii. Along the east coast of North America from Eastport, Maine, to Brownsville, Texas, the warning is accompanied by a probability. The probability gives the percent chance of the hurricane's center passing within 105 kilometers (65 miles) of a particular community. The warning is designed to give residents ample time to secure property and, if necessary, to evacuate the area.

A hurricane warning is issued for a rather large coastal area, usually about 550 kilometers (342 miles) in length. Since the average swath of hurricane damage is normally about one-third this length, much of the area is "overwarned." As a consequence, many people in a warning area feel that they are needlessly forced to evacuate. For example, approximately one million people were evacuated from low-lying Gulf coastal areas during the approach of hurricane Elena in 1985. However, this probably contributed to the fact that there were no deaths in the area of *landfall*.*

As new hurricane-prediction models are implemented, and as our understanding of the nature of hurricanes increases, improved forecasts of hurricane movement and intensification should become available.

Ample warning by the National Weather Service probably saved the lives of many people as hurricane Allen moved onshore along the south Texas coast during the morning of August 10, 1980 (Fig. 11.11). The

---

Did you know?
In the Gulf Coast states, aside from the damage and flooding that occurs with a hurricane, one of the greatest fears in the cleanup effort is the poisonous snakes that seem to find their way into various nooks and crannies.

---

*Landfall is the position along a coast where the center of a hurricane passes from ocean to land.

storm formed over the warm, tropical Atlantic and moved westward on a rampage through the Caribbean, where it killed almost 300 people and caused extensive damage. After raking the Yucatán Peninsula with 150-knot winds, Allen howled into the warm Gulf of Mexico. It reintensified and its winds increased to 160 knots. Gale-force winds reached outward for 320 kilometers (200 miles) north of its center. As it approached the south Texas coast, it was one of the greatest storms to ever enter that area. The central pressure of the storm dropped to a low of 899 millibars (26.55 inches). Up until this time, only the 1935 Labor Day storm that hit the Florida Keys with a pressure of 892 millibars (26.35 inches) was stronger.* But its path became wobbly and it stalled offshore just long enough to lose much of its intensity. It moved sluggishly inland on the morning of August 10.

Once over land, it quickly became a tropical storm with peak winds of less than 50 knots. Before weakening, however, Allen hammered the south Texas coast with winds of 100 knots. It spawned several tornadoes and swamped the area with high tides and flooding. In Corpus Christi, the storm felled trees and power lines;

*During September, 1988, hurricane Gilbert, with a pressure of 888 millibars (26.22 inches) set a low pressure record for hurricanes in the Atlantic, Carribean, and Gulf of Mexico.

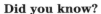

Did you know?
The huge storm surge and high winds of hurricane Camille carried several ocean-going ships over 7 miles inland near Pass Christian, Mississippi.

streets were flooded and marinas were littered with the wreckage of small boats. In all of this, only a few people lost their lives, as 200,000 were evacuated from the Texas coast many hours before the storm struck. Further inland, the remains of the storm dumped heavy rains over a wide area of southern Texas. This was a blessing for some because it marked a break in the summer-long drought; for others, it meant despair as they fled from swollen rivers that had overflowed their banks.

A costlier hurricane (in terms of loss of life and total damage) occurred on August 18, 1983, when hurricane Alicia slammed into the Texas coast just southwest of Galveston. With winds of 100 knots, a central pressure of 962 millibars, and a storm surge of at least 3 meters (10 feet), Alicia uprooted trees, knocked out skyscraper windows, tossed hundreds of mobile homes about like cardboard boxes, and flooded an extensive area. Total damage from Alicia was estimated at two billion dollars, and twenty-one persons lost their lives.

Figure 11.11
Hurricane Allen in the Gulf of Mexico on August 8, 1980. This large storm with its spiral cloud bands occupies most of the Gulf. Compare Allen's size with the smaller hurricane positioned off the west coast of Mexico.

Did you know?

As hurricane Andrew stormed through south Florida, weatherman Bryan Norcross of Miami's WTVJ-TV calmed millions of south Floridians as they anxiously watched and listened to his 3-day hurricane broadcast marathon, which for one stretch lasted 23 consecutive hours. Broadcasting from the steps of the Bunker—a concrete-walled storeroom just off the studio—Norcross became an instant hero. To show their appreciation, grateful residents showered him with letters, gifts, and even marriage proposals.

In recent years, the hurricane death toll in the United States has averaged between 50 and 100 persons, although over 200 people died in Mexico when hurricane Gilbert slammed the Gulf Coast of Mexico during September, 1988. This relatively low total is partly due to the advance warning provided by the National Weather Service and to the fact that only a few really intense storms have reached land during the past 30 years. However, as the population density continues to increase in vulnerable coastal areas, the potential for a hurricane-caused disaster continues to increase also.

Hurricane Camille (1969) stands out as one of the most intense hurricanes to reach the coastline of the United States in recent decades. With a central pressure of 905 millibars, tempestuous winds reaching 160 knots, and a storm surge more than 7 meters (23 feet) above the normal high-tide level, Camille unleashed its fury on Mississippi, destroying thousands of buildings. During its rampage, it caused an estimated one-and-a-half billion dollars in property damage and took more than 200 lives.

During September, 1989, hurricane Hugo was born as a cluster of thunderstorms became a tropical depression off the coast of Africa, southeast of the Cape Verde Islands. The storm grew in intensity, tracked westward for several days, then turned northwestward, striking the island of St. Croix with sustained winds of 125 knots. After passing over the eastern tip of Puerto Rico, this large, powerful hurricane took aim at the coastline of South Carolina. With maximum winds estimated at about 120 knots (138 miles per hour), and a central pressure near 934 millibars, Hugo made landfall near Charleston, South Carolina, about midnight on September 21st. (See Figs. 11.12 and 11.13.) The high winds and storm surge, which ranged between 2.5 and 6 meters (8 and 20 feet), hurled a thundering wall of water against the shore. This knocked out power, flooded streets, and caused widespread destruction to coastal communities (Fig. 11.14). The total damage in the United States attributed to Hugo was over $7 billion, with a death toll of 21 in the United States and 49 overall. But Hugo does not even come close to the

Figure 11.12
A color enhanced infrared satellite photograph of hurricane Hugo with its eye over the coast near Charleston, South Carolina.

Figure 11.13
Digitized radar display showing hurricane Hugo's eye and eye wall with its heavy precipitation (dark red area), as the center of the storm moves onshore.

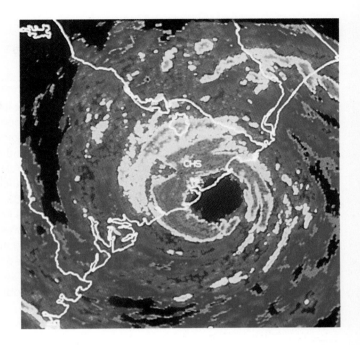

costliest hurricane on record—that dubious distinction goes to hurricane Andrew.

On August 21, 1992, as tropical storm Andrew churned westward across the Atlantic it began to weaken, prompting some forecasters to surmise that this tropical storm would never grow to hurricane strength. But Andrew moved into a region favorable for hurricane development. Warm surface water and weak wind shear allowed Andrew to intensify rapidly. And in just two days Andrew's winds increased from 45 knots to 122 knots, turning an average tropical storm into one of the most intense hurricanes to strike Florida this century.

With winds gusting to 150 knots and a powerful storm surge, Andrew made landfall south of Miami on the morning of August 24. (See Fig. 11.15.) The eye of the storm moved over Homestead, Florida. Andrew's fierce winds completely devastated the area—steel-reinforced tie beams weighing tons were torn free of townhouses and hurled as far as several blocks. The hurricane roared westward across Southern Florida, weakened slightly, then regained strength over the warm Gulf of Mexico. Surging northwestward, Andrew slammed into Louisiana with 120-knot winds.

All told, hurricane Andrew was the costliest natural disaster ever to hit the United States. It destroyed or damaged over 200,000 homes and businesses, left more than 160,000 homeless, caused over $20 billion

(a)

(b)

Figure 11.14
Beach homes at Folly Beach, South Carolina (a) before and (b) after hurricane Hugo.

Focus on a Special Topic
Modifying Hurricanes

Because of the potential destruction and loss of lives that hurricanes can inflict, attempts have been made to reduce their winds by seeding them with silver iodide. The idea is to seed the clouds just outside the eye wall with just enough artificial ice nuclei so that the latent heat given off will stimulate cloud growth in this area of the storm. These clouds, which grow at the expense of the eye wall thunderstorms, actually form a new eye wall farther away from the hurricane's center. As the storm center widens, its pressure gradient should weaken, which may cause its spiraling winds to decrease in speed. During project STORMFURY, a joint effort of the National Oceanic and Atmospheric Administration (NOAA) and the U.S. Navy, several hurricanes were seeded by aircraft. In 1963, shortly after hurricane Beulah was seeded with silver iodide, surface pressure in the eye began to rise and the region of maximum winds moved away from the storm's center. Even more encouraging results were obtained from the multiple seeding of hurricane Debbie in 1969. After one day of seeding, Debbie showed a 30 percent reduction in maximum winds. However, the question remains: Would the winds have lowered naturally had the storm not been seeded? A recent study even casts doubts upon the theoretical basis for this kind of hurricane modification because hurricanes appear to contain too little supercooled water and too much natural ice. Consequently, there are many uncertainties about the effectiveness of seeding hurricanes in an attempt to reduce their winds.

Figure 11.15
A color-enhanced infrared satellite photograph shows hurricane Andrew making landfall south of Miami, Florida, on the morning of August 24, 1992. The storm has a central pressure of 932 mb (27.52 in.) and sustained winds of 122 knots.

in damages, and took 55 lives, including 41 in Florida. Although Andrew may well be the most expensive hurricane on record, it is far from the deadliest.

Before the era of satellites and radar, catastrophic losses of life had occurred. In 1900, about 6000 people lost their lives when a hurricane slammed into Galveston, Texas, with a huge storm surge. Most of the deaths occurred in the low-lying coastal regions as flood waters pushed inland. In October, 1893, nearly 2000 people perished on the Gulf Coast of Louisiana as a giant storm surge swept that region. Spectacular losses are not confined to the Gulf Coast. Nearly 1000 people lost their lives in Charleston, South Carolina, during August of the same year. But these statistics are small compared to the more than 300,000 lives taken as a killer tropical cyclone and storm surge ravaged the coast of Bangladesh with flood waters in 1970. Again in April, 1991, a similar cyclone devastated the area with reported winds of 127 knots and a 7 meter (20 feet) storm surge. In all, the storm destroyed 1.4 million houses and killed 140,000 people and one million cattle. Unfortunately, the potential for a repeat of this type of disaster remains high in Bangladesh, as many people live along the relatively low, wide flood plain that slopes outward to the Bay. And, historically, this region is in a path frequently taken by tropical cyclones. (Look back at Fig. 11.7, p. 283.)

In an effort to estimate the possible damage a hurricane's sustained winds and storm surge could do to a coastal area, the **Saffir-Simpson scale** was developed (see Table 11.1). The scale numbers are based on actual conditions at some time during the life of the storm.

As the hurricane intensifies or weakens, the category, or scale number, is reassessed accordingly. (For some information on modifying hurricanes, see the Focus section on p. 290.)

Naming Hurricanes   In reading the previous sections, you noticed that hurricanes were assigned

Table 11.1   Saffir-Simpson Hurricane Damage-Potential Scale

Scale Number (Category)	Central Pressure		Winds		Storm Surge		Damage
	mb	in.	mi/hr	knots	ft	m	
1	≥980*	≥28.94	74–95	64–82	4–5	~1.5	damage mainly to trees, shrubbery, and unanchored mobile homes
2	965–979	28.50–28.91	96–110	83–95	6–8	~2.0–2.5	some trees blown down; major damage to exposed mobile homes; some damage to roofs of buildings
3	945–964	27.91–28.47	111–130	96–113	9–12	~2.5–4.0	foliage removed from trees; large trees blown down; mobile homes destroyed; some structural damage to small buildings
4	920–944	27.17–27.88	131–155	114–135	13–18	~4.0–5.5	all signs blown down; extensive damage to roofs, windows, and doors; complete destruction of mobile homes; flooding inland as far as 10 kilometers (6 miles); major damage to lower floors of structures near shore
5	<920	<27.17	>155	>135	>18	>5.5	severe damage to windows and doors; extensive damage to roofs of homes and industrial buildings; small buildings overturned and blown away; major damage to lower floors of all structures less than 4.5 meters (15 feet) above sea level within 500 meters (1600 feet) of shore

*Symbol > means greather than; < means less than; ≥ means equal to or greater than; ~ means approximately equal to.

Table 11.2 Retired Atlantic Hurricane Names in Alphabetical Order Since 1979	
Alicia, 1983	Frederick, 1979
Allen, 1990	Gilbert, 1988
Andrew, 1992	Gloria, 1985
Bob, 1991	Hugo, 1989
David, 1979	Joan, 1988
Elena, 1985	

War II, names like Able and Baker were used. (These names correspond to the radio code words associated with each letter of the alphabet.) This method also seemed cumbersome so, beginning in 1953, the National Weather Service began using female names to identify hurricanes. The list of names for each year was in alphabetical order, so that the name of the season's first storm began with the letter *A*, the second with *B*, and so on.

From 1953 to 1977, only female names were used. However, beginning in 1978, hurricanes in the eastern Pacific were alternately assigned female and male names, but not just English names, as Spanish and French ones were used too. This practice was started for North Atlantic hurricanes in 1979. (Remember that a storm only gets a name when it reaches tropical storm strength.) Once a storm has caused great damage and it becomes infamous, its name is retired. (See Table 11.2.) (Appendix E gives the proposed list of names for both North Atlantic and eastern Pacific hurricanes.)

names. (The name is assigned when the storm reaches tropical storm strength.) Before this practice was started, hurricanes were identified according to their latitude and longitude. This method was confusing, especially when two or more storms were present over the same ocean. To reduce the confusion, hurricanes were identified by letters of the alphabet. During World

Summary

Hurricanes are tropical cyclones with winds that exceed 64 knots (74 miles per hour) and blow counterclockwise about their centers in the Northern Hemisphere. A hurricane consists of a mass of organized thunderstorms that spiral in toward the extreme low pressure of the storm's eye. The most intense thunderstorms, the heaviest rain, and the highest winds occur outside the eye, in the region known as the eye wall. In the eye itself, the air is warm, winds are light, and skies may be broken or overcast.

Hurricanes are born over warm tropical waters where surface winds converge and thunderstorms become organized. Convergence may occur along the ITCZ, on the eastern side of an easterly wave, or along a front that has moved into the tropics from higher latitudes. If the disturbance becomes more organized, it becomes a tropical depression. If central pressures drop and surface winds increase, the depression becomes a tropical storm. At this point, the storm is given a name. Some tropical storms continue to deepen into full-fledged hurricanes and some do not.

The easterly winds in the tropics usually steer hurricanes westward. Most storms then gradually swing northwestward around the subtropical high to the north. If the storm moves into middle latitudes, the prevailing westerlies steer it northeastward. Because hurricanes derive their energy from the warm surface water and from the latent heat of condensation, they tend to dissipate rapidly when they move over cold water or over a large mass of land.

Although the high winds of a hurricane can inflict a great deal of damage, it is the huge waves and the flooding associated with the storm surge that normally cause the most destruction. The Saffir-Simpson hurricane scale was developed to estimate the potential destruction that a hurricane can cause.

Key Terms

The following terms are listed in the order they appear in the text. Define each. Doing so will aid you in reviewing the material covered in this chapter.

streamlines	eye (of hurricane)	tropical depression	hurricane watch
easterly wave	eye wall	tropical storm	hurricane warning
hurricane	trade wind inversion	storm surge	Saffir-Simpson scale
typhoon	tropical disturbance		

Review Questions

1. (a) What is an easterly wave? How do easterly waves generally move?
   (b) Are showers found on the eastern or western side of the wave?
2. Why are streamlines rather than isobars used on surface weather maps in the tropics?
3. What is the name given to a hurricane-like storm over the western North Pacific Ocean?
4. In your own words, what is a hurricane?
5. (a) Where do hurricanes derive their energy?
   (b) Would it be possible for a hurricane to form over land? Explain.
6. What conditions at the surface and aloft are most conducive for the formation of a hurricane?
7. Describe the horizontal and vertical structure of a hurricane.
8. What factors tend to weaken hurricanes?
9. Distinguish among a tropical disturbance, a tropical depression, a tropical storm, and a hurricane.
10. (a) In what ways is a hurricane different from a mid-latitude cyclone?
    (b) In what ways are they similar?

11. Why do most hurricanes move westward over tropical waters?
12. If the high winds of a hurricane are not responsible for inflicting the most damage, what is?
13. If a hurricane is moving westward at 10 knots, will the strongest winds be on its northern or southern side? Explain.
14. Explain how a storm surge forms.
15. How does a hurricane watch differ from a hurricane warning?
16. Why have hurricanes been seeded with silver iodide?
17. Give two reasons why hurricanes are more likely to strike New Jersey than Oregon.
18. Hurricanes are given names when the storm is in what stage of development?
19. Use the Saffir-Simpson hurricane scale (Table 11.1, p. 291) to determine the category of each of the following hurricanes:
    (a) hurricane Gloria in Fig. 11.8, p. 284.
    (b) hurricane Hugo in Fig. 11.12, p. 288.
    (c) hurricane Andrew in Fig. 11.15, p. 290.

Air pollution from the Navajo generating station in Arizona. (©. Steven Gottlieb, F.P.G.)

Chapter 12

Air Pollution

Contents

▲▼▲

Air pollution makes the earth a less pleasant place to live. It reduces the beauty of nature. This blight is particularly noticed in mountain areas. Views that once made the pulse beat faster because of the spectacular panorama of mountains and valleys are more often becoming shrouded in smoke. When once you almost always could see giant boulders sharply etched in the sky and the tapered arrow-heads of spired pines, you now often see a fuzzy picture of brown and green. The polluted air acts like a translucent screen pulled down by an unhappy God.

Louis J. Battan, *The Unclean Sky*

12 Every deep breath fills our lungs mostly with gaseous nitrogen and oxygen. Also inhaled, in minute quantities, may be other gases and particles, some of which could be considered pollutants. These contaminants come from car exhaust, chimneys, forest fires, factories, power plants, and other sources related to human activities.

Virtually every large city has to contend in some way with air pollution, which clouds the sky, injures plants, and damages property. Some pollutants merely have a noxious odor, whereas others can cause severe health problems. The cost is high. In the United States, for example, outdoor air pollution takes its toll in health care and lost work productivity at an annual expense that runs into *billions* of dollars. Estimates are that, worldwide, nearly one billion people in urban environments are continuously being exposed to health hazards from air pollutants.

This chapter takes a look at this serious contemporary concern. We begin by briefly examining the history of problems in this area, and then go on to explore the types and sources of air pollution, as well as the weather that can produce an unhealthful accumulation of pollutants. Finally, we investigate how air pollution influences the urban environment and also how it brings about unwanted acid precipitation.

▲▼▲

A Brief History of Air Pollution

Strictly speaking, air pollution is not a new problem. More than likely it began when humans invented fire whose smoke choked the inhabitants of poorly ventilated caves. In fact, very early accounts of air pollution characterized the phenomenon as "smoke problems," the major cause being people burning wood and coal to keep warm.

To alleviate the smoke problem in old England, King Edward I issued a proclamation in 1273 forbidding the use of sea coal, an impure form of coal that produced a great deal of soot and sulfur dioxide when burned. One person was reputedly executed for violating this decree. In spite of such restrictions, the use of coal grew as a heating fuel during the fifteenth and sixteenth centuries.

As industrialization increased, the smoke problem worsened. In 1661, the prominent scientist John Evelyn wrote an essay deploring London's filthy air. And by the

1850s, London had become notorious for its "pea-soup" fog, a thick mixture of smoke and fog that hung over the city. These fogs could be dangerous. In 1873, one was responsible for 700 deaths. Another in 1911 claimed the lives of 1150 Londoners. To describe this chronic atmospheric event, a physician, Harold Des Voeux, coined (around 1911) the word *smog*, meaning a combination of smoke and fog.

Little was done to control the burning of coal as time went by, primarily because it was extremely difficult to counter the basic attitude of the powerful industrialists: "Where there's muck, there's money." London's acute smog problem intensified. Then, during the first week of December, 1952, a major disaster struck. The winds died down over London and the fog and smoke became so thick that people walking along the street literally could not see where they were going. This particular disastrous smog lasted five days and took nearly 4000 lives, prompting parliament to pass a Clean Air Act in 1956. Additional air pollution incidents occurred in England during 1956, 1957, and 1962, but due to the strong legislative measures taken against air pollution, London's air today is much cleaner, and "pea soup" fogs are a thing of the past.

Air pollution episodes were by no means limited to Great Britain. During the winter of 1930, for instance, Belgium's highly industrialized Meuse Valley experienced an air pollution tragedy when smoke and other contaminants accumulated in a narrow steep-sided valley. The tremendous buildup of pollutants caused about 600 people to become ill, and ultimately 63 died. Not only did humans suffer, but cattle, birds, and rats fell victim to the deplorable conditions.

The industrial revolution brought air pollution to the United States, as homes and coal-burning industries belched smoke, soot, and other undesirable emissions into the air. Soon, large industrial cities, such as St. Louis and Pittsburgh (which became known as the "Smoky City"), began to feel the effects of the ever-increasing use of coal. As early as 1911, studies documented the irritating effect of smoke particles on the human respiratory system and the "depressing and devitalizing" effects of the constant darkness brought on by giant, black clouds of smoke. By 1940, the air over some cities had become so polluted that automobile headlights had to be turned on during the day.

The first major documented air pollution disaster in the United States occurred at Donora, Pennsylvania, during October, 1948, when industrial pollution became trapped in the Monongahela River Valley. During the ordeal, which lasted five days, more than twenty people died and thousands became ill.* Several times during the 1960s, air pollution levels became dangerously high over New York City. Meanwhile, on the West Coast, in cities such as Los Angeles, the ever-rising automobile population, coupled with the large petroleum processing plants, were instrumental in generating a different type of pollutant—one that forms in sunny weather and irritates the eyes. Toward the end of World War II, Los Angeles had its first (of many) smog alerts.

Air pollution episodes in Los Angeles, New York, and other large American cities led to the establishment of much stronger emission standards for industry and automobiles. The Clean Air Act of 1970, for example, empowered the Federal government to set emission standards that each state was required to enforce. The Clean Air Act was revised in 1977 and updated by congress in 1990 to include even stricter emission requirements for autos and industry. The new version of the Act also includes incentives to encourage companies to lower emissions of those pollutants contributing to the current problem of acid rain.

▲▽▲

Types and Sources of Air Pollutants

Air pollutants are airborne substances (either solids, liquids, or gases) that occur in concentrations high enough to threaten the health of people and animals, to harm vegetation and structures, or to toxify a given environment. Air pollutants come from both natural sources and human activities. Examples of natural sources include wind picking up dust and soot from the earth's surface and carrying it aloft, volcanoes belching tons of ash and dust into our atmosphere, and forest fires producing vast quantities of drifting smoke (Fig. 12.1).

Human-induced pollution enters the atmosphere from both *fixed sources* and *mobile sources*. Fixed sources encompass industrial complexes, power plants, homes, office buildings, and so forth; mobile sources include motor vehicles, jet aircrafts, and ships. Certain pollutants are called **primary air pollutants** because they enter the atmosphere directly—from

Figure 12.1
The leading edge of a smoke cloud that covered about one-third of South America during October, 1988. The smoke is from the burning of tropical rainforest debris in the Amazon River basin.

smoke stacks and tail pipes, for example. Other pollutants, known as **secondary air pollutants**, form only when a chemical reaction occurs between a primary pollutant and some other component of air, such as water vapor.

We can see in Fig. 12.2 that transportation (motor vehicles, and so on) accounts for nearly 50 percent (by weight) of the pollution across the United States, with fuel combustion from stationary (fixed) sources coming in a distant second. Although hundreds of pollutants are found in our atmosphere, most fall into five groups, which are summarized in the following section.

Did you know?
In London, England, during the severe smog episode of 1952, people wore masks over their mouths and found their way along the sidewalks by feeling the walls of buildings.

---

*Additional information about the Donora air pollution disaster is given in the Focus section on p. 309.

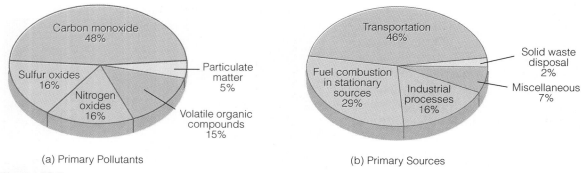

(a) Primary Pollutants

(b) Primary Sources

Figure 12.2
(a) Estimates of emissions of the primary air pollutants in the United States on a per weight basis; (b) the primary sources for the pollutants. (Data courtesy United States Environmental Protection Agency)

Principal Air Pollutants   Particulate matter represents a group of solid particles and liquid droplets that are small enough to remain suspended in the air. Collectively known as *aerosols*, this grouping includes solid particles that may irritate people but are usually not poisonous, such as soot (tiny solid carbon particles), dust, smoke, and pollen. Some of the more dangerous substances include asbestos fibers and arsenic. Tiny liquid droplets of sulfuric acid, PCBs, oil, and various pesticides are also placed into this category.

Because it often dramatically reduces visibility in urban environments, particulate matter pollution is the most noticeable (see Fig. 12.3). Some particulate matter collected in cities include iron, copper, nickel, and lead. This type of pollution can immediately influence the human respiratory system. Once inside the lungs, it can make breathing difficult, particularly for those suffering from chronic respiratory disorders. Lead particles especially are dangerous because they are absorbed into the body and accumulate in bone and soft tissues. High concentrations of lead in the human body can cause brain damage, convulsions, and death.

Industrial processes account for nearly 40 percent of the estimated 6.6 million metric tons of particulate matter emitted over the United States during the course of a year, whereas highway vehicles account

for about 17 percent. One main problem is that particulate pollution may remain in the atmosphere for some time depending on the size and the amount of precipitation that occurs. For example, larger, heavier particles with diameters greater than about 10 micrometers* (0.01 millimeters) tend to settle to the ground in about a day or so after being emitted; whereas fine, lighter particles with diameters less than 1 micrometer (0.001 millimeters) can remain suspended in the lower atmosphere for several weeks. These fine particles pose the greatest health risk, as they are small enough to penetrate the lung's natural defense mechanisms. Moreover, winds can carry these fine particles greater distances before they finally reach the surface. Rain and snow remove many of these particles from the air, even the minute ones that tend to stick to ice crystals and cloud droplets, then fall as precipitation.

Carbon monoxide (CO), a major pollutant of city air, is a colorless, odorless, poisonous gas that forms during the incomplete combustion of carbon-containing fuels. From Fig. 12.2, we can see that carbon monoxide is the most plentiful of the primary pollutants.

The Environmental Protection Agency (EPA) estimates that over 60 million metric tons of carbon monoxide enter the air annually over the United States alone—about half from highway vehicles. However, due to stricter air quality standards and the use of

Did you know?
On any given day, estimates are that about 10 million tons of solid particulate matter is suspended in our atmosphere.

---

*Particulate matter smaller than 10 micrometers are commonly referred to as PM10. In recent years, new standards set by the Environmental Protection Agency have related only to these smaller particles.

(a)

(b)

Figure 12.3
(*a*) Denver, Colorado on a clear day, and
(*b*) on a day when particulate matter and
other pollutants greatly reduce visibility.

emission-control devices, carbon monoxide levels have decreased by about 40 percent since the early 1970s.

Fortunately, carbon monoxide is quickly removed from the atmosphere by microorganisms in the soil, for even in small amounts this gas is dangerous. Hence, it poses a serious problem in poorly ventilated areas, such as highway tunnels and underground parking garages. Because carbon monoxide cannot be seen or smelled, it can kill without warning. Here's how: Normally, your cells obtain oxygen through a blood pigment called *hemoglobin*, which picks up oxygen from the lungs, combines with it, and carries it throughout your body. Unfortunately, human hemoglobin prefers carbon monoxide to oxygen so if there is too much carbon monoxide in the air you breathe, your brain will soon be starved of oxygen, and headache, fatigue, drowsiness, and even death may result.*

Sulfur dioxide (SO_2) is a colorless gas that comes primarily from the burning of sulfur-containing fossil fuels (such as coal and oil). Its primary source includes power plants, heating devices, smelters, petroleum refineries, and paper mills. However, it can enter the atmosphere naturally during volcanic eruptions and as sulfate particles from ocean spray.

Sulfur dioxide readily oxidizes to form the secondary pollutants *sulfur trioxide* (SO_3) and, in moist air, highly corrosive *sulfuric acid* (H_2SO_4). Winds can carry these particles great distances before they reach the earth as undesirable contaminants. When inhaled into the lungs, high concentrations of sulfur dioxide aggravate respiratory problems, such as asthma, bronchitis, and emphysema. Over the decade of the '80s, sulfur dioxide emissions have decreased approximately 11 percent.

Volatile organic compounds (VOC), or *hydrocarbons*, are individual organic compounds composed of hydrogen and carbon. At room temperature they occur as solids, liquids, and gases. Even though thousands of such compounds are known to exist, methane (which occurs naturally and poses no known dangers to health) is the most abundant. Other volatile organic compounds include benzene, formaldehyde, and chlorofluorocarbons. The EPA estimates that over 18 million metric tons of VOCs are emitted into the air over the United States each year, with about 34 percent of the total coming from vehicles used for transportation and about 50 percent from industrial processes.

Certain VOCs, such as benzene (an industrial solvent) and benzpyrene (a product of burning tobacco and barbecuing), are known to be *carcinogens*—cancer-causing agents. Although many VOCs are not intrinsically harmful, some will react with nitrogen oxides in the presence of sunlight to produce secondary pollutants, which are harmful to human health.

Nitrogen oxides are gases that form when some of the nitrogen in the air reacts with oxygen during the high temperature combustion of fuel. The two primary nitrogen pollutants are *nitrogen dioxide* (NO_2) and *nitric oxide* (NO), which, together, are commonly referred to as NO_x.

Although both nitric oxide and nitrogen dioxide are produced by natural bacterial action, their concentration in urban environments is between 10 and 100 times greater than in nonurban areas. In moist air, nitrogen dioxide reacts with water vapor to form corrosive nitric acid (HNO_3), a substance that adds to the problem of acid rain, which we will address later. In cities such as Los Angeles, concentrations of nitrogen dioxide may be high enough to produce a reddish-brown haze that reduces visibility.

The primary source of nitrogen oxides are motor vehicles, power plants, and waste disposal systems. High concentrations are believed to contribute to heart and lung problems, as well as lower the body's resistance to respiratory infections. Recent studies on test animals suggest that nitrogen oxides may encourage the spread of cancer. Moreover, nitrogen oxides are highly reactive gases that play a key role in producing ozone and other ingredients of photochemical smog.

As we mentioned earlier, the word **smog** originally indicated the combining of smoke and fog. Today, however, the word mainly refers to the type of smog that forms in large cities, such as Los Angeles. Because this type of smog forms when chemical reactions take place in the presence of sunlight (called *photochemical reactions*), it is termed **photochemical smog**, or *Los Angeles-type smog*. When the smog is of the smoky-foggy variety, it is usually called *London-type smog*.

The main component of photochemical smog is the gas **ozone** (O_3). Ozone is a noxious substance with an unpleasant odor that irritates eyes and the mucous membranes of the respiratory system, aggravating chronic diseases, such as asthma and bronchitis. In addition, it retards tree growth, damages crops, and attacks rubber.

---

*Should you become trapped in your car during a snowstorm and you have your engine and heater running to keep warm, roll down the window just a little. This action will allow any carbon monoxide that may have entered the car through leaks in the exhaust system to escape.

In Chapter 1, we learned that ozone forms naturally in the stratosphere through the combining of molecular oxygen and atomic oxygen. There, ozone provides a protective shield against the sun's harmful ultraviolet rays. However, near the surface, in polluted air, ozone is a secondary pollutant that is not emitted directly into the air. Rather, it forms from chemical reactions involving other pollutants, such as nitrogen oxides and volatile organic compounds (hydrocarbons).* Because certain hydrocarbons may react with many other gases, it is believed that chemical reactions involving hydrocarbons disrupt the normal mechanisms for ozone removal, thereby allowing the concentration of ozone to increase. Moreover, because sunlight is required to produce ozone, concentrations of ozone are normally higher during afternoons (Fig. 12.4) and during the summer months, when sunlight is more intense.

The reactions between hydrocarbons and oxygen can produce a substance (known as a *hydrocarbon free radical*) that can further react with oxygen and nitrogen dioxide to produce other undesirable contaminants, such as *PAN* (peroxyacetyl nitrate)—a pollutant that irritates eyes and is extremely harmful to vegetation and organic compounds. Ozone, PAN, and small amounts of other oxidating pollutants form the ingredients of photochemical smog. Instead of being specified individually, these pollutants are sometimes grouped under a single heading called *oxidants*.†

Trends in Air Pollution   Over the past decades, strides have been made in the United States to improve the quality of the air we breathe. Figure 12.5 shows the estimated emission trends over the United States for the primary pollutants. Notice that since the Clean Air Act of 1970, emissions of most pollutants have fallen off substantially, with lead showing the greatest reduction, primarily due to the gradual elimination of leaded gasoline.

Although the situation has improved, we can see from Fig. 12.5 that much more needs to be done, as large quantities of pollutants still spew into our air. In fact, many areas of the country do not conform to the standards for air quality set by the Clean Air Act of

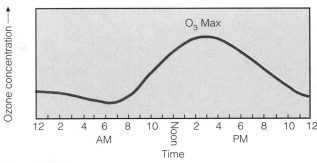

Figure 12.4
Average hourly concentrations of ozone measured at six major cities over a two-year period.

1990. Part of the problem lies in the fact that even with stricter emission laws, increasing numbers of autos (estimates are that 198 million are on the road today) and other sources can overwhelm control efforts.

Clean air standards are established by the Environmental Protection Agency. *Primary ambient* (surrounding) *air quality standards* are set to protect human health, whereas *secondary standards* protect human welfare, as measured by the effects of air pollution on visibility, crops, and buildings. Even with stronger emission laws, estimates are that, in 1992, more than 86 million Americans were breathing air that did not meet at least one of the standards.

To indicate the air quality in a particular region, the EPA developed the **pollutant standards index (PSI)**. The index includes the pollutants carbon monoxide, sulfur dioxide, nitrogen dioxide, particulate matter, and ozone. On any given day, the pollutant measuring the highest value is the one used in the index. The pollutant's measurement is then converted to a number that ranges from 0 to 500 (Table 12.1). When the pollutant's value is the same as the primary ambient air quality standard, the pollutant is assigned a PSI number of 100; when the value is twice the primary standard, its PSI number is 200, and so on. Generally, the public is given the number and the responsible pollutant.

Notice in Table 12.1 that a pollutant is considered unhealthful when its PSI value exceeds 100. When the PSI value is between 51 and 100, the air quality is described as "moderate." Although these levels may not be harmful to humans during a 24-hour period, they may exceed long-term standards. Table 12.1 also shows the health effects and the precautions that should be taken when the PSI value reaches a certain level.

---

*Very simply, during the daylight hours, ultraviolet light from the sun breaks down nitrogen dioxide (NO_2) into nitric oxide (NO) and atomic oxygen (O). The oxygen (O) is then able to combine with molecular oxygen (O_2) to produce ozone (O_3).

†An oxidant is a substance (such as ozone) whose oxygen combines chemically with another substance.

Figure 12.5
Emission estimates of six pollutants in the United States from 1940–1988. (Data courtesy of United States Environmental Protection Agency)

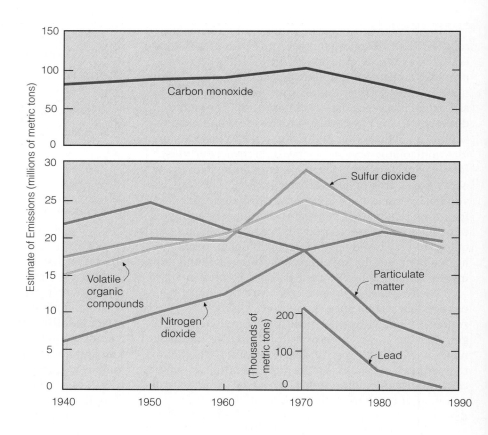

Stronger emission standards, along with cleaner fuels (such as natural gas) have made the air over our large cities cleaner today than it was years ago. For instance, in 1980, Los Angeles had 100 stage 1 smog alert days, whereas in 1991, it had 47. Moreover, total emissions of toxic chemicals spewed into the skies over the United States have been declining steadily since the EPA began its inventory of these chemicals in 1987.

▲▼▲

Air Pollution Weather

If you live in a region that periodically experiences smog, you may have noticed that these episodes often occur with clear skies, light winds, and generally warm, sunny weather. Although this may be "typical" air pollution weather, it by no means represents the only weather conditions necessary to produce high concentrations of pollutants, as we will see in the following sections.

The Role of the Wind   The wind speed plays a role in diluting pollution. When vast quantities of pollutants are spewed into the air, the wind speed determines how quickly the pollutants mix with the surrounding air and, of course, how fast they move away from their source. Strong winds tend to lower the concentration of pollutants by spreading them apart as they move downstream. Moreover, the stronger the wind, the more turbulent the air. Turbulent air produces swirling eddies that dilute the pollutants by mixing them with the cleaner surrounding air. Hence, when the wind dies down, pollutants are not readily dispersed, and they tend to become more concentrated.

The Role of Stability and Inversions   Recall from Chapter 5 that atmospheric stability determines the extent to which air will rise. Remember also that an unstable atmosphere favors vertical air currents, whereas a stable atmosphere strongly resists upward vertical motions. Consequently, smoke emitted into a stable atmosphere tends to spread horizontally, rather than mix vertically.

Table 12.1   The Pollutant Standards Index

PSI Value	Description	General Health Effects	PSI Episode Level	Cautionary Statements
0–50	Good	None		
51–100	Moderate	None		
101–199	Unhealthful	Mild aggravation of symptoms in susceptible persons, with irritation symptoms in the healthy population.		Persons with existing heart or respiratory ailments should reduce physical exertion and outdoor activity.
200–299	Very unhealthful	Significant aggravation of symptoms and decreased exercise tolerance in persons with heart or lung disease, with widespread symptoms in the healthy population.	Stage 1 Health advisory alert	Elderly and persons with existing heart or lung disease should stay indoors and reduce physical activity.
300–399	Hazardous	Premature onset of certain diseases in addition to significant aggravation of symptoms and decreased exercise tolerance in healthy persons.	Stage 2 Health advisory warning	Elderly and persons with existing diseases should stay indoors and avoid physical exertion. General population should avoid outdoor activity.
400–500	Hazardous	Premature death of ill and elderly. Healthy people will experience adverse symptoms that affect their normal activity.	Stage 3 Emergency	All persons should remain indoors, keeping windows and doors closed. All persons should minimize physical exertion and avoid traffic.

The stability of the atmosphere is determined by the way the air temperature changes with height (the lapse rate). When the measured air temperature decreases rapidly as we move up into the atmosphere, the atmosphere is unstable. If, however, the measured air temperature either decreases quite slowly as we ascend, or actually increases with height (remember that this is called an inversion), the atmosphere is stable.*

One type of very stable atmosphere usually exists during the night and early morning hours. If the sky is clear and the winds are light, the air near the ground can become much cooler than the air higher up. Recall that this situation is called a **radiation** (or **surface**) **inversion** (Chapter 3).

Figure 12.6 shows a strong radiation inversion on a clear, calm, winter night. Notice that within the stable inversion, the smoke from the shorter stack does not rise very high, but spreads out, contaminating the area around it. In the relatively unstable air above the inversion, smoke from the taller stack is able to rise and become dispersed. Since radiation inversions are often rather shallow, it should be apparent why taller chimneys have replaced many of the shorter ones (Fig. 12.7). Although these taller stacks do improve the air quality in their immediate area, they may also contribute to the acid rain problem by allowing the pollutants to be swept great distances downwind.

---

*An inversion represents an extremely stable atmosphere where warm air lies above cool air. Any air that attempts to rise into the inversion will, at some point, be cooler and heavier than the warmer air surrounding it. Hence, the inversion acts like a lid on vertical air motions.

Did you know?

In a polluted environment, a volume of air about the size of a sugar cube can contain as many as 200,000 particles.

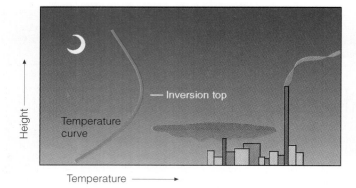

Figure 12.6
The smoke from the shorter stack is trapped within the inversion, while the smoke from the taller stack, above the inversion, rises, mixes, and disperses downwind.

As the sun rises and the surface warms, the radiation inversion normally weakens and disappears before noon. By afternoon, the atmosphere is sufficiently unstable so that, with adequate winds, pollutants are able to disperse vertically. The changing atmospheric stability, from stable in the early morning to unstable in the afternoon, can have a profound effect on the daily concentrations of pollution in certain regions. For example, on a busy city street corner, carbon monoxide levels can be considerably higher in the early morning than in the early afternoon (with the same flow of traffic). In addition, smoke plumes from chimneys will usually change during the course of a day. (Some of these changes are described in the Focus section on p. 305.)

Radiation inversions normally last just a few hours, while **subsidence inversions** may persist for

Figure 12.7
An aerial view looking down on a portion of western Pennsylvania, where a shallow radiation inversion (during the early morning) traps haze, fog, and pollutants close to the ground. The tall chimney, however, extends above the inversion and its smoke rises and disperses vertically.

Focus on an Observation
Smoke Plumes

We know that the stability of the air (especially near the surface) changes during the course of a day. These changes can influence the pollution near the ground as well as the behavior of smoke leaving a chimney. Figure 1 illustrates several different smoke plumes that can develop with adequate wind, but different types of stability.

In diagram A, it is early morning, the winds are light, and a radiation inversion extends from the surface to well above the height of the smoke stack. In this stable environment, there is little up and down motion, so the smoke spreads horizontally. When viewed from above, the smoke plume resembles the shape of a fan. For this reason, it is referred to as a *fanning smoke plume.*

Later in the morning, the surface air warms quickly and becomes unstable as the radiation inversion gradually disappears from the surface upward (diagram B). However, the air above the chimney is still stable, as indicated by the presence of the inversion. Consequently, up and down motions are confined to the region near the surface. Hence, the smoke mixes downwind, increasing the concentration of pollution at the surface— sometimes to dangerously high levels. This effect is called *fumigation.*

If daytime heating of the ground continues, the depth of atmospheric

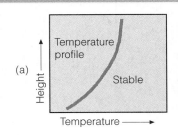

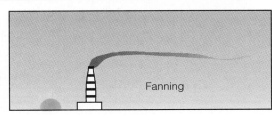

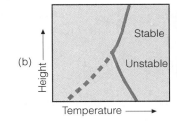

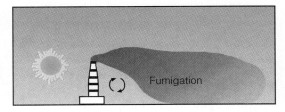

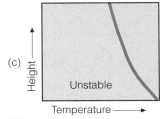

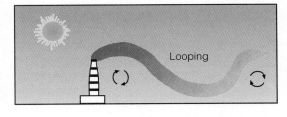

Figure 1
As the radiation inversion disappears during the course of a day (A through C), the pattern of smoke emitted from the stack changes as well.

instability increases. Notice in diagram C that the inversion has completely disappeared. Light-to-moderate winds combine with rising and sinking air to cause the smoke to move up and down in a wavy pattern, producing a *looping smoke plume.*

Thus, watching smoke plumes provides a clue to the stability of the atmosphere, and knowing the stability yields important information about the dispersion of pollutants.

several days or longer. Subsidence inversions, therefore, are the ones commonly associated with major air pollution episodes. They form as the air above a deep anticyclone slowly sinks (subsides) and warms.*

*Remember from Chapter 2 that sinking air always warms because it is being compressed by the surrounding air.

A typical temperature profile of a subsidence inversion is shown in Fig. 12.8. Notice that in the relatively unstable air beneath the inversion, the pollutants are able to mix vertically up to the inversion base. The stable air of the inversion, however, inhibits vertical mixing and acts like a lid on the pollution below,

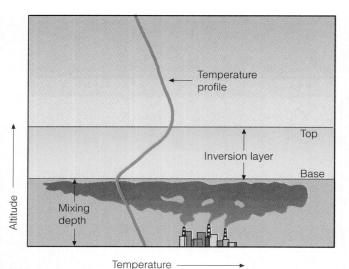

Figure 12.8
The inversion layer prevents pollutants from escaping into the air above it. If the inversion lowers, the mixing depth decreases and the pollutants are concentrated within a smaller volume.

Figure 12.9
A thick layer of polluted air is trapped in the valley. The top of the polluted air marks the base of a subsidence inversion.

preventing it from entering into the inversion. (See Fig. 12.9.)

In Fig. 12.8, the region of relatively unstable (well mixed) air that extends from the surface to the base of the inversion is referred to as the **mixing layer**. The vertical extent of the mixing layer is called the **mixing depth**. Observe that if the inversion rises, the mixing depth increases and the pollutants would be dispersed throughout a greater volume of air; if the inversion lowers, the mixing depth would decrease and the pollutants would become more concentrated, sometimes reaching unhealthy levels. Since the atmosphere tends to be most unstable in the afternoon and most stable in the early morning, we typically find the greatest mixing depth in the afternoon and the most shallow one (if one exists at all) in the early morning.

The position of the semipermanent Pacific high off the coast of California contributes greatly to the air pollution in that region. The Pacific high promotes subsiding air, which warms the air aloft. Surface winds around the high promote upwelling of ocean water (Chapter 7). Upwelling makes the surface water cool, which, in turn, cools the air above. Warm air aloft coupled with cool, surface (marine) air, together produce a strong and persistent subsidence inversion—one that exists 80 to 90 percent of the time over the city of Los Angeles between June and October, the smoggy months. The pollutants trapped within the cool marine air are occasionally swept eastward by a sea breeze. This action carries smog from the coastal regions into the interior valleys. (See Fig. 12.10.)

The Role of Topography   The shape of the landscape (topography) plays an important part in trapping pollutants. We know from Chapter 3 that, at night, cold air tends to drain downhill, where it settles into low-lying basins and valleys. The cold air can have several effects: It can strengthen a pre-existing surface inversion, and it can carry pollutants downhill from the surrounding hillsides (Fig. 12.11).

Valleys prone to pollution are those completely encased by mountains and hills. The surrounding mountains tend to block the prevailing wind. With light winds, and a shallow mixing layer, the poorly ventilated cold valley air can only slosh back and forth like a murky bowl of soup.

Air pollution concentrations in mountain valleys tend to be greatest during the colder months. During the warmer months, daytime heating can warm the

Figure 12.10
The leading edge of cool, marine air carries pollutants into Riverside, California.

sides of the valley to the point that upslope valley winds vent the pollutants upward, like a chimney. Valleys susceptible to stagnant air exist in just about all mountainous regions.

The pollution problem in several large cities is, at least, partly due to topography. For example, the city of Los Angeles is surrounded on three sides by hills and mountains. Cool marine air from off the ocean moves inland and pushes against the hills, which tend to block the air's eastward progress. Unable to rise, the cool air settles in the basin, trapping pollutants from industry and millions of autos. Baked by sunlight, the pollutants become the infamous photochemical smog. By the same token, the "mile high" city of Denver, Colorado, sits in a broad shallow basin that frequently traps both cold air and pollutants.

Severe Air Pollution Potential   The greatest potential for an episode of severe air pollution occurs

Figure 12.11
At night, cold air and pollutants drain downhill and settle in low-lying valleys.

when all of the factors mentioned in the previous sections come together simultaneously.

Following are the ingredients for a major buildup of atmospheric pollution that can change an otherwise clean atmosphere into one that resembles smog soup:

1. lots of sources of air pollution (preferably clustered close together)
2. a deep high-pressure area that becomes stationary over a region
3. light surface winds that are unable to disperse the pollutants
4. a strong subsidence inversion produced by the sinking of air aloft
5. a shallow mixing layer with poor ventilation
6. a valley where the pollutants can accumulate
7. clear skies so that radiational cooling at night will produce a surface inversion, which can cause an even greater buildup of pollutants near the ground
8. and for photochemical smog, adequate sunlight to produce secondary pollutants, such as ozone

When these conditions prevail, some of the worst air pollution disasters on record have occurred. (The Focus section on p. 309 details the 1948 Donora disaster where 17 people died within 14 hours.)

▲▽▲

Air Pollution and the Urban Environment

For more than 100 years, it has been known that cities are generally warmer than surrounding rural areas. This region of city warmth, known as the **urban heat island**, can influence the concentration of air pollution. However, before we look at its influence, let's see how the heat island actually forms.

The urban heat island is due to industrial and urban development. In rural areas, a large part of the incoming solar energy is used to evaporate water from vegetation and soil. In cities, where less vegetation and exposed soil exists, the majority of the sun's energy is absorbed by urban structures and asphalt. Hence, during warm daylight hours, less evaporative cooling in cities allows surface temperatures to rise higher than in rural areas. Additional city heat is given off by vehicles and factories, as well as by industrial and domestic heating and cooling units.

At night, the solar energy (stored as vast quantities of heat in city buildings and roads) is slowly released into the city air. The release of heat energy is retarded even more by the tall vertical city walls that do not allow infrared radiation to escape as readily as do the relatively level surfaces of the surrounding countryside. The slow release of heat tends to keep city temperatures higher than those of the faster cooling rural areas.

On clear, still nights when the heat island is pronounced, a small thermal low-pressure area forms over the city. Sometimes a light breeze—called a **country breeze**—blows from the countryside into the city. If there are major industrial areas along the city's outskirts, pollutants are carried into the heart of town, where they tend to concentrate (Fig. 12.12).

At night, the extra warmth of the city occasionally produces a shallow unstable layer near the surface. Pollutants emitted from low-level sources, such as home heating units, tend to concentrate in this shallow layer, often making it unhealthy to breathe.

Figure 12.12
On a clear, relatively calm night, a weak country breeze carries pollutants from the outskirts into the city, where they concentrate and rise due to the warmth of the city's urban heat island. This effect may produce a pollution (or dust) dome from the suburb to the center of town.

Focus on an Observation
Five Days in Donora—An Air Pollution Episode

On Tuesday morning, October 26, 1948, a cold surface anticyclone moved over the eastern half of the United States. There was nothing unusual about this high-pressure area; with a central pressure of only 1025 millibars (30.27 inches), it was not exceptionally strong (Fig. 2). Aloft, however, a large blocking-type ridge formed over the region, and the jet stream, which moves the surface pressure features along, was far to the west. Consequently, the surface anticyclone became entrenched over Pennsylvania and remained nearly stationary for five days.

The widely spaced isobars around the high-pressure system produced a weak pressure gradient and generally light winds throughout the area. These light winds, coupled with the gradual sinking of air from aloft, set the stage for a disastrous air pollution episode.

On Tuesday morning, radiation fog gradually settled over the moist ground in Donora, a small town nestled in the Monongahela Valley of western Pennsylvania. Because Donora rests on bottom land, surrounded by rolling hills, its residents were accustomed to fog, but not what was to follow.

The strong radiational cooling that formed the fog, along with the sinking air of the anticyclone, combined to produce a strong temperature inversion. Light, downslope winds spread cool air and contaminants over Donora from the community's steel mill, zinc smelter, and sulfuric acid plant.

The fog with its burden of pollutants lingered into Wednesday. Cool drainage winds during the night strengthened the inversion and added more effluents to the already filthy air. The dense fog layer blocked sunlight from reaching the ground. With essentially no surface heating, the mixing

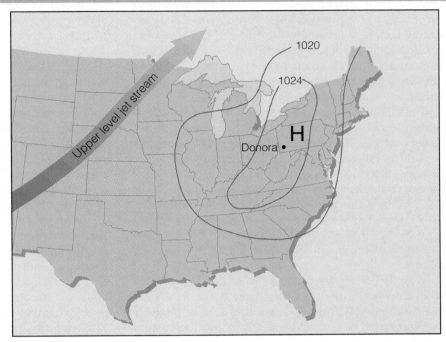

Figure 2
Surface weather map that shows a stagnant anticyclone over the eastern United States on October 26, 1948. The heavy arrow represents the position of the jet stream.

depth lowered and the pollution became more concentrated. Unable to mix and disperse both horizontally and vertically, the dirty air became confined to a shallow, stagnant layer.

Meanwhile, the factories continued to belch impurities into the air (primarily sulfur dioxide and particulate matter) from stacks no higher than 130 feet tall. The fog gradually thickened into a moist clot of smoke and water droplets. By Thursday, the visibility had decreased to the point where one could barely see across the street. At the same time, the air had a penetrating, almost sickening, smell of sulfur dioxide. At this point, a large percentage of the population became ill.

The episode reached a climax on Saturday, as seventeen deaths were reported. As the death rate mounted, alarm swept through the town. An emergency meeting was called between city officials and factory representatives to see what could be done to cut down on the emission of pollutants.

The light winds and unbreathable air persisted until, on Sunday, an approaching storm generated enough wind to vertically mix the air and disperse the pollutants. A welcome rain then cleaned the air further. All told, the episode had claimed the lives of 22 people. During the five-day period, about half of the area's 14,000 inhabitants experienced some ill effects from the pollution. Most of those affected were older people with a history of cardiac or respiratory disorders.

The constant outpouring of pollutants into the environment may actually influence the climate of a city. For example, certain pollutants reflect solar energy, thereby reducing the sunlight that reaches the surface. Some particles serve as nuclei upon which water and ice form. Water vapor condenses onto these particles, forming haze that greatly reduces visibility. Moreover, the added nuclei increases the frequency of city fog. (The impact pollutants may have on a larger scale is exceedingly complex and depends upon a number of factors, which are addressed in Chapter 14.)

Pollutants from urban areas may even affect the weather downwind from them. In a controversial study conducted at La Porte, Indiana—a city located about thirty miles downwind of the industries of south Chicago—scientists suggested that La Porte had experienced a notable increase in annual precipitation since 1925. Because this rise closely followed the increase in steel production, it was proposed that the phenomenon was due to the additional emission of particles or moisture (or both) by industries to the west of La Porte.

A study conducted in St. Louis, Missouri (the Metropolitan Meteorological Experiment, or *METROMEX*), indicated that the average annual precipitation downwind from this city increased by about 10 percent. These increases closely followed industrial development upwind. This study also demonstrated that precipitation amounts were significantly greater on weekdays (when pollution emissions were higher) than on weekends (when pollution emissions were lower). Corroborative findings have been reported for Paris, France, and for other cities as well.

Acid Deposition

Air pollution emitted from industrial areas, especially products of combustion, such as oxides of sulfur and nitrogen, can be carried many miles downwind. Either these particles and gases slowly settle to the ground in dry form (*dry deposition*) or they are removed from the air during the formation of cloud particles and then carried to the ground in rain and snow (*wet deposition*). **Acid rain** and *acid precipitation* are common terms used to describe wet deposition, while **acid deposition** encompasses both dry and wet acidic substances. How, then, do these substances become acidic?

Emissions of sulfur dioxide (SO_2) and oxides of nitrogen may settle on the local landscape, where they transform into acids as they interact with water, especially during the formation of dew or frost. The remaining airborne particles may transform into tiny dilute drops of sulfuric acid (H_2SO_4) and nitric acid (HNO_3) during a complex series of chemical reactions involving sunlight, water vapor, and other gases. These acid particles may then fall slowly to earth, or they may adhere to cloud droplets or to fog droplets, producing **acid fog**. They may even act as nuclei on which the cloud droplets begin to grow. When precipitation occurs in the cloud, it carries the acids to the ground. Because of this, precipitation is becoming increasingly acidic in many parts of the world, especially downwind of major industrial areas.

Airborne studies conducted during the middle 1980s revealed that high concentrations of acid-rain-producing pollutants can be carried great distances from their sources. In 1986, for example, scientists discovered high concentrations of pollutants hundreds of miles off the east coast of North America. It is suspected that they came from industrial East Coast cities. Although most pollutants are washed from the atmosphere during storms, some may be swept over the Atlantic, reaching places like Bermuda and Ireland. Acid rain knows no national boundaries.

Although recent studies suggest that acid precipitation may be nearly worldwide in distribution, regions noticeably affected are eastern North America, central Europe, and Scandinavia. Sweden contends that most of the sulfur emissions responsible for its acid precipitation are coming from factories in England. In some places, acid precipitation occurs naturally, such as in northern Canada, where natural fires in exposed coal beds produce tremendous quantities of sulfur dioxide.

High concentrations of acid deposition can damage plants and water resources (freshwater ecosystems seem to be particularly sensitive to changes in acidity). Concern centers chiefly on areas where interactions with the soil are unable to neutralize the acidic inputs. Studies indicate that thousands of lakes in the United States and Canada are so acidified that entire fish populations may have been adversely affected. In an attempt to reduce acidity, lime is being poured into some lakes.

Figure 12.13
Exposure to air pollution and an extensive drought have weakened many of these trees in the northern Sierra Nevada, making them susceptible to disease and insect infestation.

About a third of the trees in Germany show signs of a blight that is due, in part, to acid deposition. Apparently, acidic particles raining down on the forest floor for decades cause a chemical imbalance in the soil that, in turn, causes serious deficiencies in certain elements necessary for the tree's growth. The trees are thus weakened and become susceptible to insects and drought. The same type of processes may be affecting North American forests, but at a much slower pace (Fig. 12.13).

Also, acid deposition is eroding the foundations of structures in many cities throughout the world. In Rome, the acidity of rainfall is beginning to disfigure priceless outdoor fountain sculptures and statues. The estimated annual cost of this damage to building surfaces, monuments, and other structures is more than $2 billion.

Precipitation is naturally somewhat acidic. The carbon dioxide occurring naturally in the air dissolves in precipitation, making it slightly acidic with a pH* between 5.0 and 5.6. Consequently, precipitation is considered acidic when its pH is below about 5.0. In the northeastern United States, where emissions of sulfur dioxide are primarily responsible for the acid precipitation, typical pH values range between 4.0 and 4.5. (See Fig. 12.14.) But acid precipitation is not confined to the Northeast; the acidity of precipitation has increased rapidly during the past twenty years in the southeastern states, too. Further west, rainfall acidity

*The pH scale ranges from 0 to 14, with a value of 7 considered neutral. Values greater than 7 are alkaline and below 7 are acidic. The scale is logarithmic, which means that rain with pH 3 is 10 times more acidic than rain with pH 4 and 100 times more acidic than rain with pH 5.

also appears to be on the increase. Along the West Coast, the main cause of acid deposition appears to be the oxides of nitrogen released in automobile exhaust. Acid fog, with a pH of 1.7 in the Los Angeles area, has actually been more acidic than lemon juice.

Some scientists believe that if the United States turns more to coal-fired power plants, which are among the leading sources of sulfur oxide emissions, the acid deposition problem will worsen. In fact, there are experts who feel that the acid rain dilemma may be one of the greatest environmental problems facing the world in the near future.

In an attempt to better understand acid deposition, the National Center for Atmospheric Research (NCAR) and the Environmental Protection Agency have been working to develop computer models that better describe the many physical and chemical processes contributing to acid deposition. To deal with the acid deposition problem, the Clean Air Act of 1990 imposed a reduction in the United States' emissions of sulfur

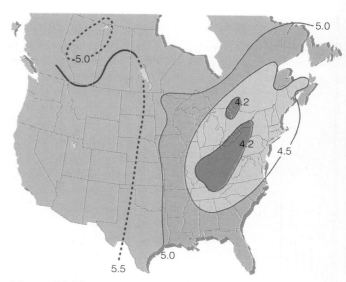

Figure 12.14
Annual average value of pH in precipitation weighted by the amount of precipitation in the United States and Canada for 1980.

dioxide and nitrogen dioxide. Canada has recently imposed new pollution control standards and set a goal of reducing industrial air pollution by 50 percent.

Summary

In this chapter, we found that air pollution has plagued humanity for centuries. Air pollution problems began when people tried to keep warm by burning wood and coal. These problems worsened during the industrial revolution as coal became the primary fuel for both homes and industry. Even though many American cities do not meet all of the air quality standards set by the federal Clean Air Act of 1990, the air over our large cities is cleaner today than it was 50 years ago due to stricter emission standards and cleaner fuels.

We examined the types and sources of air pollution and found that primary air pollutants enter the atmosphere directly, whereas secondary pollutants form by chemical reactions that involve other pollutants. The secondary pollutant ozone is the main ingredient of photochemical smog—a smog that irritates the eyes and forms in the presence of sunlight.

We looked at air pollution weather and found that most air pollution episodes occur when the winds are light, the mixing layer is shallow, the air is stable, and a strong inversion exists, as a high pressure area stalls over a region.

We observed that urban environments tend to be warmer and more polluted than the rural areas that surround them. We saw that pollution from industrial areas can modify environments downwind of them, as pollution that is swept into the air may transform into acid precipitation. Acid deposition knows no national boundaries—the pollution of one country becomes the acid rain of another.

Key Terms

The following terms are listed in the order they appear in the text. Define each.
Doing so will aid you in reviewing the material covered in this chapter.

air pollutants
primary air pollutants
secondary air pollutants
particulate matter
carbon monoxide (CO)
sulfur dioxide (SO₂)

volatile organic
   compounds (VOC)
nitrogen oxides
smog
photochemical smog
ozone

pollutant standards
   index (PSI)
radiation (surface)
   inversion
subsidence inversion
mixing layer

mixing depth
urban heat island
country breeze
acid rain
acid deposition
acid fog

Review Questions

1. What are some of the main sources of air pollution?
2. How do primary air pollutants differ from secondary air pollutants?
3. List a few of the substances that fall under the category of particulate matter.
4. Why do the very fine particles of particulate pollution pose the greatest risk to human health?
5. List two ways particulate pollution is removed from the atmosphere.
6. Describe the primary sources and some of the health problems associated with each of the following pollutants:
   (a) carbon monoxide (CO)
   (b) sulfur dioxide (SO₂)
   (c) volatile organic compounds (VOC)
   (d) nitrogen oxides
   (e) ozone
7. Describe how London-type smog differs from Los Angeles-type smog.
8. (a) What is photochemical smog and how does it form?
   (b) What is the main component of photochemical smog?
   (c) Why is photochemical smog more prevalent during the summer and early fall?
9. Why do many of the pollutants in Fig. 12.5 show a dramatic drop after 1970?
10. (a) How is the PSI scale a measure of air quality?
   (b) On the PSI scale, how would air be described if it had a PSI value of 250 for ozone?
   (c) What would be the general health effects, and what precautions should a person take, with a PSI value of 250 for ozone?

11. Explain why light winds are more conducive to air pollution conditions than strong winds.
12. How is an accumulation of air pollutants related to the stability of the atmosphere?
13. Why is it that polluted air and inversions seem to go hand in hand?
14. Major air pollution episodes are mainly associated with radiation inversions or subsidence inversions? Why?
15. Give two reasons why taller smoke stacks are replacing shorter ones.
16. Explain how the mixing depth changes during the course of a day.
17. For least-polluted conditions, what would be the best time of day for a farmer to burn agricultural debris? Explain your reasoning.
18. Explain why most severe episodes of air pollution are associated with high pressure areas.
19. How does topography influence the concentration of pollutants in cities such as Los Angeles and Denver? In mountainous terrain?
20. What are the main causes of the urban heat island?
21. How can a country breeze make the heart of a large city more polluted?
22. How can pollution play a role in influencing the precipitation downwind of certain large industrial complexes?
23. Why is acid deposition considered a serious problem in many regions of the world? How does precipitation become acidic?

Ten thousand years ago, additional water from precipitation and melting glaciers formed giant Lake Bonneville—a lake that covered over nineteen thousand square miles of northwestern Utah. Since then, the climate has changed. Evaporation in this semi-arid region has gradually lowered the lake until, today, most of the landscape is a flat, salt-covered bed that occasionally becomes dotted with puddles after a summer rain shower. (Photo by author)

Chapter 13

Climate Change

Contents

▲▼▲

A change in our climate however is taking place very sensibly. Both heats and colds are becoming much more moderate within the memory even of the middle-aged. Snows are less frequent and less deep. They do not often lie, below the mountains, more than one, two, or three days, and very rarely a week. They are remembered to have been formerly frequent, deep, and of long continuance. The elderly inform me the earth used to be covered with snow about three months in every year. The rivers, which then seldom failed to freeze over in the course of the winter, scarcely ever do now. This change has produced an unfortunate fluctuation between heat and cold, in the spring of the year, which is very fatal to fruits. In an interval of twenty-eight years, there was no instance of fruit killed by the frost in the neighborhood of Monticello. The accumulated snows of the winter remaining to be dissolved all together in the spring, produced those overflowings of our rivers, so frequent then, and so rare now.

Thomas Jefferson, *Notes on the State of Virginia*, 1781   ▲

13

The climate is always changing. Evidence shows that climate has changed in the past and nothing suggests that it will not continue to do so. Some climate changes are small and go practically unnoticed from one year to the next. Others, such as a persistent drought or a delay in the annual monsoon rains, can adversely affect the lives of millions. Even small changes can have an adverse effect when averaged over many years, as grasslands once used for grazing gradually become uninhabited deserts. In this chapter, we will investigate not only how the global climate has changed, but also some of the theories as to why it has changed.

▲▼▲

The Earth's Changing Climate

Not only is the earth's climate always changing, but a mere 18,000 years ago the earth was in the grip of a cold spell, with *alpine glaciers* extending their ice fingers down river valleys and huge ice sheets (*continental glaciers*) covering vast areas of North America and Europe. The ice measured thousands of feet thick and extended as far south as New York and the Ohio River Valley. Perhaps the glaciers advanced ten times during the last two million years, only to retreat. In the warmer periods, between glacier advances, average global tem-

peratures were slightly higher than at present. Hence, some scientists believe we are still in an ice age, but in the comparatively warmer part of it.

Presently, glaciers cover only about 10 percent of the earth's land surface. The total volume of ice over the face of the earth amounts to a little more than 25 million cubic kilometers. If global temperatures were to rise enough so that all of this ice melted, the level of the ocean would rise about 65 meters (213 feet). (See Fig. 13.1.) Imagine the catastrophic results: Many major cities (such as New York, Tokyo, and London) would be inundated. Even a rise in global temperature of several degrees Celsius might be enough to raise sea level by about half a meter or so, flooding coastal lowlands.

Did you know?
A 3-foot rise in average sea level would push shorelines back about 100 feet.

How has the climate changed, and what evidence is there to suggest that indeed it has changed? These are some of the questions we will address in the next several sections.

Determining Past Climates   The study of the geological evidence left behind by advancing and retreating glaciers is one factor suggesting that global climate

Figure 13.1
If all the ice locked up in glaciers were to melt, estimates are that this coastal area of North America would be under 65 meters (or about 200 feet) of water. Even a relatively small 1-meter (3-foot) rise in sea level would threaten half of the world's population with rising seas.

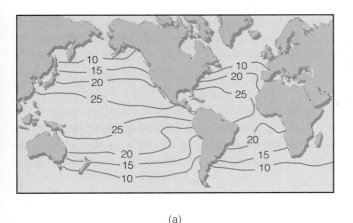

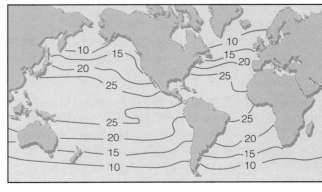

(a)   (b)

Figure 13.2
(a) Sea surface isotherms (°C) during August 18,000 years ago and (b) during August today.

has undergone slow but continuous changes. To reconstruct past climates, scientists must examine and then carefully piece together all the available evidence. Unfortunately, the evidence only gives a general understanding of what the climate was like. For example, fossil pollen of a tundra plant collected in a layer of sediment in New England and dated to be 12,000 years old suggests that the climate of that region was much colder than it is today.

Other biological evidence of climatic change comes from the study of annual growth rings of trees, called **dendrochronology**. As a tree grows, it produces a layer of wood cells under its bark. When the tree is cut down, each year's growth appears as a ring. The changes in thickness of the rings indicate climatic changes that may have taken place from one year to the next. The growth of tree rings has been correlated with precipitation and temperature patterns for hundreds of years into the past in various regions of the world.

Still other evidence of global climatic change comes from core samples taken from ocean floor sediments and ice from Greenland. A multiuniversity research project known as CLIMAP (*C*limate: *l*ong-range *i*nvestigation *m*apping *a*nd *p*rediction) studied the past million years of global climate. Thousands of feet of ocean sediment obtained with a hollow-centered drill were analyzed. The sediment contains the remains of calcium carbonate shells of organisms that once lived near the surface. Because certain organisms live within a narrow range of temperature, the distribution and

type of organisms within the sediment indicate the surface water temperature.

In addition, the oxygen-isotope* ratio of these shells provide information about the sequence of glacier advances. For example, most of the oxygen in sea water is composed of 8 protons and 8 neutrons in its nucleus, giving it an atomic weight of 16. However, about one out of every thousand oxygen atoms contains an extra 2 neutrons, giving it an atomic weight of 18. When ocean water evaporates, the heavy oxygen 18 tends to be left behind. Consequently, during periods of glacier advance, more oxygen 16 is stored on land as ice, and the oceans, which contain less water, have a higher concentration of oxygen 18. Since the shells of marine organisms are constructed from the oxygen atoms existing in ocean water, determining the ratio of oxygen 18 to oxygen 16 within these shells yields information about how the climate may have altered in the past. A higher ratio of oxygen 18 to oxygen 16 in the ocean sediment record suggests a colder climate, while a lower ratio suggests a warmer climate. Using data such as these, the CLIMAP project was able to reconstruct the earth's surface ocean temperature for various times during the past (Fig. 13.2).

Vertical ice cores extracted from glaciers in Antarctica and Greenland provide additional information on past temperature patterns. Glaciers form over

---

*Isotopes are atoms whose nuclei have the same number of protons but different numbers of neutrons.

land where temperatures are sufficiently low so that,
during the course of a year, more snow falls than will
melt. Successive snow accumulations over many years
compact the snow, which slowly recrystallizes into ice.
Under the influence of gravity, the ice begins to move,
and a glacier is born. Since ice is composed of hydrogen and oxygen, examining the oxygen-isotope ratio in
ancient cores provides a past record of temperature
trends. Generally, the colder the air when the snow fell,
the richer the concentration of oxygen 16 in the core.

Other data have been used to reconstruct past climates, such as:

1. analysis of sulfuric acid in ice cores as well as
   trapped air bubbles containing CO_2, and methane
2. natural records of lake-bottom sediment and soil deposits
3. the study of pollen in deep ice caves, soil deposits,
   and sea sediments
4. certain geologic evidence (ancient coal beds, sand
   dunes, and fossils)
5. documents concerning droughts, floods, and crop
   yields

Despite all of these data, our knowledge about
past climates is still incomplete. Now that we have reviewed *how* the climatologist gains information about
the past, let's look at *what* this information reveals.

Climate Through the Ages   Throughout much of
the earth's history, the global climate was probably between 8°C and 15°C (14°F and 27°F) warmer than it is
today. During most of this time, the polar regions were
free of ice. These comparatively warm conditions, however, were interrupted by several periods of glaciation.
Geologic evidence suggests that one glacial period occurred about 700 million years ago (m.y.a.) and another
about 300 m.y.a. The most recent one—the *Pleistocene
epoch* or, simply, the **Ice Age**—began about 2 m.y.a.
Let's summarize the climatic conditions that led up to
the Pleistocene.

About 65 m.y.a., the earth was warmer than it is
now; polar ice caps did not exist. Beginning about 55

m.y.a., the earth entered a long cooling trend. After millions of years, polar ice appeared. As average temperatures continued to lower, the ice grew thicker, and by
about 10 m.y.a. a deep blanket of ice covered the
Antarctic. Meanwhile, snow and ice began to accumulate in high mountain valleys of the Northern Hemisphere, and alpine, or valley, glaciers soon appeared.

About 2 m.y.a., continental glaciers appeared in
the Northern Hemisphere, marking the beginning of the
Pleistocene epoch. The Pleistocene, however, was not
a period of continuous glaciation but a time when glaciers alternately advanced and retreated (melted back)
over large portions of North America and Europe. Between the glacial advances were warmer periods called
interglacial periods, which lasted for 10,000 years or
more.

The most recent North American glaciers reached
their maximum thickness and extent about 18,000–
22,000 years ago (y.a.). At that time, the sea level was
perhaps 85 meters (280 feet) lower than it is now. The
lower sea level exposed vast areas of land, such as the
Bering land bridge (a strip of land that connected
Siberia to Alaska), which allowed human migration
from Asia to North America.

The ice began to retreat about 14,000 y.a., and by
about 8000 y.a. the continental ice sheets over North
America were gone. From about 7000–5000 y.a., the climate was probably about 1°C warmer than at present.
Because this period favored the development of plants,
it is called the **climatic optimum**. Then, about 5000
y.a., a cooling trend set in, during which extensive
alpine glaciers returned, but not continental ice sheets.

Climate During the Last 1000 Years   About 1000
y.a., the Northern Hemisphere was relatively warm and
dry. During this time, vineyards flourished and wine
was produced in England, indicating warm, dry summers and the absence of cold springs. It was during this
tranquil period of several hundred years that the
Vikings colonized Iceland and Greenland.

Sometime around A.D. 1200, the mild climate of
western Europe began to show extreme variations. For
several hundred years the climate grew stormy. Both
great floods and great droughts occurred. Extremely
cold winters were followed by relatively warm ones.
During the cold spells, the English vineyards and the
Viking settlements suffered. Europe experienced several famines during the 1300s.

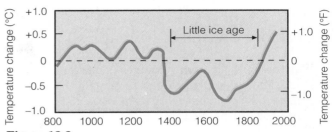

Figure 13.3
The average temperature variations of eastern Europe for about the last 1200 years.

Around 1400 to 1550, the climate moderated (Fig. 13.3). However, starting in the middle 1550s, the average temperature began to drop. This cooling trend (which continued for almost 300 years) is known as the **Little Ice Age**. During this time, alpine glaciers increased in size and advanced down river canyons. Winters were long and severe, summers short and wet. The vineyards in England vanished and farming became impossible in the more northerly latitudes. Cut off from the rest of the world by an advancing ice pack, the Viking colony in Greenland perished.

During the Little Ice Age, one particular year stands out: 1816. In Europe that year, bad weather contributed to a poor wheat crop, and famine spread across the land. In North America, unusual blasts of cold arctic air moved through Canada and the northeastern United States between May and September. The cold spells brought heavy snow in June and killing frosts in July and August. In the warmer days that followed each cold snap, farmers replanted, only to have another cold outbreak damage the planting. The year 1816 has come to be known as "the year without a summer" or "eighteen hundred and froze-to-death." The unusually cold summer was followed by a bitterly cold winter.

In the late 1800s, the average temperature in the Northern Hemisphere began to rise. (See Fig. 13.4.) From about 1900 until about 1940, the average temperature of the lower atmosphere rose nearly 0.5°C. Following the warmer period, the Northern Hemisphere began to cool slightly over the next twenty-five years or so. In the late 1960s and early 1970s, the cooling trend ended over most of the Northern Hemisphere. During the 1970s and into the 1980s, the average yearly temperature showed considerable fluctuation from year to year and from region to region, with the overall trend pointing to warming. The decade of the 1980s has been the warmest of this century so far. Several studies, in fact, indicate that the earth's surface may have warmed by about 0.6°C (about 1°F) over the last 100 years.

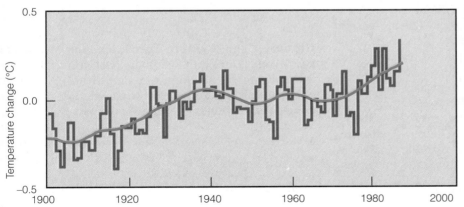

Year

Figure 13.4
Changes in the average global surface air temperature from 1900 to 1987. The zero line represents the average surface air temperature from 1950 to 1979. [From P. D. Jones et al., "Evidence for global warming in the past decade," *Nature* (28 April 1988) 332: 790. Reprinted with permission.]

▲▼▲

Possible Causes of Climatic Change

Why the earth's climate changes is not totally under-stood. Many theories attempt to explain the changing climate, but no single theory alone can satisfactorily account for *all* the climatic variations of the geologic past.

Why hasn't the riddle of a fluctuating climate been solved? One major problem facing any comprehensive theory is the intricate interrelationship of the elements involved. For example, if temperature changes, many other elements may be altered as well. The interactions among the atmosphere, the oceans, and the ice are extremely complex and the number of possible interactions among these systems is enormous. No climatic element within the system is isolated from the others. With this in mind, we will first investigate a warmer and a cooler world to see how feedback systems work; then we will consider some of the current theories of climatic change.

Climate Change and Feedback Mechanisms   In Chapter 2, we learned that the earth-atmosphere system is in a delicate balance between incoming and outgoing energy. If this balance is upset, even slightly, global climate can undergo a series of complicated changes.

Let's assume that the earth-atmosphere system has been disturbed to the point that the earth has entered a slow warming trend. (At this point we need not worry about why—that comes later.) Over the years the temperature slowly rises, and water from the oceans rapidly evaporates into the warmer air. The increased quantity of water vapor absorbs more of the earth's infrared energy, thus strengthening the atmospheric greenhouse effect.* This raises the air temperature even more, which, in turn, further increases the evaporation rate. The greenhouse effect becomes even stronger and the air temperature rises even more. If left unchecked, the temperature would increase until the oceans boiled away. Such a chain reaction is called a *runaway greenhouse effect*. This situation is known as

a **positive feedback mechanism** because the initial increase in temperature is reinforced by the other processes that occur. (Additional information on a runaway greenhouse effect on Venus is given in the Focus section on p. 321.)

The earth-atmosphere system has a number of checks and balances that help it readjust itself into a new equilibrium. Hence, a runaway greenhouse effect is not very likely on earth. Helping to counteract the positive feedback mechanisms are **negative feedback mechanisms**—those that tend to weaken the interactions among the variables rather than reinforce them. Let's look at an example of how a negative feedback mechanism might work on a warming planet. As the air warms and becomes more moist, there is a greater likelihood of convection and, perhaps, an increase in global cloudiness. A greater cloud cover reflects a larger percentage of incoming sunlight. Because there is less solar energy to warm the ground, the warming is slowed, or perhaps even reversed.

How do feedback systems affect a slow global cooling trend over hundreds or even thousands of years? Lower temperatures would allow for a greater snow cover in middle and high latitudes. Because snow has a high albedo (reflectivity), much of the incident sunlight would be reflected back to space. Less sunlight absorbed at the surface would cause a further drop in temperature. This, in turn, would increase the snow cover, which would lower the temperature even more. This positive feedback mechanism, if left unchecked, would produce a *runaway ice age*. A runaway ice age is not likely because the earth-atmosphere system would respond to the cooling with other feedback mechanisms.

Climate Change, Plate Tectonics, and Mountain Building   During the geologic past, the earth's surface has undergone extensive modifications. One involves the slow shifting of the continents and the ocean floors. This motion is explained in the widely acclaimed

---

*The earth's atmospheric greenhouse effect is due mainly to the absorption and re-emission of infrared radiation by gases such as water vapor and CO_2. Refer back to Chapter 2 for additional information on this topic.

Did you know?

If all the snow that normally falls over central Canada were to stay on the ground and not melt (even during the summer), it would take nearly 3000 years to build an ice sheet comparable to the one that existed there 18,000 years ago.

Focus on a Special Topic
A Runaway Greenhouse Effect on Venus

Our closest planetary neighbor, Venus, is about the same size as Earth. Venus is slightly closer to the sun, so compared to Earth, its average surface temperature should be slightly warmer. However, observations reveal that the surface temperature of Venus is not slightly warmer—it is scorching hot, averaging about 480°C (900°F). The cause for these high temperatures is a positive feedback mechanism that some scientists refer to as a *runaway greenhouse effect.*

Unlike Earth, the atmosphere of Venus is almost entirely CO_2 with minor amounts of other gases such as water vapor, sulfur dioxide, and nitrogen. The CO_2 probably originated in much the same way as it did in the Earth's early atmosphere—through volcanic outgassing of CO_2, water vapor, and hydrogen compounds from the planet's hot interior. As the Earth's atmosphere cooled, however, its water vapor condensed into clouds that produced vast amounts of liquid water, which filled the basins to form the seas. Much of the CO_2 dissolved in the ocean water, and through chemi-

Figure 1
A runaway greenhouse effect on Venus has left this planet with a mean surface temperature of 480°C (900°F).

cal and biological processes became carbonate rocks. Plants evolved that further removed CO_2 and, during photosynthesis, enriched the Earth's atmosphere with oxygen.

On Venus, the story is different. Being closer to the sun, Venus was warmer. In the warmer air, the water vapor probably did not condense, but remained as a vapor to enhance the

greenhouse effect. The lack of oceans on Venus meant that its CO_2 was to remain in its atmosphere. Infrared energy from the surface tried to penetrate the Venusian atmosphere, but it was absorbed and reradiated back. Volcanoes continued to spew CO_2 and water vapor into the atmosphere. More CO_2 meant more warming and the runaway positive feedback mechanism was underway.*

Eventually, the outgoing energy from the surface balanced the incoming energy (mainly from the atmosphere), but not until the average surface temperature reached an unbearable 480°C. Although scientists feel that this type of runaway greenhouse effect will not occur on earth, it is something worth pondering as we continue to release greenhouse gases into our atmosphere.

---

*On Venus, at some point, energetic rays from the sun probably separated the water vapor into hydrogen and oxygen. The lighter hydrogen more than likely escaped from the hot atmosphere while the heavier oxygen became trapped in surface rocks and minerals.

---

theory of plate tectonics (formerly called the *theory of continental drift*). According to this theory, the earth's outer shell is composed of huge plates that fit together like pieces of a jigsaw puzzle. The plates, which slide over a partially molten zone below them, move in relation to one another. Continents are embedded in the plates and move along like luggage riding piggyback on a conveyer belt. The rate of motion is extremely slow, only a few centimeters per year.

Besides providing insights into many geological processes, plate tectonics also helps to explain past climates. For example, we find glacial features near sea level in Africa today. This fact suggests that the area underwent a period of glaciation hundreds of millions of years ago. Were temperatures at low elevations near the equator ever cold enough to produce ice sheets? Probably not. The ice sheets formed when this land mass was located at a much higher latitude. Over the

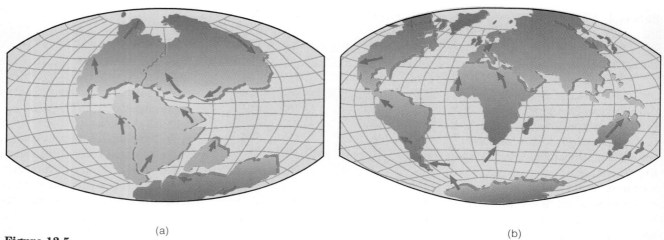

(a)   (b)

Figure 13.5
Geographical distribution of (a) land masses about 180 million years ago, and (b) today.
Arrows show the relative direction of continental movement.

many millions of years since then, the land has slowly moved to its present position. Along the same line, we can see how the fossil remains of tropical vegetation can be found under layers of ice in polar regions today.

According to plate tectonics, the now existing continents were at one time joined together in a single huge continent, which broke apart. Its pieces slowly moved across the face of the earth, thus changing the distribution of continents and ocean basins (Fig. 13.5). Some scientists feel that, when land masses are concentrated in middle and high latitudes, ice sheets are more likely to form. During these times, there is a greater likelihood that more sunlight will be reflected back into space and that the positive feedback mechanism mentioned earlier will develop.

The various arrangements of the continents may also influence the path of ocean currents. This would alter the transport of heat from low to high latitudes and change both the global wind system and the climate in middle and high latitudes. As an example, suppose that plate movement "pinches off" a rather large body of high latitude ocean water such that the transport of warm water into the region is cut off. In winter, the surface water would eventually freeze over with ice. This freezing would, in turn, reduce the amount of sensible and latent heat given up to the atmosphere. Furthermore, the ice allows snow to accumulate on top of it, thereby setting up conditions that could lead to even lower temperatures.

There are other mechanisms by which tectonic processes* may influence climate. In Fig. 13.6, notice that the formation of oceanic plates (plates that lie beneath the ocean) begins at a *ridge*, where dense, molten material from inside the earth wells up to the surface, forming new sea floor material as it hardens. Spreading (on the order of several centimeters a year) takes place at the ridge center, where two oceanic plates move away from one another. When an oceanic plate encounters a lighter continental plate it responds by diving under it, in a process called *subduction*. Heat and pressure then melt a portion of the subducting rock, which usually consists of volcanic rock and calcium-rich ocean sediment. The molten rock may then gradually work its way to the surface, producing volcanic eruptions that spew water vapor and carbon dioxide into the atmosphere. The release of these gases (called *degassing*) usually takes place at other locations as well, for instance, at ridges where new crustal rock is forming.

Some scientists speculate that climatic change, which takes place over millions of years, might be related to the rate at which the plates move and, hence, related to the amount of CO_2 in the air. For example, during times of rapid spreading, an increase in volcanic activity vents large quantities of CO_2 into the atmo-

---

*Tectonic processes are large-scale processes that deform the earth's crust.

sphere, which may enhance the atmospheric greenhouse effect, causing global temperatures to rise. However, as global temperatures climb, a negative feedback kicks in, as increasing temperatures promote the chemical weathering of rocks on the continents—a process that removes CO_2 from the atmosphere.

Millions of years later, when spreading rates decrease, less volcanic activity means less degassing, and less CO_2. As CO_2 levels drop further due to chemical weathering of rocks on the land (and to perhaps biological activity in the oceans), the greenhouse effect weakens, causing global temperatures to drop. The accumulation of ice and snow over portions of the continents may promote additional cooling by reflecting more sunlight back to space. The cooling, however, will not go unchecked, as lower temperatures retard the chemical weathering of rocks and the depletion of atmospheric CO_2.

A chain of volcanic mountains forming above a subduction zone may disrupt the air flow over them, especially if their orientation is north-south. By the same token, mountain building that occurs when two continental plates collide (like that which presumably formed the Himalayan Mountains and Tibetan highlands) can have a marked influence on global circulation patterns and, hence, on the climate of an entire hemisphere.

Did you know?
About 100 million years ago, when dinosaurs ruled the planet, the earth's temperature was between 18° and 27°F warmer than today's, and the concentration of CO_2 in the atmosphere was much higher.

Up to now, we have examined how climatic variations can take place over millions of years due to the movement of continents and the associated restructuring of landmasses, mountains, and oceans. We will now turn our attention to variations in the earth's orbit that may account for climatic fluctuations that take place on a time scale of tens of thousands of years.

Climate Change and Variations in the Earth's Orbit   A popular theory ascribing climatic changes to variations in the earth's orbit is the **Milankovitch theory**, named for the Yugoslavian astronomer Milutin Milankovitch, who first proposed the idea in the 1930s. The basic premise of this theory is that, as the earth travels through space, three separate cyclic movements combine to produce variations in the amount of solar energy that falls on the earth.

The first cycle deals with changes in the shape (**eccentricity**) of the earth's orbit as the earth revolves

Figure 13.6
The earth is composed of a series of moving plates. The rate at which plates move (spread) may influence global climate. During times of rapid spreading, increased volcanic activity may promote global warming by enriching the CO_2 content of the atmosphere.

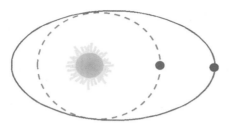

Figure 13.7
For the earth's orbit to stretch from nearly a circle (dashed line) to an elliptical orbit (solid line) and back again takes nearly 100,000 years. (Diagram is not to scale.)

about the sun. Notice in Fig. 13.7 that the earth's orbit changes from being elliptical to being nearly circular. To go from less elliptical to more elliptical and back again takes about 100,000 years. The greater the eccentricity of the orbit, the greater the variation in solar energy received at the top of the atmosphere between the earth's closest and farthest approach to the sun.

Presently, we are in a period of low eccentricity. The earth is closer to the sun in January and farther away in July (see Chapter 2). The difference in distance (which only amounts to about 3 percent) is responsible for a nearly 7 percent increase in the solar energy received at the top of the atmosphere from July to January. When the difference in distance is 9 percent (a highly eccentric orbit), the difference in solar energy

received will be on the order of 20 percent. In addition, the more eccentric orbit will change the length of seasons in each hemisphere by changing the length of time between the vernal and autumnal equinoxes.

The second cycle takes into account the fact that, as the earth rotates on its axis, it wobbles like a spinning top. This wobble, known as the **precession** of the earth's axis, occurs in a cycle of about 23,000 years. Presently, the earth is closer to the sun in January and farther away in July. Due to precession, the reverse will be true in about 11,000 years (see Fig. 13.8). In about 23,000 years we will be back to where we are today. This means, of course, that if everything else remains the same, 11,000 years from now seasonal variations in the Northern Hemisphere should be greater than at present. The opposite would be true for the Southern Hemisphere.

The third cycle takes about 41,000 years to complete and relates to the changes in tilt (**obliquity**) of the earth as it orbits the sun. Presently, the earth's orbital tilt is 23½°, but during the 41,000-year cycle the tilt varies from about 22° to 24½°. The smaller the tilt, the less seasonal variation there is between summer and winter in middle and high latitudes. Thus, winters tend to be milder and summers cooler. During the warmer winters, more snow would probably fall in polar regions due to the air's increased capacity for water vapor. And during the cooler summers, less snow would melt. As a consequence, the periods of smaller tilt would tend to promote the formation of glaciers in high

Figure 13.8
(a) Like a spinning top, the earth's axis of rotation slowly moves and traces out the path of a cone in space. (b) Presently the earth is closer to the sun in January, when the Northern Hemisphere experiences winter. (c) In about 11,000 years, due to precession, the earth will be closer to the sun in July, when the Northern Hemisphere experiences summer.

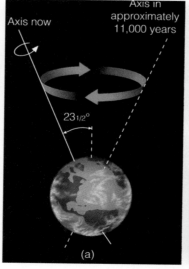

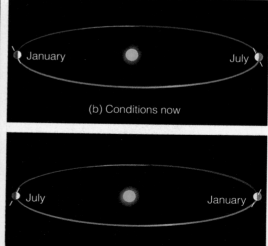

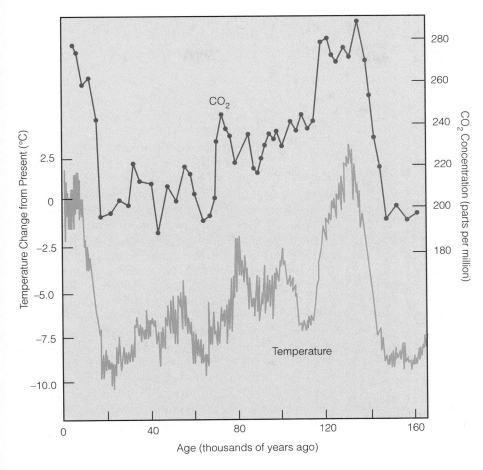

Figure 13.9
Analysis of trapped bubbles of ancient air in the polar ice sheet at the Vostok station in Antarctica reveals that over the past 160,000 years, CO_2 levels (upper curve) correlate well with air temperature changes (bottom curve). Temperatures are derived from the analysis of oxygen isotopes. Note that CO_2 levels were about 30 percent lower and Antarctic temperatures about 10°C (18°F) lower during the colder glacial periods. [From J. M. Barnola et al., "Vostok ice cure provides 160,000-year record of atmospheric CO_2," *Nature* (1 Oct., 1987) 329: 410. Reprinted with permission.]

latitudes. In fact, when all of the cycles are taken into account, the present trend should be toward a *cooler climate* over the Northern Hemisphere, with extensive glaciation.

In summary, the Milankovitch cycles that combine to produce variations in solar radiation received at the earth's surface include:

1. changes in the shape (*eccentricity*) of the earth's orbit about the sun
2. *precession* of the earth's axis of rotation, or wobbling
3. changes in the tilt (*obliquity*) of the earth's axis

In the 1970s, scientists of the CLIMAP project found strong evidence in deep-ocean sediments that variations in climate during the past several hundred thousand years were closely associated with the Milankovitch cycles. Recent studies have even strengthened this premise. For example, studies conclude that during the past 800,000 years, ice sheets have peaked about every 100,000 years. This conclusion corresponds naturally to variations in the earth's eccentricity. Superimposed on this situation are smaller ice advances that show up at intervals of about 41,000 years and 23,000 years. It appears, then, that eccentricity is the forcing factor—the external cause—for the frequency of glaciation, as it appears to control the severity of the climatic variation.

But orbital changes alone are probably not totally responsible for ice buildup and retreat. Recent evidence (from trapped air bubbles in the ice sheets of Greenland and Antarctica representing thousands of years of snow accumulation) reveal that CO_2 levels were about 30 percent lower during colder glacial periods than during warmer interglacial periods. (See Fig. 13.9.) This knowledge suggests that lower atmospheric CO_2 levels may have had the effect of amplifying the cooling initiated by the orbital changes. Likewise, increasing CO₂ levels at the end of the glacial period may have accounted for the rapid melting of the ice sheets. Just why atmospheric CO_2 levels have varied as glaciers expanded and contracted stirs up much debate, but it appears to be due to changes in biological activity taking place in the oceans.

Perhaps, also, changing levels of CO_2 indicate a shift in ocean circulation patterns. Such shifts, brought on by changes in precipitation and evaporation rates, may alter the distribution of heat energy around the world. Alteration wrought in this manner could, in turn, affect the global circulation of winds, which may explain why alpine glaciers in the Southern Hemisphere expanded and contracted in tune with Northern Hemisphere glaciers during the last ice age, even though the Southern Hemisphere (according to the Milankovitch cycles) was not in an orbital position for glaciation.

Still other factors may work in conjunction with the earth's orbital changes to explain the temperature variations between glacial and interglacial periods. Some of these are:

1. the amount of dust in the atmosphere
2. the reflectivity of the ice sheets
3. the concentration of other trace gases, such as methane
4. the changing characteristics of clouds
5. the rebounding of land, having been depressed by ice

Hence, the Milankovitch cycles, in association with other natural factors, may explain the advance and retreat of ice over periods of 10,000 to 100,000 years. But what caused the Ice Age to begin in the first place? And why have periods of glaciation been so infrequent during geologic time? The Milankovitch theory does not attempt to answer these questions.

Climate Change and Atmospheric Particles   Tiny liquid and solid particles (called *aerosols*) that enter the atmosphere from both human and natural sources can remain suspended in the lower atmosphere—the troposphere—for several days. The impact of these particles on climate is exceedingly complex and depends upon a number of factors, such as size, shape, color, vertical distribution above the surface, water content of the surface, and surface reflectivity (albedo). For instance, where the particles are bright in relation to the surface below, the amount of sunlight reflected and scattered back to space is likely to increase, causing air

temperatures to lower. On the other hand, in regions where the particles are relatively dark compared to the surface below, greater absorption of sunlight by the particles could cause air temperatures aloft to rise.

To further complicate the picture, studies show that, even though certain particles may slightly lower the amount of sunlight reaching the surface, these same particles can selectively absorb outgoing infrared radiation from the surface. This process, of course, enhances the atmospheric greenhouse effect and causes an overall warming of the air. Computer models indicate that the pollutant-laden haze layer (comprised mostly of soot), which apparently originates from industrial complexes in Europe and northern Asia and drifts over the Arctic during the winter, may be responsible for the atmospheric warming in that region.

Although the net effect of aerosols in climate in the lower atmosphere is not fully understood, scientists have recently gained a better understanding of how human-generated sulfate particles influence climate. Sulfate particles (produced mainly from fossil fuel combustion) tend to concentrate in regions where they are emitted. Usually lasting only a few days in the atmosphere, these tiny aerosols tend to reflect incoming solar energy, thereby slightly lowering daytime surface air temperatures. Moreover, sulfate particles serve as cloud condensation nuclei. Additional cloud cover can further reduce incoming sunlight and promote even lower daytime temperatures.

Volcanic eruptions can have a definitive impact on climate. During volcanic eruptions, fine particles of ash and dust (as well as gases) can be ejected into the stratosphere (see Fig. 13.10). Scientists agree that the volcanic eruptions having the greatest impact on climate are those rich in sulfur gases. These gases, over a period of about two months, combine with water vapor in the presence of sunlight to produce tiny, bright sulfuric acid particles that grow in size, forming a dense layer of haze. As heavier particles fall out of the stratosphere, new particles form. And so the haze layer may reside in the stratosphere for several years, absorbing and reflecting back to space a portion of the sun's in-

Did you know?
Only one year after the eruption of the Philippine volcano, Mount Pinatubo, in 1991, the average global temperature had cooled by about 1°F.

coming energy. This reduction in sunlight may cause a decrease in average global surface temperature, especially in the hemisphere where the eruption occurs. Both observation and the latest numerical climate models suggest that the cooling response to volcanic eruptions is greatest during the winter.

The models also predict that large volcanic eruptions rich in sulfur, such as El Chichón in 1982 and Mount Pinatubo in 1991, can lower surface temperatures by about 0.2°C or more for a few years afterwards. The exact amount of cooling is often difficult to determine, however, as the average hemispheric temperature usually varies from year to year. But it is interesting to note that, over much of North America, the summer of 1992 was exceptionally cool, just one year after Mount Pinatubo ejected an estimated twenty million tons of sulfur dioxide into the stratosphere.

An infamous cold spell often linked to volcanic activity occurred during the year 1816, "the year without a summer" mentioned earlier. Apparently, a rather stable longwave pattern in the atmosphere produced unseasonably cold summer weather over eastern North America and western Europe. The cold weather followed the massive eruption in 1815 of Mount Tambora in Indonesia. In addition to this, major volcanic eruptions occurred in the four years preceding Tambora. If, indeed, the cold weather pattern was brought on by volcanic eruptions, it was probably an accumulation of several volcanoes loading the stratosphere with particles—particles that probably remained there for several years.

In an attempt to correlate sulfur-rich volcanic eruptions with long-term trends in global climate, scientists are measuring the acidity of annual ice layers in Greenland and Antarctica. Generally, the greater the concentration of sulfuric acid particles in the atmosphere, the greater the acidity of the ice layer. Relatively acidic ice has been uncovered from about A.D. 1350 to about 1700, a time that corresponds to the Little Ice Age. Such findings suggest that sulfur-rich volcanic eruptions may have played an important role in triggering this comparatively cool period and, perhaps, other cool periods during the geologic past.

If the Northern Hemisphere were to cool by several degrees Celsius, what immediate effect might this have on weather and climate patterns? Some numerical climate models predict that a cooler Northern Hemisphere would cause a greater contrast in temperature between high and low latitudes that, in turn, would

Figure 13.10
Large volcanic eruptions rich in sulfur may have an effect on climate. As sulfur gases in the stratosphere transform into tiny bright sulfuric acid particles, they prevent a portion of the sun's energy from reaching the surface.

cause the jet stream to move farther south than normal, especially in summer. In order to transfer heat from low to high latitudes, the upper-level flow and the jet stream would circle the world in broad loops that meander. Well-defined troughs and ridges would form in the flow and remain implanted over a given area for weeks at a time. Vigorous storms would develop in regions near an upper-level trough, while warm and comparatively dry conditions would prevail in areas influenced by a ridge. The total weather pattern would be one of extreme variability, with droughts and floods

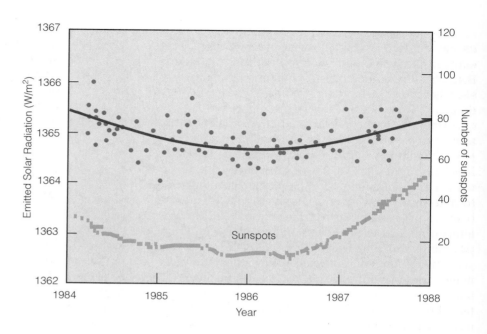

Figure 13.11
Changes in solar energy output (upper curve) as measured by the *Earth Radiation Budget Satellite*. Bottom curve represents the yearly average number of sunspots. [From V. Ramanathan, B. R. Barstrom, and E. F. Harrison, "Climate and the Earth's Radiation Budget," *Physics Today* (May, 1989), Fig. 5, p. 27. Reprinted with permission.]

occurring at the same time in different parts of the world.

Alpine glaciers would slowly advance down river valleys and, in response to this, the sea level might begin to lower. And for crops grown in middle and high latitudes, there would be a greater likelihood of a late spring or early fall freeze.

Climate Change and Variations in Solar Output
In the past, it was thought that solar energy does not vary by more than a fraction of a percent over many years. However, recent measurements made by sophisticated radiation detectors aboard satellites suggest that the sun's energy output may vary considerably more than was thought. Moreover, the sun's total output decreased by about 0.1 percent from 1981 to 1986, after which it began to increase. This variation in solar output appears to be linked to sunspot activity. (See Fig. 13.11.)

Sunspots are huge magnetic storms that show up as cooler (darker) regions on the sun's surface. They occur in cycles, with the number and size reaching a maximum approximately every eleven years. The decrease in solar energy observed in 1986 corresponded to a period of minimum sunspots. Evidently, the greater number of bright areas (*faculae*) around the sunspots radiate more energy, which offsets the effect of the dark spots.

It appears that the eleven-year sunspot cycle has not always prevailed. Apparently, between 1645 and 1715, during the period known as the **Maunder minimum**,* there were few, if any, sunspots. It is interesting to note that the minimum occurred during the coldest stage of the Little Ice Age—a time when global mean temperature decreased by about 0.5°C over the long-term average. Some scientists suggest that a reduction in solar brightness was, in part, responsible for this cold spell. Numerical climate models predict that a change in solar output of only 0.5 percent per century could alter the earth's climate. And, according to one model, a decrease in solar energy of 1 percent per century would lower the earth's average temperature by 1°C.

In an attempt to better understand the sun's behavior, solar researchers are examining stars that are similar in age and mass to our sun. Recent observations suggest that, in some of these stars, energy output may vary by as much as 0.4 percent, leading some scientists to speculate that changes in the sun's brightness might account for part of the global warming during the last century.

---

*This period is named after E. W. Maunder, the British solar astronomer who first discovered the low sunspot period sometime in the late 1880s.

The sun's magnetic field varies with sunspot activity and actually reverses every 11 years. Because it takes 22 years to return to its original state, the sun's *magnetic cycle* is 22 years, rather than 11. Some researchers point to the fact that periodic 20-year droughts on the Great Plains of the United States seem to correlate with this 22-year solar cycle. More recently, scientists have found a relationship between the 11-year sunspot cycle and weather patterns across the Northern Hemisphere. It appears that winter warmings might be related to variations in sunspots and to a pattern of reversing stratospheric winds over the tropics. How such a small change in solar output could bring about such a large variation in weather remains a tantalizing mystery.

To sum up, fluctuations in solar output may account for climatic changes over time scales of decades and centuries. To date, many theories have been proposed linking solar variations to climate change, but none have been proven. However, instruments aboard satellites and solar telescopes on the earth are monitoring the sun to observe how its energy output may vary. Because many years of data are needed, it may be some time before we fully comprehend the relationship between solar activity and climate change on earth.

In previous sections, we saw how increasing levels of CO_2 may have contributed to changes in global climate spanning thousands and even millions of years. Today, we may be undertaking a global scientific experiment by injecting vast quantities of CO_2 into our atmosphere without really knowing the long-term consequences. The next section describes how CO_2 and other trace gases may be enhancing the earth's greenhouse effect, producing global warming.

Carbon Dioxide, the Greenhouse Effect, and Recent Global Warming

We know from Chapter 2 that CO_2 is a greenhouse gas that strongly absorbs infrared radiation and plays a major role in warming the lower atmosphere. We also know that CO_2 has been increasing steadily in the atmosphere, primarily due to the burning of fossil fuels. (See Fig. 1.3, p. 5.) However, deforestation may also be adding to this increase as tropical rain forests are removed and replaced with less efficient plants. Currently, the annual average of CO_2 is about 355 parts per million, and present estimates are that this value may double sometime in the next century.

Most numerical model experiments (mathematical models that simulate climate) predict that a doubling of CO_2 will result in a global warming of surface air between about 2°C and 5°C (Fig. 13.13). There are,

Figure 13.12
The sun's energy output (brightness) changes slightly over a period of years. Such changes may cause variations in the earth's climate system.

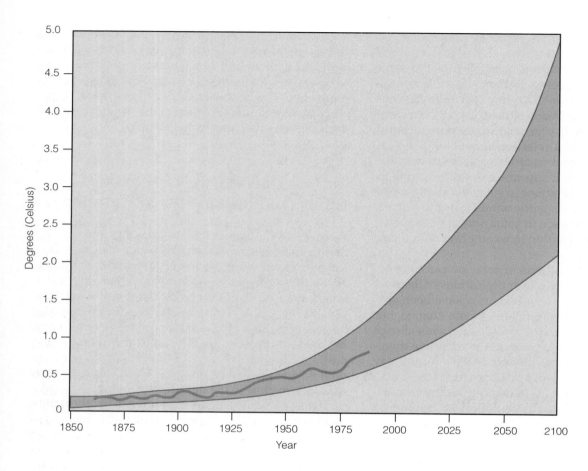

Figure 13.13
Projected warming of the earth's surface by various computer models due to increasing
levels of greenhouse gases. The range of the warming (orange shade) shows that the models
predict a warming of between about 2°C and 5°C by the year 2100. If these predictions are
correct, mean global temperatures within the next 100 years will be higher than they have
been in the previous 150,000 years. The red line shows the change in earth's mean surface
temperature from about 1860 to 1990. (Data from Intergovernmental Panel on Climate
Change and National Academy of Sciences.)

however, uncertainties in the models. For example, if
deforestation is mainly responsible for the rising
levels of CO₂ (rather than the burning of fossil fuels),
the increase in CO₂ will occur much more slowly than
predicted, and global warming will not be as great. In
addition, researchers need a better understanding of
the process by which CO₂ is removed from the atmosphere.

To complicate the picture, trace gases such as
methane (CH₄), nitrous oxide (N₂O), and chlorofluorocarbons (CFCs), all of which readily absorb infrared radiation, have been increasing in concentration over the

past century.* Collectively, these gases are about equal
to CO₂ in their ability to enhance the atmospheric
greenhouse effect. Moreover, the models predict that
rising ocean temperatures will cause an increase in
evaporation rates and, hence, an increase in atmospheric *water vapor*, which is the most potent greenhouse gas. The added water vapor produces a positive
feedback by accelerating the temperature rise. Without this feedback, most models predict that doubling

*Refer back to Chapter 1 and to Table 1.1, p. 3, for additional infor-
mation on the concentration of these gases.

the current values of CO_2 will produce a warming of only about 1°C or 2°C.

Oceans, Clouds, and Global Warming The oceans play a major role in the climate system, yet the exact effect they will have on rising levels of CO_2 and global warming is uncertain. The oceans are huge storehouses for CO_2. Microscopic plants (phytoplankton) extract CO_2 from the atmosphere during photosynthesis and store some of it below the ocean surface when they die. Would a warmer earth trigger a larger blooming of these tiny plants, in effect reducing CO_2 in the atmosphere? Or, would a gradual rise in ocean temperature increase the amount of CO_2 in the air due to the fact that warmer oceans can't hold as much CO_2 as colder ones?* Furthermore, the oceans have a large capacity for storing heat energy. Thus, as they slowly warm, they should retard the rate at which the atmosphere warms. Overall, the response of ocean temperatures, ocean circulations, and sea ice to global warming will probably determine the global pattern and speed of climate change.

If the atmosphere's water vapor content increases as ocean temperatures rise, so might global cloudiness. How, then, would clouds—which come in a variety of shapes and sizes and form at different altitudes—affect the climate system? Clouds reflect incoming sunlight back to space, a process that tends to cool the climate, but they also absorb infrared radiation from the earth, which tends to warm it. Just how the climate will respond to changes in cloudiness will probably depend on the type of clouds that form and their physical properties, such as liquid water (or ice) content and droplet size distribution. For example, high, thin cirriform clouds (composed mostly of ice) tend to promote a net warming effect: They allow a good deal of sunlight to pass through (which warms the earth's surface), yet because they are cold, they warm the atmosphere by absorbing more infrared radiation from the earth than they emit.

Low stratified clouds, on the other hand, tend to promote a net cooling effect. Composed mostly of water droplets, they reflect much of the sun's incoming energy, and, because their tops are relatively warm, they radiate away much of the infrared energy they receive

from the earth. Satellite data from the Earth Radiation Budget Experiment confirm that, overall, clouds have a *net cooling effect* on our planet, which means that, without clouds, our atmosphere would be warmer. So, if global temperatures rise, an increase in cloudiness at all levels might offset some of the warming by providing a negative feedback on the climate system.

Some scientists even speculate that an increase in towering cumuliform clouds, brought on by enhanced convection, will promote a negative feedback on global warming. They contend that as cumulus clouds develop, much of their water vapor will condense and fall to the surface as rain, leaving the upper part of the clouds relatively dry. Additionally, sinking air filling the space around the clouds produces warmer and dryer air aloft. Less water vapor, they feel, will diminish the effect of greenhouse warming.

To illustrate the effect that cloud properties might have on climate models, researchers at the British Meteorological Office altered the representation of clouds in their model. Initially, the model projected a global temperature rise of about 5°C, accompanying a doubling of atmospheric CO_2. However, when water clouds replaced ice clouds in the simulation, the projected temperature rise was less than 2°C. It is no wonder, then, that a study conducted in 1989 with 14 climate models showed good agreement on how the global climate would respond if current values of CO_2 were doubled under clear skies. But when clouds were incorporated into the models, the models did not agree, and, in fact, varied greatly over a wide range.

Possible Consequences of Global Warming Some climatic models predict that, if average global temperatures increase by about 3°C or 4°C, the jet stream will weaken and global winds will shift from their "normal" position. The added surface warmth will enhance evaporation, which will lead to a greater worldwide average precipitation. However, the shifting upper-level winds might reduce precipitation over certain areas, especially middle latitude interior continental regions. This, in turn, would put added stress on important agricultural areas (especially those in the western United States) that depend greatly on irrigation water from streams and reservoirs. Moreover, if the reduction in rainfall occurs in summer (as some models propose), agricultural productivity might be severely impaired over the Central Plains of North America and elsewhere.

---

*This fact has caused a few scientists to suggest that the rise in CO_2 levels over the last century is the result of a warmer ocean, rather than the burning of more fossil fuels.

Figure 13.14
Oceans and clouds play an important part in the earth's climate system. How they will respond to increasing global temperatures is not clear. Oceans may well add water vapor to the atmosphere, which might promote warming, by enhancing the greenhouse effect. Clouds, especially low ones, might provide a negative feedback on global warming, as clouds overall have a net cooling effect on climate.

Other consequences of global warming might be a rise in the sea level of about one-half meter or so, as alpine glaciers recede, polar ice melts, and the oceans expand as they slowly warm. Rising ocean levels might have a damaging influence on coastal marine life. In addition, coastal groundwater supplies might become contaminated with salt water.

Did you know?
If the world warmed by 7°F, estimates are that in Washington, D.C., the number of days when the air temperature exceeded 90°F would increase from an average of 35 per year to 85 per year.

In polar regions, where the warming should be several times greater than in middle and low latitudes (Fig. 13.15), scientists initially turned their attention to ice sheets (especially the west Antarctic ice sheet), to see if shrinkage of the ice might signal a global warming trend due to increasing CO_2 levels. But in polar regions, as elsewhere around the globe, rising temperatures produce complex interactions among temperature, precipitation, and wind patterns. Consequently, it is now believed that as temperatures rise in south polar regions, more snow will fall in the warmer (but still cold) air, causing snow and ice to build up over the continent of Antarctica.

A few scientists even suggest that increased levels of CO_2 in the atmosphere will have some positive consequences. The higher level of CO_2 will act as a "fertilizer" for some plants, accelerating their growth. Increased plant growth consumes more CO_2, which might retard the increasing rate of CO_2 in the environment. Other scientists feel that the increased plant growth might force some insects to eat more, resulting in a net loss of vegetation. There is concern also that a major increase in CO_2 might upset the balance of nature, with some plant species becoming so dominant that others are eliminated. In cold climates where crops are now grown only marginally, the warming effect may actually increase crop yield.

Forests might also feel the effect of a warmer planet. Trees that grow in a climate zone defined mainly by temperature may become especially hard hit as rising temperatures place them in an inhospitable environment. In addition, they may become, in their weakened state, more susceptible to insects and disease.

The effect that increasing levels of CO_2 might have on the upper atmosphere is not totally clear. However, climate models suggest that while the lower atmosphere (troposphere) steadily warms, the upper atmosphere (stratosphere, mesosphere, and thermosphere) will cool. The cooling is brought on by the additional molecules of CO_2 (and other trace gases) emitting more infrared radiation to space than they receive in turn. Recent calculations indicate that a cooler stratosphere would retard the rate of ozone destruction. (See Chapter 1 for more information on stratospheric ozone.)

Is the Warming Real?   Earlier we learned that, since the beginning of this century, the earth's surface appears to have warmed by about 0.6°C. (See Fig. 13.4 and 13.13.) There are, however, uncertainties in the temperature record. For example, during this time, recording stations have moved, techniques for measuring temperature have varied, and overall there is a sparcity of marine observing stations. Moreover, there is a tendency for urbanization (especially in developed nations) to artificially raise average temperatures as cities grow in size—the urban *heat island effect*. (See Chapter 12.) When urban warming is taken into account and improved sea surface temperatures are incorporated into the data, it appears that global temperatures over the last 100 years may have risen between 0.3°C and 0.6°C.

Did you know?
As Columbia, Maryland, grew in population from 200 to 20,000 from 1968 to 1975, its average nightime air temperature, during the winter, increased by about 14°F.

Is the warming trend due to increasing levels of CO_2 and other trace gases? A few scientists contend that it is. Most believe that it is too early to tell. Some point to the fact that many mathematical climate models predict that, due to increasing levels of greenhouse gases, average global temperatures should have risen by at least 1°C, instead of less than 1°C, as observed

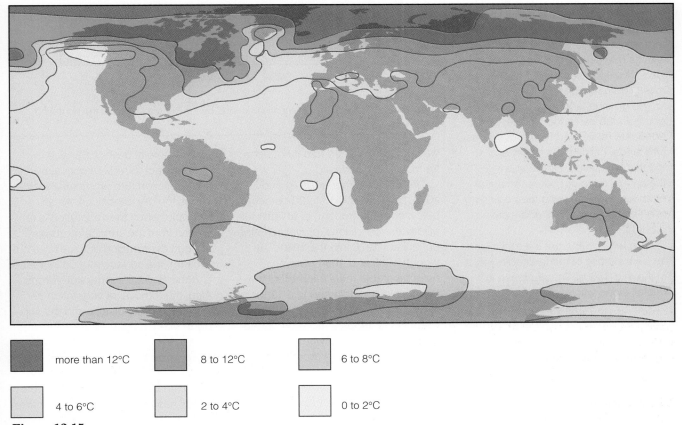

	more than 12°C		8 to 12°C		6 to 8°C
	4 to 6°C		2 to 4°C		0 to 2°C

Figure 13.15
Projected changes in surface air temperature averaged for the months of December, January, and February, due to a doubling of CO_2 as simulated by the Geophysical Fluids Dynamic Laboratory model (GFHI). Notice that, during the Northern Hemisphere winter, the greatest warming is projected for the polar latitudes.

Focus on a Special Topic
The Sahel—An Example of Climatic Variability and Human Existence

The Sahel is in North Africa, located between about 14° and 18°N latitude. (See Fig. 2.) Bounded on the north by the dry Sahara and on the south by the grasslands of the Sudan, the Sahel is a semi-arid region of variable rainfall. Precipitation totals may exceed 50 cm (20 in.) in the southern portion while in the north, rainfall is scanty. Yearly rainfall amounts are also variable as a year with adequate rainfall can be followed by a dry one.

During the winter, the Sahel is dry but, as summer approaches, the intertropical convergence zone (ITCZ) with its rain usually moves into the region. The inhabitants of the Sahel are mostly nomadic people who migrate to find grazing land for their cattle and goats. In the early and middle 1960s, adequate rainfall led to improved pasture lands; herds grew larger and so did the population. However, in 1968, the annual rains did not reach as far north as usual, marking the beginning of a series of dry years and a severe drought.

Rain fell in 1969, but the totals were far below those of the favorable years in the mid-1960s. The decrease in rainfall along with overgrazing turned thousands of square kilometers of pasture into barren wasteland. By 1973, when the severe drought reached its climax, rainfall totals were 50 percent of the long-term average, and perhaps 50 percent of the cattle and goats had died. The Sahara Desert had migrated southward into the northern fringes of the region, and a great famine had taken the lives of more than 100,000 people. Many more of the 2 million or so inhabitants

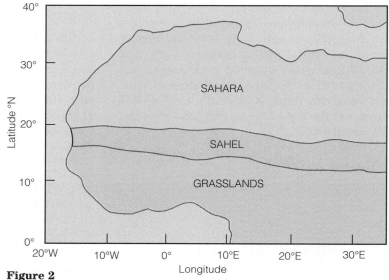

Figure 2
The semi-arid Sahel of North Africa is bounded by the Sahara Desert to the north and grasslands to the south.

would have perished had it not been for massive outside aid.

The rains returned in 1974 and 1975, but were still 15 to 20 percent below normal. Drought, or at least relatively dry conditions, prevailed in the Sahel from 1976 into the 1980s. Although 1983 and 1984 were dry, more favorable rains returned to the region in the middle and late 1980s.

Studies suggest that the wetter years of the 1950s and 1960s appear to be due to the northward displacement of the ITCZ. The drought, however, appears to be more related to the intensity of rain that falls during the so-called rainy season. Some scientists feel that the lack of intense rain is due to a *biogeophysical feedback mechanism* wherein less rainfall and reduced vegetation cover modify the surface and promote a positive

feedback relationship: Surface changes act to reduce convective activity, which in turn promotes or reinforces the dry conditions. As an example, when the vegetation is removed from the surface (perhaps through overgrazing or excessive cultivation), the surface albedo increases and the surface temperature drops. Also, the removal of vegetation reduces the soil moisture and the supply of latent heat to the atmosphere.

Is this a long-term fluctuation in climate, or do the abundant rains of the 1980s suggest that the more favorable climate of the early and middle 1960s has returned? Will the Sahara continue to migrate southward? And if global temperatures rise into the next century, how will precipitation patterns change? At present, we have no answers.

(Fig. 13.13). Others feel that the warming is statistical in nature and falls within the earth's *natural variability* of climate change.

In an attempt to find a signal that suggests that greenhouse gases may have already altered earth's climate, researchers examined rainfall and temperature data within the contiguous United States over a period dating back to 1895. But they could find no overall trend in either rainfall or temperature. Another study examined satellite measurements to see if the lower atmosphere warmed between 1979 and 1988, only a few years in the long-term record. But the satellite data could detect no *significant* rise in global temperatures during this short period.

A few scientists even contend that the problem of global warming is overstated. They believe that there are too many uncertainties in the climate models to adequately represent the highly complex atmosphere, especially with respect to clouds and oceans. Some feel that several warm years during the 1980s (especially 1983 and 1987) were the result of El Niño/Southern Oscillation events (Chapter 7), rather than increasing levels of greenhouse gases.

With levels of CO_2 increasing by about 25 percent over the past 100 years, why has the observed increase in global temperature been so small? Some researchers feel that the large heat capacity of the ocean is delaying the warming of the atmosphere. Others suggest that the cooling witnessed for about 30 years (beginning in the 1940s) may have been due to industrial emissions of sulfate particles that increased the reflectivity of clouds and haze. Also, other factors—such as volcanic eruptions that inject large quantities of dust into the stratosphere and changes in the radiation output of the sun—may be countering the warming.

In Perspective   As the battle rages among scientists over greenhouse warming, modification of the earth's surface, taking place right now, could potentially influence the immediate climate of certain regions. For example, studies show that about half the rainfall in the Amazon River Basin is returned to the atmosphere through evaporation and through transpiration from the leaves of trees. Consequently, clearing large areas of tropical rain forests in South America to create open areas for farms and cattle ranges will most likely cause a decrease in evaporative cooling. This, in turn, could lead to a warming in that area of at least several degrees Celsius. In turn, the reflectivity of the deforested area will change. Similar changes in albedo result from the overgrazing and excessive cultivation of grasslands in semi-arid regions, causing an increase in desert conditions (a process known as **desertification**).

Currently, billions of acres of the world's range and cropland are affected by desertification. Annually, millions of acres are reduced to a state of near or complete uselessness. The main cause is overgrazing, although overcultivation, poor irrigation practices, and deforestation also play a role. The effect this will have on climate, as surface albedos increase and more dust is swept into the air, is uncertain. (For a look at how a modified surface influences the inhabitants of a region in Africa, read the Focus section on p. 334.)

Summary

In this chapter, we examined some of the possible causes of climate change, noting that the problem is extremely complex, as a change in one variable in the climate system almost immediately changes other variables. One theory suggests that the shifting of the continents, along with volcanic activity and mountain building, may account for variations in climate that take place over millions of years.

The Milankovitch theory proposes that alternating glacial and interglacial episodes during the past two million years are the result of small variations in the tilt of the earth's axis and in the geometry of the earth's orbit around the sun. Another theory suggests that certain cooler periods in the geologic past may have been caused by volcanic eruptions rich in sulfur. Still another theory postulates that climatic variations on earth might be due to variations in the sun's energy output.

We examined how sophisticated climate models project that the earth's surface will warm by at least several degrees Celsius over the next century as increasing levels of CO_2 and other trace gases enhance the atmospheric greenhouse effect. We learned that some scientists are skeptical as to whether these predictions are accurate, based on the fact that we presently have a limited understanding of how clouds and oceans will respond to a warmer earth.

Even today, modification of the earth's surface may be altering the climate, as overgrazing and deforestation are changing the earth's surface albedo and, in certain regions, rendering the land useless.

Finally, we learned that recent studies indicate that the world is in a slight warming trend. Will the climate slowly continue to warm, or will it warm at an accelerated rate due to increasing concentrations of greenhouse gases? Are we in a natural cycle where the temperature will soon level off, then slowly drop? In centuries to come, will the Northern Hemisphere enter a cooler period as predicted by the Milankovitch cycles? Certainly, climate has changed in the past, but how much and how quickly will it change in the future? Unfortunately, at present, we have no answers to these important questions. However, as we learn more about our complicated and imperfectly understood atmosphere, we will gain better insight into what the future may have in store for the climate of our globe.

Key Terms

The following terms are listed in the order they appear in the text. Define each. Doing so will aid you in reviewing the material covered in this chapter.

dendrochronology
Ice Age
climatic optimum
Little Ice Age
positive feedback
   mechanism

negative feedback
   mechanism
theory of plate tectonics
Milankovitch theory
eccentricity (of earth's
   orbit)

precession (of earth's
   axis of rotation)
obliquity (of earth's axis)

Maunder minimum
desertification

Review Questions

1. Describe some of the methods scientists use to determine past climatic conditions.
2. How does the overall climate of the world today compare with the so-called "normal" climate throughout earth's history?
3. Explain how the changing climate influenced the formation of the Bering land bridge.
4. When did the Little Ice Age occur?
5. Is a runaway greenhouse effect a positive or negative feedback mechanism? Explain.
6. How does the theory of plate tectonics explain climate change over periods of millions of years?
7. Describe the Milankovitch Theory of climate change by explaining how each of the three cycles alters the amount of solar energy reaching the earth.

8. Given the analysis of air bubbles trapped in polar ice during the past 160,000 years, were CO_2 levels generally higher or lower during colder glacial periods?
9. Would volcanic eruptions rich in sulfur tend to produce higher or lower temperatures at the earth's surface? Explain.
10. Explain how variations in the sun's energy output might influence global climate.
11. As the number of sunspots increases, will the sun's energy output slightly increase or decrease?
12. Most climate models predict that, if CO_2 values double, surface air temperatures will rise by as much as 5°C. What other greenhouse gas *must* also increase in concentration in order for this condition to occur?

13. (a) Describe how clouds influence the climate system.

(b) Which clouds would tend to promote surface cooling: high clouds or low clouds?

14. List the negative (and the potential positive) aspects that increasing levels of CO_2 might have on the atmosphere and its inhabitants.

15. What are some of the uncertainties in the temperature record for the past 100 years?

16. Even though CO_2 concentrations have risen dramatically over the past 100 years, how do scientists explain the fact that global temperatures have risen only slightly?

17. Describe the biogeophysical feedback mechanism that may be responsible for the lack of intense rain in the African Sahel.

The cool summer climate near the top of the Minarets in the Sierra Nevada allows winter snow to survive in sheltered coves and valleys. Afternoon thunderstorms that build along the windward slopes produce ample summer rains. However, just a few hundred miles to the east and several thousand feet lower in elevation, the climate is hot (with afternoon summer temperatures hovering near 40°C or 104°F) and dry (as the region is in the rain shadow of this mountain range). (Photo: T. Ansel Toney)

Chapter 14

Global Climate

Contents

The climate is unbearable . . . At noon today the highest temperature measured was –33°C. We really feel that it is late in the season. The days are growing shorter, the sun is low and gives no warmth, katabatic winds blow continuously from the south with gales and drifting snow. The inner walls of the tent are like glazed parchment with several millimeters thick ice-armour . . . Every night several centimeters of frost accumulate on the walls, and each time you inadvertently touch the tent cloth a shower of ice crystals falls down on your face and melts. In the night huge patches of frost from my breath spread around the opening of my sleeping bag and melt in the morning. The shoulder part of the sleeping bag facing the tent-side is permeated with frost and ice, and crackles when I roll up the bag . . . For several weeks now my fingers have been permanently tender with numb fingertips and blistering at the nails after repeated frostbites. All food is frozen to ice and it takes ages to thaw out everything before being able to eat. At the depot we could not cut the ham, but had to chop it in pieces with a spade. Then we threw ourselves hungrily at the chunks and chewed with the ice crackling between our teeth. You have to be careful with what you put in your mouth. The other day I put a piece of chocolate from an outer pocket directly in my mouth and promptly got frostbite with blistering of the palate.

Ove Wilson. (Quoted in David M. Gates, *Man and His Environment*.)

14 Our opening comes from a report by Norwegian scientists on their encounter with one of nature's cruelest climates—that of Antarctica. The experience of the scientists illustrates the profound effect that climate can have on even ordinary events, such as eating a piece of chocolate. Though we may not always think about it, climate profoundly affects nearly everything in the middle latitudes, too. For instance, it influences our housing, clothing, the shape of landscapes, agriculture, how we feel and live, and even where we reside, as most people will choose to live on a sunny hillside rather than in a cold, dark, and foggy river basin. Entire civilizations have flourished in favorable climates and have moved away from, or perished in, unfavorable ones. We learned early in this text that climate is the average of the day-to-day weather over a long duration. But the concept of climate is much larger than what happens with the weather everyday, for it encompasses, among other things, the daily and seasonal extremes of weather within specified areas.

When we speak of climate, then, we must be careful to specify the spatial location we are talking about. For example, the Chamber of Commerce of a rural town may boast that its community has mild winters with air temperatures seldom below freezing. This may be true several feet above the ground in an instrument shelter, but near the ground the temperature may drop below freezing on many winter nights. This small climatic region near or on the ground is referred to as a **microclimate**. Because a much greater extreme in daily air temperatures exists near the ground than several feet above, the microclimate for small plants is far more harsh than the thermometer in an instrument shelter would indicate.

When we examine the climate of a small area of the earth's surface, we are looking at the **mesoclimate**. The size of the area may range from a few acres to several square miles. Mesoclimate includes regions such as forests, valleys, beaches, and towns. The climate of a much larger area, such as a state or a country, is called **macroclimate**. The climate extending over the entire earth is often referred to as *global climate*.

In this chapter, we will concentrate on the larger scales of climate. We will begin with the factors that regulate global climate, then we will discuss how climates are classified. Finally, we will examine the different types of climate.

▲▼▲

A World with Many Climates

The world is rich in climatic types. From the teeming tropical jungles to the frigid polar "wastelands," there seems to be an almost endless variety of climatic regions. The factors that produce the climate in any given place—the **climatic controls**—are the same that produce our day-to-day weather. Briefly, the controls are the:

1. intensity of sunshine and its variation with latitude
2. distribution of land and water
3. ocean currents
4. prevailing winds
5. positions of high- and low-pressure areas
6. mountain barriers
7. altitude

We can ascertain the effect these controls have on climate by observing the global patterns of two weather elements—temperature and precipitation.

Global Temperatures   Figure 14.1 shows mean annual temperatures for the world. To eliminate the distorting effect of topography, the temperatures are corrected to sea level.* Notice that in both hemispheres the isotherms are oriented east-west, reflecting the fact that locations at the same latitude receive nearly the same amount of solar energy. In addition, the annual solar heat that each latitude receives decreases from low to high latitude; hence, annual temperatures tend to decrease from equatorial toward polar regions.†

The bending of the isotherms along the coastal margins is due in part to the unequal heating and cooling properties of land and water, and to ocean currents and upwelling. For example, along the west coast of North and South America, ocean currents transport cool water equatorward. In addition to this, the wind in both regions blows toward the equator, parallel to the coast. This situation favors upwelling of cold water (Chapter 7), which cools the coastal margins. In the area of the eastern North Atlantic Ocean (north of

---

*This correction is made by adding to each station above sea level an amount of temperature that would correspond to the normal (standard) temperature lapse rate of 6.5°C per 1000 meters (3.6°F per 1000 feet).

†Average global temperatures for January and July are given in Figs. 3.7 and 3.8, respectively, on pp. 62 and 63.

40°N), the poleward bending of the isotherms is due to the Gulf Stream and the North Atlantic Drift, which carry warm water northward. The highest mean temperatures occur in the subtropical deserts of the Northern Hemisphere. Here, the sinking air associated with the subtropical anticyclones produces generally clear skies and low humidity. In summer, the high sun beating down upon a relatively barren landscape produces scorching heat.

The lowest mean temperatures occur in the Antarctic. During part of the year, the sun is below the horizon; when it is above the horizon, it is low in the sky and its rays do not effectively warm the surface. Consequently, the land remains snow- and ice-covered year-round. The snow and ice reflect perhaps 80 percent of the sunlight that reaches the surface. Much of the unreflected solar energy is used to transform the ice and snow into water vapor. The relatively dry air and the Antarctic's high elevation permit rapid radiational cooling during the dark winter months, producing extremely cold surface air.

Did you know?
The warm water of the Gulf Stream helps to keep the average winter temperature in Bergen, Norway (latitude 60°N), about a degree fahrenheit warmer than the average winter temperature in Philadelphia, Pennsylvania (latitude 40°N).

Global Precipitation   Figure 14.2 shows the worldwide general pattern of annual precipitation, which varies from place to place. There are, however, certain regions that stand out as being wet or dry. For example, equatorial regions are typically wet, while the subtropics and the polar regions are relatively dry. The global distribution of precipitation is closely tied to the general circulation of the atmosphere (Chapter 7) and to the distribution of mountain ranges and high plateaus.

Figure 14.3 shows in simplified form how the general circulation influences the north-to-south distribution of precipitation to be expected on a uniformly

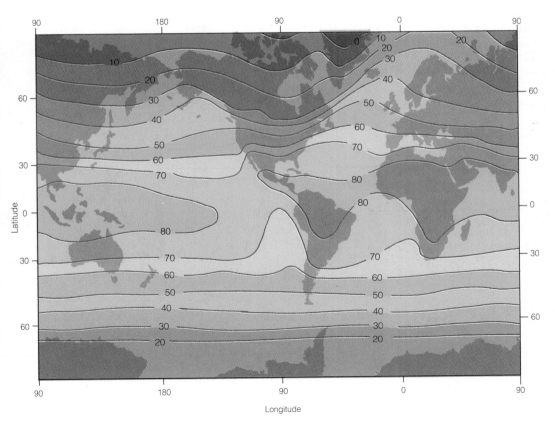

Figure 14.1
Average annual sea level temperatures throughout the world (°F).

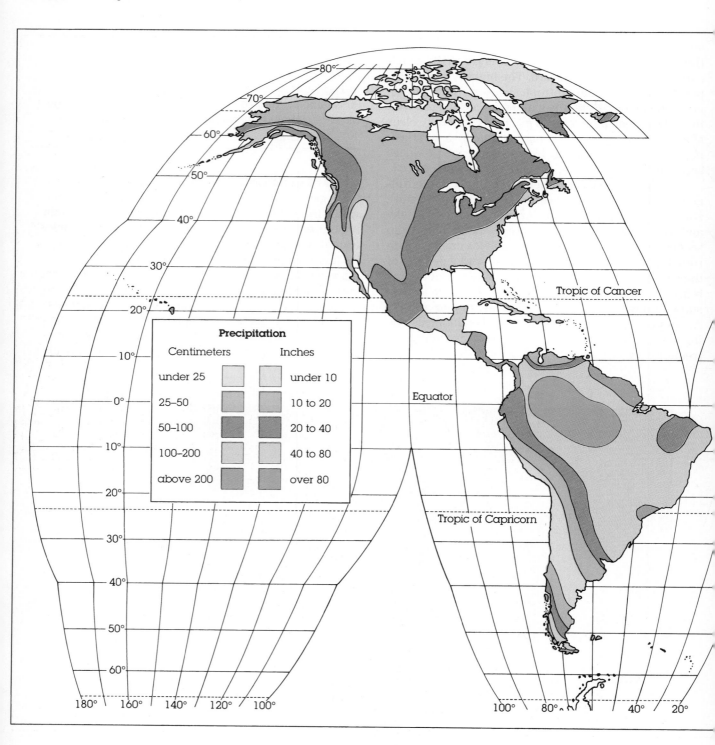

Figure 14.2
Annual global pattern of precipitation.

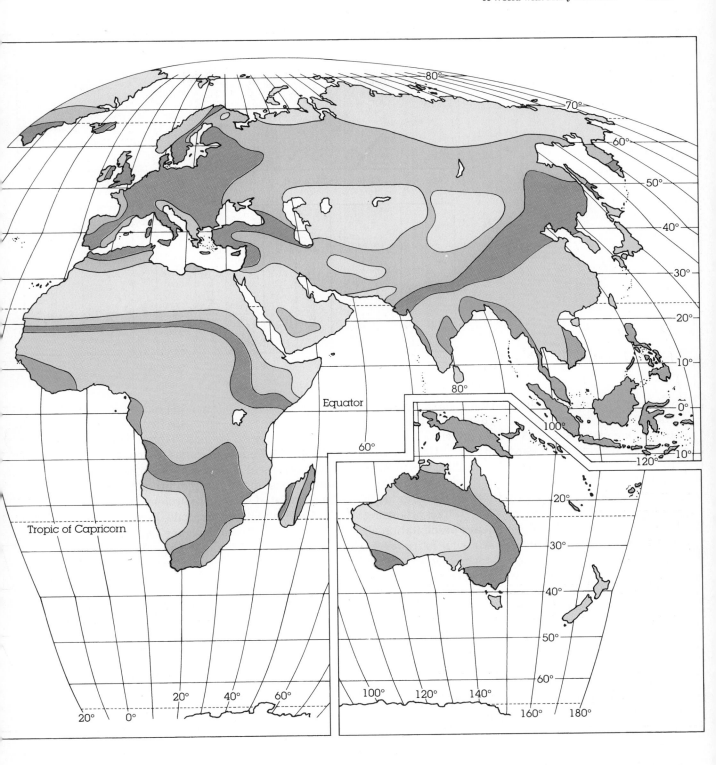

Figure 14.3
A vertical cross section along a line running north to south illustrates the main global regions of rising and sinking air and how each region influences precipitation.

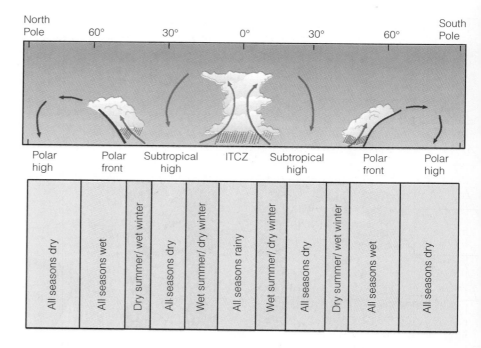

water-covered earth. Precipitation is most abundant where the air rises; least abundant where it sinks. Hence, one expects a great deal of precipitation in the tropics and along the polar front, and little near subtropical highs and at the poles. Let's look at this in more detail.

In tropical regions, the trade winds converge along the Intertropical Convergence Zone (ITCZ), producing rising air, towering clouds, and heavy precipitation all year long. Poleward of the equator, near latitude 30°, the sinking air of the subtropical highs produces a "dry belt" around the globe. The Sahara Desert of North Africa is in this region. Here, annual rainfall is exceedingly light and varies considerably from year to year. Because the major wind belts and pressure systems shift with the season—northward in July and southward in January—the area between the rainy tropics and the dry subtropics is influenced by both the ITCZ and the subtropical highs.

In the cold air of the polar regions there is little moisture, so there is little precipitation. Winter storms drop light, powdery snow that remains on the ground for a long time because of the low evaporation rates. In summer, a ridge of high pressure tends to block storm systems that would otherwise travel into the area; hence, precipitation in polar regions is meager in all seasons.

There are exceptions to this idealized pattern. For example, in middle latitudes the migrating position of the subtropical anticyclones also has an effect on the west-to-east distribution of precipitation. The sinking air associated with these systems is more strongly developed on their eastern side. Hence, the air along the eastern side of an anticyclone tends to be more stable; it is also drier, as cooler air moves equatorward because of the circulating winds around these systems. In addition, along coastlines, cold upwelling water cools the surface air even more, adding to the air's stability. Consequently, in summer, when the Pacific high moves to a position centered off the California coast, a strong, stable subsidence inversion forms above coastal regions. With the strong inversion and the fact that the anticyclone tends to steer storms to the north, central and southern California areas experience little, if any, rainfall during the summer months.

On the western side of subtropical highs, the air is less stable and more moist, as warmer air moves poleward. In summer, over the North Atlantic, the Bermuda high pumps moist tropical air northward from the Gulf of Mexico into the eastern two-thirds of the United

States. The humid air is conditionally unstable to begin with, and by the time it moves over the heated ground, it becomes even more unstable. If conditions are right, the moist air will rise and condense into cumulus clouds, which may build into towering thunderstorms.

In winter, the subtropical North Pacific high moves south, allowing storms traveling across the ocean to penetrate the western states, bringing much needed rainfall to California after a long, dry summer. The Bermuda high also moves south in winter. Across much of the United States, intense winter storms develop and travel eastward, frequently dumping heavy precipitation as they go. Usually, however, the heaviest precipitation is concentrated in the eastern states, as moisture from the Gulf of Mexico moves northward ahead of these systems. Therefore, cities on the plains typically receive more rainfall in summer, those on the west coast have maximum precipitation in winter, while cities in the midwest and east usually have abundant precipitation all year long.

Mountain ranges disrupt the idealized pattern of global precipitation (1) by promoting convection (because their slopes are warmer than the surrounding air) and (2) by forcing air to rise along their upwind (windward) slopes (*orographic uplift*). Consequently, the windward side of mountains tends to be "wet." As air descends and warms along the downwind (leeward)

side, there is less likelihood of clouds and precipitation. Thus, the leeward side of mountains tends to be "dry." As Chapter 5 points out, a region on the leeward side of a mountain where precipitation is noticeably less is called a *rain shadow*.

A good example of the rain shadow effect occurs in the northwestern part of Washington State. Situated on the western (windward) side at the base of the Olympic Mountains, the Hoh River Valley receives an average 380 centimeters (150 inches) of precipitation. On the eastern (leeward) side of this range, only about sixty miles from the Hoh rain forest, the mean annual precipitation is less than 43 centimeters (17 inches), and irrigation is necessary to grow certain crops. Figure 14.4 shows a classic example of how topography produces several rainshadow effects. (Additional information on precipitation extremes is given in the Focus section on p. 346.)

Figure 14.4
The effect of topography on average annual precipitation along a line running from the Pacific Ocean through central California into western Nevada.

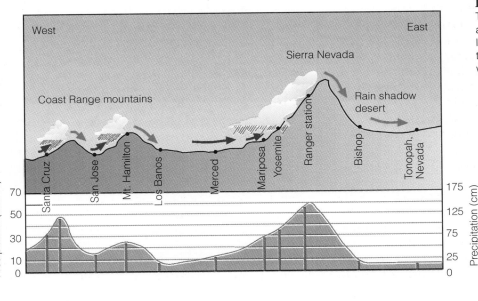

Focus on a Special Topic
Precipitation Extremes

Most of the "rainiest" places in the world are located on the windward side of mountains. For example, Mount Waialeale on the island of Kauai, Hawaii, has the greatest annual average rainfall on record: 1168 centimeters (460 inches). Cherrapunji, on the crest of the southern slopes of the Khasi Hills in northeastern India, receives an average of 1080 centimeters (425 inches) of rainfall each year, the majority of which falls between April and October. Cherrapunji once received 380 centimeters (150 inches) of rain in just five days.

Record rainfall amounts are often associated with tropical storms. On the island of La Réunion (about four hundred miles east of Madagascar in the Indian Ocean), a tropical cyclone dumped 135 centimeters (53 inches) of rain on Belouve in twelve hours. Heavy rains of short duration often occur with severe thunderstorms that move slowly or stall over a region. On July 4, 1956, 3 centimeters (1.2 inches) of rain fell from a thunderstorm on Unionville, Maryland, in one minute.

Snowfalls tend to be heavier where cool, moist air rises along the windward slopes of mountains. One of the snowiest places in North America is located at the Paradise Ranger Station in Mt. Rainier National Park, Washington. Situated at over five thousand feet in elevation, this station receives an average 1575 centimeters (620 inches) of snow annually. However, a record 2850 centimeters (1122 inches) was received during the winter of 1971–1972.

As we noted earlier, the driest regions of the world lie in the frigid, polar region, the leeward side of mountains, and in the belt of subtropical high pressure, between 15° and 30° latitude. Arica in northern Chile holds the world record for lowest annual rainfall, 0.08 centimeters (0.03 inches). In the United States, Death Valley, California, averages only 4.3 centimeters (1.68 inches) of precipitation annually. Fig. 1 gives additional information on world precipitation records.

Figure 1
Some precipitation records throughout the world.

KEY TO MAP

❶	World's greatest annual average rainfall	460 inches	Mt. Waialeale, Hawaii
❷	Greatest 1-month rainfall total	366 inches	Cherrapunji, India, July, 1861
❸	Greatest 12-hour rainfall total	53 inches	Belouve, La Réunion Island, February 28, 1964
❹	Greatest 24-hour rainfall total in United States	43 inches	Alvin, Texas, July 25, 1979
❺	Greatest 42-minute rainfall total	12 inches	Holt, Missouri, June 22, 1947
❻	Greatest 1-minute rainfall total in United States	1.2 inches	Unionville, Maryland, July 4, 1956
❼	Lowest annual average rainfall in Northern Hemisphere	1.2 inches	Bataques, Mexico
❽	Lowest annual average rainfall in the world	0.03 inches	Arica, Chile
❾	Greatest annual snowfall in United States	1122 inches	Paradise Ranger Station, Mt. Rainier, Washington, 1971–1972
❿	Greatest snowfall in 1 month	390 inches	Tamarack, California, January, 1911
⓫	Greatest snowfall in 24 hours	76 inches	Silverlake, Boulder, Colorado, April 14–15, 1921

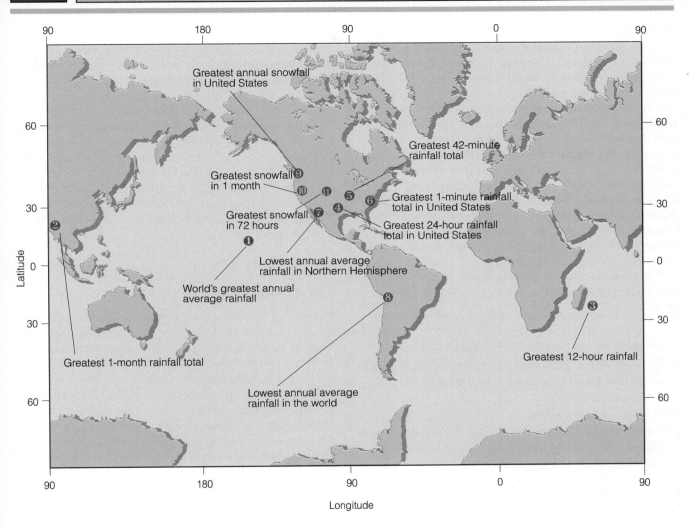

Climatic Classification— the Köppen System

The climatic controls interact to produce such a wide array of different climates that no two places experience exactly the same climate. However, the similarity of climates within a given area allows us to divide the earth into climatic regions.

A widely used classification of world climates based on the annual and monthly averages of temperature and precipitation was devised by the famous German scientist Waldimir Köppen (1846–1940). Initially published in 1918, the original **Köppen classification system** has since been modified and refined. Faced with the lack of adequate observing stations throughout the world, Köppen related the distribution and type of native vegetation to the various climates. In this way, climatic boundaries could be approximated where no climatological data were available.

Köppen's scheme employs five major climatic types; each type is designated by a capital letter:

A *Tropical moist climates*: All months have an average temperature above 18°C (64°F). Since all months are warm, there is no real winter season.

B *Dry climates*: Deficient precipitation most of the year. Potential evaporation and transpiration exceed precipitation.

C *Moist mid-latitude climates with mild winters*: Warm-to-hot summers with mild winters. The average temperature of the coldest month is below 18°C (64°F) and above –3°C (27°F).

D *Moist mid-latitude climates with severe winters*: Warm summers and cold winters. The average temperature of the warmest month exceeds 10°C (50°F), and the coldest monthly average drops below –3°C (27°F).

E *Polar climates*: Extremely cold winters and summers. The average temperature of the warmest month is below 10°C (50°F). Since all months are cold, there is no real summer season.

Each group contains subregions that describe special regional characteristics, such as seasonal changes in temperature and precipitation. (See Table 14.1.) In mountainous country, where rapid changes in elevation bring about sharp changes in climatic type, delineating the climatic regions is impossible. These regions are designated by the letter H, for highland climates.

Köppen's system has been criticized primarily because his boundaries (which relate vegetation to monthly temperature and precipitation values) do not correspond to the natural boundaries of each climatic zone. In addition, the Köppen system implies that there is a sharp boundary between climatic zones, when in reality there is a gradual transition.

The Köppen system has been revised several times, most notably by the German climatologist Rudolf Geiger, who worked with Köppen on amending the climatic boundaries of certain regions. A popular modification of the Köppen system was developed by the American climatologist Glenn T. Trewartha, who redefined some of the climatic types and altered the climatic world map by putting more emphasis on the lengths of growing seasons and average summer temperatures.

The Global Pattern of Climate

Figure 14.5 displays how the major climatic regions of the world are distributed, based mainly on the work of Köppen. (The major climatic types along with their subdivisions are given in Table 14.1.) We will first examine humid tropical climates in low latitudes and then we'll look at middle latitude and polar climates. Bear in mind that each climatic region has many subregions of local climatic differences wrought by such factors as topography, elevation, and large bodies of water. Remember, too, that boundaries of climatic regions represent gradual transitions. Thus, the major climatic characteristics of a given region are best observed away from its periphery.

Tropical Moist Climates (Group A)

General characteristics: year-round warm temperatures (all months have a mean temperature above 18°C, or 64°F); abundant rainfall (typical annual average exceeds 150 centimeters, or 59 inches)

Extent: northward and southward from the equator to about latitude 15° to 25°

Major types (based on seasonal distribution of rainfall): *tropical wet* (Af), *tropical monsoon* (Am), and *tropical wet and dry* (Aw)

Table 14.1　Köppen's Climatic Classification System

Letter Symbol 1st	2nd	3rd	Climatic Characteristics	Criteria
A			Humid tropical	All months have an average temperature of 18°C (64°F) or higher
	f		Tropical wet (rain forest)	Wet all seasons; all months have at least 6 cm (2.4 in.) of rainfall
	w		Tropical wet and dry (savanna)	Winter dry season; rainfall in driest month is less than 6 cm (2.4 in.) and less than 10 – P/25 (P is mean annual rainfall in cm)
	m		Tropical monsoon	Short dry season; rainfall in driest month is less than 6 cm (2.4 in.) but equal to or greater than 10 – P/25
B			Dry	Potential evaporation and transpiration exceed precipitation. The dry/ humid boundary is defined by the following formulas: $p = 2t + 28$ when 70% or more of rain falls in warmer 6 months (dry winter) $p = 2t$ when 70% or more of rain falls in cooler 6 months (dry summer) $p = 2t + 14$ when neither half year has 70% or more of rain (p is the mean annual precipitation in cm and t is the mean annual temperature in °C)*
	S		Semiarid (steppe)	The BS/BW boundary is exactly ½ the dry/humid boundary
	W		Arid (desert)	
		h	Hot and dry	Mean annual temperature is 18°C (64°F) or higher
		k	Cool and dry	Mean annual temperature is below 18°C (64°F)
C			Moist with mild winters	Average temperature of coolest month is below 18°C (64°F) and above –3°C (27°F)
	w		Dry winters	Average rainfall of wettest summer month at least 10 times as much as in driest winter month
	s		Dry summers	Average rainfall of driest summer month less than 4 cm (1.6 in.); average rainfall of wettest winter month at least 3 times as much as in driest summer month
	f		Wet all seasons	Criteria for w and s cannot be met
		a	Summers long and hot	Average temperature of warmest month above 22°C (72°F); at least 4 months with average above 10°C (50°F)
		b	Summers long and cool	Average temperature of all months below 22°C (72°F); at least 4 months with average above 10°C (50°F)
		c	Summers short and cool	Average temperature of all months below 22°C (72°F); 1 to 3 months with average above 10°C (50°F)
D			Moist with cold winters	Average temperature of coldest month is –3°C (27°F) or below; average temperature of warmest month is greater than 10°C (50°F)
	w		Dry winters	Same as under C
	s		Dry summers	Same as under C
	f		Wet all seasons	Same as under C
		a	Summers long and hot	Same as under C
		b	Summers long and cool	Same as under C
		c	Summers short and cool	Same as under C
		d	Summers short and cool; winters severe	Average temperature of coldest month is –38°C (–36°F) or below
E			Polar climates	Average temperature of warmest month is below 10°C (50°F)
	T		Tundra	Average temperature of warmest month is greater than 0°C (32°F) but less than 10°C (50°F)
	F		Ice cap	Average temperature of warmest month is 0°C (32°F) or below

*The dry/humid boundary is defined in English units as: $p = 0.44t$ –3 (dry winter); $p = 0.44t$ –14 (dry summer); and $p = 0.44t$ –8.6 (rainfall evenly distributed). Where p is mean annual rainfall in inches and t is mean annual temperature in °F.

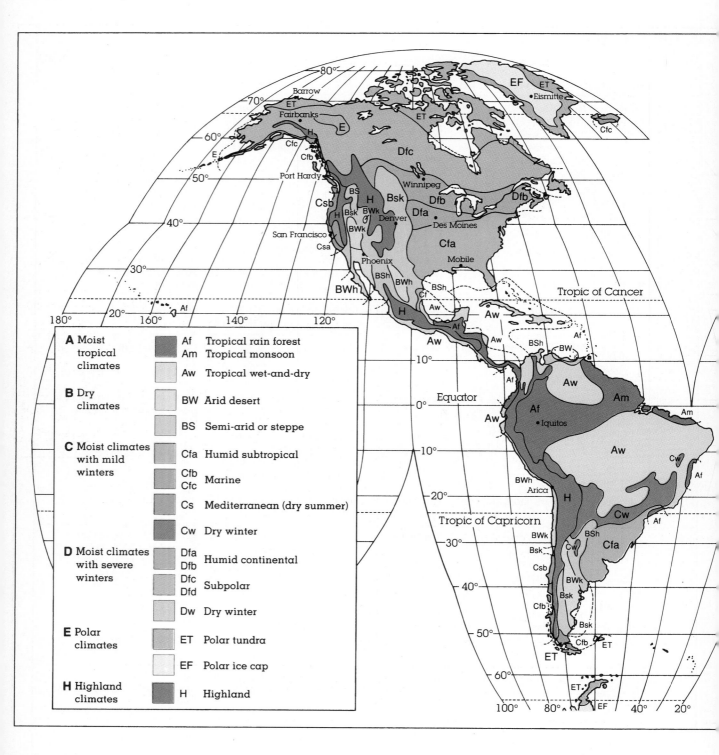

Figure 14.5
Worldwide distribution of climatic regions (after Köppen).

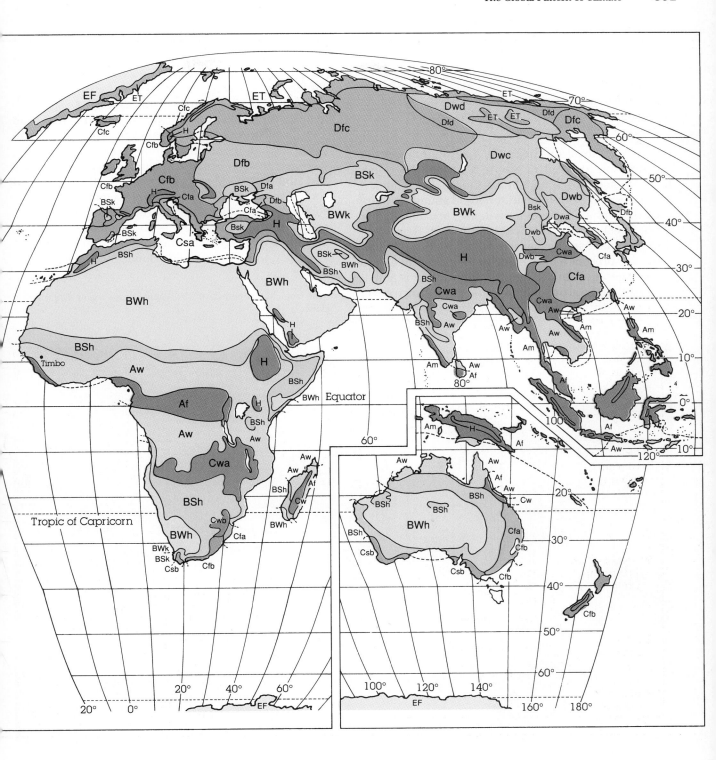

Did you know?
Hot and humid Belem, Brazil—a city situated near the equator with a tropical wet climate—had an all time record high temperature of 98°F, exactly 1°F *less* than the highest temperature ever measured in Fairbanks, Alaska.

At low elevations near the equator, in particular the Amazon lowland of South America, the Congo River Basin of Africa, and the East Indies from Sumatra to New Guinea, high temperatures and abundant yearly rainfall combine to produce a dense, broadleaf, evergreen forest called a **tropical rain forest**. Here, many different plant species, each adapted to differing light intensity, present a crudely layered appearance of diverse vegetation. In the forest, little sunlight is able to penetrate to the ground through the thick crown cover. As a result, little plant growth is found on the forest floor. However, at the edge of the forest, or where a clearing has been made, abundant sunlight allows for the growth of tangled shrubs and vines, producing an almost impenetrable jungle (see Fig. 14.6).

Within the **tropical wet climate*** (Af), seasonal temperature variations are small (normally less than 3°C) because the noon sun is always high and the number of daylight hours is relatively constant. However, there is a greater variation in temperature between day and night than there is between the warmest and coolest months. This is why people remark that winter comes to the tropics at night. The weather here is monotonous and sultry. There is little change in temperature from one day to the next. Furthermore, almost every day, towering cumulus clouds form and produce

---

*The tropical wet climate is also known as the tropical rain forest climate.

Figure 14.6
Tropical rain forest near Iquitos, Peru. (Climatic information for this region is presented in Fig. 14.7.)

heavy, localized showers by early afternoon. As evening approaches, the showers usually end and skies clear. Typical annual rainfall totals are greater than 150 centimeters (59 inches) and, in some cases, especially along the windward side of hills and mountains, the total may exceed 400 centimeters (157 inches).

The high humidity and cloud cover tend to keep maximum temperatures from reaching extremely high values. In fact, summer afternoon temperatures are normally higher in middle latitudes than here. Nighttime cooling can produce saturation and, hence, a blanket of dew and—occasionally—fog covers the ground.

An example of a station with a tropical wet climate (Af) is Iquitos, Peru (Fig. 14.7). Located near the equator (latitude 4°S), in the low basin of the upper Amazon River, Iquitos has an average annual temperature of 25°C (77°F), with an annual temperature range of only 2.2°C (4°F). Notice also that the monthly rainfall totals vary more than do the monthly temperatures. This is due primarily to the migrating position of the Intertropical Convergence Zone and its associated windflow patterns. Although monthly precipitation totals vary considerably, the average for each month exceeds 6 cm, and consequently no month is considered deficient of rainfall.

Köppen classified tropical wet regions, where the monthly precipitation totals drop below 6 centimeters (2.4 inches) for perhaps one or two months, as **tropical monsoon climates** (Am). Here, yearly rainfall totals are similar to those of the tropical wet climate, usually exceeding 150 centimeters a year. Because the dry season is brief and copious rains fall throughout the rest of the year, there is sufficient soil moisture to maintain the tropical rain forest through the short dry period. Tropical monsoon climates can be seen in Fig. 14.5 along the coasts of Southeast Asia, India, and in northeastern South America.

Poleward of the tropical wet region, total annual rainfall diminishes, and there is a gradual transition from the tropical wet climate to the **tropical wet-and-dry climate** (Aw), where a distinct dry season prevails. Even though the annual precipitation usually exceeds 100 centimeters (40 inches), the dry season, where the monthly rainfall is less than 6 centimeters (2.4 inches), lasts for more than two months. Because tropical rain forests cannot survive this "drought," the jungle gradually gives way to tall, coarse **savanna grass**, scattered with low, drought-resistant deciduous trees (Fig. 14.8). The dry season occurs during the winter (low sun

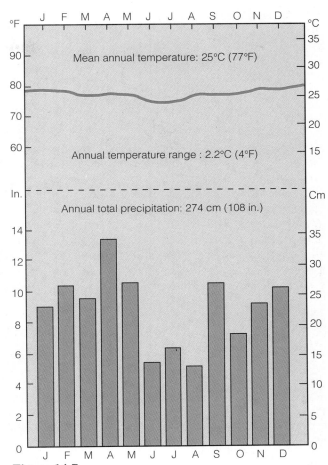

Figure 14.7
Temperature and precipitation data for Iquitos, Peru, latitude 4°S. A station with a tropical wet climate (Af).

period), when the region is under the influence of the subtropical highs. In summer, the ITCZ moves poleward, bringing with it heavy precipitation, usually in the form of showers. Rainfall is enhanced by slow moving shallow lows that move through the region.

Tropical wet-and-dry climates not only receive less total rainfall than the tropical wet climates, but the rain that does occur is much less reliable, as the total rainfall often fluctuates widely from one year to the next. In the course of a single year, for example, destructive floods may be followed by serious droughts. As with tropical wet regions, the daily range of temperature usually exceeds the annual range, but the climate here is much less monotonous. There is a cool season in

Figure 14.8
Baobob and acacia illustrate typical trees of the East African grassland savanna, a region
with a tropical wet-and-dry climate (Aw).

winter when the maximum temperature averages 30°C
to 32°C (86°F to 90°F). At night, the low humidity and
clear skies allow for rapid radiational cooling and, by
early morning, minimum temperatures drop to 20°C
(68°F) or below.

From Fig. 14.5, we can see that the principal areas
having a tropical wet-and-dry climate (Aw) are those located in western Central America, in the region both
north and south of the Amazon Basin (South America),
in southcentral and eastern Africa, in parts of India and
Southeast Asia, and in northern Australia. In many areas (especially within India and Southeast Asia), the
marked variation in precipitation is associated with the
monsoon—the seasonal reversal of winds.

As we saw in Chapter 7, the monsoon circulation is
due in part to differential heating between land masses
and oceans. During winter in the Northern Hemisphere,
winds blow outward, away from a cold, shallow high-

pressure area centered over continental Siberia. These
downslope, relatively dry northeasterly winds from the
interior provide India and Southeast Asia with generally fair weather and the dry season. In summer, the
windflow pattern reverses as air flows into a developing
thermal low over the continental interior. The humid
air from the water rises and condenses, resulting in
heavy rain and the wet season.

An example of a station with a tropical wet-and-
dry climate (Aw) is given in Fig. 14.9. Located at latitude 11°N in west Africa, Timbo, Guinea, receives an
annual average 163 centimeters (64 inches) of rainfall.
Notice that the rainy season is during the summer
when the ITCZ has migrated to its most northern position. Note also that practically no rain falls during the
months of December, January, and February, when the
region comes under the domination of the subtropical
high-pressure area and its sinking air.

The monthly temperature patterns at Timbo are characteristic of most tropical wet-and-dry climates. As spring approaches, the noon sun is slightly higher, and the more intense sunshine produces greater surface heating and higher afternoon temperatures—usually above 32°C (90°F) and occasionally above 38°C (100°F)—creating hot, dry desert-like conditions. After this brief hot season, a persistent cloud cover and the evaporation of rain tends to lower the temperature during the summer. The warm, muggy weather of summer often resembles that of the tropical wet climate (Af). The rainy summer is followed by a warm, relatively dry period, with afternoon temperatures usually climbing above 30°C (86°F).

Poleward of the tropical wet-and-dry climate, the dry season becomes more severe. Clumps of trees are more isolated and the grasses dominate the landscape. When the potential annual water loss through evaporation and transpiration exceeds the annual water gain from precipitation, the climate is described as dry.

Dry Climates (Group B)

General characteristics: deficient precipitation most of the year; potential evaporation and transpiration (i.e., the evaporation of water from plant surfaces) exceed precipitation

Extent: the subtropical deserts extend from roughly 20° to 30° latitude; in large continental regions of the middle latitudes, often surrounded by mountains

Major types: arid (BW)—the "true desert"—and semi-arid (BS)

A quick glance at Fig. 14.5 reveals that, according to Köppen, the dry regions of the world occupy more land area (about 26 percent) than any other major climatic type. Within these dry regions, a deficiency of water exists. Here, the potential annual loss of water through evaporation is greater than the annual water gained through precipitation. Thus, classifying a climate as dry depends not only on precipitation totals but also on temperature, which greatly influences evaporation. For example, 35 centimeters (14 inches) of precipitation in a hot climate will support only sparse vegetation, while the same amount of precipitation in northcentral Canada will support a conifer forest. In addition, a region with a low annual rainfall total is more likely to be classified as dry if the majority of pre-

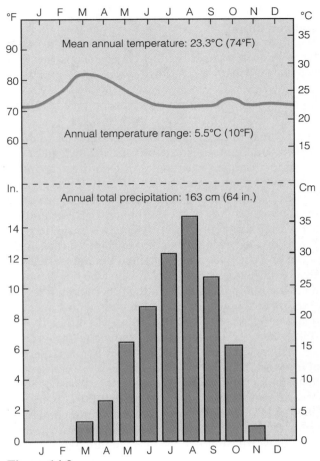

Figure 14.9
Climatic data for Timbo, Guinea, latitude 11°N. A station with a tropical wet-and-dry climate (Aw).

cipitation is concentrated during the warm summer months, when evaporation rates are greater.

Precipitation in a dry climate is both meager and irregular. Typically, the lower the average annual rainfall, the greater its variability. For example, a station that reports an annual rainfall of 5 centimeters (2 inches) may actually measure no rainfall for two years; then, in a single downpour, it receives 10 centimeters (4 inches).

The major dry regions of the world can be divided into two primary categories. The first includes the area of the subtropics (between latitude 15° and 30°), where the sinking air of the subtropical anticyclones produces generally clear skies. The second is found in the continental areas of the middle latitudes. Here, far removed from a source of moisture, areas are deprived of precipitation. Dryness here is often accentuated by mountain ranges that produce a rainshadow effect.

Köppen divided dry climates into two types based on their degree of dryness: the *arid* (BW)* and the *semi-arid*, or steppe (BS). These two climatic types can be divided even further. For example, if the climate is hot and dry with a mean annual temperature above 18°C (64°F), it is either BWh or BSh (the h is for *heiss*, meaning hot in German). On the other hand, if the climate is cold (in winter, that is) and dry with a mean annual temperature below 18°C, then it is either BWk or BSk (where the k is for *kalt*, meaning cold in German).

The **arid climates** (BW) occupy about 12 percent of the world's land area. From Fig. 14.5, we can see that this climatic type is found along the west coast of South America and Africa and over much of the interior of Australia. Notice, also, that a swath of arid climate extends from northwest Africa all the way into central Asia. In North America, the arid climate extends from northern Mexico into the southern interior of the United States and northward along the leeward slopes of the Sierra Nevada. This region includes both the Sonoran and Mojave deserts and the Great Basin.

The southern desert region of North America is dry because it is dominated by the subtropical high most of the year, and winter storm systems tend to weaken before they move into the area. The northern region is in the rain shadow of the Sierra Nevada. These regions are deficient in precipitation all year long, with many stations receiving less than 13 centimeters (5 inches) annually. As noted earlier, the rain that does fall is spotty, often in the form of scattered summer afternoon showers. Some of these showers can be downpours that change a gentle gully into a raging torrent of water. More often than not, however, the rain evaporates into the dry air before ever reaching the ground,

*The letter W is for Wüste, the German word for desert.

and what is seen are rainstreamers (virga) dangling beneath the clouds.

Contrary to popular belief, few deserts are completely without vegetation. Although meager, the vegetation that does exist must depend on the infrequent rains. Thus, most of the native plants are **xerophytes** —those capable of surviving prolonged periods of drought. (See Fig. 14.10.) Such vegetation includes various forms of cacti and short-lived plants that spring up during the rainy periods.

In low-latitude deserts (BWh), intense sunlight produces scorching heat on the parched landscape. Here, air temperatures are as high as anywhere in the world. Maximum daytime readings during the summer can exceed 50°C (122°F), although 40°C to 45°C (104°F to 113°F) are more common. In the middle of the day, the relative humidity is usually between 5 and 25 percent. At night, the air's relatively low water vapor content allows for rapid radiational cooling. Minimum temperatures often drop below 25°C (77°F). Thus, arid climates have large daily temperature ranges, often between 15°C and 25°C (27°F and 45°F) and occasionally higher.

During the winter, temperatures are more moderate, and minimums may, on occasion, drop below freezing. The variation in temperature from summer to winter produces large annual temperature ranges. We can see this in the climate record for Phoenix, Arizona (Fig. 14.11), a city in the southwestern United States with a BWh climate. Notice that the annual temperature average in Phoenix is 22°C (72°F), and that the average temperature of the warmest month (July) reaches a sizzling 32°C (90°F). As we would expect, rainfall is meager in all months. There is, however, a slight maximum in July and August. This is due to the summer monsoon, when more humid, southerly winds are likely to sweep over the region and develop into afternoon showers and thunderstorms.

In middle latitude deserts (BWk), average annual temperatures are lower. Summers are typically warm to hot, with afternoon temperatures frequently reaching

40°C (104°F). Winters are usually extremely cold, with minimum temperatures sometimes dropping below −35°C (−31°F). Many of these deserts lie in the rainshadow of an extensive mountain chain, such as the Cascade Mountains and Sierra Nevada in North America, the Himalayan Mountains in Asia, and the Andes in South America. The meager precipitation that falls comes from an occasional summer shower or a passing mid-latitude cyclone in winter.

Again, refer to Fig. 14.5 and notice that around the margins of the arid regions, where rainfall amounts are greater, the climate gradually changes into **semi-arid** (BS). This region is called **steppe** and typically has short bunch grass, scattered low bushes, trees, or sagebrush (Fig. 14.12). In North America, this climatic region includes most of the Great Plains, the southern coastal sections of California, and the northern valleys of the Great Basin. As in the arid region, northern areas experience lower winter temperatures and more

Figure 14.11
Climatic data for Phoenix, Arizona, latitude 33.5°N. A station with an arid climate (BWh).

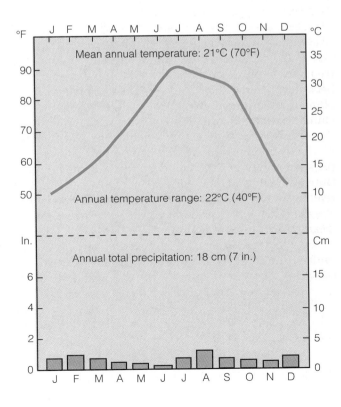

Mean annual temperature: 21°C (70°F)

Annual temperature range: 22°C (40°F)

Annual total precipitation: 18 cm (7 in.)

Figure 14.12
Cumulus clouds forming over the steppe grasslands of western
North America, a region with a semi-arid climate (BS).

frequent snowfalls. Annual precipitation is generally
between 20 and 40 centimeters (8 and 16 inches). The
climatic record for Denver, Colorado (Fig. 14.13), ex-
emplifies the semi-arid (BSk) climate.

    As average rainfall amounts increase, the climate
gradually changes to one that is more humid. Hence,
the semi-arid (steppe) climate marks the transition between the arid and the humid climatic regions. (Before
reading about moist climates, you may wish to read the
Focus section on p. 359 about deserts that experience
drizzle but little rainfall.)

Moist Subtropical Mid-latitude Climates (Group C)

General characteristics: humid with mild winters (i.e.,
average temperature of the coldest month below 18°C,
or 64°F, and above –3°C, or 27°F)

Figure 14.13
Climatic data for Denver, Colorado, latitude 40°N. A station with a
semi-arid climate (BSk).

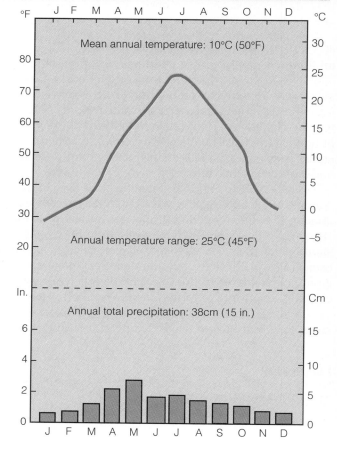

Mean annual temperature: 10°C (50°F)

Annual temperature range: 25°C (45°F)

Annual total precipitation: 38cm (15 in.)

Focus on a Special Topic
A Desert with Clouds and Drizzle

We already know that not all deserts are hot. By the same token, not all deserts are sunny. In fact, some coastal deserts experience considerable cloudiness, especially low stratus and fog.

Amazingly, these coastal deserts are some of the driest places on earth. They include the Atacama Desert of Chile and Peru, the coastal Sahara Desert of northwest Africa, the Namib Desert of southwestern Africa, and a portion of the Sonoran Desert in Baja, California (see Fig. 2). On the Atacama Desert, for example, some regions go without measurable rainfall for decades. And Arica, in northern Chile, has an annual rainfall of only 0.08 centimeters (0.03 inches).

The cause of this aridity is, in part, due to the fact that each region is adjacent to a large body of relatively cool water. Notice in Fig. 2 that these deserts are located along the western coastal margins of continents, where a subtropical high-pressure area causes prevailing winds to move cool water from higher latitudes along the coast. In addition, these winds help to accentuate the water's coldness by initiating *upwelling*—the rising of cold water from lower levels. The combination of these conditions tends to produce coastal water temperatures between 10°C and 15°C (50°F and 59°F), which is quite cool for such low latitudes. As surface air sweeps across the cold water, it is chilled to

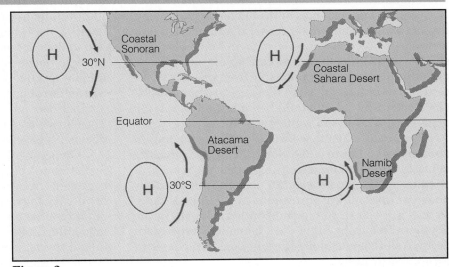

Figure 2
Location of coastal deserts that experience frequent fog, drizzle, and low clouds.

its dew point, often producing a blanket of fog and low clouds, from which drizzle falls. The drizzle, however, accounts for very little rainfall. In most regions, it is only enough to dampen the streets with a mere trace of precipitation.

As the cool, stable air moves inland, it warms, and the water droplets evaporate. Hence, most of the cloudiness and drizzle is found along the immediate coast. Although the relative humidity of this air is high, the dew-point temperature is comparatively low (often near that of the coastal surface water). Inland, further warming causes the air to rise. However, a stable subsidence inversion, associated

with the subtropical highs, inhibits vertical motions by capping the rising air, causing it to drift back toward the ocean, where it sinks, completing a rather strong sea breeze circulation. The position of the subtropical highs, which tend to remain almost stationary, plays an additional role by preventing the Intertropical Convergence Zone with its rising, unstable air from entering the region.

And so we have a desert with clouds and drizzle—a desert that owes its existence, in part, to its proximity to rather cold ocean water and, in part, to the position and air motions of the subtropical highs.

Extent: on the eastern and western regions of most continents, from about 25° to 40° latitude

Major types: humid subtropical (Cfa), marine (Cfb), and dry-summer subtropical, or Mediterranean (Cs)

The Group C climates of the middle latitudes have distinct summer and winter seasons. Additionally, they have ample precipitation to keep them from being classified as dry. Although winters can be cold, and air temperatures can change appreciably from one day to the next, no month has a mean temperature below −3°C (27°F), for if it did, it would be classified as a D climate—one with severe winters.

The first C climate we will consider is the **humid subtropical climate** (Cfa). Notice in Fig. 14.5, that Cfa climates are found principally along the east coasts of continents, roughly between 25° and 40° latitude. They dominate the southeastern section of the United States, as well as eastern China and southern Japan. In the Southern Hemisphere, they are found in southeastern South America and along the southeastern coasts of Africa and Australia.

A trademark of the humid subtropical climate is its hot, muggy summers. This sultry summer weather occurs because Cfa climates are located on the western side of subtropical highs, where maritime tropical air from lower latitudes is swept poleward into these regions. Generally, summer dew-point temperatures are high (often exceeding 23°C, or 73°F) and so is the relative humidity, even during the middle of the day. The high humidity combines with the high air temperature (usually above 32°C, or 90°F) to produce more oppressive conditions than are found in equatorial regions. Summer morning low temperatures often range between 21°C and 27°C (70°F and 81°F). Occasionally, a weak summer cool front will bring temporary relief from the sweltering conditions. However, devastating heat waves, sometimes lasting many weeks, can occur when an upper-level ridge moves over the area.

Winters tend to be relatively mild, especially in the lower latitudes, where air temperatures rarely dip much below freezing. Poleward regions experience winters that are colder and harsher. Here, frost, snow, and ice storms are more common, but heavy snowfalls are rare. Winter weather can be quite changeable, as almost summerlike conditions can give way to cold rain and wind in a matter of hours when a middle latitude storm and its accompanying fronts pass through the region.

Humid subtropical climates experience adequate and fairly well-distributed precipitation throughout the year, with typical annual averages between 80 and 165 centimeters (31 and 65 inches). In summer, when air-mass thunderstorms are common, much of the precipitation falls as afternoon showers. Tropical storms entering the United States and China can substantially add to their summer and autumn rainfall totals. Winter precipitation most often occurs with eastward-trekking middle latitude storms. In the southeastern United States, the abundant rainfall supports a thick pine forest that becomes mixed with oak at higher latitudes. The climate data for Mobile, Alabama, a city with a Cfa climate, is given in Fig. 14.14.

Glance back at Fig. 14.5 and observe that C climates extend poleward along the western side of most continents from about latitude 40° to 60°. These regions are dominated by prevailing winds from the ocean that moderate the climate, keeping winters considerably milder than stations located at the same latitude farther inland. In addition to this, summers are quite cool. When the summer season is both short and cool, the climate is designated as Cfc.* Equatorward, where summers are longer (but still cool), the climate is classified as *west coast marine*, or simply **marine**, Cfb.

Where mountains parallel the coastline, such as along the west coasts of North and South America, the marine influence is restricted to narrow belts. Unobstructed by high mountains, prevailing westerly winds pump ocean air over much of western Europe and thus provide this region with a marine climate (Cfb).

During much of the year, marine climates are characterized by low clouds, fog, and drizzle. The ocean's influence produces adequate precipitation in all months, with much of it falling as light or moderate rain associated with maritime polar air masses. Snow does fall, but frequently it turns to slush after only a day or so. In some locations, topography greatly enhances precipitation totals.

Along the northwest coast of North America, rainfall amounts decrease in summer. This phenomenon is caused by the northward migration of the subtropical Pacific high, which is located southwest of this region. The summer decrease in rainfall can be seen by examining the climatic record of Port Hardy (Fig. 14.15), a station situated along the coast of Canada's Vancouver

---

*Table 14.1, p. 349, details the necessary criteria for each climatic type.

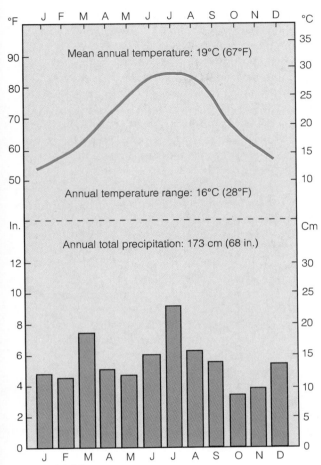

Figure 14.14
Climatic data for Mobile, Alabama, latitude 30°N. A station with a humid subtropical climate (Cfa).

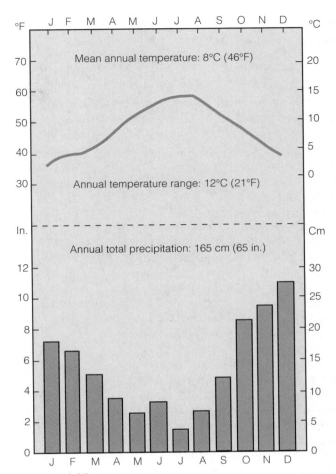

Figure 14.15
Climatic data for Port Hardy, Canada, latitude 51°N. A station with a marine climate (Cfb).

Island. The data illustrate another important characteristic of marine climates: the low annual temperature range for such a high-latitude station. The ocean's influence keeps daily temperature ranges low as well. In this climate type, it rains on many days and when it is not raining, skies are usually overcast. The heavy rains produce a dense forest of Douglas fir.

Moving equatorward of marine climates, the influence of the subtropical highs becomes greater, and the summer dry period more pronounced. Gradually, the climate changes from marine to one of **dry-summer subtropical** (Cs), or **Mediterranean**, because it also borders the coastal areas of the Mediterranean Sea. Along the west coast of North America, Portland, Ore-

gon, because it has rather dry summers, marks the transition between the marine climate and the dry-summer subtropical climate to the south.

The extreme summer aridity of the Mediterranean climate, which in California may exist for five months, is caused by the sinking air of the subtropical highs. In addition, these anticyclones divert summer storm

Did you know?
In the humid, subtropical climate of Washington, D.C., the highest temperature ever measured during January was 79°F and the lowest, −18°F.

systems poleward. During the winter, when the sub-tropical highs move equatorward, mid-latitude storms from the ocean frequent the region, bringing with them much needed rainfall. Consequently, Mediterranean climates are characterized by mild, wet winters, and mild-to-hot, dry summers.

Where surface winds parallel the coast, upwelling of cold water helps keep the water itself and the air

above it cool all summer long. In these coastal areas, which are often shrouded in low clouds and fog, the climate is called *coastal Mediterranean* (Csb). Here, summer daytime maximum temperatures usually reach about 21°C (70°F), while overnight lows often drop below 15°C (59°F). Inland, away from the ocean's influence, summers are hot and winters are a little cooler than coastal areas. In this *interior Mediterranean climate* (Csa), summer afternoon temperatures usually climb above 34°C (93°F) and occasionally above 40°C (104°F).

Figure 14.16 contrasts the coastal Mediterranean climate of San Francisco, California, with the interior Mediterranean climate of Sacramento, California.

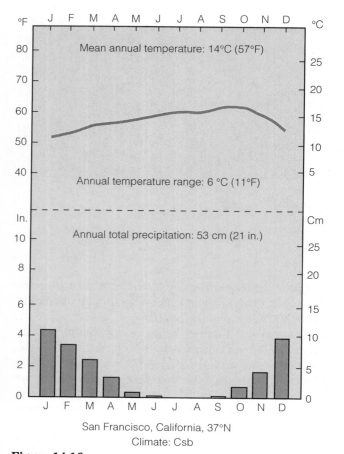

San Francisco, California, 37°N
Climate: Csb

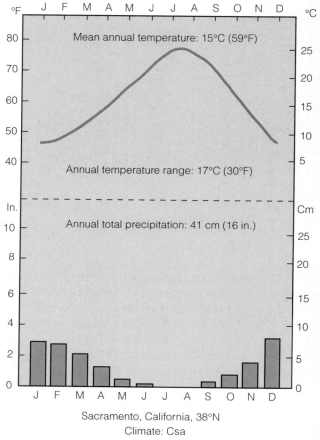

Sacramento, California, 38°N
Climate: Csa

Figure 14.16

Comparison of a coastal Mediterranean climate, Csb (San Francisco, at left) with an interior Mediterranean climate, Csa (Sacramento, at right).

Figure 14.17
In the Mediterranean-type climates of North America, typical chaparral vegetation includes chamise, manzanita, and digger pine.

While Sacramento is only about eighty miles inland from San Francisco, Sacramento's average July temperature is 9°C (16°F) higher. As we would expect, Sacramento's annual temperature range is considerably higher, too. Although Sacramento and San Francisco both experience an occasional frost, snow in these areas is a rarity.

In Mediterranean climates, yearly precipitation amounts range between 30 and 90 centimeters (12 and 35 inches). However, much more precipitation falls on surrounding hillsides and mountains. Because of the summer dryness, the land supports only a scrubby type of low-growing woody plants and trees called *chaparral* (Fig. 14.17).

At this point, we should note that summers are not as dry along the Mediterranean Sea as they are along the west coast of North America. Moreover, coastal Mediterranean areas are also warmer, due to the lack of upwelling in the Mediterranean Sea.

Before leaving our discussion of C climates, note that when the dry season is in winter, the climate is classified as Cw. Over northern India and parts of China, the relatively dry winters are the result of northerly winds from continental regions circulating southward around the cold Siberian high. Many lower-latitude regions with a Cw climate would be tropical if it were not for the fact they are too high in elevation and, consequently, too cool to be designated as tropical.

Moist Continental Climates (Group D)

General characteristics: warm-to-cool summers and cold winters (i.e., average temperature of warmest month exceeds 10°C, or 50°F, and the coldest monthly average drops below –3°C, or 27°F); winters are severe with snowstorms, blustery winds, bitter cold; climate controlled by large continent

Extent: north of moist subtropical mid-latitude climates

Major types: humid continental with hot summers (Dfa), humid continental with cool summers (Dfb), and subpolar (Dfc)

The D climates are controlled by large land masses. Therefore, they are found only in the North-ern Hemisphere. Look at the climate map, Fig. 14.5 on pp. 350–351, and notice that D climates extend across North America and Eurasia, from about latitude 40°N to almost 70°N. In general, they are characterized by cold winters and warm-to-cool summers.

As we know, for a station to have a D climate, the average temperature of its coldest month must dip below –3°C (27°F). This is not an arbitrary number. Köppen found that, in Europe, this temperature marked the southern limit of persistent snow cover in winter.* Hence, D climates experience a great deal of winter snow that stays on the ground for extended periods. When the temperature drops to a point such that no month has an average temperature of 10°C (50°F), the climate is classified as polar (E). Köppen found that the average monthly temperature of 10°C tended to represent the minimum temperature required for tree growth. So no matter how cold it gets in a D climate (and winters can get extremely cold), there is enough summer warmth to support the growth of trees.

There are two basic types of D climates: the **humid continental** (Dfa and Dfb) and the **subpolar** (Dfc). Humid continental climates are observed from about latitude 40°N to 50°N (60°N in Europe). Here, precipitation is adequate and fairly evenly distributed throughout the year, although interior stations experience maximum precipitation in summer. Annual precipitation totals usually range from 50 to 100 centimeters (20 to 40 inches). Native vegetation in the wetter regions includes forests of spruce, fir, pine, and oak. In autumn, nature's pageantry unveils itself as the leaves of deciduous trees turn brilliant shades of red, orange, and yellow (Fig. 14.18).

Humid continental climates are subdivided on the basis of summer temperatures. Where summers are long and hot,† the climate is described as *humid continental with hot summers* (Dfa). Here summers are often hot and humid, especially in the southern regions. Midday temperatures often exceed 32°C (90°F) and occasionally 40°C (104°F). Summer nights are usually

Figure 14.18
The leaves of deciduous trees burst into brilliant color during autumn over the countryside of Vermont, a region with a humid continental climate.

---

*In North America, studies suggest that an average monthly temperature of 0°C (32°F) or below for the coldest month seems to correspond better to persistent winter snow cover.

†Hot means that the average temperature of the warmest month is above 22°C (72°F) and at least four months have a monthly mean temperature above 10°C (50°F).

warm and humid, as well. The frost-free season normally lasts from five to six months, long enough to grow a wide variety of crops. Winters tend to be windy, cold, and snowy. Farther north, where summers are shorter and not as hot,‡ the climate is described as *humid continental with long cool summers* (Dfb). In Dfb climates, summers are not only cooler but much less humid. Temperatures may exceed 35°C (95°F) for a time, but extended hot spells lasting many weeks are rare. The frost-free season is shorter than in the Dfa climate, and normally lasts between three and five months. Winters are long, cold, and windy. It is not uncommon for temperatures to drop below –30°C (–22°F) and stay below –18°C (0°F) for days and sometimes weeks. Autumn is short, with winter often arriving right on the heels of summer. Spring, too, is short, as late spring snowstorms are common, especially in the more northern latitudes.

Figure 14.19 compares the Dfa climate of Des Moines, Iowa, with the Dfb climate of Winnipeg, Canada. Notice that both cities experience a large annual temperature range. This is characteristic of climates located in the northern interior of continents. In fact, as we move poleward, the annual temperature range increases. In Des Moines, it is 31°C (56°F), while about six hundred miles to the north in Winnipeg, it is 38°C (68°F). The summer precipitation maximum expected for these interior continental locations shows up well. Most of the summer rain is in the form of air-mass convective showers, although an occasional weak frontal system can produce more widespread precipitation, as can a cluster of thunderstorms—the Mesoscale Convective Complex described in Chapter 10. The weather in both climatic types can be quite changeable, especially in winter, when a brief warm spell is replaced by blustery winds and temperatures plummeting well below –30°C.

When winters are severe and summers short and cool, with only one to three months having a mean temperature exceeding 10°C (50°F), the climate is described as *subpolar* (Dfc). From Fig. 14.5, we can see that, in North America, this climate occurs in a broad belt across Canada and Alaska; in Eurasia, it stretches from Norway over much of Siberia. The exceedingly

low temperatures of winter account for these areas being the primary source regions for continental polar and arctic air masses. Extremely cold winters coupled with cool summers produce large annual temperature ranges, as exemplified by the climate data in Fig. 14.20 for Fairbanks, Alaska.

Precipitation is comparatively light in the subpolar climates, especially in the interior regions, with most places receiving less than 50 centimeters (20 inches) annually. A good percentage of the precipitation falls when weak cyclonic storms move through the region in summer. The total snowfall is usually not large but the cold air prevents melting, so snow stays on the ground for months at a time. Because of the low temperatures, there is a low annual rate of evaporation that ensures adequate moisture to support the boreal forests of conifers and birches known as **taiga** (Fig. 14.21). Hence, the subpolar climate is known also as a *boreal climate* and as a *taiga climate*.

In the taiga region of northern Siberia and Asia, where the average temperature of the coldest month drops to a frigid –38°C (–36°F) or below, the climate is designated Dfd. Where the winters are considered dry, the climate is designated Dwd.

Polar Climates (Group E)

General characteristics: year-round low temperatures (i.e., average temperature of the warmest month is below 10°C, or 50°F)

Extent: northern coastal areas of North America and Eurasia; Greenland and Antarctica

Major types: polar tundra (ET) and polar ice caps (EF)

In the **polar tundra** (ET), the average temperature of the warmest month is below 10°C (50°F), but above freezing. (See Fig. 14.22, the climate data for Barrow, Alaska.) Here, the ground is permanently frozen to depths of hundreds of meters, a condition known as **permafrost**. Summer weather, however, is just warm enough to thaw out the upper meter or so of soil. Hence, during the summer, the tundra turns swampy and muddy. Annual precipitation on the tundra is meager, with most stations receiving less than 20 centimeters (8 inches). In lower latitudes, this would constitute a desert, but in the cold polar regions evaporation rates are very low and moisture remains adequate. Because of the extremely short growing season, *tundra vegeta-*

---

‡"Not as hot" means that the average temperature of the warmest month is below 22°C (72°F) and at least four months have a monthly mean temperature above 10°C (50°F).

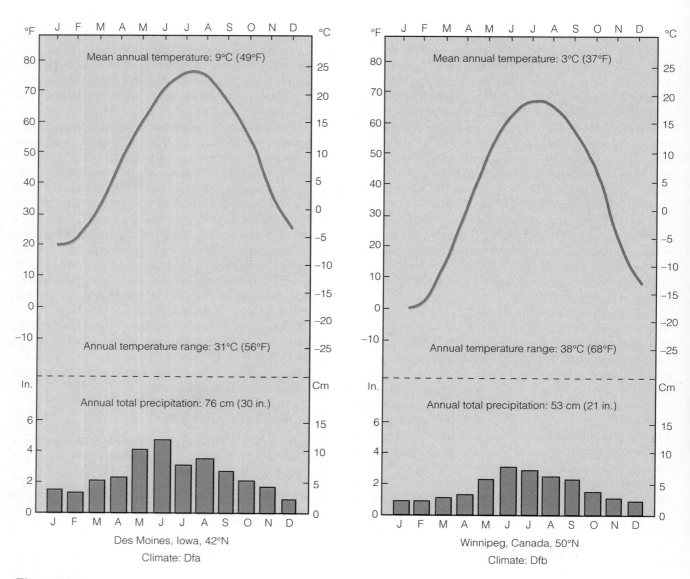

Figure 14.19
Comparison of a humid continental hot summer climate, Dfa (Des Moines, at left), with a humid continental cool summer climate, Dfb (Winnipeg, at right).

tion consists of mosses, lichens, dwarf trees, and scattered woody vegetation, fully grown and only several inches tall (Fig. 14.23).

Even though summer days are long, the sun is never very high above the horizon. Additionally, some of the sunlight that reaches the surface is reflected by snow and ice, while some is used to melt the frozen soil. Consequently, in spite of the long hours of daylight, summers are quite cool. The cool summers and the extremely cold winters produce large annual temperature ranges.

When the average temperature for every month

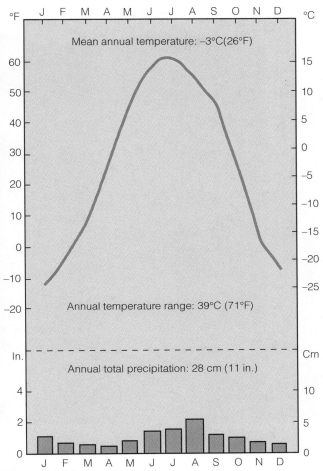

Figure 14.20
Climatic data for Fairbanks, Alaska, latitude 65°N. A station with a subpolar climate (Dfc).

Figure 14.21
Coniferous forests (taiga) such as this occur where winter temperatures are low and precipitation is abundant.

drops below freezing, plant growth is impossible, and the region is perpetually covered with snow and ice. This climatic type is known as **polar ice cap** (EF). It occupies the interior ice sheets of Greenland and Antarctica, where the depth of ice in some places measures thousands of meters. In this region, temperatures are never much above freezing, even during the middle of "summer." The coldest places in the world are located here. Precipitation is extremely meager with many places receiving less than 10 centimeters (4 inches) annually. Most precipitation falls as snow dur-

ing the "warmer" summer. Strong downslope winds frequently whip the snow about, adding to the climate's harshness. The data in Fig. 14.24 for Eismitte, Greenland, illustrate the severity of an EF climate.

Did you know?
If all the snow and ice in Greenland were piled onto the island of Manhattan, it would produce a mountain of ice more than 31,000 miles high.

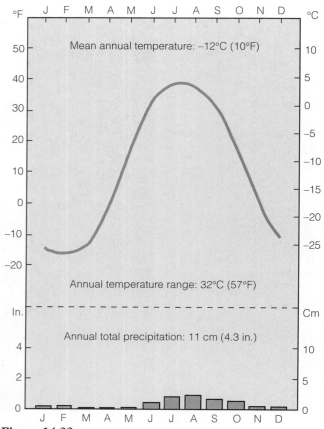

Figure 14.22
Climatic data for Barrow, Alaska, latitude 71°N. A station with a polar tundra climate (ET).

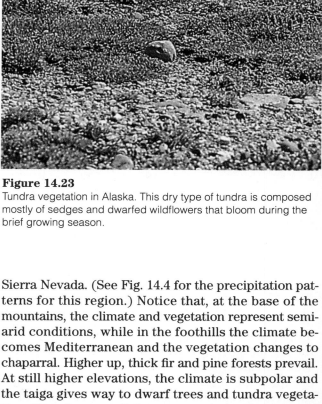

Figure 14.23
Tundra vegetation in Alaska. This dry type of tundra is composed mostly of sedges and dwarfed wildflowers that bloom during the brief growing season.

Highland Climates (Group H)   It is not necessary to visit the polar regions to experience a polar climate. Because temperature decreases with altitude, climatic changes experienced when climbing one thousand feet in elevation are about equivalent in high latitudes to horizontal changes experienced when traveling two hundred miles northward—a distance equal to about 3° latitude. Therefore, when ascending a high mountain one can travel through many climatic regions in a relatively short distance.

Figure 14.25 shows how the climate and vegetation change along the western slopes of the central

Sierra Nevada. (See Fig. 14.4 for the precipitation patterns for this region.) Notice that, at the base of the mountains, the climate and vegetation represent semi-arid conditions, while in the foothills the climate becomes Mediterranean and the vegetation changes to chaparral. Higher up, thick fir and pine forests prevail. At still higher elevations, the climate is subpolar and the taiga gives way to dwarf trees and tundra vegetation. Near the summit there are permanent patches of ice and snow, with some small glaciers nestled in protected areas. Hence, in less than 13,000 vertical feet, the climate has changed from semi-arid to polar.

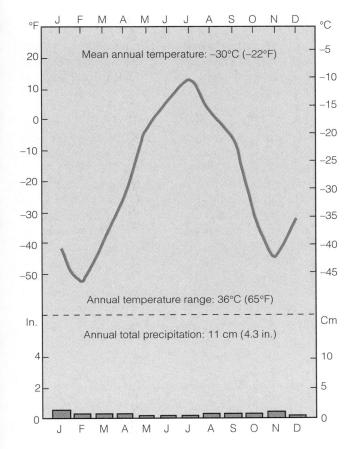

Mean annual temperature: –30°C (–22°F)

Annual temperature range: 36°C (65°F)

Annual total precipitation: 11 cm (4.3 in.)

Figure 14.24
Climatic data for Eismitte, Greenland, latitude 71°N. Located in the interior of Greenland, at an elevation of almost ten thousand feet above sea level, Eismitte has a polar ice cap climate (EF).

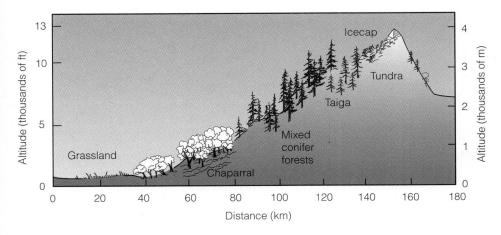

Icecap

Tundra

Taiga

Mixed conifer forests

Grassland

Chaparral

Figure 14.25
Vertical view of changing vegetation and climate due to elevation in the central Sierra Nevada.

Summary

In this chapter, we examined global temperature and precipitation patterns, as well as the various climatic regions throughout the world. Tropical climates are found in low latitudes, where the noon sun is always high, day and night are of nearly equal length, every month is warm, and no real winter season exists. Some of the rainiest places in the world exist in the tropics, especially where warm, humid air rises upslope along mountain ranges.

Dry climates prevail where potential evaporation and transpiration exceed precipitation. Some deserts, such as the Sahara, are mainly the result of sinking air associated with the subtropical highs, while others, due to the rain shadow effect, are found on the leeward side of mountains. Many deserts form in response to both of these effects.

Middle latitudes are characterized by a distinct winter and summer season. Winters tend to be milder in lower latitudes and more severe in higher latitudes.

Along the east coast of some continents, summers tend to be hot and humid as moist air sweeps poleward around the subtropical highs. The air often rises and condenses into afternoon thunderstorms in this humid subtropical climate. The west coast of many continents tends to be drier, especially in summer, as the combination of cool ocean water and sinking air of the subtropical high inhibit to a large degree the formation of cumuliform clouds.

In the middle of large continents, such as North America and Eurasia, summers are usually wetter than winters. Winter temperatures are generally lower than those experienced in coastal regions. As one moves northward, summers become shorter and winters longer and colder. Polar climates prevail at high latitudes, where winters are severe and there is no real summer. When ascending a high mountain, one can travel through many climatic zones in a relatively short distance.

Key Terms

The following terms are listed in the order they appear in the text. Define each.
Doing so will aid you in reviewing the material covered in this chapter.

microclimate	tropical monsoon	semi-arid climate	humid continental
mesoclimate	climate	steppe	climate
macroclimate	tropical wet-and-dry	humid subtropical	subpolar climate
climatic controls	climate	climate	taiga
Köppen classification	savanna grass	marine climate	polar tundra climate
system	arid climate	dry-summer subtropical	permafrost
tropical rain forest	xerophytes	(Mediterranean)	polar ice cap climate
tropical wet climate			

Review Questions

1. Distinguish among microclimate, mesoclimate, and macroclimate.
2. Describe as many factors as you can that influence the global pattern of precipitation.
3. Explain why, in North America, precipitation typically is a maximum along the West Coast in winter, a maximum on the Central Plains in summer, and fairly evenly distributed between summer and winter along the East Coast.

4. According to Köppen's climatic system (Fig. 14.5, p. 350), what major climatic type is most abundant (a) in North America, (b) in South America, (c) throughout the world?
5. What constitutes a dry climate?
6. What climatic information did Köppen use in classifying climates?
7. On a blank map of the world, roughly outline Köppen's major climatic regions.

8. In which climatic region would each of the following be observed: tropical rain forest, xerophytes, steppe, taiga, tundra, and savanna?

9. What are the controlling factors (the major climatic controls) that produce the following climatic regions: (a) tropical wet and dry; (b) Mediterranean; (c) marine; (d) humid subtropical; (e) subpolar; (f) polar ice cap?

10. Why do marine climates (Cs) usually lie along the west coast of continents?

11. Why are large annual temperature ranges characteristic of D-type climates?

12. Why are D-type climates only found in the Northern Hemisphere?

13. Explain why a tropical rain forest climate will support a tropical rain forest, while a tropical wet-and-dry climate will not.

14. What is the primary distinction between a Cfa and a Dfa climate?

15. Explain how arid deserts can be found adjacent to oceans.

16. Why did Köppen use the 10°C (50°F) average temperature for July to distinguish between D and E climates?

17. What accounts for the existence of a BWk climate in the western Great Basin of North America?

18. Barrow, Alaska, receives a mere 11 centimeters (about 4.3 inches) of precipitation annually. Explain why its climate is not classified as arid or semi-arid.

19. Explain why subpolar climates are also known as boreal climates and taiga climates.

Sunlight reflected and scattered by air molecules and cloud particles produces a dramatic, colorful sunset. (Photo by Pat Badalamente Zawadzki)

Chapter 15

Light, Color, and Atmospheric Optics

Contents

▲▼▲

On a crisp day in 1818, a ship with full sails entered unknown waters near Baffin Island in Canada. Passengers James and John Ross, English brothers, had come to find the elusive "Northwest Passage," the waterway linking the Atlantic and Pacific oceans. Their hopes were dashed, however, for blocking their path was a towering mountain range. Disappointed, they returned to report that the Northwest Passage did not exist. Admiral Perry arrived 75 years later, naming the same barrier "Crocker land."

Speculation about Crocker land continued until the American Museum of Natural History sent Donald MacMillan to solve the mystery of the uncharted mountain range in 1913. At first, the journey was disappointing, for MacMillan found only vast stretches of open water. Finally, Crocker land was sighted, but it was 200 miles farther west from where Perry had encountered it. MacMillan sailed as far as possible, dropped anchor, and set out on foot with a small crew of men.

As the team moved toward the mountains, the mountains seemed to move away. Puzzled, they treked onward until huge mountains surrounded them. But in the next instant, when the sun dropped below the horizon, as if by magic, the mountains dissolved into the cold arctic twilight. Dumbfounded, the men looked around only to see ice in all directions—not a mountain was in sight. They had been the victims of one of nature's greatest practical jokes, for Crocker land was a mirage.

▲

15

The sky is full of visual events. Optical illusions (mirages) can appear as towering mountains or wet roadways. In clear weather, the sky can appear blue, while the horizon appears milky white. Sunrises and sunsets can fill the sky with brilliant shades of pink, red, orange, and purple. At night, the sky is black, except for the light from the stars, planets, and the moon. The moon's size and color seem to vary during the night, and the stars twinkle. To understand what we see in the sky, we will take a closer look at sunlight, examining how it interacts with the atmosphere to produce an array of atmospheric visuals.

White and Colors

We know from Chapter 2 that nearly half of the solar radiation that reaches the atmosphere is in the form of visible light. As sunlight enters the atmosphere, it is either absorbed, reflected, scattered, or transmitted on through. How objects at the surface respond to this energy depends on their general nature (color, density, composition) and the wavelength of light that strikes them. How do we see? Why do we see various colors? What kind of visual effects do we observe because of the interaction between light and matter? In particular, what can we *see* when light interacts with our atmosphere?

We perceive light because electromagnetic waves stimulate antenna-like nerve endings in the retina of the human eye. These antennae are of two types—*rods* and *cones*. The rods respond to all wavelengths of visible light and give us the ability to distinguish light from dark. If people possessed rod-type receptors only, then only black and white vision would be possible. The cones respond to specific wavelengths of visible light. The cones fire an impulse through the nervous system to the brain, and we perceive this impulse as the sensation of color. (Color blindness is caused by missing or malfunctioning cones.) Wavelengths of radiation shorter or longer than those of visible light do not stimulate color vision in humans.

White light is perceived when all visible wavelengths strike the cones of the eye with nearly equal intensity. Because the sun radiates almost half of its energy as visible light, all visible wavelengths from the midday sun reach the cones, and the sun usually appears white. A star that is cooler than our sun radiates most of its energy at slightly longer wavelengths; therefore, it appears redder. On the other hand, a star much hotter than our sun radiates more energy at shorter wavelengths and thus appears bluer. A star whose temperature is about the same as the sun's appears white.

Objects that are not hot enough to produce radiation at visible wavelengths can still have color. Everyday objects we see as red are those that absorb all visible radiation except red. The red light is reflected from the object to our eyes. Blue objects have blue light returning from them, since they absorb all visible wavelengths except blue. Some surfaces absorb all visible wavelengths and reflect no light at all. Since no radiation strikes the rods or cones, these surfaces appear black. Therefore, when we see colors, we know that light must be reaching our eyes.

White Clouds and Scattered Light

One exciting feature of the atmosphere can be experienced when we watch the underside of a puffy, growing cumulus cloud change color from white to dark gray or black. When we see this change happen, our first thought is usually, "It's going to rain." Why is the cloud initially white? Why does it change color? To answer these questions, let's examine the concept of *scattering*.

When sunlight bounces off a surface at the same angle at which it strikes the surface, we say that the light is **reflected**, and call this phenomenon *reflection*. There are various constituents of the atmosphere, however, that tend to deflect sunlight from its path and send it out in all directions. We know from Chapter 2 that radiation reflected in this way is said to be **scattered**. (Scattered light is also called *diffuse light*.) During the scattering process, no energy is gained or lost and, therefore, no temperature changes occur. In the atmosphere, scattering is usually caused by small objects, such as air molecules, fine particles of dust, water molecules, and some pollutants. Just as the ball in a pinball machine bounces off the pins in many directions, so solar radiation is knocked about by small particles in the atmosphere.

Typical cloud droplets are large enough to effectively scatter all wavelengths of visible radiation more or less equally. Clouds, even small ones, are optically thick, meaning that very little unscattered light gets through them. These same clouds are poor absorbers of sunlight. Hence, when we look at a cloud, it appears white because countless cloud droplets scatter all wavelengths of visible sunlight in all directions (Fig. 15.1).

As a cloud grows larger and taller, more sunlight is reflected from it and less light can penetrate all the way through it. In fact, relatively little light penetrates a cloud whose thickness is several thousand feet. Since little sunlight reaches the underside of the cloud, little light is scattered, and the cloud base appears dark. At the same time, if droplets near the cloud base grow larger, they become less effective scatterers and better absorbers. As a result, the meager amount of visible light that does reach this part of the cloud is absorbed rather than scattered, which makes the cloud appear even darker. These same cloud droplets may even grow large and heavy enough to fall to the earth as rain. From a casual observation of clouds, we know that dark, threatening ones frequently produce rain. Now we know why they appear so dark.

Blue Skies and Hazy Days

The sky appears blue because light that stimulates the sensation of blue color is reaching the retina of the eye. How does this happen?

Individual air molecules are much smaller than cloud droplets—their diameters are small even when compared with the wavelength of visible light. Each air molecule of oxygen and nitrogen is a *selective scatterer* in that each scatters shorter waves of visible light much more effectively than longer waves.

As sunlight enters the atmosphere, the shorter visible wavelengths of violet, blue, and green are scattered more by atmospheric gases than are the longer wavelengths of yellow, orange, and especially red. (Violet light is scattered about 16 times more than red light.) As we view the sky, the scattered waves of violet, blue, and green strike the eye from all directions. Because our eyes are more sensitive to blue light, these waves, viewed together, produce the sensation of blue coming from all around us. (See Fig. 15.2.) Therefore, when we look at the sky it appears blue. (Earth, by the way, is not the only planet with a colorful sky. On Mars, dust in the air turns the sky red at midday and purple at sunset.)

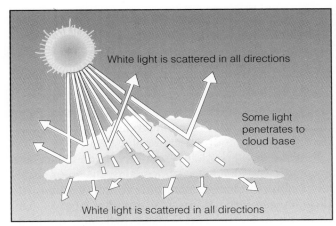

Figure 15.1
Since tiny cloud droplets scatter visible light in all directions, light from many billions of droplets turns a cloud white.

The selective scattering of blue light by air molecules and very small particles can make distant mountains appear blue, such as the Blue Ridge Mountains of Virginia (Fig. 15.4) and the Blue Mountains of Australia. In some places, a *blue haze* may cover the landscape, even in areas far removed from human contamination. Although its cause is still controversial, the blue haze appears to be the result of a particular process. Extremely tiny particles (hydrocarbons called *terpenes*) are released by vegetation to combine chemically with

Figure 15.2
The sky appears blue because billions of air molecules selectively scatter the shorter wavelengths of visible light more effectively than the longer ones. This causes us to see blue light coming from all directions.

small amounts of ozone, which may filter down from the stratosphere. This reaction produces extremely tiny particles that selectively scatter blue light.

When small particles, such as fine dust and salt, become suspended in the atmosphere, the color of the sky begins to change from blue to milky white. Although these particles are small, they are large enough to scatter all wavelengths of visible light fairly evenly in all directions. When our eyes are bombarded by all wavelengths of visible light, the sky appears milky white, and we call the day "hazy." If the humidity is high enough, soluble particles (nuclei) will "pick up" water vapor and grow into haze particles. Thus, the color of the sky gives us a hint about how much material is suspended in the air: the more particles, the more scattering, and the whiter the sky becomes. On top of a high mountain, when we are above many of these haze particles, the sky usually appears a deep blue.

Haze can scatter light from the rising or setting sun, so that we see bright lightbeams, or **crepuscular rays**, radiating across the sky. A similar effect occurs when the sun shines through a break in a layer of clouds (Fig. 15.5). Dust, tiny water droplets, or haze in the air beneath the clouds scatters sunlight, making that region of the sky appear bright with rays. Because these rays seem to reach downward from clouds, some people will remark that the "sun is drawing up water." In England, this same phenomenon is referred to as "Jacob's ladder." No matter what name these sunbeams go by, it is the scattering of sunlight by particles in the atmosphere that makes them visible.

Red Suns and Blue Moons

At midday, the sun seems intensely white, while at sunset it usually appears to be yellow, orange, or red. At noon, when the sun is high in the sky, light from the sun is most intense—all wavelengths of visible light are able to reach the eye with about equal intensity, and the sun appears white. (Looking directly at the sun, espe-

Figure 15.3
Blue skies and white clouds. The selective scattering of blue light by air molecules produces the blue sky, while the scattering of all wavelengths of visible light by liquid cloud droplets produces the white clouds.

Figure 15.4
The Blue Ridge Mountains in Virginia. The blue haze is caused by the scattering of blue light
by extremely small particles—smaller than the wavelengths of visible light.

Figure 15.5
The scattering of sunlight by dust and haze produces these white bands of crepuscular rays.

cially during this time of day, can cause irreparable damage to the eye. Normally, we get only glimpses or impressions of the sun out of the corner of our eye.)

Near sunrise or sunset, however, the rays coming directly from the sun strike the atmosphere at a low angle. They must pass through much more atmosphere than at any other time during the day. (When the sun is 4° above the horizon, sunlight must pass through an atmosphere more than twelve times thicker than when the sun is directly overhead.) By the time sunlight has penetrated this large amount of air, most of the shorter waves of visible light have been scattered away by the air molecules. Just about the only waves from a setting sun that make it on through the atmosphere on a fairly direct path are the yellow, orange, and red. Upon reaching the eye, these waves produce a sunset that is bright yellow-orange. (See Fig. 15.6.)

Bright, yellow-orange sunsets only occur when the atmosphere is fairly clean, such as after a recent rain. If the atmosphere contains many fine particles whose diameters are a little larger than air molecules, slightly longer (yellow) waves also would be scattered away. Only orange and red waves would penetrate through to the eye, and the sun would appear red-orange. When

the atmosphere becomes loaded with particles, only the longest red wavelengths are able to penetrate the atmosphere, and we see a red sun.

Natural events may produce red sunrises and sunsets. Over the oceans, for example, the scattering characteristics of small, suspended salt particles and water vapor are responsible for the brilliant red suns observed from the beach. Moreover, major volcanic eruptions send vast amounts of dust and ash high into the atmosphere. These fine particles, moved by the winds aloft, circle the globe, producing beautiful sunrises and sunsets for months and even years. There were beautiful ruddy sunsets in many parts of the Northern Hemisphere after the the eruption of the Mexican volcano El Chichón in 1982 and the Philippine volcano Mount Pinatubo in 1991.

Occasionally, the atmosphere becomes so laden with dust, smoke, and pollutants that even red waves are unable to pierce the filthy air. An eerie effect then occurs. Because no visible waves enter the eye, the sun literally disappears before it reaches the horizon!

The scattering of light by large quantities of atmospheric particles can cause some rather unusual sights. If the volcanic ash, dust, or smoke particles are

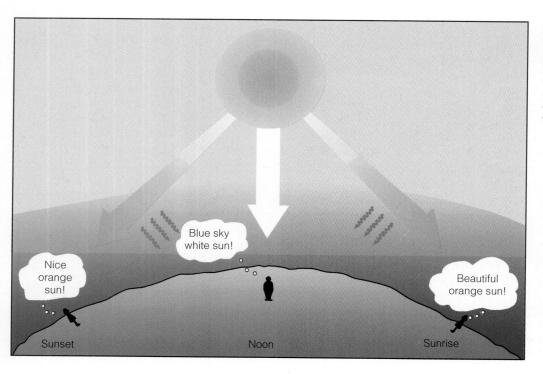

Figure 15.6
Because of the selective scattering by a thick section of atmosphere, the sun at sunrise and sunset appears either yellow, orange, or red. At noon, it is usually white.

roughly uniform in size, they can selectively scatter the sun's rays. Even at noon, various colored suns have appeared: orange suns, green suns, and even blue suns. For blue suns to appear, the size of the suspended particles must be similar to the wavelength of visible light. When these particles are present they tend to scatter red light more than blue, which causes a bluing of the sun. Although rare, the same phenomenon can happen to moonlight, making the moon appear blue; thus, the expression "once in a blue moon."

In summary, the scattering of light by small particles in the atmosphere causes many familiar effects: white clouds, blue skies, hazy skies, crepuscular rays, and colorful sunsets. In the absence of any scattering, we would simply see a white sun against a black sky—not an attractive alternative.

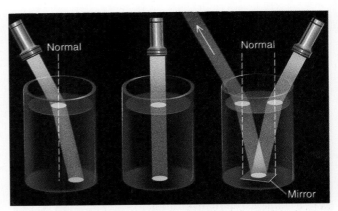

Figure 15.7
The behavior of light as it enters and leaves a more-dense substance, such as water.

Twinkling, Twilight, and the Green Flash

Light that passes through a substance is said to be *transmitted*. Upon entering a denser substance, transmitted light slows in speed. If it enters the substance at an angle, the light's path also bends. This bending is called **refraction**. The amount of refraction depends primarily on two factors: the density of the material and the angle at which the light enters the material.

Refraction can be demonstrated in a darkened room by shining a flashlight into a beaker of water (Fig. 15.7). If the light is held directly above the water so that the beam strikes the surface of the water straight on, no bending occurs. But, if the light enters the water at some angle, it bends toward the *normal*, which is the dashed line in the diagram running perpendicular to the air-water boundary. (The normal is simply a line that intersects any surface at a right angle. We use it as a reference to see how much bending occurs as light enters and leaves various substances.) A small mirror on the bottom of the beaker reflects the light upward. This reflected light bends away from the normal as it re-enters the air. We can summarize these observations as follows: *Light that travels from a less-dense to a more-dense medium loses speed and bends toward the normal, whereas light that enters a less-dense medium increases in speed and bends away from the normal.*

The refraction of light within the atmosphere causes a variety of visual effects. At night, for exam-

ple, the light from the stars that we see directly above us is not bent, but starlight that enters the earth's atmosphere at an angle is bent (Fig. 15.8). In fact, a star whose light enters the atmosphere just above the horizon has more atmosphere to penetrate and is thus refracted the most. By the time this "bent" starlight reaches our eyes, the star appears to be higher than it actually is because our eyes cannot detect that the light path is bent. We see light coming from a particular direction and interpret the star to be in that direction. So, the next time you take a midnight stroll, point to any star near the horizon and remember: this is where the star appears to be. To point to the star's true position, you would have to lower your arm just a tiny bit.

As starlight enters the atmosphere, it often passes through regions of differing air density. Each of these regions deflects and bends the tiny beam of starlight, constantly changing the apparent position of the star. This causes the star to appear to *twinkle* or flicker, a condition known as **scintillation**. Planets, being much closer to us, appear larger, and usually do not twinkle because their size is greater than the angle at which their light deviates as it penetrates the atmosphere. Planets sometimes twinkle, however, when they are near the horizon, where the bending of their light is greatest.

Did you know?
The moon has no atmosphere. Consequently, its sky color is always black and there is no twilight.

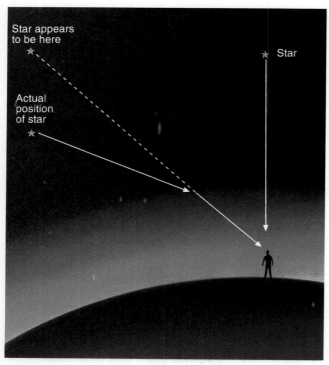

Figure 15.8
Due to the bending of starlight by the atmosphere, stars not directly overhead appear to be higher than they really are.

The refraction of light by the atmosphere has some other interesting consequences. For example, the atmosphere gradually bends the rays from a rising or setting sun or moon. Because light rays from the lower part of the sun (or moon) are bent more than those from the upper part, the sun appears to flatten out on the horizon, taking on an elliptical shape. Also, since light is bent most on the horizon, the sun and moon both appear to be higher than they really are (Fig. 15.9). Consequently, they both rise about two minutes earlier and set about two minutes later than they would if there were no atmosphere.

You may have noticed that on clear days the sky is often bright for some time after the sun sets. The atmosphere refracts and scatters sunlight to our eyes, even though the sun itself has disappeared from our view. **Twilight** is the name given to the time after sunset (and immediately before sunrise) when the sky remains illuminated and allows outdoor activities to continue without artificial lighting.

The length of twilight depends on season and latitude. During the summer in middle latitudes, twilight adds about thirty minutes of light to each morning and evening for outdoor activities. The duration of twilight increases with increasing latitude, especially in summer. At high latitudes during the summer, morning and evening twilight may converge, producing a *white night*—a nightlong twilight.

The color of twilight can change when particles of volcanic ash and dust are present in the upper atmosphere. For example, after the volcanic eruptions of El Chichón in 1982 and Mount Pinatubo in 1991, brilliant red twilights occurred over parts of North America as sulfur-rich volcanic particles scattered red light from the setting (and rising) sun. (See Fig. 15.10.)

In general, without the atmosphere, there would be no refraction or scattering, and the sun would rise

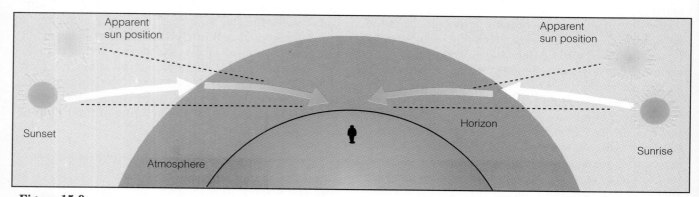

Figure 15.9
The bending of sunlight by the atmosphere causes the sun to rise about two minutes earlier, and set about two minutes later, than it would otherwise.

Figure 15.10
Bright twilight sky over California produced by the sulfur–rich particles from the volcano
Mount Pinatubo during September, 1992.

later and set earlier than it now does. Instead of twilight, darkness would arrive immediately when the sun disappears below the horizon. Imagine the number of sandlot baseball games that would be called because of instant darkness.

Occasionally, a flash of green light—called the **green flash**—may be seen near the upper rim of a rising or setting sun (Fig. 15.11). Remember from our earlier discussion that, when the sun is near the horizon, its light must penetrate a thick section of atmosphere. This thick atmosphere refracts sunlight, with purple and blue light bending the most, and red light the least. Because of this bending, more blue light should appear along the top of the sun. But because the atmosphere selectively scatters blue light, very little reaches us, and we see green light instead.

Usually, the green light is too faint to see with the human eye. However, under certain atmospheric conditions, such as when the surface air is very hot or when an upper-level inversion exists, the green light is magnified by the atmosphere. When this happens, a momentary flash of green light appears, often just before the sun disappears from view.

The flash usually lasts about a second, although in polar regions it can last longer. Here, the sun slowly

Did you know?
An old Scottish legend says that if you see the green flash you will not err in matters of love.

Figure 15.11
The very light green on the upper rim of the sun is the green flash. Also, observe how the atmosphere makes the sun appear to flatten on the horizon into an elliptical shape.

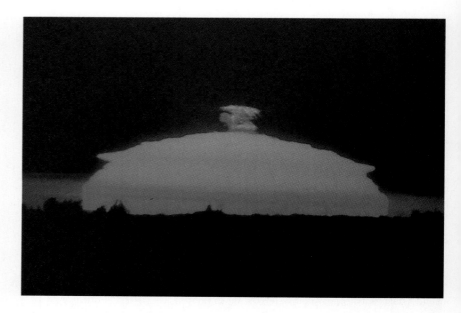

changes in elevation and the flash may exist for many minutes. Members of Admiral Byrd's expedition in the south polar region reported seeing the green flash for 35 minutes in September as the sun slowly rose above the horizon, marking the end of the long winter.

The Mirage: Seeing Is Not Believing

In the atmosphere, when an object appears to be displaced from its true position, we call this phenomenon a **mirage**. A mirage is not a figment of the imagination—our minds are not playing tricks on us, but the atmosphere is.

Atmospheric mirages are created by light passing through and being bent by air layers of different densities. Such changes in air density are usually caused by sharp changes in air temperature. The greater the rate of temperature change, the greater the light rays are bent. For example, on a warm, sunny day, black road surfaces absorb a great deal of solar energy and be-

Did you know?
On December 14, 1890, a mirage—lasting several hours—gave the residents of Saint Vincent, Minnesota, a clear view of cattle nearly eight miles away.

come very hot. Air in contact with these hot surfaces warms by conduction and, because air is a poor thermal conductor, we find much cooler air only a few meters higher. On hot days, these road surfaces often appear wet. (See Fig. 15.12.) Such "puddles" disappear as we approach them, and advancing cars seem to swim in them. Yet, we know the road is dry. The apparent wet pavement above a road is the result of blue sky-light refracting up into our eyes as it travels through air of different densities. A similar type of mirage occurs in deserts during the hot summer. Many thirsty travelers have been disappointed to find that what appeared to be a water hole was in actuality hot desert sand.

Sometimes, these "watery" surfaces appear to *shimmer*. The shimmering results as rising and sinking air near the ground constantly change the air density. As light moves through these regions, its path also changes, causing the shimmering effect.

When the air near the ground is much warmer than the air above, objects may not only appear to be lower than they really are, but also (often) inverted. These mirages are called **inferior** (lower) **mirages**. The tree in Fig. 15.13 certainly doesn't grow upside down. So why does it look that way? It appears to be inverted because light reflected from the top of the tree moves outward in all directions. Rays that enter the hot, less-dense air above the sand are refracted upward, entering the eye

Figure 15.12
The road in the photo appears wet because blue skylight is bending up into the camera as the light passes through air of different densities.

from below. The brain is fooled into thinking that these rays came from below the ground, which makes the tree appear upside down. Some light from the top of the tree travels directly toward the eye through air of nearly constant density and, therefore, bends very little. These rays reach the eye "straight-on," and the tree appears upright. Hence, off in the distance, we see a tree and its upsidedown image beneath it. (Some of the trees in Fig. 15.12 show this effect.)

The atmosphere can play optical jokes on us in extremely cold areas, too. In polar regions, air next to a snow surface can be much colder than the air many meters above. Because the air in this cold layer is very dense, light from distant objects entering it bends toward the normal in such a way that the objects can appear to be shifted upward. This phenomenon is called a **superior** (upward) **mirage**. Figure 15.14 shows the conditions favorable for a superior mirage. (A special type of mirage, the *Fata Morgana*, is described in the Focus section on p. 384.)

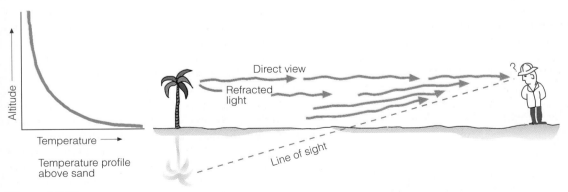

Temperature profile
above sand

Figure 15.13
Inferior mirage over hot desert sand.

Focus on an Observation
The *Fata Morgana*

A special type of superior mirage is the *Fata Morgana*, a mirage that transforms a fairly uniform horizon into one of vertical walls and columns with spires. (See Fig. 1.) According to legend, *Fata Morgana* (Italian for fairy Morgan) was the half-sister of King Arthur. Morgan, who was said to live in a crystal palace beneath the water, had magical powers that could build fantastic castles out of thin air. Looking across the Straits of Messina (between Italy and Sicily), residents of Reggio, Italy, on occasion would see buildings, castles, and sometimes whole cities appear, only to vanish again in minutes. The *Fata Morgana* is observed where the air temperature increases with height above the surface, slowly at first, then more rapidly, then slowly again. Consequently, mirages like the *Fata Morgana* are frequently seen where warm air rests above a cold surface, such as above large bodies of water and in polar regions.

Figure 1
The *Fata Morgana* mirage over water.

Halos, Sundogs, and Sun Pillars

A ring of light encircling and extending outward from the sun or moon is called a **halo**. Such a display is produced when sunlight or moonlight is refracted as it passes through ice crystals. Hence, the presence of a halo indicates that *cirriform clouds* are present. (See Chapter 4.)

The most common type of halo is the 22° halo—a ring of light 22° from the sun or moon.* Such a halo forms when tiny suspended ice crystals become randomly oriented as air molecules constantly bump against them. The refraction of light rays through these

---

*Extend your arm and spread your fingers apart. An angle of 22° is about the distance from the tip of the thumb to the tip of the little finger.

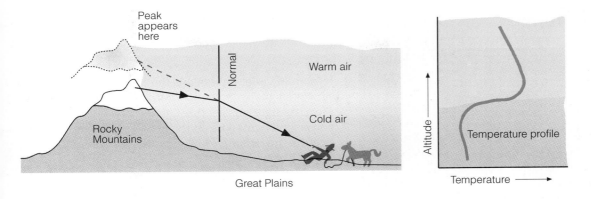

Figure 15.14
The formation of a superior mirage. When cold air lies close to the surface with warm air aloft, light from distant mountains is refracted away from the normal as it enters the cold air. This causes an observer on the ground to see mountains higher and closer than they really are.

crystals forms a halo like the one shown in Fig. 15.15. Less common is the 46° halo, which forms in a similar fashion to the 22° halo. (See Fig. 15.16.) With the 46° halo, however, the light is refracted through hexagonal column-type ice crystals that have the shape of tiny pencils.

A halo is usually seen as a bright, white ring, but there are refraction effects that can cause it to have color. To understand this, we must first examine refraction more closely.

When white light passes through a glass prism, it is refracted and split into a spectrum of visible colors (Fig. 15.17). Each wavelength of light is slowed by the glass, but each is slowed a little differently. Because longer wavelengths (red) slow the least and shorter wavelengths (violet) slow the most, red light bends the

Figure 15.15
A 22° halo around the sun, produced by the refraction of sunlight through ice crystals.

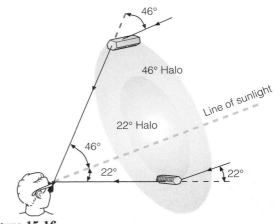

Figure 15.16
The formation of a 22° and a 46° halo.

Figure 15.17
Refraction and dispersion of light through a glass prism.

least, and violet light bends the most. The breaking up of white light by "selective" refraction is called **dispersion**. As light passes through ice crystals, dispersion causes red light to be on the inside of the halo and blue light on the outside.

When hexagonal platelike ice crystals are present in the air, they tend to fall slowly and orient themselves horizontally (Fig. 15.18). (The horizontal orientation of these ice crystals prevents a ring halo.) In this position, the ice crystals act as small prisms, refracting and dispersing sunlight that passes through them. If the sun is near the horizon in such a configuration that it, ice crystals, and observer are all in the same horizontal plane, the observer will see a pair of brightly colored spots, one on either side of the sun. These colored spots are

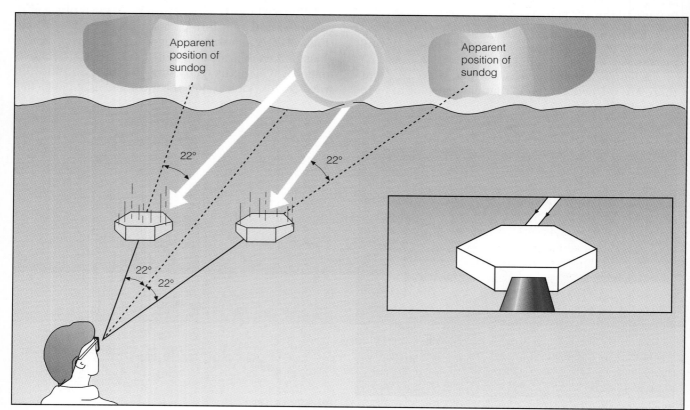

Figure 15.18
Platelike ice crystals falling with their flat surfaces parallel to the earth produce sundogs.

Figure 15.19
A sundog photographed at Bartlett Cove, Glacier Bay, Alaska.

called **sundogs**, *mock suns*, or *parhelia* (meaning "with the sun"). (See Fig. 15.19.) The colors usually grade from red (bent least) on the inside closest to the sun to blue (bent more) on the outside.

While sundogs and halos are caused by *refraction* of sunlight *through* ice crystals, **sun pillars** are caused by *reflection* of sunlight *off* ice crystals. Sun pillars appear most often at sunrise or sunset as a vertical shaft of light extending upward or downward from the sun. (See Fig. 15.20.) Pillars may form as hexagonal plate-like ice crystals fall with their flat bases oriented horizontally. As the tiny crystals fall in still air, they tilt from side to side like a falling leaf. This allows sunlight to reflect off the tipped surfaces of the crystals, producing a relatively bright area in the sky above the sun. Pillars may also form as sunlight reflects off hexagonal pencil-shaped ice crystals that fall with their long axes oriented horizontally. As these crystals fall, they can rotate about their horizontal axes, producing many orientations that reflect sunlight. So, look for sun pillars

Figure 15.20
A sun pillar produced by the reflection of sunlight off ice crystals.

when the sun is low on the horizon and cirriform (ice crystal) clouds are present.

Rainbows

Now we come to one of the most spectacular light shows observed on the earth—the rainbow. **Rainbows** occur when rain is falling in one part of the sky, and the sun is shining in another. (Rainbows also may form by the sprays from waterfalls and water sprinklers.) To see the rainbow, we must face the falling rain with the sun at our backs. Look at Fig. 15.21 closely and note that, when we see a rainbow in the evening, we are facing east toward the rainshower. Behind us—in the west—it is clear. Because clouds tend to move from west to east in middle latitudes, the clear skies in the west suggest that the showers will give way to clearing. However, when we see a rainbow in the morning, we are facing west, toward the rainshower. It is a good bet that the clouds and showers will move toward us and

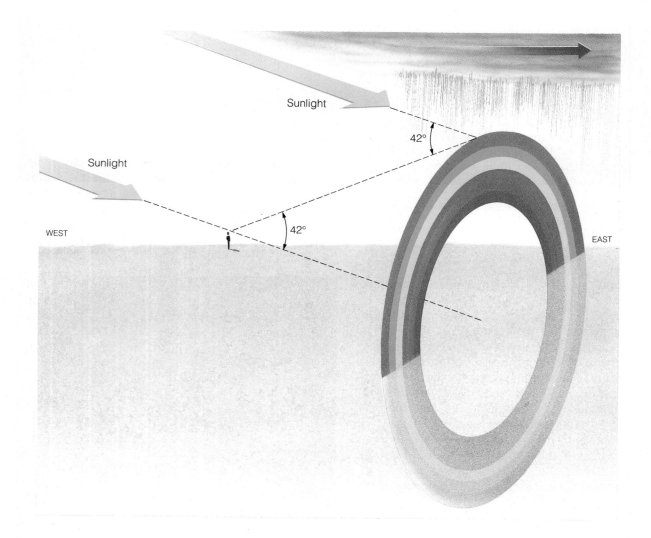

Figure 15.21
When you observe a rainbow, the sun is always to your back. In middle latitudes a rainbow in the evening indicates that clearing weather is ahead.

it will rain soon. These observations explain why the following weather rhyme became popular:

> Rainbow in morning, sailors take warning
> Rainbow at night, a sailor's delight.*

When we look at a rainbow we are looking at sunlight that has entered the falling drops, and, in effect, has been redirected back toward our eyes. Exactly how this process happens requires some discussion.

As sunlight enters a raindrop, it slows and bends, with violet light refracting the most and red light the least. (Fig. 15.22.) Although most of this light passes right on through the drop and is not seen by us, some of it strikes the backside of the drop at such an angle that it is reflected within the drop. The angle at which this occurs is called the *critical angle*. For water, this angle is 48°. Light that strikes the back of a raindrop at an angle exceeding the critical angle, bounces off the back of the drop and is *internally reflected* toward our eyes. Because each light ray bends differently from the rest, each ray emerges from the drop at a slightly different angle. For red light, the angle is 42° from the beam of sunlight; for violet light, it is 40°. (See Fig.

---

*This rhyme is often used with the words "red sky" in the place of rainbow. The red sky makes sense when we consider that it is the result of red light from a rising or setting sun being reflected from the underside of clouds above us. In the morning, a red sky indicates that it is clear to the east and cloudy to the west. A red sky in the evening suggests the opposite.

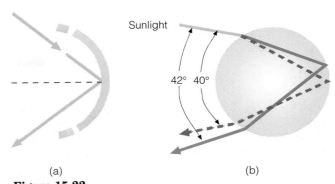

(a)   (b)

Figure 15.22
Sunlight internally reflected and dispersed by a raindrop. (a) The light ray is internally reflected only when it strikes the backside of the drop at an angle greater than the critical angle for water. (b) Refraction of the light as it enters the drop causes the point of reflection (on the back of the drop) to be different for each color. Hence, the colors are separated from each other when the light emerges from the raindrop.

15.22b.) The light leaving the drop is, therefore, dispersed into a spectrum of colors from red to violet. Since we see only a single color from each drop, it takes myriads of raindrops (each refracting and reflecting light back to our eyes at slightly different angles) to produce the brilliant colors of a *primary rainbow*.

Figure 15.22 might lead us erroneously to believe that red light should be at the bottom of the bow and violet at the top. A more careful observation of the behavior of light leaving two drops (Fig. 15.23) shows us

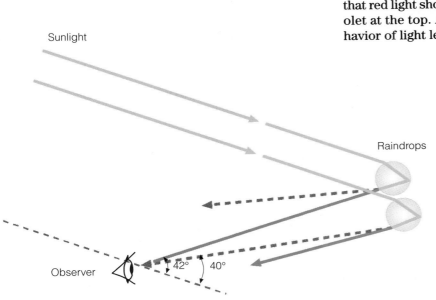

Figure 15.23
The formation of a primary rainbow. The observer sees red light from the upper drop and violet light from the lower drop.

Figure 15.24
A primary and a secondary rainbow.

why the reverse is true. When violet light from the *lower drop* reaches an observer's eye, red light from the same drop is incident elsewhere, toward the waist. Notice that red light reaches the observer's eye from the *higher drop*. Because the color red comes from higher drops and the color violet from lower drops, the colors of a primary rainbow change from red on the outside (top) to violet on the inside (bottom).

Frequently, a larger second (secondary) rainbow with its colors reversed can be seen above the primary bow. (See Fig. 15.24.) Usually this secondary bow is much fainter than the primary one. The *secondary bow* is caused when sunlight enters the raindrops at an angle that allows the light to make two internal reflections in each drop. Each reflection weakens the light intensity and makes the bow dimmer. Figure 15.25 shows that the color reversals—with red now at the bottom

Did you know?
During the summer, rainbows are occasionally seen after a thunderstorm. Because of this fact, the Shoshone Indians viewed the rainbow as a giant serpent that would sometimes rub its back on the icy sky and hurl pieces of ice (hail) to the ground.

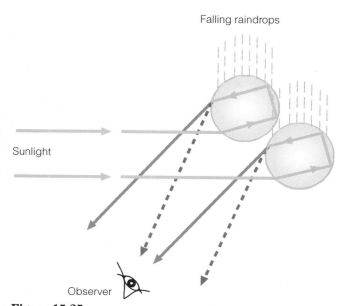

Figure 15.25
Two internal reflections are responsible for the weaker secondary rainbow.

Focus on an Observation
Coronas and Cloud Iridescence

When the moon is seen through a thin veil of clouds composed of tiny spherical water droplets, a bright ring of light, called a *corona* (meaning crown), may appear to rest on the moon (Fig. 2). The same effect can occur with the sun, but, due to the sun's brightness, it is usually difficult to see.

The corona is due to *diffraction*—the bending of light as it passes *around* objects. To understand the corona, imagine water waves moving around a small stone in a pond. As the waves spread around the stone, the trough of one wave may meet the crest of another wave. This situation results in the waves canceling each other, thus producing calm water. Where two crests come together, they produce a much larger wave. The

Figure 2
Corona around the moon, resulting from the diffraction of light by tiny liquid cloud droplets of uniform size.

same thing happens when light passes around tiny cloud droplets. Where light waves come together, we see bright light; where they cancel each other, we see darkness. Sometimes, the corona appears white, with alternating bands of light and dark. On other occasions, the rings have color. The colors of the corona appear when the cloud droplets (or any kind of small particles) are of uniform size.

When different size droplets exist within a cloud, the corona becomes distorted and irregular. Sometimes the cloud exhibits patches of color, often pastel shades of pink, blue, or green. These bright areas produced by diffraction are called *iridescence*. Cloud iridescence is most often seen within 20° of the sun.

Figure 3
Cloud iridescence.

and violet on top—are due to the way the light emerges from each drop after going through two internal reflections.

As you look at a rainbow, keep in mind that only one ray of light is able to enter your eye from each drop. Everytime you move, whether it be up, down, or sideways, the rainbow moves with you. The reason why this happens is that, with every movement, light from different raindrops enters your eye. The bow you see is not exactly the same rainbow that the person standing next to you sees. In effect, each of us has a personal rainbow to ponder and enjoy!

Summary

The scattering of sunlight in the atmosphere can produce a variety of atmospheric visuals, from hazy days and blue skies to crepuscular rays and blue moons. Refraction of light by the atmosphere causes stars near the horizon to appear higher than they really are. It also causes the sun and moon to rise earlier and set later than they otherwise would. Under certain atmospheric conditions, the amplification of green light near the upper rim of a rising or setting sun produces the illusive green flash.

Mirages form when refraction of light displaces objects from their true positions. Inferior mirages cause objects to appear lower than they really are, while superior mirages displace objects upward.

Halos and sundogs form from the refraction of light through ice crystals. Sun pillars are the result of sunlight reflecting off gently falling ice crystals. The refraction, reflection, and dispersion of light in raindrops create a rainbow. To see a rainbow, the sun must be to your back, and rain must be falling in front of you.

Key Terms

The following terms are listed in the order they appear in the text. Define each. Doing so will aid you in reviewing the material covered in this chapter.

reflected light	scintillation	inferior mirage	sundog (or parhelia)
scattered light	twilight	superior mirage	sun pillar
crepuscular rays	green flash	halo	rainbow
refraction (of light)	mirage	dispersion (of light)	

Review Questions

1. (a) Why are cumulus clouds normally white?
    (b) Why do the undersides of building cumulus clouds frequently change color from white to dark gray or even black?
2. Explain why the sky is blue during the day and black at night.
3. What can make a setting (or rising) sun appear red?
4. If the earth had no atmosphere, what color would the sky be during the day?
5. Why, on a hazy day, does the horizon appear white?
6. What process (refraction or scattering) produces crepuscular rays?
7. Why do stars "twinkle"?
8. How does refraction of light differ from reflection of light?
9. How long does twilight last on the moon? (Hint: the moon has no atmosphere.)
10. At what time of day would you expect to observe the green flash?
11. How does light bend as it enters a more-dense substance at an angle? How does it bend upon leaving the more-dense substance? Make a sketch to illustrate your answer.

12. On a clear, dry, warm day, why do dark road surfaces frequently appear wet?

13. What atmospheric conditions are necessary for an inferior mirage? A superior mirage?

14. (a) Describe how a halo forms.
    (b) How is the formation of a halo different from that of a sundog?

15. Would you expect to see a ringed halo if the sky contains a few wispy cirrus clouds? Explain.

16. What process is believed to be mainly responsible for the formation of sun pillars: refraction, reflection, or scattering?

17. Explain the rhyme below:
    Rainbow in morning, joggers take warning.
    Rainbow at night (evening), jogger's delight.

18. With the aid of a diagram, describe why the sun must be at a person's back in order for that person to observe a rainbow.

19. Why are secondary rainbows much dimmer than primary rainbows?

20. How does the appearance of a halo differ from that of a corona?

Units, Conversions, and Abbreviations

Length

1 kilometer (km)	=	1000 meters (m)
	=	3281 feet (ft)
	=	0.62 mile (mi)
1 mile (mi)	=	5280 feet (ft)
	=	1609 meters (m)
	=	1.61 kilometers (km)
1 centimeter (cm)	=	0.39 inch (in.)
	=	0.01 meter (m)
	=	10 millimeters (mm)
1 inch (in.)	=	2.54 cm
	=	0.08 ft
1 meter (m)	=	100 cm
	=	3.28 ft
	=	39.37 in.
1 micrometer (μm)	=	0.0001 cm
	=	0.000001 m
1 degree latitude	=	111 km
	=	60 nautical mi
	=	69 statute mi

Area

1 square centimeter (cm^2)	=	0.15 $in.^2$
1 square inch ($in.^2$)	=	6.45 cm^2
1 square meter (m^2)	=	10.76 ft^2
1 square foot (ft^2)	=	0.09 m^2

Volume

1 cubic centimeter (cm^3)	=	0.06 $in.^3$
1 cubic inch ($in.^3$)	=	16.39 cm^3
1 liter (l)	=	1000 cm^3
	=	0.264 gallon (gal) U.S.

Speed

1 knot	=	1 nautical mi/hr
	=	1.15 statute mi/hr
	=	0.51 m/sec
	=	1.85 km/hr
1 mile per hour (mi/hr)	=	0.87 knot
	=	0.45 m/sec
	=	1.61 km/hr
1 kilometer per hour (km/hr)	=	0.54 knot
	=	0.62 mi/hr
	=	0.28 m/sec
1 meter per second (m/sec)	=	1.94 knots
	=	2.24 mi/hr
	=	3.60 km/hr

Force

1 dyne	=	1 gram centimeter per second per second
	=	2.2481×10^{-6} pound (lb)
1 newton (N)	=	1 kilogram meter per second per second
	=	10^5 dynes
	=	0.2248 lb

Mass

1 gram (g)	=	0.035 ounce
	=	0.002 lb
1 kilogram (kg)	=	1000 g
	=	2.2 lb

Energy

1 erg	=	1 dyne per cm
	=	2.388×10^{-8} cal
1 joule (J)	=	1 newton meter
	=	0.239 cal
	=	10^7 erg
1 calorie (cal)	=	4.186 J
	=	4.186×10^7 erg

Pressure

1 millibar (mb)	=	1000 dynes/cm^2
	=	0.75 millimeter of mercury (mm Hg)
	=	0.02953 inch of mercury (in. Hg)
	=	0.01450 pound per square inch (lb/in.^{2})
	=	100 pascals (Pa)
1 standard atmosphere	=	1013.25 mb
	=	760 mm Hg
	=	29.92 in. Hg
	=	14.7 lb/in.^{2}
1 inch of mercury	=	33.865 mb
1 millimeter of mercury	=	1.3332 mb
	=	0.01 mb
1 pascal	=	1 N/m^2

Power

1 watt (W)	=	1 J/sec
	=	14.3353 cal/min
1 cal/min	=	0.06973 W
1 horse power (hp)	=	746 W

Powers of Ten

Prefix

nano	one-billionth	=	10^{-9}	=	0.000000001
micro	one-millionth	=	10^{-6}	=	0.000001
milli	one-thousandth	=	10^{-3}	=	0.001
centi	one-hundredth	=	10^{-2}	=	0.01
deci	one-tenth	=	10^{-1}	=	0.1
hecto	one hundred	=	10^{2}	=	100
kilo	one thousand	=	10^{3}	=	1000
mega	one million	=	10^{6}	=	1,000,000
giga	one billion	=	10^{9}	=	1,000,000,000

Temperature $°C = 5/9 \ (°F -32)$

To convert degrees Fahrenheit (°F) to degrees Celsius (°C): Subtract 32 degrees from °F, then divide by 1.8

To convert degrees Celsius (°C) to degrees Fahrenheit (°F): Multiply °C by 1.8, then add 32 degrees

To convert degrees Celsius (°C) to Kelvins (K): Add 273 to Celsius temperature, as

$$K = °C + 273.$$

Table A.1   Temperature Conversions

°F	°C	°F	°C	°F	°C	°F	°C	°F	°C	°F	°C	°F	°C	°F	°C
−40	−40	−20	−28.9	0	−17.8	20	−6.7	40	4.4	60	15.6	80	26.7	100	37.8
−39	−39.4	−19	−28.3	1	−17.2	21	−6.1	41	5.0	61	16.1	81	27.2	101	38.3
−38	−38.9	−18	−27.8	2	−16.7	22	−5.6	42	5.6	62	16.8	82	27.8	102	38.9
−37	−38.3	−17	−27.2	3	−16.1	23	−5.0	43	6.1	63	17.2	83	28.3	103	39.4
−36	−37.8	−16	−26.7	4	−15.6	24	−4.4	44	6.7	64	17.8	84	28.9	104	40.0
−35	−37.2	−15	−26.1	5	−15.0	25	−3.9	45	7.2	65	18.3	85	29.4	105	40.6
−34	−36.7	−14	−25.6	6	−14.4	26	−3.3	46	7.8	66	18.9	86	30.0	106	41.1
−33	−36.1	−13	−25.0	7	−13.9	27	−2.8	47	8.3	67	19.4	87	30.6	107	41.7
−32	−35.6	−12	−24.4	8	−13.3	28	−2.2	48	8.9	68	20.0	88	31.1	108	42.2
−31	−35.0	−11	−23.9	9	−12.8	29	−1.7	49	9.4	69	20.6	89	31.7	109	42.8
−30	−34.4	−10	−23.3	10	−12.2	30	−1.1	50	10.0	70	21.1	90	32.2	110	43.3
−29	−33.9	−9	−22.8	11	−11.7	31	−0.6	51	10.6	71	21.7	91	32.8	111	43.9
−28	−33.3	−8	−22.2	12	−11.1	32	0.0	52	11.1	72	22.2	92	33.3	112	44.4
−27	−32.8	−7	−21.7	13	−10.6	33	0.6	53	11.7	73	22.8	93	33.9	113	45.0
−26	−32.2	−6	−21.1	14	−10.0	34	1.1	54	12.2	74	23.3	94	34.4	114	45.6
−25	−31.7	−5	−20.6	15	−9.4	35	1.7	55	12.8	75	23.9	95	35.0	115	46.1
−24	−31.1	−4	−20.0	16	−8.9	36	2.2	56	13.3	76	24.4	96	35.6	116	46.7
−23	−30.6	−3	−19.4	17	−8.3	37	2.8	57	13.9	77	25.0	97	36.1	117	47.2
−22	−30.0	−2	−18.9	18	−7.8	38	3.3	58	14.4	78	25.6	98	36.7	118	47.8
−21	−29.4	−1	−18.3	19	−7.2	39	3.9	59	15.0	79	26.1	99	37.2	119	48.3

Table A.2   SI Units* and Their Symbols

Quantity	Name	Units	Symbol
length	meter	m	m
mass	kilogram	kg	kg
time	second	sec	sec
temperature	Kelvin	K	K
density	kilogram per cubic meter	kg/m^3	kg/m^3
speed	meter per second	m/sec	m/sec
force	newton	$m \cdot kg/sec^2$	N
pressure	pascal	N/m^2	Pa
energy	joule	$N \cdot m$	J
power	watt	J/sec	W

*SI stands for Système International, which is the international system of units and symbols.

Appendix B

Equations and Constants

Gas Law (Equation of State)

The relationship among air pressure, air density, and air temperature can be expressed by

Pressure = density × temperature × constant.

This relationship, often called the gas law (or equation of state), can be expressed in symbolic form as:

$$p = \rho R T$$

where p is air pressure, ρ is air density, R is a constant, and T is air temperature.

Units/Constants
p = pressure in N/m^2 (SI)
ρ = density (kg/m^3)
T = temperature (K)
R = 287 J/kg · K (SI) or
R = 2.87 × 10^6 erg/g · K

Stefan-Boltzmann Law

The Stefan-Boltzmann law is a law of radiation. It states that all objects with temperatures above absolute zero emit radiation at a rate proportional to the fourth power of their absolute temperature. It is expressed mathematically as:

$$E = \sigma T^4$$

where E is the maximum rate of radiation emitted each second per unit surface area, T is the object's surface temperature, and σ is a constant.

Units/Constants
E = radiation emitted in W/m^2 (SI)
σ = 5.67 × 10^{-8} W/m^2 · K^4 (SI) or
σ = 5.67 × 10^{-5} erg/cm^2 · K^4 · sec
T = temperature (K)

Wien's Law

Wien's law relates an object's maximum emitted wavelength of radiation to the object's temperature. It states that the wavelength of maximum emitted radiation by an object is inversely proportional to the object's absolute temperature. In symbolic form, it is written as:

$$\lambda_{max} = \frac{w}{T}$$

where λ_{max} is the wavelength at which maximum radiation emission occurs, T is the object's temperature, and w is a constant.

Units/Constants		
λ_{max}	=	wavelength (micrometers)
w	=	0.2897 μm K
T	=	temperature (K)

Geostrophic Wind Equation

The geostrophic wind equation gives an approximation of the wind speed above the level of friction, where the wind blows parallel to the isobars or contours. The equation is expressed mathematically as:

$$V_g = \frac{1}{2\Omega \sin\phi \rho} \frac{\Delta p}{d}$$

where V_g is the geostrophic wind, Ω is a constant (twice the earth's angular spin), $\sin\phi$ is a trigonometric function that takes into account the variation of latitude (ϕ), ρ is the air density, Δp is the pressure difference between two places on the map some horizontal distance (d) apart.

Units/Constants		
V_g	=	geostrophic wind (m/sec)
Ω	=	7.29×10^{-5} radian*/sec
ϕ	=	latitude
ρ	=	air density (kg/m³)
d	=	distance (meters)
Δp	=	pressure difference (newton/m²)

*2π radians equal 360°.

Hydrostatic Equation

The hydrostatic equation relates to how quickly the air pressure decreases in a column of air above the surface. The equation tells us that the rate at which the air pressure decreases with height is equal to the air density times the acceleration of gravity. In symbolic form, it is written as:

$$\frac{\Delta p}{\Delta z} = -\rho g$$

where Δp is the decrease in pressure along a small change in height Δz, ρ is the air density, and g is the force of gravity.

Units/Constants		
Δp	=	pressure difference (newton/m²)
Δz	=	change in height (meters)
ρ	=	air density (kg/m³)
g	=	force of gravity (9.8m/sec²)

▲▼▲
Relative Humidity

The relative humidity of the air can be expressed as:

$$RH = \frac{e}{e_s} \times 100\%.$$

To determine e and e_s when the air temperature and dew-point temperature are known, consult Table B.1. Simply read the value adjacent to the air temperature and obtain e_s; read the value adjacent to the dew-point temperature and obtain e.

Units/Constants		
e	=	actual vapor pressure (millibars)
e_s	=	saturation vapor pressure (millibars)
RH	=	relative humidity (percent)

Table B.1 Saturation Vapor Pressures over Water for Various Air Temperatures

Temperature (0°C)	Vapor Pressure (millibars)	Temperature (0°C)	Vapor Pressure (millibars)
−30	0.5	10	12.3
−25	0.8	15	17.0
−20	1.2	20	23.4
−15	1.9	25	31.7
−10	2.9	30	42.4
−5	4.2	35	56.2
0	6.1	40	73.8
5	8.7	45	95.8

Weather Symbols and the Station Model

Simplified Surface-Station Model

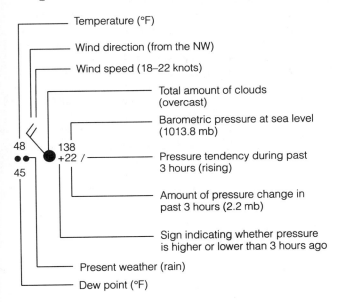

- Temperature (°F)
- Wind direction (from the NW)
- Wind speed (18–22 knots)
- Total amount of clouds (overcast)
- Barometric pressure at sea level (1013.8 mb)
- Pressure tendency during past 3 hours (rising)
- Amount of pressure change in past 3 hours (2.2 mb)
- Sign indicating whether pressure is higher or lower than 3 hours ago
- Present weather (rain)
- Dew point (°F)

48
•• 138
45 +22 /

Total Sky Cover

◯	No clouds
◔	Less than one-tenth or one-tenth
◔	Two-tenths or three-tenths
◕	Four-tenths
◑	Five-tenths
⊖	Six-tenths
◕	Seven-tenths or eight-tenths
◉	Nine-tenths or overcast with openings
●	Completely overcast
⊗	Sky obscured

Upper-Air Model (500 mb)

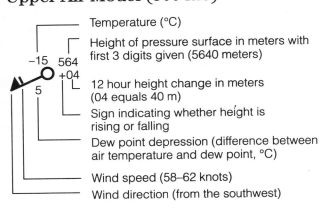

- Temperature (°C)
- Height of pressure surface in meters with first 3 digits given (5640 meters)
- 12 hour height change in meters (04 equals 40 m)
- Sign indicating whether height is rising or falling
- Dew point depression (difference between air temperature and dew point, °C)
- Wind speed (58–62 knots)
- Wind direction (from the southwest)

−15 564
5   +04

Common Weather Symbols

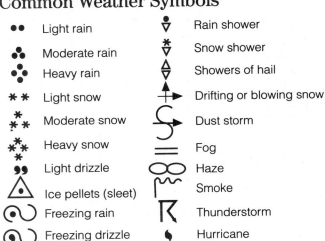

••	Light rain		Rain shower
⦂•	Moderate rain		Snow shower
•••	Heavy rain		Showers of hail
∗ ∗	Light snow		Drifting or blowing snow
∗∗∗	Moderate snow		Dust storm
∗∗∗	Heavy snow		Fog
,,	Light drizzle		Haze
△	Ice pellets (sleet)		Smoke
∿	Freezing rain		Thunderstorm
∿	Freezing drizzle		Hurricane

401

Wind Entries

	Miles (statute) per hour	Knots	Kilometers per Hour
Calm	Calm	Calm	Calm
	1–2	1–2	1–3
	3–8	3–7	4–13
	9–14	8–12	14–19
	15–20	13–17	20–32
	21–25	18–22	33–40
	26–31	23–27	41–50
	32–37	28–32	51–60
	38–43	33–37	61–69
	44–49	38–42	70–79
	50–54	43–47	80–87
	55–60	48–52	88–96
	61–66	53–57	97–106
	67–71	58–62	107–114
	72–77	63–67	115–124
	78–83	68–72	125–134
	84–89	73–77	135–143
	119–123	103–107	144–198

Pressure Tendency

Rising, then falling

Rising, then steady; or rising, then rising more slowly

Rising steadily or unsteadily

Falling or steady, then rising; or rising, then rising more quickly

⎫ Barometer now higher than 3 hours ago

Steady, same as 3 hours ago

Falling, then rising, same or lower than 3 hours ago

Falling, then steady; or falling, then falling more slowly

Falling steadily, or unsteadily

Steady or rising, then falling; or falling, then falling more quickly

⎫ Barometer now lower than 3 hours ago

Front Symbols

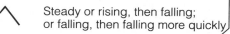 Cold front (surface)

Warm front (surface)

Occluded front (surface)

 Stationary front (surface)

 Warm front (aloft)

 Cold front (aloft)

—— •• —— Squall line

Appendix D

Humidity and Dew-Point Tables

To obtain the dew point (or relative humidity), simply read down the temperature column and then over to the wet-bulb depression. For example, in Table D.1, a temperature of 10°C with a wet-bulb depression of 3°C produces a dew-point temperature of 4°C.

Table D.1 Dew-Point Temperature (°C)

Air (Dry-Bulb) Temperature (°C)	Wet-Bulb Depression (Dry-Bulb Temperature Minus Wet-Bulb Temperature) (°C)																
	0.5	1.0	1.5	2.0	2.5	3.0	3.5	4.0	4.5	5.0	7.5	10.0	12.5	15.0	17.5	20.0
−20	−25	−33														
−17.5	−21	−27	−38													
−15	−19	−23	−28													
−12.5	−15	−18	−22	−29												
−10	−12	−14	−18	−21	−27	−36										
−7.5	−9	−11	−14	−17	−20	−26	−34									
−5	−7	−8	−10	−13	−16	−19	−24	−31								
−2.5	−4	−6	−7	−9	−11	−14	−17	−22	−28	−41						
0	−1	−3	−4	−6	−8	−10	−12	−15	−19	−24						
2.5	1	0	−1	−3	−4	−6	−8	−10	−13	−16						
5	4	3	2	0	−1	−3	−4	−6	−8	−10	−48					
7.5	6	6	4	3	2	1	−1	−2	−4	−6	−22					
10	9	8	7	6	5	4	2	1	0	−2	−13					
12.5	12	11	10	9	8	7	6	4	3	2	−7	−28				
15	14	13	12	12	11	10	9	8	7	5	−2	−14				
17.5	17	16	15	14	13	12	12	11	10	8	2	−7	−35			
20	19	18	18	17	16	15	14	14	13	12	6	−1	−15			
22.5	22	21	20	20	19	18	17	16	16	15	10	3	−6	−38		
25	24	24	23	22	21	21	20	19	18	18	13	7	0	−14		
27.5	27	26	26	25	24	23	23	22	21	20	16	11	5	−5	−32	
30	29	29	28	27	27	26	25	25	24	23	19	14	9	2	−11	
32.5	32	31	31	30	29	29	28	27	26	26	22	18	13	7	−2	
35	34	34	33	32	32	31	31	30	29	28	25	21	16	11	4	
37.5	37	36	36	35	34	34	33	32	32	31	28	24	20	15	9	0
40	39	39	38	38	37	36	36	35	34	34	30	27	23	18	13	6
42.5	42	41	41	40	40	39	38	38	37	36	33	30	26	22	17	11
45	44	44	43	43	42	42	41	40	40	39	36	33	29	25	21	15
47.5	47	46	46	45	45	44	44	43	42	42	39	35	32	28	24	19
50	49	49	48	48	47	47	46	45	45	44	41	38	35	31	28	23

Table D.2   Relative Humidity (Percent)

Air (Dry-Bulb) Temperature (°C)	Wet-Bulb Depression (Dry-Bulb Temperature Minus Wet-Bulb Temperature) (°C)																	
	0.5	1.0	1.5	2.0	2.5	3.0	3.5	4.0	4.5	5.0	7.5	10.0	12.5	15.0	17.5	20.0	22.5	25.0
−20	70	41	11															
−17.5	75	51	26	2														
−15	79	58	38	18														
−12.5	82	65	47	30	13													
−10	85	69	54	39	24	10												
−7.5	87	73	60	48	35	22	10											
−5	88	77	66	54	43	32	21	11	1									
−2.5	90	80	70	60	50	42	37	22	12	3								
0	91	82	73	65	56	47	39	31	23	15								
2.5	92	84	76	68	61	53	46	38	31	24								
5	93	86	78	71	65	58	51	45	38	32	1							
7.5	93	87	80	74	68	62	56	50	44	38	11							
10	94	88	82	76	71	65	60	54	49	44	19							
12.5	94	89	84	78	73	68	63	58	53	48	25	4						
15	95	90	85	80	75	70	66	61	57	52	31	12						
17.5	95	90	86	81	77	72	68	64	60	55	36	18	2					
20	95	91	87	82	78	74	70	66	62	58	40	24	8					
22.5	96	92	87	83	80	76	72	68	64	61	44	28	14	1				
25	96	92	88	84	81	77	73	70	66	63	47	32	19	7				
27.5	96	92	89	85	82	78	75	71	68	65	50	36	23	12	1			
30	96	93	89	86	82	79	76	73	70	67	52	39	27	16	6			
32.5	97	93	90	86	83	80	77	74	71	68	54	42	30	20	11	1		
35	97	93	90	87	84	81	78	75	72	69	56	44	33	23	14	6		
37.5	97	94	91	87	85	82	79	76	73	70	58	46	36	26	18	10	3	
40	97	94	91	88	85	82	79	77	74	72	59	48	38	29	21	13	6	2
42.5	97	94	91	88	86	83	80	78	75	72	61	50	40	31	23	16	9	6
45	97	94	91	89	86	83	81	78	76	73	62	51	42	33	26	18	12	6
47.5	97	94	92	89	86	84	81	79	76	74	63	53	44	35	28	21	15	9
50	97	95	92	89	87	84	82	79	77	75	64	54	45	37	30	23	17	11

Table D.3 Dew-Point Temperature (°F)

Wet-Bulb Depression (Dry-Bulb Temperature Minus Wet-Bulb Temperature (°F)

Air (Dry-Bulb) Temperature (°F)	1	2	3	4	5	6	7	8	9	10	11	12	13	14	15	16	17	18	19	20	25	30	35	40
0	−7	−20																						
5	−1	−9	−24																					
10	5	−2	−10	−27																				
15	11	6	0	−9	−26																			
20	16	12	8	2	−7	−21																		
25	22	19	15	10	5	−3	−15	−51																
30	27	25	21	18	14	8	2	−7	−25															
35	33	30	28	25	21	17	13	7	0	−11														
40	38	35	33	30	28	25	21	18	13	7	−1	−14												
45	43	41	38	36	34	31	28	25	22	18	13	7	−1	−14										
50	48	46	44	42	40	37	34	32	29	26	22	18	13	8	0	−13								
55	53	51	50	48	45	43	41	38	36	33	30	27	24	20	15	9	1	−12						
60	58	57	55	53	51	49	47	45	43	40	38	35	32	29	25	21	17	11	4	−8				
65	63	62	60	59	57	55	53	51	49	47	45	42	40	37	34	31	27	24	19	14				
70	69	67	65	64	62	61	59	57	55	53	51	49	47	44	42	39	36	33	30	26	−11			
75	74	72	71	69	68	66	64	63	61	59	57	55	54	51	49	47	44	42	39	36	15			
80	79	77	76	74	73	72	70	68	67	65	63	62	60	58	56	54	52	50	47	44	28			
85	84	82	81	80	78	77	75	74	72	71	69	68	66	64	62	61	59	57	54	52	39	19		
90	89	87	86	85	83	82	81	79	78	76	75	73	72	70	69	67	65	63	61	59	48	32		
95	94	93	91	90	89	87	86	85	83	81	80	79	78	76	74	73	71	70	68	66	56	43	24	
100	99	98	96	95	94	93	91	90	89	87	86	85	83	82	80	79	77	76	74	72	63	52	37	12
105	104	103	101	100	99	98	96	95	94	93	91	90	89	87	86	84	83	82	80	78	70	61	48	30
110	109	108	106	105	104	103	102	100	99	98	97	95	94	93	91	90	89	87	86	84	77	68	57	43
115	114	113	112	110	109	108	107	106	104	103	102	101	99	98	97	96	94	93	92	90	83	75	65	54
120	119	118	117	115	114	113	112	111	110	108	107	106	105	104	102	101	100	98	97	96	89	81	73	63

Table D.4 Relative Humidity (Percent)

Air (Dry-Bulb) Temperature (°F)	Wet-Bulb Depression (Dry-Bulb Temperature Minus Wet-Bulb Temperature) (°F)																							
	1	2	3	4	5	6	7	8	9	10	11	12	13	14	15	16	17	18	19	20	25	30	35	40
0	67	33	1																					
5	73	46	20																					
10	78	56	34	13																				
15	82	64	46	29	11																			
20	85	70	55	40	26	12																		
25	87	74	62	49	37	25	13	1																
30	89	78	67	56	46	36	26	16	6															
35	91	81	72	63	54	45	36	27	19	10	2													
40	92	83	75	68	60	52	45	37	29	22	15	7												
45	93	86	78	71	64	57	51	44	38	31	25	18	12	6										
50	93	87	80	74	67	61	55	49	43	38	32	27	21	16	10	5								
55	94	88	82	76	70	65	59	54	49	43	38	33	28	23	19	14	9	5						
60	94	89	83	78	73	68	63	58	53	48	43	39	34	30	26	21	17	13	9	5				
65	95	90	85	80	75	70	66	61	56	52	48	44	39	35	31	27	24	20	16	12				
70	95	90	86	81	77	72	68	64	59	55	51	48	44	40	36	33	29	25	22	19	3			
75	96	91	86	82	78	74	70	66	62	58	54	51	47	44	40	37	34	30	27	24	9			
80	96	91	87	83	79	75	72	68	64	61	57	54	50	47	44	41	38	35	32	29	15	3		
85	96	92	88	84	80	76	73	69	66	62	59	56	52	49	46	43	41	38	35	32	20	8		
90	96	92	89	85	81	78	74	71	68	65	61	58	55	52	49	47	44	41	39	36	24	13	3	
95	96	93	89	85	82	79	75	72	69	66	63	60	57	54	51	49	46	43	41	38	27	17	7	1
100	96	93	89	86	83	80	77	73	70	68	65	62	59	56	54	51	49	46	44	41	30	21	12	4
105	97	93	90	87	83	80	77	74	71	69	66	63	60	58	55	53	50	48	46	43	33	23	15	7
110	97	93	90	87	84	81	78	75	73	70	67	65	62	60	57	55	52	50	48	46	36	26	18	11
115	97	94	91	88	85	82	79	76	74	71	68	66	63	61	58	56	54	52	49	47	37	28	21	13
120	97	94	91	88	85	82	80	77	74	72	69	67	65	62	60	58	55	53	51	49	40	31	23	17

Hurricane Names

Names for Hurricanes

North Atlantic Hurricane Names				Eastern Pacific Hurricane Names			
1993	**1994**	**1995**	**1996**	**1993**	**1994**	**1995**	**1996**
Arlene	Alberto	Allison	Arthur	Adrian	Aletta	Adolph	Alma
Bret	Beryl	Barry	Bertha	Beatriz	Bud	Barbara	Boris
Cindy	Chris	Chantal	Cesar	Calvin	Carlotta	Cosme	Christina
Dennis	Debby	Dean	Diana	Dora	Daniel	Dalilia	Douglas
Emily	Ernesto	Erin	Edouard	Eugene	Emilia	Erick	Elida
Floyd	Florence	Felix	Fran	Fernanda	Fabio	Flossie	Fausto
Gert	Gordon	Gabrielle	Gustav	Greg	Gilma	Gil	Genevieve
Harvey	Helene	Humberto	Hortense	Hilary	Hector	Henrietta	Herman
Irene	Isaac	Iris	Isidore	Irwin	Iva	Ismael	Iselle
Jose	Joyce	Jerry	Josephine	Jova	John	Juliette	Julio
Katrina	Keith	Karen	Klaus	Knut	Kristy	Kiko	Kenna
Lenny	Leslie	Luis	Lili	Lidia	Lane	Lorena	Lowell
Maria	Michael	Marily	Marco	Max	Miriam	Manuel	Marie
Nate	Nadine	Noel	Nana	Norma	Norman	Narda	Norbert
Ophelia	Oscar	Opal	Omar	Otis	Olivia	Octave	Odile
Philippe	Patty	Pablo	Paloma	Pilar	Paul	Priscilla	Polo
Rita	Rafael	Roxanne	Rene	Ramon	Rosa	Raymond	Rachel
Stan	Sandy	Sebastien	Sally	Selma	Sergio	Sonia	Simon
Tammy	Tony	Tanya	Teddy	Todd	Tara	Tico	Trudy
Vince	Valerie	Van	Vicky	Veronica	Vicente	Velma	Vance
Wilma	William	Wendy	Wilfred	Wiley	Willa	Winnie	Wallis
				Xina	Xavier		
				York	Yolanda		
				Zelda	Zeke		

Heat Index (HI) Table

Table F.1 Air Temperature (°F) and Relative Humidity Are Combined to Determine an Apparent Temperature or Heat Index (HI). An Air Temperature of 95°F with a Relative Humidity of 55 Percent Produces an Apparent Temperature (HI) of 110°F

Relative Humidity (%)

Air Temperature (°F)	0	5	10	15	20	25	30	35	40	45	50	55	60	65	70	75	80	85	90	95	100
140	125																				
135	120	128																			
130	117	122	131																		
125	111	116	123	131	141																
120	107	111	116	123	130	139	148														
115	103	107	111	115	120	127	135	143	151												
110	99	102	105	108	112	117	123	130	137	143	150										
105	95	97	100	102	105	109	113	118	123	129	135	142	149								
100	91	93	95	97	99	101	104	107	110	115	120	126	132	138	144						
95	87	88	90	91	93	94	96	98	101	104	107	110	114	119	124	130	136				
90	83	84	85	86	87	88	90	91	93	95	96	98	100	102	106	109	113	117	122		
85	78	79	80	81	82	83	84	85	86	87	88	89	90	91	93	95	97	99	102	105	108
80	73	74	75	76	77	77	78	79	79	80	81	81	82	83	85	86	86	87	88	89	91
75	69	69	70	71	72	72	73	73	74	74	75	75	76	76	77	77	78	78	79	79	80
70	64	64	65	65	66	66	67	67	68	68	69	69	70	70	70	70	71	71	71	71	72

Heat Index (or apparent temperature)

Appendix G

Beaufort Wind Scale

Beaufort Number	Description	Wind Speed			Observations
		mi/hr	*knots*	*km/hr*	
0	Calm	0–1	0–1	0–2	Smoke rises vertically
1	Light air	1–3	1–3	2–6	Direction of wind shown by drifting smoke, but not by wind vanes
2	Slight breeze	4–7	4–6	7–11	Wind felt on face; leaves rustle; wind vanes moved by wind; flags stir
3	Gentle breeze	8–12	7–10	12–19	Leaves and small twigs move; wind will extend light flag
4	Moderate breeze	13–18	11–16	20–29	Wind raises dust and loose paper; small branches move; flags flap
5	Fresh breeze	19–24	17–21	30–39	Small trees with leaves begin to sway; flags ripple
6	Strong breeze	25–31	22–27	40–50	Large tree branches in motion; whistling heard in telegraph wires; umbrellas used with difficulty
7	High wind	32–38	28–33	51–61	Whole trees in motion; inconvenience felt walking against wind; flags extend
8	Gale	39–46	34–40	62–74	Wind breaks twigs off trees; walking is difficult
9	Strong gale	47–54	41–47	75–87	Slight structural damage occurs (signs and antennas blown down)
10	Whole gale	55–63	48–55	88–101	Trees uprooted; considerable damage occurs
11	Storm	64–74	56–64	102–119	Winds produce widespread damage
12	Hurricane	≥ 75	≥ 65	≥ 120	Winds produce extensive damage

Table G.1 Estimating Wind Speed from Surface Observations

Standard Atmosphere

Altitude				Pressure	Temperature		Density
meters	feet	km	mi	millibars	°C	°F	kg/m³
0	0	(0.0)	(0.0)	1013.25	15.0	(59.0)	1.225
500	1,640	(0.5)	(0.3)	954.61	11.8	(53.2)	1.167
1,000	3,280	(1.0)	(0.6)	898.76	8.5	(47.3)	1.112
1,500	4,921	(1.5)	(0.9)	845.59	5.3	(41.5)	1.058
2,000	6,562	(2.0)	(1.2)	795.01	2.0	(35.6)	1.007
2,500	8,202	(2.5)	(1.5)	746.91	−1.2	(29.8)	0.957
3,000	9,842	(3.0)	(1.9)	701.21	−4.5	(23.9)	0.909
3,500	11,483	(3.5)	(2.2)	657.80	−7.7	(18.1)	0.863
4,000	13,123	(4.0)	(2.5)	616.60	−11.0	(12.2)	0.819
4,500	14,764	(4.5)	(2.8)	577.52	−14.2	(6.4)	0.777
5,000	16,404	(5.0)	(3.1)	540.48	−17.5	(0.5)	0.736
5,500	18,045	(5.5)	(3.4)	505.39	−20.7	(−5.3)	0.697
6,000	19,685	(6.0)	(3.7)	472.17	−24.0	(−11.2)	0.660
6,500	21,325	(6.5)	(4.0)	440.75	−27.2	(−17.0)	0.624
7,000	22,965	(7.0)	(4.3)	411.05	−30.4	(−22.7)	0.590
7,500	24,606	(7.5)	(4.7)	382.99	−33.7	(−28.7)	0.557
8,000	26,247	(8.0)	(5.0)	356.51	−36.9	(−34.4)	0.526
8,500	27,887	(8.5)	(5.3)	331.54	−40.2	(−40.4)	0.496
9,000	29,528	(9.0)	(5.6)	308.00	−43.4	(−46.1)	0.467
9,500	31,168	(9.5)	(5.9)	285.84	−46.6	(−51.9)	0.440
10,000	32,808	(10.0)	(6.2)	264.99	−49.9	(−57.8)	0.413
11,000	36,089	(11.0)	(6.8)	226.99	−56.4	(−69.5)	0.365
12,000	39,370	(12.0)	(7.5)	193.99	−56.5	(−69.7)	0.312
13,000	42,651	(13.0)	(8.1)	165.79	−56.5	(−69.7)	0.267
14,000	45,932	(14.0)	(8.7)	141.70	−56.5	(−69.7)	0.228
15,000	49,213	(15.0)	(9.3)	121.11	−56.5	(−69.7)	0.195
16,000	52,493	(16.0)	(9.9)	103.52	−56.5	(−69.7)	0.166
17,000	55,774	(17.0)	(10.6)	88.497	−56.5	(−69.7)	0.142
18,000	59,055	(18.0)	(11.2)	75.652	−56.5	(−69.7)	0.122
19,000	62,336	(19.0)	(11.8)	64.674	−56.5	(−69.7)	0.104
20,000	65,617	(20.0)	(12.4)	55.293	−56.5	(−69.7)	0.089
25,000	82,021	(25)	(15.5)	25.492	−51.6	(−60.9)	0.040
30,000	98,425	(30)	(18.6)	11.970	−46.6	(−51.9)	0.018
35,000	114,829	(35)	(21.7)	5.746	−36.6	(−33.9)	0.008
40,000	131,234	(40)	(24.9)	2.871	−22.8	(−9.0)	0.004
45,000	147,638	(45)	(28.0)	1.491	−9.0	(15.8)	0.002
50,000	164,042	(50)	(31.1)	0.798	−2.5	(27.5)	0.001
60,000	196,850	(60)	(37.3)	0.220	−26.1	(−15.0)	0.0003
70,000	229,659	(70)	(43.5)	0.052	−53.6	(−64.5)	0.00008
80,000	262,467	(80)	(49.7)	0.010	−74.5	(−102.1)	0.00002

Additional Reading Material

Periodicals

Selected nontechnical periodicals that contain articles on weather and climate.

Bulletin of the American Meteorological Society. Monthly. The American Meteorological Society, 45 Beacon St., Boston, Mass. 02108.

Meteorological Magazine. Monthly. British Meteorological Office, British Information Services, 845 Third Avenue, New York, New York.

National Weather Digest. Quarterly. National Weather Association, 4400 Stamp Road, Room 404, Marlow Heights, MD 20031. (Deals mainly with weather forecasting.)

NOAA. Bimonthly. Office of Public Affairs, National Oceanic and Atmospheric Administration, Rockville, MD 20852.

Weather. Monthly. Royal Meteorological Society, James Glaisher House, Grenville Place, Bracknell, Berkshire, England.

Weatherwise. Bimonthly. Heldref Publications, 4000 Albermarle St., N.W., Washington, D.C. 20016.

Selected Technical Periodicals

EOS—Transaction of the American Geophysical Union. American Geophysical Union (AGS), Washington, D.C.

Journal of Applied Meteorology. American Meteorological Society (AMS), Boston, Mass.

Journal of Atmospheric and Oceanic Technology. American Meteorological Society (AMS), Boston, Mass.

Journal of Atmospheric Science. AMS, Boston, Mass.

Journal of Climate. American Meteorological Society (AMS), Boston, Mass.

Journal of Geophysical Research. American Geophysical Union, Washington, D.C.

Monthly Weather Review. AMS, Boston, Mass.

Weather and Forecasting. AMS, Boston, Mass.

Additional periodicals that frequently contain articles of meteorological interest.

American Scientist. Bimonthly. Sigma Xi, the Scientific Research Society, Inc., New Haven, Conn.

Science. Weekly. American Association for the Advancement of Science, Washington, D.C.

Scientific American. Monthly. Scientific American Inc., New York, New York.

Smithsonian. Monthly. The Smithsonian Association, Washington, D.C.

Books

The titles listed below may be drawn upon for additional information. Many are written at the introductory level. Those that are more advanced are marked with an asterisk.

Anderson, Bette R. *Weather in the West*, American West Publishing Co., Palo Alto, CA, 1975.

Anthes, R. A. *Tropical Cyclones: Their Evolution, Structure, and Effect*, American Meteorological Society, Boston, Mass., 1982.

Bach, W. *Atmospheric Pollution*, McGraw-Hill, New York, 1972.

Battan, Louis J. *Harvesting the Clouds*, Anchor Books, Doubleday and Co., Garden City, N.Y., 1961.

———. *The Nature of Violent Storms*, Anchor Books, Doubleday and Co., Garden City, N.Y., 1961.

———. *The Unclean Sky*, Anchor Books, Doubleday and Co., Garden City, N.Y., 1966.

Bohren, Craig F. *Clouds in a Glass of Beer: Simple Experiments in Atmospheric Physics*, John Wiley, New York, 1987.

———. *What Light Through Yonder Window Breaks?*, John Wiley, New York, 1991.

Burroughs, William J. *Watching the World's Weather*, Cambridge University Press, New York, 1991.

*Byers, H. R. *General Meteorology*, McGraw-Hill, New York, 1974.

Cotton, W. R. and R. A. Anthes. *Storm and Cloud Dynamics*, Academic Press, New York, 1989.

Cotton, William R. *Storms*, ASTeR Press, Fort Collins, Co., 1990.

Craig, R. A. *The Edge of Space*, Anchor Books, Doubleday and Co., Garden City, N.Y., 1968.

Eagleman, Joe R. *Air Pollution Meteorology*, Trimedia Publishing Co., Lenexa, KS., 1991.

Edinger, James G. *Watching for the Wind*, Anchor Books, Doubleday and Co., Garden City, N.Y., 1967.

*Fleagle, Robert G. and Joost A. Businger. *An Introduction to Atmospheric Physics* (2nd ed.), Academic Press, New York, 1980.

Fujita, T. T. *The Downburst—Microburst and Macroburst*, The University of Chicago Press, Chicago, 1985.

Geiger, R. *The Climate Near the Ground*, Harvard University Press, Cambridge, Mass., 1965.

Goody, Richard M. and C. G. Walker. *Atmospheres*, Prentice-Hall, Englewood Cliffs, N.J., 1972.

Greenler, Robert. *Rainbows, Halos and Glories*, Cambridge University Press, New York, 1980.

Hartmann, William K. *Astronomy: The Cosmic Journey* (3rd ed.), Wadsworth, Belmont, CA, 1984.

Hewitt, Paul G. *Conceptual Physics: A New Introduction to Your Environment* (6th ed.), Little, Brown, Boston, Mass., 1988.

Holdgate, M. F. *A Perspective of Environmental Pollution*, Cambridge University Press, New York, 1979.

*Holton, James R. *An Introduction to Dynamic Meteorology* (2nd ed.), Academic Press, New York, 1979.

Houghton, David D. *Handbook of Applied Meteorology*, John Wiley & Sons, (through AMS, Boston, Mass.), 1985.

Hughes, Patrick. *American Weather Stories*, U.S. Department of Commerce, NOAA, Washington, D.C., 1976.

Imbrie, John, and K. P. Imbrie. *Ice Ages: Solving the Mystery*, Enslow Publishers, Short Hills, N.J., 1979.

Inadvertent Climate Modification. Report of the Study of Man's Impact on Climate (SMIC), The MIT Press, Cambridge, Mass., 1971.

International Cloud Atlas. World Meteorological Organization, Geneva, Switzerland, 1987.

Keen, Richard A. *Skywatch: The Western Weather Guide*. Fulcrum Incorporated, Golden, CO, 1987.

———. *Skywatch East: A Weather Guide*, Fulcrum Incorporated, Golden, CO, 1992.

Kellogg, W. W. and M. Mead, Eds. *The Atmosphere: Endangered and Endangering*. U.S. Government Printing Office, Washington, D.C., 1977.

Kessler, Edwin. *Thunderstorm Morphology and Dynamics* (2nd ed.), University of Oklahoma Press, Norman, OK, 1986.

Kocin, Paul J. and L. W. Uccellini. *Snowstorms along the Northeastern Coast of the United States: 1955 to 1985*, American Meteorological Society, Boston, Mass., 1990.

Kotsch, William J. *Weather for the Mariner* (2nd ed.), Naval Institute Press, Annapolis, MD, 1977.

Ladurie, E. L. R. *Times of Feast, Times of Famine: A History of Climate since the Year 1000*, Doubleday and Co., Garden City, N.Y., 1971.

Landsberg, H. E. *Weather and Health*, Anchor Books, Doubleday and Co., Garden City, N.Y., 1969.

Lowry, W. D. *Weather and Life: An Introduction to Biometeorology*, Oregon State University Book Store, Inc., Corvallis, Oregon, 1968.

*Ludlam, F. H. *Clouds and Storms: The Behavior and Effects of Water in the Atmosphere*, The Pennsylvania State University Press (through AMS, Boston, Mass.), 1980.

Ludlum, D. M. *The Country Journal New England Weather Book*, Houghton Mifflin, Boston, Mass., 1976.

———. *Early American Winters: 1604–1870*, American Meteorological Society, Boston, Mass., 1967.

———. *Weather Record Book*, Weatherwise, Inc., Princeton, N.J., 1971.

Mason, B. J. *Clouds, Rain and Rainmaking* (2nd ed.), Cambridge University Press, New York, 1975.

Mather, J. R. *Climatology: Fundamentals and Applications*, McGraw-Hill, New York, 1974.

Middleton, W. E. K. *The Invention of the Meteorological Instruments*, Johns Hopkins University Press, Baltimore, MD, 1969.

National Academy Press. *Acid Deposition—Atmospheric Processes in Eastern North America*, Washington, D.C., 1983.

———. *Causes and Effects of Stratospheric Ozone Reduction: An Update*, Washington, D.C., 1982.

———. *Changing Climate: A Report of the Carbon Dioxide Assessment Committee*, Washington, D.C., 1983.

———. *The Effects on the Atmosphere of a Major Nuclear Exchange*, Washington, D.C., 1985.

———. *Ozone Depletion, Greenhouse Gases, and Climate Change*, Washington, D.C., 1989.

———. *Solar Variability, Weather, and Climate*, Washington, D.C., 1982.

National Research Council. *Severe Storms: Prediction, Detection, and Warning*, National Academy of Sciences, Washington, D.C., 1977.

*Palmen, E. and C. W. Newton. *Atmospheric Circulation Systems*, Academic Press, New York, 1969.

Reiter, E. R. *Jet Streams: How Do They Affect Our Weather?* Anchor Books, Doubleday and Co., Garden City, N.Y., 1967.

Roberts, Walter Orr and Henry Landsford. *The Climate Mandate*, W H. Freeman and Co., San Francisco, 1979.

*Rogers, R. R. *A Short Course in Cloud Physics* (3rd ed.), Pergamon Press, Oxford, England, 1989.

Schaeter, Vincent J. and John A. Day. *A Field Guide to the Atmosphere*, Houghton Mifflin (through AMS, Boston, Mass.), 1981.

Schroeder, Mark S. and Charles C. Buck. *Fire Weather*, U.S. Department of Agriculture, Washington, D.C., 1970.

Simpson, Robert H. and Herbert Riehl. *The Hurricane and Its Impact*, Louisiana State University Press, Baton Rouge, La., 1981.

Stanford, John L. *Tornado: Accounts of Tornadoes in Iowa*, Iowa State University Press, Ames, Iowa, 1977.

U.S. Department of Commerce, NOAA. *Aviation Weather*, U.S. Department of Agriculture, Washington, D.C., 1975.

*Wallace, J. M. and P. V. Hobbs. *Atmospheric Science: An Introductory Survey*, Academic Press, New York, 1977.

Glossary

Absolute humidity   The mass of water vapor in a given volume of air. It represents the density of water vapor in the air.

Absolute zero   A temperature reading of –273°C, –460°F, or 0K. Theoretically, there is no molecular motion at this temperature.

Absolutely stable air   An atmospheric condition that exists when a lifted parcel of air is colder than the air around it.

Absolutely unstable air   An atmospheric condition that exists when a lifted parcel of air is warmer than the air around it.

Accretion   The growth of a precipitation particle by the collision of an ice crystal or snowflake with a supercooled liquid droplet that freezes upon impact.

Acid deposition   The depositing of acidic particles (usually sulfuric acid and nitric acid) at the earth's surface. Acid deposition occurs in dry form (*dry deposition*) or wet form (*wet deposition*). Acid rain and acid precipitation often denote wet deposition. (*See* Acid rain.)

Acid fog   *See* Acid rain.

Acid rain   Cloud droplets or raindrops combining with gaseous pollutants, such as oxides of sulfur and nitrogen, to make falling rain (or snow) acidic—pH less than about 5.0. If fog droplets combine with such pollutants it becomes *acid fog*.

Actual vapor pressure   *See* Vapor pressure.

Adiabatic process   A process that takes place without a transfer of heat between the system (such as an air parcel) and its surroundings. In an adiabatic process, compression always results in warming, and expansion results in cooling.

Advection   The horizontal transfer of any atmospheric property by the wind.

Advection fog   Occurs when warm, moist air moves over a cold surface and the air cools to below its dew point.

Aerosols   Tiny suspended solid particles (dust, smoke, etc.) or liquid droplets that enter the atmosphere from either natural or human (anthropogenic) sources, such as the burning of fossil fuels.

Aerovane   A wind instrument that indicates or records both wind speed and wind direction.

AFOS   Acronym for *A*utomation of *F*ield *O*perations and *S*ervices. Electronic-computerized system that displays weather information on TV-type consoles.

Air density   *See* Density.

Air mass   A large body of air that has similar horizontal temperature and moisture characteristics.

Air mass (ordinary) thunderstorm   A thunderstorm produced by local convection within an unstable air mass.

Air mass weather   A persistent type of weather that may last for several days (up to a week or more). It occurs when an area comes under the influence of a particular air mass.

Air parcel   *See* Parcel of air.

Air pollutants   Solid, liquid, or gaseous airborne substances that occur in concentrations high enough to threaten the health of people and animals, to harm vegetation and structures, or to toxify a given environment.

Air pressure (atmospheric pressure)   The pressure exerted by the weight of air above a given point, usually expressed in millibars (mb) or inches of mercury (Hg).

Albedo   The percent of radiation returning from a surface compared to that which strikes it.

Aleutian low   The subpolar low-pressure area that is centered near the Aleutian Islands on charts that show mean sea level pressure.

Altimeter   An instrument that indicates the altitude of an object above a fixed level. Pressure altimeters use an aneroid barometer with a scale graduated in altitude instead of pressure.

Altocumulus   A middle cloud, usually white or gray. Often occurs in layers or patches with wavy, rounded masses or rolls.

Altostratus   A middle cloud composed of gray or bluish sheets or layers of uniform appearance. In the thinner regions, the sun or moon usually appears dimly visible.

Analogue method of forecasting   A forecast made by comparison of past large-scale synoptic weather patterns that resemble a given (usually current) situation in its essential characteristics.

Analysis   The drawing and interpretation of the patterns of various weather elements on a surface or upper-air chart.

Anemometer   An instrument designed to measure wind speed.

Aneroid barometer   An instrument designed to measure atmospheric pressure. It contains no liquid.

Annual range of temperature   The difference between the warmest and coldest months at any given location.

Anticyclone An area of high pressure around which the wind blows clockwise in the Northern Hemisphere and counterclockwise in the Southern Hemisphere.

Apparent temperature What the air temperature "feels like" for various combinations of air temperature and relative humidity.

Arcus cloud *See* Roll cloud.

Arid climate An extremely dry climate—drier than the semi-arid climate. Often referred to as a "true desert" climate.

ASOS Acronym for *Automated Surface Observing Systems*. A system designed to provide continuous information on wind, temperature, pressure, cloud base height, and runway visibility at selected airports.

Atmosphere The envelope of gases that surround a planet and are held to it by the planet's gravitational attraction. The earth's atmosphere is mainly nitrogen and oxygen.

Atmospheric greenhouse effect The warming of an atmosphere by its absorbing and reemitting infrared radiation while allowing shortwave radiation to pass on through. The gases mainly responsible for the earth's atmospheric greenhouse effect are water vapor and carbon dioxide. Also called the *greenhouse effect.*

Atmospheric models Simulation of the atmosphere's behavior by mathematical equations or by physical models.

Atmospheric stagnation A condition of light winds and poor vertical mixing that can lead to a high concentration of pollutants. Air stagnations are most often associated with fair weather, an inversion, and the sinking air of a high-pressure area.

Atmospheric window The wavelength range between 8 and 11 micrometers in which little absorption of infrared radiation takes place.

Aurora Glowing light display in the nighttime sky caused by excited gases in the upper atmosphere giving off light. In the Northern Hemisphere it is called the *aurora borealis* (northern lights); in the Southern Hemisphere, the *aurora australis* (southern lights).

Autumnal equinox The equinox at which the sun approaches the Southern Hemisphere and passes directly over the equator. Occurs around September 23.

AWIPS Acronym for *Advanced Weather Interactive Processing System*. New computerized system that integrates and processes data received at a Weather Forecasting Office from NEXRAD, ASOS, and analysis and guidance products prepared by the National Meteorological Center.

Back-door cold front A cold front moving south or southwest along the Atlantic seaboard of the United States.

Backing wind A wind that changes direction in a counterclockwise sense (e.g., north to northwest to west).

Ball lightning A rare form of lightning that may consist of a reddish, luminous ball of electricity or charged air.

Barograph A recording barometer.

Barometer An instrument that measures atmospheric pressure. The two most common barometers are the *mercury barometer* and the *aneroid barometer.*

Bergeron process *See* Ice crystal process.

Bermuda high *See* Subtropical high.

Billow clouds Broad, nearly parallel lines of clouds oriented at right angles to the wind.

Bimetallic thermometer A temperature-measuring device usually consisting of two dissimilar metals that expand and contract differentially as the temperature changes.

Black body A hypothetical object that absorbs all of the radiation that strikes it. It also emits radiation at a maximum rate for its given temperature.

Blizzard A severe weather condition characterized by low temperatures and strong winds (greater than 35 mi/hr) bearing a great amount of snow either falling or blowing. When these conditions continue after the falling snow has ended, it is termed a *ground blizzard.*

Boulder winds Fast-flowing, local downslope winds that may attain speeds of 100 knots or more. They are especially strong along the eastern foothills of the Rocky Mountains near Boulder, Colorado.

Buys-Ballot's law A law describing the relationship between the wind direction and the pressure distribution. In the Northern Hemisphere, if you stand with your back to the wind, lower pressure will be to your left. In the Southern Hemisphere, if you stand with your back to the wind, lower pressure will be to your right.

California current The ocean current that flows southward along the west coast of the United States from about Washington to Baja California.

Cap cloud *See* Pileus cloud.

Carbon dioxide (CO_2) A colorless, odorless gas whose concentration is about 0.035 percent (355 ppm) in a volume of air near sea level. It is a selective absorber of infrared radiation and, consequently, it is important in the earth's atmospheric greenhouse effect. Solid CO_2 is called *dry ice.*

Carbon monoxide (CO) A colorless, odorless, toxic gas that forms during the incomplete combustion of carbon-containing fuels.

Celsius scale A temperature scale where zero is assigned to the temperature where water freezes and 100 to the temperature where water boils (at sea level).

Centripetal acceleration The inward-directed acceleration on a particle moving in a curved path.

Centripetal force The radial force required to keep an object moving in a circular path. It is directed toward the center of that curved path.

Chinook wall cloud A bank of clouds over the Rocky Mountains that signifies the approach of a chinook.

Chinook wind A warm, dry wind on the eastern side of the Rocky Mountains. In the Alps, this wind is called a *foehn*.

Cirrocumulus A high cloud that appears as a white patch of clouds without shadows. It consists of very small elements in the form of grains or ripples.

Cirrostratus High, thin, sheetlike clouds, composed of ice crystals. They frequently cover the entire sky and often produce a halo.

Cirrus A high cloud composed of ice crystals in the form of thin, white, featherlike clouds in patches, filaments, or narrow bands.

Clear air turbulence (CAT) Turbulence encountered by aircraft flying through cloudless skies. Thermals, wind shear, and jet streams can each be a factor in producing CAT.

Clear ice A layer of ice that appears transparent because of its homogeneous structure and small number and size of air pockets.

Climate The accumulation of daily and seasonal weather events over a long period of time.

Climatic controls The relatively permanent factors that govern the general nature of the climate of a region.

Climatic optimum A period in geological history (about 7000 to 5000 years ago) when temperatures were warmer than at present.

Climatological forecast A weather forecast, usually a month or more in the future, which is based upon the climate of a region rather than upon current weather conditions.

Cloudburst Any sudden and heavy rain shower.

Cloud seeding The introduction of artificial substances (usually silver iodide or dry ice) into a cloud for the purpose of either modifying its development or increasing its precipitation.

Coalescence The merging of cloud droplets into a single larger droplet.

Cold front A transition zone where a cold air mass advances and replaces a warm air mass.

Cold occlusion *See* Occluded front.

Cold wave A rapid fall in temperature within 24 hours that often requires increased protection for agriculture, industry, commerce, and human activities.

Computer enhancement A process where the temperatures of radiating surfaces are assigned different shades of gray (or different colors) on an infrared picture. This allows specific features to be more clearly delineated.

Condensation The process by which water vapor becomes a liquid.

Condensation level The level above the surface marking the base of a cumuliform cloud.

Condensation nuclei Tiny particles upon whose surfaces condensation of water vapor begins in the atmosphere.

Conditionally unstable air An atmospheric condition that exists when the environmental lapse rate is between the dry adiabatic rate and the moist adiabatic rate. Also called *conditional instability*.

Conduction The transfer of heat by molecular activity from one substance to another, or through a substance. Transfer is always from warmer to colder regions.

Continental arctic air mass An air mass characterized by extremely low temperatures and very dry air.

Continental polar air mass An air mass characterized by low temperatures and dry air. Not as cold as arctic air masses.

Continental tropical air mass An air mass characterized by high temperatures and low humidity.

Contour line A line that connects points of equal elevation above a reference level, most often sea level.

Contrail (condensation trail) A cloudlike streamer frequently seen forming behind aircraft flying in clear, cold, humid air.

Controls of temperature The main factors that cause variations in temperature from one place to another.

Convection Motions in a fluid that result in the transport and mixing of the fluid's properties. In meteorology, convection usually refers to atmospheric motions that are predominantly vertical, such as rising air currents due to surface heating. The rising of heated surface air and the sinking of cooler air aloft is often called *free convection*. (Compare with *forced convection*.)

Convergence An atmospheric condition that exists when the winds cause a horizontal net inflow of air into a specified region.

Cooling degree-day A form of degree-day used in estimating the amount of energy necessary to reduce the effective temperature of warm air. A cooling degree-day is a day on which the average temperature is one degree above a desired base temperature.

Coriolis force An apparent force observed on any free-moving object in a rotating system. On the earth this deflective force results from the earth's rotation and causes moving particles (including the wind) to deflect to the right in the Northern Hemisphere and to the left in the Southern Hemisphere.

Corona (optic) A series of colored rings concentrically surrounding the disk of the sun or moon. Smaller than the halo, the corona is caused by the diffraction of light around small water droplets of uniform size.

Country breeze A light breeze that blows into a city from the surrounding countryside. It is best observed on clear nights when the urban heat island is most pronounced.

Crepuscular rays Alternating light and dark bands of light that appear to fan out from the sun's position, usually at twilight.

Cumulonimbus An exceptionally dense and vertically developed cloud, often with a top in the shape of an anvil. The cloud is frequently accompanied by heavy showers, lightning, thunder, and sometimes hail. It is also known as a *thunderstorm cloud*.

Cumulus   A cloud in the form of individual, detached domes or towers that are usually dense and well defined. It has a flat base with a bulging upper part that often resembles cauliflower. Cumulus clouds of fair weather are called *cumulus humilis*. Those that exhibit much vertical growth are called *cumulus congestus* or *towering cumulus*.

Cumulus stage   The initial stage in the development of an air mass thunderstorm in which rising, warm, humid air develops into a cumulus cloud.

Cyclogenesis   The development or strengthening of middle latitude (extratropical) cyclones.

Cyclone   An area of low pressure around which the winds blow counterclockwise in the Northern Hemisphere and clockwise in the Southern Hemisphere.

Daily range of temperature   The difference between the maximum and minimun temperatures for any given day.

Dart leader   The discharge of electrons that proceeds intermittently toward the ground along the same ionized channel taken by the initial lightning stroke.

Dendrochronology   The analysis of the annual growth rings of trees as a means of interpreting past climatic conditions.

Density   The ratio of the mass of a substance to the volume occupied by it. Air density is usually expressed as g/cm^3 or kg/m^3.

Deposition   A process that occurs in subfreezing air when water vapor changes directly to ice without becoming a liquid first.

Desertification   A general increase in the desert conditions of a region.

Dew   Water that has condensed onto objects near the ground when their temperatures have fallen below the dew point of the surface air.

Dew cell   An instrument used to determine the dew-point temperature.

Dew point (dew-point temperature)   The temperature to which air must be cooled (at constant pressure and constant water vapor content) for saturation to occur.

Diffraction   The bending of light around objects, such as cloud and fog droplets, producing fringes of light and dark or colored bands.

Dispersion   The separation of white light into its different component wavelengths.

Dissipating stage   The final stage in the development of an air mass thunderstorm when downdrafts exist throughout the cumulonimbus cloud.

Divergence   An atmospheric condition that exists when the winds cause a horizontal net outflow of air from a specific region.

Doldrums   The region near the equator that is characterized by low pressure and light, shifting winds.

Doppler lidar   The use of light beams to determine the velocity of objects such as dust and falling rain by taking into account the *Doppler shift*.

Doppler radar   A radar that determines the velocity of falling precipitation either toward or away from the radar unit by taking into account the *Doppler shift*.

Doppler shift (effect)   The change in the frequency of waves that occurs when the emitter or the observer is moving toward or away from the other.

Downburst   A severe localized downdraft that can be experienced beneath a severe thunderstorm.

Drizzle   Small water drops between 0.2 and 0.5 millimeters in diameter that fall slowly and reduce visibility more than light rain.

Drought   A period of abnormally dry weather sufficiently long enough to cause serious effects on agriculture and other activities in the affected area.

Dry adiabatic rate   The rate of change of temperature in a rising or descending unsaturated air parcel. The rate of adiabatic cooling or warming is about 10°C per 1000 meters (5.5°F per 1000 ft).

Dry bulb temperature   The air temperature measured by the dry-bulb thermometer of a psychrometer.

Dry climate   A climate deficient in precipitation where annual potential evaporation and transpiration exceed precipitation.

Dry line   A boundary that separates warm, dry air from warm, moist air. It usually represents a zone of instability along which thunderstorms form.

Dry-summer subtropical climate   A climate characterized by mild, wet winters and warm to hot, dry summers. Typically located between 30 and 45 degrees latitude on the western side of continents. Also called *Mediterranean climate*.

Dust devil (or whirlwind)   A small but rapidly rotating wind made visible by the dust, sand, and debris it picks up from the surface. It develops best on clear, dry, hot afternoons.

Easterly wave   A migratory wavelike disturbance in the tropical easterlies. Easterly waves occasionally intensify into tropical cyclones.

Eccentricity (of the earth's orbit)   The deviation of the earth's orbit from elliptical to nearly circular.

Eddy   A small volume of air (or any fluid) that behaves differently from the larger flow in which it exists.

Electrical thermometers   Thermometers that use elements that convert energy from one form to another (transducers). Common electrical thermometers include the electrical resistance thermometer, thermocouple, and thermistor.

Electromagnetic waves   *See* Radiant energy.

El Niño   An extensive ocean warming that begins along the coast of Peru and Ecuador. Major El Niño events occur once every 3 to 7 years as a current of nutrient-poor tropical water moves southward along the west coast of South America.

Energy The property of a system that generally enables it to do work. Some forms of energy are kinetic, radiant, potential, chemical, electric, and magnetic.

Ensemble forecasting A method of weather forecasting where several forecast charts are examined to see how well they agree (or disagree) in predicting the weather.

Entrainment The mixing of environmental air into a preexisting air current or cloud so that the environmental air becomes part of the current or cloud.

Environmental lapse rate The rate of decrease of air temperature with elevation. It is most often measured with a radiosonde.

Evaporation The process by which a liquid changes into a gas.

Evaporation (mixing) fog Fog produced when sufficient water vapor is added to the air by evaporation, and the moist air mixes with relatively drier air. The two common types are *steam fog*, which forms when cold air moves over warm water, and *frontal fog*, which forms as warm raindrops evaporate in a cool air mass.

Exosphere The outermost portion of the atmosphere.

Extratropical cyclone A cyclonic storm that most often forms along a front in middle and high latitudes. Also called a *middle latitude storm*, a *depression*, and a *low*. It is not a tropical storm or hurricane.

Eye A region in the center of a hurricane (tropical storm) where the winds are light and skies are clear to partly cloudy.

Eye wall A wall of dense thunderstorms that surrounds the eye of a hurricane.

Fahrenheit scale A temperature scale where 32 is assigned to the temperature where water freezes and 212 to the temperature where water boils (at sea level).

Fall streaks Falling ice crystals that evaporate before reaching the ground.

Fall wind A strong, cold katabatic wind that blows downslope off snow-covered plateaus.

Fata morgana A complex mirage that is characterized by objects being distorted in such a way as to appear as castle-like features.

Feedback mechanism A process whereby an initial change in an atmospheric process will tend to either reinforce the process (*positive feedback*) or weaken the process (*negative feedback*).

Flash flood A flood that rises and falls quite rapidly with little or no advance warning, usually as the result of intense rainfall over a relatively small area.

Foehn *See* Chinook wind.

Fog A cloud with its base at the earth's surface. It reduces visibility to below about 0.6 miles or 1 kilometer.

Forced convection On a small scale, a form of mechanical stirring taking place when twisting eddies of air are able to mix hot surface air with the cooler air above. On a larger scale, it can be induced by the lifting of warm air along a front (*frontal uplift*) or along a topographic barrier (*orographic uplift*).

Free convection *See* Convection.

Freeze A condition occurring over a widespread area when the surface air temperature remains below freezing for a sufficient time to damage certain agricultural crops. A freeze most often occurs as cold air is advected into a region, causing freezing conditions to exist in a deep layer of surface air. Also called *advection frost*.

Freezing rain and freezing drizzle Rain or drizzle that falls in liquid form and then freezes upon striking a cold object or ground. Both can produce a coating of ice on objects which is called *glaze*.

Friction layer The atmospheric layer near the surface usually extending up to about 1 km (3300 ft) where the wind is influenced by friction of the earth's surface and objects on it.

Front The transition zone between two distinct air masses.

Frontal fog *See* Evaporation fog.

Frontal thunderstorms Thunderstorms that form in response to forced convection (forced lifting) along a front. Most go through a cycle similar to those of air-mass thunderstorms.

Frontal wave A wavelike deformation along a front in the lower levels of the atmosphere. Those that develop into storms are termed *unstable waves*, while those that do not are called *stable waves*.

Frost (also called hoarfrost) A covering of ice produced by deposition on exposed surfaces when the air temperature falls below the frost point.

Frostbite The partial freezing of exposed parts of the body, causing injury to the skin and sometimes to deeper tissues.

Frost point The temperature at which the air becomes saturated with respect to ice when cooled at constant pressure and constant water vapor content.

Frozen dew The transformation of liquid dew into tiny beads of ice when the air temperature drops below freezing.

Fujita scale A scale developed by T. Theodore Fujita for classifying tornadoes according to the damage they cause and their rotational wind speed.

Funnel cloud A rotating conelike cloud that extends downward from the base of a thunderstorm. When it reaches the surface it is called a *tornado*.

General circulation of the atmosphere Large-scale atmospheric motions over the entire earth.

Geostationary satellite A satellite that orbits the earth at the same rate that the earth rotates and thus remains over a fixed place above the equator.

Geostrophic wind A theoretical horizontal wind blowing in a straight path, parallel to the isobars or contours, at a constant speed. The geostrophic wind results when the Coriolis force exactly balances the horizontal pressure gradient force.

Global scale   The largest scale of atmospheric motion. Also called the *planetary scale.*

Gradient wind   A theoretical wind that blows parallel to curved isobars or contours.

Graupel   Ice particles between 2 and 5 millimeters in diameter that form in a cloud often by the process of accretion. Snowflakes that become rounded pellets due to riming are called *graupel* or *snow pellets.*

Green flash   A small green color that occasionally appears on the upper part of the sun as it rises or sets.

Greenhouse effect   *See* Atmospheric greenhouse effect.

Ground fog   *See* Radiation fog.

Growing degree-day   A form of the degree-day used as a guide for crop planting and for estimating crop maturity dates.

Gulf stream   A warm, swift, narrow ocean current flowing along the east coast of the United States.

Gust front   A boundary that separates a cold downdraft of a thunderstorm from warm, humid surface air. On the surface its passage resembles that of a cold front.

Haboob   A dust or sandstorm that forms as cold downdrafts from a thunderstorm turbulently lift dust and sand into the air.

Hadley cell   A thermal circulation proposed by George Hadley to explain the movement of the trade winds. It consists of rising air near the equator and sinking air near 30° latitude.

Hailstones   Transparent or partially opaque particles of ice that range in size from that of a pea to that of golf balls.

Halos   Rings or arcs that encircle the sun or moon when seen through an ice crystal cloud or a sky filled with falling ice crystals. Halos are produced by refraction of light.

Haze   Fine dry or wet dust or salt particles dispersed through a portion of the atmosphere. Individually these are not visible but cumulatively they will diminish visibility.

Heat   A form of energy transferred between systems by virtue of their temperature differences.

Heat index (HI)   An index that combines air temperature and relative humidity to determine an apparent temperature—how hot it actually feels.

Heating degree-day   A form of the degree-day used as an index for fuel consumption.

Heat lightning   Distant lightning that illuminates the sky but is too far away for its thunder to be heard.

Heatstroke   A physical condition induced by a person's overexposure to high air temperatures, especially when accompanied by high humidity.

Heterosphere   The region of the atmosphere above about 85 km where the composition of the air varies with height.

High   *See* Anticyclone.

High inversion fog   A fog that lifts above the surface but does not completely dissipate because of a strong inversion (usually subsidence) that exists above the fog layer.

Homosphere   The region of the atmosphere below about 85 km where the composition of the air remains fairly constant.

Hook-shape echo   The shape of an echo on a radar screen that indicates the possible presence of a tornado.

Horse latitudes   The belt of latitude at about 30° to 35° where winds are predominantly light and weather is hot and dry.

Humid continental climate   A climate characterized by severe winters and mild to warm summers with adequate annual precipitation. Typically located over large continental areas in the Northern Hemisphere between about 40° and 70° latitude.

Humidity   A general term that refers to the air's water vapor content. (*See* Relative humidity.)

Humid subtropical climate   A climate characterized by hot, muggy summers, cool to cold winters and abundant precipitation throughout the year.

Hurricane   A severe tropical cyclone having winds in excess of 64 knots (74 mi/hr).

Hurricane warning   A warning given when it is likely that a hurricane will strike an area within 24 hours.

Hurricane watch   A hurricane watch indicates that a hurricane poses a threat to an area (often within several days) and residents of the watch area should be prepared.

Hydrologic cycle   A model that illustrates the movement and exchange of water among the earth, atmosphere, and oceans.

Hydrostatic equilibrium   The state of the atmosphere when there is a balance between the vertical pressure gradient force and the downward pull of gravity.

Hygrometer   An instrument designed to measure the air's water vapor content. The sensing part of the instrument can be hair (*hair hygrometer*), a plate coated with carbon (*electrical hygrometer*), or an infrared sensor (*infrared hygrometer*).

Hypothermia   The deterioration in one's mental and physical condition brought on by a rapid lowering of human body temperature.

Hypoxia   A condition experienced by humans when the brain does not receive sufficient oxygen.

Ice Age   *See* Pleistocene epoch.

Ice crystal process   A process that produces precipitation. The process involves tiny ice crystals in a supercooled cloud growing larger at the expense of the surrounding liquid droplets. Also called the *Bergeron process.*

Ice fog   A type of fog composed of tiny suspended ice particles that forms at very low temperatures.

Icelandic low   The subpolar low-pressure area that is centered near Iceland on charts that show mean sea level pressure.

Ice nuclei   Particles that act as nuclei for the formation of ice crystals in the atmosphere.

Ice pellets   *See* Sleet.

Indian summer   An unseasonably warm spell with clear skies near the middle of autumn. Usually follows a substantial period of cool weather.

Inferior mirage   *See* Mirage.

Infrared radiation   Electromagnetic radiation with wavelengths between about 0.7 and 1000 micrometers. This radiation is longer than visible radiation but shorter than microwave radiation.

Infrared radiometer   An instrument designed to measure the intensity of infrared radiation emitted by an object. Also called *infrared sensor*.

Insolation   The *in*coming *sol*ar radi*ation* that reaches the earth and the atmosphere.

Instrument shelter   A boxlike wooden structure designed to protect weather instruments from direct sunshine and precipitation.

Interglacial period   A time interval of relatively mild climate during the Ice Age when continental ice sheets were absent or limited in extent to Greenland and the Antarctic.

Intertropical convergence zone (ITCZ)   The boundary zone separating the northeast trade winds of the Northern Hemisphere from the southeast trade winds of the Southern Hemisphere.

Inversion   An increase in air temperature with height.

Ion   An electrically charged atom, molecule, or particle.

Ionosphere   An electrified region of the upper atmosphere where fairly large concentrations of ions and free electrons exist.

Iridescence   Brilliant spots or borders of colors, most often red and green, observed in clouds up to about 30° from the sun.

Isobar   A line connecting points of equal pressure.

Isobaric map   A chart that shows variables such as temperature and wind on a constant pressure surface. Variations in height (which represent the same thing as variations in pressure) are usually shown by lines of equal height (contour lines).

Isobaric surface   A surface along which the atmospheric pressure is everywhere equal.

Isotach   A line connecting points of equal wind speed.

Isotherm   A line connecting points of equal temperature.

Jet maximum   A region of high wind speed that moves through the axis of a jet stream. Also called a *jet streak*.

Jet stream   Relatively strong winds concentrated within a narrow band in the atmosphere.

Katabatic (fall) wind   Any wind blowing downslope. It is usually cold.

Kelvin   A unit of temperature. A Kelvin is denoted by K and 1 K equals 1°C. Zero Kelvin is absolute zero, or –273.15°C.

Kelvin scale   A temperature scale with zero degrees equal to the theoretical temperature at which all molecular motion ceases. Also called the *absolute scale*. The units are sometimes called "degrees Kelvin"; however, the correct SI terminology is "Kelvins," abbreviated K.

Kinetic energy   The energy within a body that is a result of its motion.

Knot   A unit of speed equal to 1 nautical mile per hour. 1 knot equals 1.15 mi/hr.

Köppen classification system   A system for classifying climates developed by W. Köppen that is based mainly on annual and monthly averages of temperature and precipitation.

Lake breeze   A wind blowing onshore from the surface of a lake.

Lake-effect snows   Localized snowstorms that form on the downwind side of a lake. Such storms are common in late fall and early winter near the Great Lakes as cold, dry air picks up moisture and warmth from the unfrozen bodies of water.

Land breeze   A coastal breeze that blows from land to sea, usually at night.

Lapse rate   The rate at which an atmospheric variable (usually temperature) decreases with height. (*See* Environmental lapse rate.)

Latent heat   The heat that is either released or absorbed by a unit mass of a substance when it undergoes a change of state, such as during evaporation, condensation, or sublimation.

Leeside low   Storm systems (extratropical cyclones) that form on the downwind (lee) side of a mountain chain. In the United States leeside lows frequently form on the eastern side of the Rockies and Sierra Nevada.

Lenticular cloud   A cloud in the shape of a lens.

Lightning   A visible electrical discharge produced by thunderstorms.

Little Ice Age   The period from about 1550 to 1850 when average global temperatures were lower, and alpine glaciers increased in size and advanced down mountain canyons.

Longwave radiation   A term most often used to describe the infrared energy emitted by the earth and the atmosphere.

Longwaves in the westerlies   A wave in the major belt of westerlies characterized by a long length (thousands of kilometers) and significant amplitude. Also called *Rossby waves*.

Low   *See* Extratropical cyclone.

Low-level jet streams   Jet streams that typically form near the earth's surface below an altitude of about 2 kilometers and usually attain speeds of less than 60 knots.

Macroclimate The general climate of a large area, such as a country.

Macroscale The normal meteorological synoptic scale for obtaining weather information. It can cover an area ranging from the size of a continent to the entire globe.

Mammatus clouds Clouds that look like pouches hanging from the underside of a cloud.

Marine climate A climate controlled largely by the ocean. The ocean's influence keep winters relatively mild and summers cool.

Maritime air Moist air whose characteristics were developed over an extensive body of water.

Maritime polar air mass An air mass characterized by low temperatures and high humidity.

Maritime tropical air mass An air mass characterized by high temperatures and high humidity.

Mature thunderstorm The second stage in the three-stage cycle of an air-mass thunderstorm. This stage is characterized by heavy showers, lightning, thunder, and violent vertical motions inside cumulonimbus clouds.

Maunder minimum A period from about 1645 to 1715 when few, if any, sunspots were observed.

Maximum thermometer A thermometer with a small constriction just above the bulb. It is designed to measure the maximum air temperature.

Mean annual temperature The average temperature at any given location for the entire year.

Mean daily temperature The average of the highest and lowest temperature for a 24-hour period.

Mediterranean climate *See* Dry-summer subtropical climate.

Meridional flow A type of atmospheric circulation pattern in which the north-south component of the wind is pronounced.

Mesoclimate The climate of an area ranging in size from a few acres to several square kilometers.

Mesocyclone A vertical column of cyclonically rotating air within a severe thunderstorm.

Mesohigh A relatively small area of high atmospheric pressure that forms beneath a thunderstorm.

Mesopause The top of the mesosphere. The boundary between the mesosphere and the thermosphere, usually near 85 kilometers.

Mesoscale The scale of meteorological phenomena that range in size from a few kilometers to about 100 kilometers. It includes local winds, thunderstorms, and tornadoes.

Mesoscale convective complex (MCC) A large organized convective weather system comprised of a number of individual thunderstorms. The size of an MCC can be 1000 times larger than an individual air-mass thunderstorm.

Mesosphere The atmospheric layer between the stratosphere and the thermosphere. Located at an average elevation between 50 and 80 kilometers above the earth's surface.

Meteorology The study of the atmosphere and atmospheric phenomena as well as the atmosphere's interaction with the earth's surface, oceans, and life in general.

Microburst A strong localized downdraft (downburst) less than 4 km wide that occurs beneath thunderstorms.

Microclimate The climate structure of the air space near the surface of the earth.

Micrometer (μm) A unit of length equal to one-millionth of a meter.

Microscale The smallest scale of atmospheric motions.

Middle latitude cyclones *See* Extratropical cyclone.

Milankovitch theory A theory proposed by Milutin Milankovitch in the 1930s suggesting that changes in the earth's orbit were responsible for climatic changes and the ice ages.

Millibar (mb) A unit for expressing atmospheric pressure. Sea level pressure is normally close to 1013 mb.

Minimum thermometer A thermometer designed to measure the minimum air temperature during a desired time period.

Mirage A refraction phenomenon that makes an object appear to be displaced from its true position. When an object appears higher than it actually is, it is called a *superior mirage*. When an object appears lower than it actually is, it is an *inferior mirage*.

Mixing depth The vertical extent of the mixing layer.

Mixing layer The unstable atmospheric layer that extends from the surface up to the base of an inversion. Within this layer the air is well stirred.

Mixing ratio The ratio of the mass of water vapor in a given volume of air to the mass of dry air.

Moist adiabatic rate The rate of change of temperature in a rising or descending saturated air parcel. The rate of cooling or warming varies but a common value of 6°C per 1000 meters (3.3°F per 1000 ft) is used.

Molecule A collection of atoms held together by chemical forces.

Monsoon wind system A wind system that reverses direction between winter and summer. Usually the wind blows from land to sea in winter and from sea to land in summer.

Mountain and valley breeze A local wind system of a mountain valley that blows downhill (*mountain breeze*) at night and uphill (*valley breeze*) during the day.

Multicell storms Thunderstorms in a line, each of which may be in a different stage of development.

Nacreous clouds Clouds of unknown composition that have a soft, pearly luster and that form at altitudes about 25 to 30 km above the earth's surface. They are also called *mother-of-pearl clouds*.

Negative feedback mechanism *See* Feedback mechanism.

NEXRAD   An acronym for *Nex*t Generation Weather *Rad*ar. It is also known as *WSR 88D*.

Nimbostratus   A dark, gray cloud characterized by more or less continuously falling precipitation. It is not accompanied by lightning, thunder, or hail.

Nitrogen oxides   Gases that form when nitrogen in the air reacts with oxygen.

Noctilucent clouds   Wavy, thin, bluish-white clouds that are best seen at twilight in polar latitudes. They form at altitudes about 80 to 90 km above the surface.

Nocturnal inversion   *See* Radiation inversion.

Northeaster   A name given to a strong, steady wind from the northeast that is accompanied by rain and inclement weather. It often develops when a storm system moves northeastward along the coast of North America.

Northern lights   *See* Aurora.

Nowcasting   Short-term weather forecasts varying from minutes up to a few hours.

Numerical weather prediction (NWP)   Forecasting the weather based upon the solutions of mathematical equations by high-speed computers.

Obliquity (of the earth's axis)   The tilt of the earth's axis. It represents the angle from the perpendicular to the plane of the earth's orbit.

Occluded front (occlusion)   A complex frontal system that ideally forms when a cold front overtakes a warm front. When the air behind the occluded front is colder than the air ahead of it, the front is called a *cold occlusion*. When the air behind the occluded front is milder than the air ahead of it, it is called a *warm occlusion*.

Offshore wind   A breeze that blows from the land out over the water. Opposite of an onshore wind.

Onshore wind   A breeze that blows from the water onto the land. Opposite of an offshore wind.

Open wave   The stage of development of a wave cyclone (mid-latitude cyclonic storm) where a cold front and warm front exist, but no occluded front.

Orchard heaters   Oil heaters placed in orchards that generate heat and promote convective circulations to protect fruit trees from damaging low temperatures. Also called *smudge pots*.

Orographic uplift   The lifting of air over a topographic barrier. Clouds that form in this lifting process are called *orographic clouds*.

Outgassing   The release of gases dissolved in hot, molten rock.

Overrunning   A condition that occurs when air moves up and over another layer of air.

Ozone (O_3)   An almost colorless gaseous form of oxygen with an odor similar to weak chlorine. The highest natural concentration is found in the stratosphere.

Ozone hole   A sharp drop in ozone concentration observed over the Antarctic during the spring.

Pacific high   *See* Subtropical high.

Parcel of air   An imaginary small body of air a few meters wide that is used to explain the behavior of air.

Parhelia   *See* Sundog.

Particulate matter   Solid particles or liquid droplets that are small enough to remain suspended in the air. Also called *aerosols*.

Permafrost   A layer of soil beneath the earth's surface that remains frozen throughout the year.

Persistence forecast   A forecast that the future weather condition will be the same as the present condition.

Photochemical smog   *See* Smog.

Photon   A discrete quantity of energy that can be thought of as a packet of electromagnetic radiation traveling at the speed of light.

Pileus cloud   A smooth cloud in the form of a cap. Occurs above, or is attached to, the top of a cumuliform cloud.

Plate tectonics   The theory that the earth's surface down to about 100 km is divided into a number of plates that move relative to one another across the surface of the earth. Once referred to as continental drift.

Pleistocene Epoch (or Ice Age)   The most recent period of extensive continental glaciation that saw large portions of North America and Europe covered with ice. It began about 2 million years ago and ended about 10,000 years ago.

Polar easterlies   A shallow body of easterly winds located at high latitudes poleward of the subpolar low.

Polar front   A semipermanent, semicontinuous front that separates tropical air masses from polar air masses.

Polar front jet stream   The jet stream that is associated with the polar front in middle and high latitudes. It is usually located at altitudes between 9 and 12 kilometers (29,500 and 39,000 feet).

Polar front theory   A theory developed by a group of Scandinavian meteorologists that explains the formation, development, and overall life history of cyclonic storms that form along the polar front.

Polar ice cap climate   A climate characterized by extreme cold, as every month has an average temperature below freezing.

Polar orbiting satellite   A satellite whose orbit closely parallels the earth's meridian lines and thus crosses the polar regions on each orbit.

Polar tundra climate   A climate characterized by extremely cold winters and cool summers, as the average temperature of the warmest month climbs above freezing but remains below 10°C (50°F).

Pollutants   Any gaseous, chemical, or organic matter that contaminates the atmosphere, soil, or water.

Pollutant Standards Index (PSI)   An index of air quality that provides daily air pollution concentrations. Intervals on the scale relate to potential health effects.

Positive feedback mechanism   *See* Feedback mechanism.

Potential energy   The energy that a body possesses by virtue of its position with respect to other bodies in the field of gravity.

Precession (of the earth's axis of rotation)   The wobble of the earth's axis of rotation that traces out the path of a cone over a period of about 23,000 years.

Precipitation   Any form of water particles—liquid or solid—that falls from the atmosphere and reaches the ground.

Pressure gradient   The rate of change of pressure over some horizontal distance. On the same chart, when the isobars are close together, the pressure gradient is steep. When the isobars are far apart, the pressure gradient is weak.

Pressure gradient force   The force due to differences in pressure within the atmosphere that causes air to move and, hence, the wind to blow. It is directly proportional to the pressure gradient.

Pressure tendency   The rate of change of atmospheric pressure within a specified period of time, most often three hours. Same as *barometric tendency.*

Prevailing westerlies   *See* Westerlies.

Prevailing wind   The wind direction most frequently observed during a given period.

Primary air pollutants   Air pollutants that enter the atmosphere directly.

Probability forecast   A forecast of the probability of occurrence of one or more of a mutually exclusive set of weather conditions.

Prognostic chart (prog)   A chart showing expected or forecasted conditions, such as pressure patterns, frontal positions, contour height patterns, and so on.

Psychrometer   An instrument used to measure the water vapor content of the air. It consists of two thermometers (dry bulb and wet bulb). After whirling the instrument, the dew point and relative humidity can be obtained with the aid of tables.

Radar   An electronic instrument used to detect objects (such as falling precipitation) by their ability to reflect and scatter microwaves back to a receiver.

Radiant energy (radiation)   Energy propagated in the form of electromagnetic waves. These waves do not need molecules to propagate them, and in a vacuum they travel at nearly 300,000 kilometers per second.

Radiational cooling   The process by which the earth's surface and adjacent air cool by emitting infrared radiation.

Radiation fog   Fog produced over land when radiational cooling reduces the air temperature to or below its dew point. It is also known as *ground fog* and *valley fog.*

Radiation inversion   An increase in temperature with height due to radiational cooling of the earth's surface. Also called a *nocturnal inversion.*

Radiative equilibrium temperature   The temperature achieved when an object, behaving as a black body, is absorbing and emitting radiation at equal rates.

Radiometer   An instrument designed to measure the intensity of radiation (usually infrared) emitted by an object.

Radiosonde   A balloon-borne instrument that measures and transmits pressure, temperature, and humidity to a ground-based receiving station.

Rain   Precipitation in the form of liquid water drops that have diameters greater than that of drizzle.

Rainbow   An arc of concentric colored bands that spans a section of the sky when rain is present and the sun is positioned at the observer's back.

Rain gauge   An instrument designed to measure the amount of rain that falls during a given time interval.

Rain shadow   The region on the leeside of a mountain where the precipitation is noticeably less than on the windward side.

Rawinsonde observation   A radiosonde observation that includes wind data.

Reflection   The process whereby a surface turns back a portion of the radiation that strikes it.

Refraction   The bending of light as it passes from one medium to another.

Relative humidity   The ratio of the amount of water vapor actually in the air compared to the amount of water vapor the air can hold at that particular temperature and pressure. The ratio of the air's actual vapor pressure to its saturation vapor pressure.

Return stroke   The luminous lightning stroke that propagates upward from the earth to the base of a cloud.

Ridge   An elongated area of high atmospheric pressure.

Rime ice   A white, granular deposit of ice formed by the freezing of water drops when they come in contact with an object.

Riming   *See* Accretion.

Roll cloud   A dense, roll-shaped cloud attached to the lower front part of the main cloud. It often forms with thunderstorms along the leading edge of a gust front. Also called an *arcus cloud.*

Rossby waves   *See* Longwaves in the westerlies.

Rotor cloud   A turbulent cumuliform type of cloud that forms on the leeward side of large mountain ranges. The air in the cloud rotates about an axis parallel to the range.

Rotors   Turbulent eddies that form downwind of a mountain chain, creating hazardous flying conditions.

Saffir-Simpson scale   A scale relating a hurricane's central pressure and winds to the possible damage it is capable of inflicting.

St. Elmo's fire   A bright electric discharge that is projected from objects (usually pointed) when they are in a strong electric field, such as during a thunderstorm.

Santa Ana wind   A warm, dry wind that blows into southern California from the east off the elevated desert plateau. Its warmth is derived from compressional heating.

Saturation (of air)   An atmospheric condition whereby the level of water vapor is the maximum possible at the existing temperature and pressure.

Saturation vapor pressure   The maximum amount of water vapor necessary to keep moist air in equilibrium with a surface of pure water or ice. It represents the maximum amount of water vapor that the air can hold at any given temperature and pressure.

Savanna   A tropical or subtropical region of grassland and drought-resistant vegetation. Typically found in tropical wet-and-dry climates.

Scales of motion   The hierarchy of atmospheric circulations from tiny gusts to giant storms.

Scattering   The process by which small particles in the atmosphere deflect radiation from its path into different directions.

Scintillation   The apparent twinkling of a star due to its light passing through regions of differing air densities in the atmosphere.

Sea breeze   A coastal local wind that at the surface blows from the ocean onto the land.

Sea level pressure   The atmospheric pressure at mean sea level.

Secondary air pollutants   Pollutants that form when a chemical reaction occurs between a primary air pollutant and some other component of air.

Selective absorbers   Substances such as water vapor, carbon dioxide, clouds, and snow that absorb radiation only at particular wavelengths.

Semi-arid climate   A dry climate where potential evaporation and transpiration exceed precipitation. Not as dry as the arid climate. Typical vegetation is short grass.

Semipermanent highs and lows   Areas of high pressure (anticyclones) and low pressure (extratropical cyclones) that tend to persist at a particular latitude belt throughout the year. In the Northern Hemisphere, typically they shift slightly northward in summer and slightly southward in winter.

Sensible heat   The heat we can feel and measure with a thermometer.

Sensible temperature   The sensation of temperature that the human body feels in contrast to the actual temperature of the environment as measured with a thermometer.

Severe thunderstorms   Intense thunderstorms capable of producing heavy showers, flash floods, hail, strong and gusty surface winds, and tornadoes.

Sheet lightning   A fairly bright lightning flash from distant thunderstorms that illuminates a portion of the cloud.

Shortwave (in the atmosphere)   A small wave that moves around longwaves in the same direction as the air flow in the middle and upper troposphere. Shortwaves are also called *shortwave troughs*.

Shortwave radiation   A term most often used to describe the radiant energy emitted from the sun, in the visible and near ultraviolet wavelengths.

Shower   Intermittent precipitation from a cumuliform cloud, usually of short duration but often heavy.

Siberian high   A strong, shallow area of high pressure that forms over Siberia in winter.

Sleet   A type of precipitation consisting of transparent pellets of ice 5 millimeters or less in diameter. Same as *ice pellets*.

Smog   Originally smog meant a mixture of smoke and fog. Today, smog means air that has restricted visibility due to pollution, or pollution formed in the presence of sunlight—*photochemical smog*.

Smog front (also smoke front)   The leading edge of a sea breeze that is contaminated with smoke or pollutants.

Snow   A solid form of precipitation composed of ice crystals in complex hexagonal form.

Snowflake   An aggregate of ice crystals that falls from a cloud.

Snow flurries   Light showers of snow that fall intermittently.

Snow grains   Precipitation in the form of very small, opaque grains of ice. The solid equivalent of drizzle.

Snow pellets   White, opaque, approximately round ice particles between 2 and 5 millimeters in diameter that form in a cloud either from the sticking together of ice crystals or from the process of accretion.

Snow squall (shower)   An intermittent heavy shower of snow that greatly reduces visibility.

Solar constant   The rate at which solar energy is received on a surface at the outer edge of the atmosphere perpendicular to the sun's rays when the earth is at a mean distance from the sun. The value of the solar constant is about two calories per square centimeter per minute or about 1376 W/m^2 in the SI system of measurement.

Sounding   An upper-air observation, such as a radiosonde observation. A vertical profile of an atmospheric variable such as temperature or winds.

Source regions   Regions where air masses originate and acquire their properties of temperature and moisture.

Southern oscillation   The reversal of surface air pressure at opposite ends of the tropical Pacific Ocean that occur during major El Niño events.

Specific heat   The ratio of the heat absorbed (or released) by the unit mass of the system to the corresponding temperature rise (or fall).

Specific humidity The ratio of the mass of water vapor in a given parcel to the total mass of air in the parcel.

Squall line Any nonfrontal line or band of active thunderstorms.

Stable air *See* Absolutely stable air.

Standard atmosphere A hypothetical vertical distribution of atmospheric temperature, pressure, and density in which the air is assumed to obey the gas law and the hydrostatic equation. The lapse rate of temperature in the troposphere is taken as 6.5°C/1000 m or 3.6°F/1000 ft.

Standard atmospheric pressure A pressure of 1013.25 millibars (mb), 29.92 inches of mercury (Hg), 760 millimeters (mm) of mercury, 14.7 pounds per square inch (lb/in.²), 101,325 pascals (Pa).

Stationary front A front that is nearly stationary with winds blowing almost parallel and from opposite directions on each side of the front.

Station pressure The actual air pressure computed at the observing station.

Steady-state forecast A weather prediction based on the past movement of surface weather systems. It assumes that the systems will move in the same direction and at approximately the same speed as they have been moving. Also called *trend forecasting*.

Steam fog *See* Evaporation (mixing) fog.

Steppe An area of grass-covered, treeless plains that has a semi-arid climate.

Stepped leader An initial discharge of electrons that proceeds intermittently toward the ground in a series of steps in a cloud-to-ground lightning stroke.

Storm surge An abnormal rise of the sea along a shore; primarily due to the winds of a storm, especially a hurricane.

Stratocumulus A low cloud, predominantly stratiform, with low, lumpy, rounded masses, often with blue sky between them.

Stratosphere The layer of the atmosphere above the troposphere and below the mesosphere (between 10 km and 50 km), generally characterized by an increase in temperature with height.

Stratospheric polar night jet A jet stream that forms near the top of the stratosphere over polar latitudes during the winter months.

Stratus A low, gray cloud layer with a rather uniform base whose precipitation is most commonly drizzle.

Streamline A line that shows the wind flow pattern.

Sublimation The process whereby ice changes directly into water vapor without melting.

Subpolar climate A climate observed in the Northern Hemisphere that borders the polar climate. It is characterized by severely cold winters and short, cool summers. Also known as *taiga climate* and *boreal climate*.

Subpolar low A belt of low pressure located between 50° and 70° latitude. In the Northern Hemisphere, this "belt" consists of the *Aleutian low* in the North Pacific and the *Ice-landic low* in the North Atlantic. In the Southern Hemisphere, it exists around the periphery of the Antarctic continent.

Subsidence The slow sinking of air, usually associated with high-pressure areas.

Subsidence inversion A temperature inversion produced by compressional warming—the adiabatic warming of a layer of sinking air.

Subtropical high A semipermanent high in the subtropical high-pressure belt centered near 30° latitude. The *Bermuda high* is located over the Atlantic Ocean off the east coast of North America. The *Pacific high* is located off the west coast of North America.

Subtropical jet stream The jet stream typically found between 20° and 30° latitude at altitudes between 12 and 14 km.

Suction vortices Small, rapidly rotating whirls perhaps 10 meters in diameter that are found within large tornadoes.

Sulfur dioxide (SO₂) A colorless gas that forms primarily in the burning of sulfur-containing fossil fuels.

Summer solstice Approximately June 22 in the Northern Hemisphere when the sun is highest in the sky and directly overhead at latitude 23½°N, the Tropic of Cancer.

Sundog A colored luminous spot produced by refraction of light through ice crystals that appears on either side of the sun. Also called *parhelia*.

Sun pillar A vertical streak of light extending above (or below) the sun. It is produced by the reflection of sunlight off ice crystals.

Sunspots Relatively cooler areas on the sun's surface. They represent regions of an extremely high magnetic field.

Supercell storm An enormous severe thunderstorm whose updrafts and downdrafts are nearly in balance, allowing it to maintain itself for several hours. It can produce large hail and tornadoes.

Supercooled cloud droplets Liquid cloud droplets observed at temperatures below freezing.

Superior mirage *See* Mirage.

Supersaturated air A condition that occurs in the atmosphere when the relative humidity is greater than 100 percent.

Surface inversion *See* Radiation inversion.

Synoptic scale The typical weather map scale that shows features such as high- and low-pressure areas and fronts over a distance spanning a continent.

Taiga (boreal forest) The open northern part of the coniferous forest. Taiga also refers to subpolar climate.

Tcu An abbreviation sometimes used to denote a towering cumulus cloud (cumulus congestus).

Temperature The degree of hotness or coldness of a substance as measured by a thermometer. It is also a measure of the average speed or kinetic energy of the atoms and molecules in a substance.

Temperature inversion An increase in air temperature with height.

Thermal   A small, rising parcel of warm air produced when the earth's surface is heated unevenly.

Thermal belts   Horizontal zones of vegetation found along hillsides that are primarily the result of vertical temperature variations.

Thermal circulations   Air flow resulting primarily from the heating and cooling of air.

Thermal lows and thermal highs   Areas of low and high pressure that are shallow in vertical extent and are produced primarily by surface temperatures.

Thermal tides   Atmospheric pressure variations due to the uneven heating of the atmosphere by the sun.

Thermistor   An electrical resistance device used in the measurement of temperature.

Thermograph   An instrument that measures and records air temperature.

Thermometer   An instrument for measuring temperature. The most common are liquid-in-glass, which have a sealed glass tube attached to a glass bulb filled with liquid.

Thermosphere   The atmospheric layer above the mesosphere (above about 85 km) where the temperature increases rapidly with height.

Thunder   The sound due to rapidly expanding gases along the channel of a lightning discharge.

Thunderstorm   A local storm produced by cumulonimbus clouds. Always accompanied by lightning and thunder.

Tornado   An intense, rotating column of air that protrudes from a cumulonimbus cloud in the shape of a funnel or a rope and touches the ground. (*See* Funnel cloud.)

Tornado outbreak   A series of tornadoes that forms within a particular region—a region that may include several states. Often associated with widespread damage and destruction.

Tornado warning   A warning issued when a tornado has actually been observed either visually or on a radar screen. It is also issued when the formation of tornadoes is imminent.

Tornado watch   A forecast issued to alert the public that tornadoes may develop within a specified area.

Trace (of precipitation)   An amount of precipitation less than 0.01 inch (0.025 cm).

Trade wind inversion   A temperature inversion frequently found in the subtropics over the eastern portions of the tropical oceans.

Trade winds   The winds that occupy most of the tropics and blow from the subtropical highs to the equatorial low.

Transpiration   The process by which water in plants is transferred as water vapor to the atmosphere.

Tropical depression   A mass of thunderstorms and clouds generally with a cyclonic wind circulation of between 20 and 34 knots.

Tropical disturbance   An organized mass of thunderstorms with a slight cyclonic wind circulation of less than 20 knots.

Tropical easterly jet   A jet stream that forms on the equatorward side of the subtropical highs near 15 kilometers.

Tropical monsoon climate   A tropical climate with a brief dry period of perhaps one or two months.

Tropical rain forest   A type of forest consisting mainly of lofty trees and a dense undergrowth near the ground.

Tropical storm   Organized thunderstorms with a cyclonic wind circulation between 35 and 64 knots.

Tropical wet-and-dry climate   A tropical climate poleward of the tropical wet climate where a distinct dry season occurs, often lasting for two months or more.

Tropical wet climate   A tropical climate with sufficient rainfall to produce a dense tropical rain forest.

Tropopause   The boundary between the troposphere and the stratosphere.

Troposphere   The layer of the atmosphere extending from the earth's surface up to the tropopause (about 11 km above the ground).

Trough   An elongated area of low atmospheric pressure.

Turbulence   Any irregular or disturbed flow in the atmosphere that produces gusts and eddies.

Twilight   The time at the beginning of the day immediately before sunrise and at the end of the day after sunset when the sky remains illuminated.

Typhoon   A hurricane that forms over the western Pacific Ocean.

Ultraviolet radiation   Electromagnetic radiation with wavelengths longer than X-rays but shorter than visible light.

Unstable air   *See* Absolutely unstable air.

Upslope fog   Fog formed as moist, stable air flows upward over a topographic barrier.

Upslope precipitation   Precipitation that forms due to moist, stable air gradually rising along an elevated plain. Upslope precipitation is common over the western Great Plains, especially east of the Rocky Mountains.

Upwelling   The rising of water (usually cold) toward the surface from the deeper regions of a body of water.

Urban heat island   The increased air temperatures in urban areas as contrasted to the cooler surrounding rural areas.

Valley breeze   *See* Mountain breeze.

Valley fog   *See* Radiation fog.

Vapor pressure   The pressure exerted by the water vapor molecules in a given volume of air.

Vernal equinox   The equinox at which the sun approaches the Northern Hemisphere and passes directly over the equator. Occurs around March 20.

Virga   Precipitation that falls from a cloud but evaporates before reaching the ground. (*See* Fall streaks.)

Visible radiation (light)   Radiation with a wavelength between 0.4 and 0.7 micrometers.

Visibility   The greatest distance an observer can see and identify prominent objects.

Volatile organic compounds (VOC)   Organic compounds composed of hydrogen and carbon. Also called *hydrocarbons*.

Wall cloud   An area of rotating clouds that extends beneath a severe thunderstorm and from which a funnel cloud may appear.

Warm-core low   A low-pressure area that is warmer at its center than at its periphery. Tropical cyclones exhibit this temperature pattern.

Warm front   A front that moves in such a way that warm air replaces cold air.

Warm occlusion   *See* Occluded front.

Warm sector   The region of warm air within a wave cyclone that lies between a retreating warm front and an advancing cold front.

Water equivalent   The depth of water that would result from the melting of a snow sample. Typically about 10 inches of snow will melt to 1 inch of water, producing a water equivalent of 10 to 1.

Waterspout   A column of rotating wind over water that has characteristics of a dust devil and tornado.

Water vapor   Water in a vapor (gaseous) form. Also called *moisture*.

Wave cyclone   An extratropical cyclone that forms and moves along a front. The circulation of winds about the cyclone tends to produce a wavelike deformation on the front.

Wavelength   The distance between successive crests, troughs, or identical parts of a wave.

Weather   The condition of the atmosphere at any particular time and place.

Weather elements   The elements of *air temperature, air pressure, humidity, clouds, precipitation, visibility,* and *wind* that determine the present state of the atmosphere, the weather.

Weather type forecasting   A forecasting method where weather patterns are categorized into similar groups or types.

Weather types   Certain weather patterns categorized into similar groups. Used as an aid in weather prediction.

Weather warning   A forecast indicating that hazardous weather is either imminent or actually occurring within the specified forecast area.

Weather watch   A forecast indicating that atmospheric conditions are favorable for hazardous weather to occur over a particular region during a specified time period.

Westerlies   The dominant westerly winds that blow in the middle latitudes on the poleward side of the subtropical high-pressure areas.

Wet-bulb depression   The difference in degrees between the air temperature (dry-bulb temperature) and the wet-bulb temperature.

Wet-bulb temperature   The lowest temperature that can be obtained by evaporating water into the air.

Whirlwinds   *See* Dust devils.

Wind   Air in motion relative to the earth's surface.

Wind-chill factor   The cooling effect of any combination of temperature and wind, expressed as the loss of body heat. Also called *wind-chill index*.

Wind direction   The direction *from which* the wind is blowing.

Wind machines   Fans placed in orchards for the purpose of mixing cold surface air with warmer air above.

Wind profiler   A Doppler radar capable of measuring the turbulent eddies that move with the wind. Because of this, it is able to provide a vertical picture of wind speed and wind direction.

Wind rose   A diagram that shows the percent of time that the wind blows from different directions at a given location over a given time.

Wind shear   The rate of change of wind speed or wind direction over a given distance.

Wind vane   An instrument used to indicate wind direction.

Windward side   The side of an object facing into the wind.

Winter solstice   Approximately December 22 in the Northern Hemisphere when the sun is lowest in the sky and directly overhead at latitude 23½°S, the Tropic of Capricorn.

Xerophytes   Drought-resistant vegetation.

Zonal wind flow   A wind that has a predominate west-to-east component.

Index

429

Chapter 10

Figs. 10.2, 10.3   Photos by author.
Fig. 10.6   Photo by Major Richard F. Picanso, USAF (Air Weather Service).
Figs. 10.7, 10.8   Photos by author.
Fig. 10.11   NOAA photo.
Fig. 10.12   After *Thunderstorms*, Vol. 2, U.S. Government Printing Office, Washington, D.C., 1982.
Fig. 10.16   Photo by Mike Spurgat.
Fig. 2   Photo by E. Philip Krider, University of Arizona.
Fig. 3   Photo by D. Baumhefner, NCAR.
Fig. 10.18   Photo by Eric J. Lance, Walnut Grove, MN.
Fig. 10.22   Photo © Wade Balzer, Weatherstock.
Fig. 10.25   Photo © Howard B. Bluestein.
Fig. 10.26   NOAA photo.
Fig. 10.27   National Center for Atmospheric Research/ National Science Foundation.
Fig. 10.28   Photo © Howard B. Bluestein.
Fig. 10.29   Courtesy Joseph H. Golden, NOAA.

Chapter 11

Fig. 11.2   NOAA photo.
Fig. 11.4   Photo © Comstock.
Fig. 11.6   NOAA photo.
Fig. 11.9   Photo © M. Laca, Weatherstock.
Fig. 11.11   NOAA photo.
Figs. 11.12, 11.13, 11.14a, 11.14b   Courtesy NOAA/ National Weather Service.
Fig. 11.15   Courtesy NOAA.

Chapter 12

Figs. 12.3a, 12.3b   Courtesy John Day and the University of Colorado Health Sciences Center.
Fig. 12.7   Photo by author.
Fig. 12.9   Photo by J. L. Medeiros.
Fig. 12.10   Photo by Jim Edwards, Riverside Press Enterprise.
Fig. 12.13   Photo by author.
Fig. 12.14   After United States and Canada Work Group No. 2, 1982.

Chapter 13

Fig. 13.1   Photo by author.
Fig. 1   NASA photo.
Fig. 13.10   Photo by Ron Tingley.
Figs. 13.12, 13.13   Photos by author.
Fig. 13.15   Data courtesy IPCC Scientific Assessment.

Chapter 14

Figs. 14.6, 14.8, 14.10   Photos by J. L. Medeiros.
Fig. 14.12   Photo by author.
Fig. 14.17   Photo by J. L. Medeiros.
Fig. 14.18   Photo by author.
Figs. 14.21, 14.23   Photos by J. L. Medeiros.

Chapter 15

Figs. 15.3, 15.4, 15.5, 15.10   Photos by author.
Fig. 15.11   Photo © Pekka Parviainen.
Fig. 15.12   Photo by author.
Fig. 1   Photo by Pekka Parviainen.
Fig. 2   Photo by author.
Fig. 3   Photo by Pekka Parviainen.
Fig. 15.15   Photo courtesy T. Ansel Toney.
Fig. 15.17   Science Source/Photo Researchers, Inc.
Fig. 15.19   Photo by J. L. Medeiros.
Fig. 15.20   From *Rainbows, Halos, and Glories* by Robert Greenler, Cambridge University Press, 1980; photo by author.
Fig. 15.24   Photo by author.